Enhancing Soil Health to Mitigate Soil Degradation

Special Issue Editors

Douglas L. Karlen
Charles W. Rice

MDPI

Guest Editors

Douglas L. Karlen
National Laboratory for Agriculture
and the Environment
USA

Charles W. Rice
Kansas State University,
Department of Agronomy
USA

Editorial Office
MDPI AG
St. Alban-Anlage 66 Basel,
Switzerland

This edition is a reprint of the Special Issue published online in the open access journal *Sustainability* (ISSN 2071-1050) from 2014–2015 (available at: http://www.mdpi.com/journal/sustainability/special_issues/soil-degradation).

For citation purposes, cite each article independently as indicated on the article page online and as indicated below:

Author 1; Author 2; Author 3 etc. Article title. *Journal Name.* **Year**. Article number/page range.

ISBN 978-3-03842-358-4 (Pbk)
ISBN 978-3-03842-359-1 (PDF)

Table of Contents

About the Guest Editors

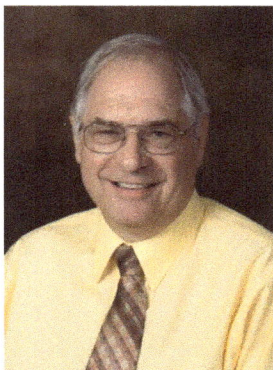

Douglas L. Karlen is a research soil scientist with the USDA-Agricultural Research Service (ARS). Growing up on a small dairy and hog farm near Monticello, Wisconsin, he was very active in 4-H which provided an introduction to the ARS through a "Citizenship Short Course" in Washington D.C. With a strong interest in science and desire to pursue a research career focused on agriculture, he chose the University of Wisconsin, Madison, for his undergraduate studies. Soon he found Soil Science as a discipline that fulfilled his desires. Karlen received his B.S. in Soil Science in 1973, his M.S. in Soil Science at Michigan State University in 1975, and his Ph.D. in Agronomy at Kansas State University in 1978. He joined the ARS and was stationed in Florence Carolina until 1988 when he transferred to Ames, Iowa where the National Soil Tilth Laboratory (NSTL) was being established. Now known as the National Laboratory for Agriculture and the Environment (NLAE), Karlen has continued his career focusing on sustainable agriculture, soil quality/soil health, and production of sustainable feedstock for bioenergy and/or other bio-products. Karlen's initial research focus was on soil fertility and plant nutrition relationships, but gradually evolved into a landscape approach to sustainable land use. During the 1990s, he became an international leader for soil quality/soil health research and technology transfer. He is a Fellow of the American Society of Agronomy (ASA), Crop Science Society of America (CSSA), Soil and Water Conservation Society (SWCS) and was recognized in 2015 as recipient of the Hugh Hammond Bennett award for national and international leadership in natural resources conservation.

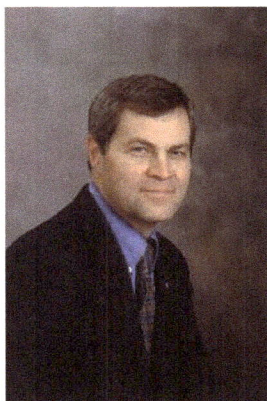

Charles W. Rice, Ph. D., is a university distinguished professor of soil microbiology. He has conducted long-term research on soil organic matter dynamics, nitrogen transformations and microbial ecology. Rice grew up in Yorkville, Illinois where he became involved in many aspects of 4-H. Rice began his undergraduate work in biology but following an honors geography water resource class, he switched his major and ultimately received his B.S. in Geography from Northern Illinois University. He then completed his Masters and Doctorate from the University of Kentucky before joining the Agronomy Department at Kansas State University in Manhattan, KS. Rice specializes in soil microbiology, carbon cycling, and climate change. His extensive research has allowed him to gain helpful insight in order to help his students. Internationally, Rice was a member of the United Nations' Intergovernmental Panel on Climate Change that received the Nobel Peace Prize in 2007. He is a Fellow of the ASA, SSSA, and American Association for the Advancement of Science (AAAS). He is past president of the SSSA and currently serves as the Chair of the National Academies Board on Agriculture and Natural Resources of the National Academies of Sciences, Engineering and Medicine.

Preface to "Enhancing Soil Health to Mitigate Soil Degradation"

In his book entitled *Out of the Earth: Civilization and the Life of the Soil* our friend and colleague Dr. Daniel Hillel included an eloquent quote attributed to Plato which stated that:

> "What now remains of the formerly rich land is like the skeleton of a sick man, with all the fat and soft earth having wasted away and only the bare framework remaining. Formerly, many of the mountains were arable. The plains that were full of rich soil are now marshes. Hills that were once covered with forests and produced abundant pasture now produce only food for bees. Once the land was enriched by yearly rains, which were not lost, as they are now, by flowing from the bare land into the sea. The soil was deep, it absorbed and kept the water in the loamy soil, and the water that soaked into the hills fed springs and running streams everywhere. Now the abandoned shrines at spots where formerly there were springs attest that our description of the land is true."

Written nearly 5000 years ago, these words clearly indicate that humankind has been forewarned regarding the fragility of our soil, water, and air resources. Unfortunately, we still have many lessons to learn. Our hope is that this collection of contributions from around the world will help inspire many to continue to strive for improved soil management practices that will protect and preserve our precious resource while still meeting the rising demand for food, feed, fiber, and fuel as our population increases beyond 9 billion.

Douglas L. Karlen
Charles W. Rice
Guest Editors

sustainability

MDPI

Editorial

Soil Degradation: Will Humankind Ever Learn?

Douglas L. Karlen [1,†,*] **and Charles W. Rice** [2,†]

1   Research Soil Scientist, USDA-Agricultural Research Service, National Laboratory for Agriculture and the Environment (NLAE), 2110 University Boulevard, Ames, IA 50011-3120, USA

2   Department of Agronomy, Kansas State University, 2701 Throckmorton Ctr., Manhattan, KS 66506, USA; cwrice@ksu.edu

*   Author to whom correspondence should be addressed; Doug.Karlen@ars.usda.gov; Tel.: +1-515-294-3336; Fax: +1-515-294-8125.

†   These authors contributed equally to this work.

Academic Editor: Marc A. Rosen

Received: 10 July 2015; Accepted: 3 September 2015; Published: 11 September 2015

Abstract: Soil degradation is a global problem caused by many factors including excessive tillage, inappropriate crop rotations, excessive grazing or crop residue removal, deforestation, mining, construction and urban sprawl. To meet the needs of an expanding global population, it is essential for humankind to recognize and understand that improving soil health by adopting sustainable agricultural and land management practices is the best solution for mitigating and reversing current soil degradation trends. This research editorial is intended to provide an overview for this Special Issue of *Sustainability* that examines the global problem of soil degradation through reviews and recent research studies addressing soil health in Africa, Australia, China, Europe, India, North and South America, and Russia. Two common factors—soil erosion and depletion of soil organic matter (SOM)—emerge as consistent indicators of how "the thin layer covering the planet that stands between us and starvation" is being degraded. Soil degradation is not a new problem but failing to acknowledge, mitigate, and remediate the multiple factors leading to it is no longer a viable option for humankind. We optimistically conclude that the most promising strategies to mitigate soil degradation are to select appropriate land uses and improve soil management practices so that SOM is increased, soil biology is enhanced, and all forms of erosion are reduced. Collectively, these actions will enable humankind to "take care of the soil so it can take care of us".

Keywords: soil health; soil quality; sustainable intensification; soil biology; erosion; soil organic matter; carbon sequestration

1. Introduction

This research editorial is intended to establish the context and provide a broad overview for the Special Issue of *Sustainability* entitled "Enhancing Soil Health to Mitigate Soil Degradation" that was initiated in 2014 to document both the magnitude and global prevalence of soil degradation. Our goals for the Special Issue were to: (1) help illustrate various factors contributing to the problem of soil degradation; (2) identify past and current impacts of soil degradation in countries around the world; and (3) suggest soil health strategies that could be used to protect our fragile soil resources.

As our global population marches steadily toward projections of 9.5 billion in 2050, natural and human induced soil degradation, if not mitigated, will undoubtedly increase the potential for negative impacts such as disease and malnutrition [1]. Currently, those problems are most severe in mountainous, tropical latitude areas of Central and South America where natural or environmentally induced soil degradation (e.g., landslides) is prevalent, and in Africa, which unlike Asia, has not been able to capitalize on benefits associated with the traditional "green revolution" even though the rate of adoption of improved crop varieties was equivalent to the rate in other developing regions around the world. In Africa, depletion of soil nutrients and poor water management are the major limiting factors, not the lack of improved crop varieties. Several recent studies confirm that no matter how good genetic improvement is, crops cannot grow well without sufficient nitrogen (N), phosphorus (P), and other essential plant nutrients. Sanchez and Swaminathan [1] also concluded that the root cause for many of Africa's malnutrition and subsequent social problems stems from a catastrophic "crisis in soil health." For years, Africa's small-scale farmers have removed large quantities of nutrients from their soils without returning them through either manure or fertilizer sources. The removal of almost all crop residues has also resulted in decreased SOM, impaired soil biological activities, weakening of soil structure, and impaired water dynamics—*i.e.*, infiltration, retention and release for plant growth. Within the Special Issue, Tully *et al.* [2] focus on soil degradation in sub-Saharan Africa (SSA), which we fully recognize is just one area within a vast continent that is struggling to mitigate the problem. Similarly, as stated above, but not adequately addressed, there are several in countries in Central America, especially the Andean ones, that are also struggling against soil degradation. In many of them, the combined impacts of agricultural activity, climate change, and extreme environmental events have had severe consequences that illustrate the global prevalence of severe soil degradation problems.

Unfortunately, soil degradation is not a new problem for humankind. Greek and Roman philosophers were well aware of the importance of soil health to agricultural prosperity and demonstrated this understanding in their treatises on farm management more than 2500 years ago. An example from Hillel [3] is an account whereby Plato has Critias proclaim:

> "What now remains of the formerly rich land is like the skeleton of a sick man, with all the fat and soft earth having wasted away and only the bare framework remaining. Formerly, many of the mountains were arable. The plains that were full of rich soil are now marshes. Hills that were once covered with forests and produced abundant pasture now produce only food for bees. Once the land was enriched by yearly rains, which were not lost as they are now, by flowing from the bare land into the sea. The soil was deep, it absorbed and kept the water in the loamy soil, and the water that soaked into the hills fed springs and running streams everywhere. Now the abandoned shrines at spots where formerly there were springs attest that our description of the land is true."

Hillel [3] provides many other references from current and historical times that address concerns related to the health of soil and land resources. Similar concerns regarding humanity's history of poor soil resource management, apparent lack of concern for soil health, and consequences of our negligence can be found in writings of Lowdermilk [4], Montgomery [5], and Larson, who often stated that soil is "the thin layer covering the planet that stands between us and starvation" [6].

2. Global Soil Degradation Perspectives

Several national research councils and advisory boards have published strategic papers related to soil resources. All reports agree that the state of our soils is deteriorating and that there is an urgent need to improve soil health (e.g., the National Research Council (NRC), the Royal Society of Chemistry (London) (RSC), the German Advisory Council on Global Change (WBGU). The most significant threats to soils around the world are:

(1)   Erosion (wind and water)
(2)   Loss of SOM (also referred to as carbon, soil carbon, or soil organic matter)
(3)   Nutrient imbalance
(4)   Salinization
(5)   Surface sealing
(6)   Loss of soil biodiversity
(7)   Contamination
(8)   Acidification

(9)   Compaction

(10)   Waterlogging

The degree of severity, geographic extent and interaction between these threats are diverse and complex. Degradation of soil results in the loss of critical functions and ecosystem services. These functions and services include production of food, feed, fuel, and fiber ensuring sufficient supplies of clean water, providing a platform for the built environment, acting as a buffer against extreme climatic events, supporting biodiversity, and providing the largest terrestrial store of carbon and nutrients [7].

Degrading soils cover approximately 24% of the global land area (35 million km^2 or 3500 million ha) [8]. Furthermore, the new global Land Cover Share-database [9] shows that croplands cover 13% of the global land surface and that grasslands, which are often used for grazing, cover another 13%. While agriculture is not the sole cause of soil degradation it is a dominant factor. Around 1000 AD cropland and pasture accounted for 1% to 2% of the ice free land area [10]. As the human population expanded, 2% to 4% of the land was in agriculture by 1700 (Figure 1). By 1900, expansion occurred into North America and by 2000, intensive agricultural practices expanded further with significant population increases in South America, Africa and Asia that have resulted in current day totals (Figure 2).

Anthropogenic Biomes of the World, Version 2, 1700: Global

Anthropogenic Biomes

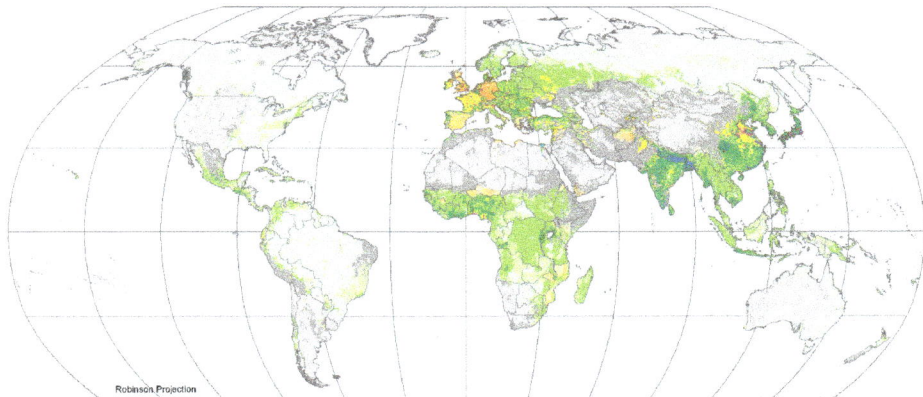

Figure 1. Global land use projected for 1700 [11].

One of the key drivers of soil degradation is land use change (LUC). An example of how LUC is affecting soil structure in Brazil is shown in Figure 3. Due to increasing global demand for bioenergy feedstock, native biomes such as the Cerado are being converted first to pasture and then to sugarcane (*Saccharum officinarum*). Soil health assessment techniques such as Visual Evaluation of Soil Structure (VESS) [12,13] are being used to document how LUC is affecting soil physical quality so that better soil and crop management practices can be implemented. Specifically, those pictures (Figure 3) show a relatively well-aggregated soil under native vegetation. Transition to pasture maintains an organic-matter enriched surface, but begins to show signs of compaction due to hoof traffic often associated with over-grazing. Finally, due to tillage and very heavy equipment associated with sugarcane production, soil structure is further degraded until productivity diminishes [14] or the site is restored through deep and aggressive tillage, but that in turn leads to accelerated decomposition of SOM.

Anthropogenic Biomes of the World, Version 2, 2000: Global
Anthropogenic Biomes

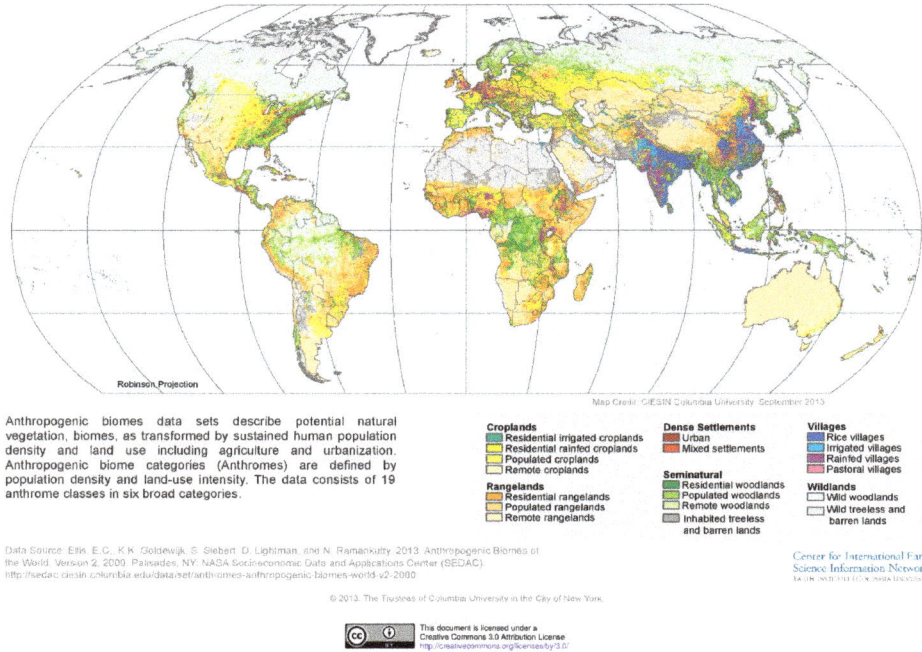

Anthropogenic biomes data sets describe potential natural vegetation, biomes, as transformed by sustained human population density and land use including agriculture and urbanization. Anthropogenic biome categories (Anthromes) are defined by population density and land-use intensity. The data consists of 19 anthrome classes in six broad categories.

Croplands
- Residential irrigated croplands
- Residential rainfed croplands
- Populated croplands
- Remote croplands

Rangelands
- Residential rangelands
- Populated rangelands
- Remote rangelands

Dense Settlements
- Urban
- Mixed settlements

Seminatural
- Residential woodlands
- Populated woodlands
- Remote woodlands
- Inhabited treeless and barren lands

Villages
- Rice villages
- Irrigated villages
- Rainfed villages
- Pastoral villages

Wildlands
- Wild woodlands
- Wild treeless and barren lands

Figure 2. Global land use projected for 2000 [15].

Soil change is a direct result of both population and economic growth. The world's population surpassed 7.3 billion in 2015 and is projected by the United Nations to reach 8.4 billion by 2030 and 9.6 billion by 2050. Estimates of global food demand, based on population growth and dietary shifts associated with economic growth, indicate that production will need to increase by 40% to 70% by 2050 [16]. Amazingly, from the period between 1961 and 2000 food production increased by 146% while the amount of land in agriculture production increased by only 8%. This was achieved through increased inputs of nutrients and water. However this intensification puts pressure on the soil resources and raises the question of whether additional food needs can be met without further degradation? Traditionally, there has been no consistent monitoring of either the extent or type of soil degradation, but FAO and the Global Soil Partnership [10] are expected to release a comprehensive State of the World Soil Resources Report in 2015. It is expected that this report will be the first of a series to document soil change.

| Native Vegetation | Pasture | Sugarcane |

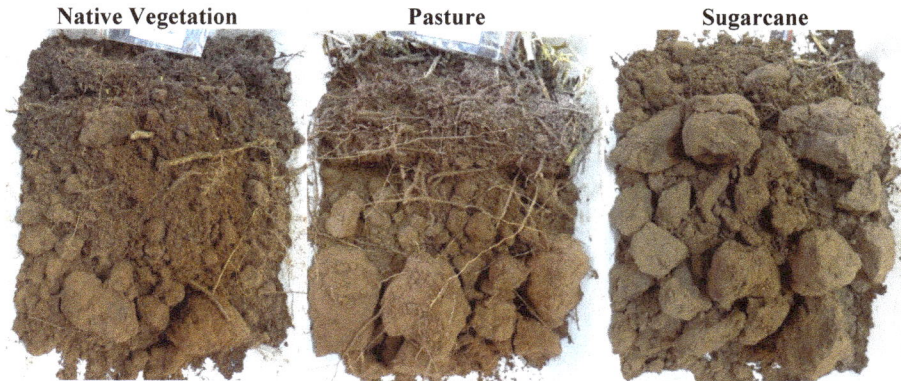

Figure 3. Land use change effects on the surface soil structure of an Oxisol in Brazil [14]. At U.S Department of Agriculture (USDA), National Laboratory for Agriculture and the Environment, Ames—IA, USA.

Another indication that public awareness of the importance of soil resources is finally increasing is that 2015 has been proclaimed as the International Year of Soils by the United Nations. As a result, many efforts are being undertaken to increase public awareness of the importance of soil [17,18]. We suggest the papers contained in the Special Issue of *Sustainability* entitled "Enhancing Soil Health to Mitigate Soil Degradation" will contribute significantly to those efforts by highlighting the global need for more awareness and understanding of how complex soils are and how valuable they are to humankind.

3. Geographic Consistency in Soil Degradation

"Enhancing Soil Health to Mitigate Soil Degradation" includes in-depth reviews and research results from Africa, Australia, China, Europe, India, North and South America, and Russia that provide a very clear, strong and consistent message that soil erosion and depletion of soil organic matter are two key indicators of and therefore intervention points for reversing soil degradation.

Key points identified by the various authors are briefly summarized below to entice potential readers explore the problem of soil degradation and potential strategies for its remediation more fully in the regions that are most important to them. For example, many people consider the U.S. Dust Bowl to have been the first major event associated with soil degradation in North America [19], but in reality, the catastrophic wind erosion associated with that event was preceded by soil degradation due to water erosion in the southern U.S. and nutrient depletion in many New England areas. One of the most significant effects of the Dust Bowl, however, was that it provided the impetus to improve agricultural management of soils being degraded by both wind and water erosion. Currently, North American soils experience soil degradation due to compaction, salinization, acidification, and contamination by anthropogenic compounds, but wind- and water-induced soil erosion that results in deteriorated physical properties, nutrient losses, and reshaped, potentially unworkable, field surface conditions remains the predominant driver. A close second, however, is the loss of soil organic matter which decreases aggregate stability, weakens soil structure, and negatively impacts soil water availability to crops. The magnitude of SOM loss is documented by the fact that in North America soil organic matter concentrations are only about 50% of the level they were when land was converted from forests or prairies to farmland. In Bolivia, Paraguay, Uruguay, Argentina and southern Brazil [20], unprecedented adoption of no-tillage, as well as improved soil fertility and plant genetics have significantly increased yields and altered the role of these countries in helping meet an increasing global food and feed demand. However, various land use changes such as converting pasture to continuous soybean

[*Glycine max* (L.) Merr.] or sugarcane have contributed to various levels of soil degradation through wind and water erosion. In Western Europe [21], the extent and causes of chemical, physical, and biological soil degradation and soil loss vary greatly, but agriculture and forestry are the main causes for physical degradation, erosion, and loss of SOM. In Eastern Europe [22], a diverse topography along with deforestation, changing climatic conditions, long-term human settlement, overuse of agricultural lands without sustainable planning, cultural difficulties in accepting conservative land management practices and wrong political decisions have all increased the vulnerability of many soils to degradation and resulted in a serious decline in their functional capacity. Once again, the predominant causes of soil degradation were water and wind erosion, organic matter depletion, salinity, acidification, crusting and sealing, and compaction.

Soil degradation in India [23] is estimated to be affecting 147 Mha which is extremely serious considering that country supports 18% of the world's human population and 15% of the animal population on just 2.4% of the global land area. The causes for this degradation are both natural and human induced. Natural causes include earthquakes, tsunamis, droughts, avalanches, landslides, volcanic eruptions, floods, tornadoes, and wildfires; while human-induced soil degradation results from land clearing and deforestation, inappropriate agricultural practices, improper management of industrial effluents and wastes, over-grazing, careless management of forests, surface mining, urban sprawl, and commercial/industrial development. Inappropriate agricultural practices include excessive tillage, use of heavy machinery, excessive and unbalanced use of inorganic fertilizers, poor irrigation and water management techniques, pesticide overuse, inadequate crop residue and/or organic carbon inputs, and poor crop cycle planning. Contributions from both China [24] and Russia [25] focus primarily on wind erosion. To decrease soil loss and enhance local ecosystems, the Chinese government has been encouraging residents to reduce wind-induced soil degradation through a series of national policies and several ecological projects. These measures include conservation tillage, windbreak networks, checkerboard barriers, afforestation, and grassland enclosures. As a result, the aeolian degradation of land in many regions of arid and semiarid northern China are being controlled. In Russia, extensive cultivation of Chernozems that were some of the most naturally fertile soils in the world, with thick A horizons, lost a significant amount of the original organic matter stocks and had become much less productive by the second half of the 19th century. Restoration programs focused on planting windbreaks were implemented to rehabilitate and remediate the degraded soils. These practices protected cropland from wind and water erosion, improved the microclimate for crop growth, and provided new refugia for wild animal and plant habitats. During the last several decades, these windbreaks have begun to be viewed as ecosystems with great potential for atmospheric carbon sequestration, which plays a positive role in climate change mitigation while also improving soil quality by increasing soil organic matter concentrations. Soil degradation in Australia [26] is also dominated by soil erosion. The authors present evidence for three key phases of soil degradation since European settlement and show a clear link between inappropriate agricultural practices and soil degradation. Fortunately, modern agricultural practices are significantly reducing erosion losses. The contribution from Ethiopia [27] pointed out that many of their soils have been exhausted for several decades due to over exploitation and mismanagement. Using the Soil Management Assessment Framework (SMAF), they then showed that implementation of agro-forestry practices resulted in improved water entry, movement and availability through increased water-stable aggregation, soil carbon and nitrogen. The second contribution from Africa [2] concluded that the primary cause of soil degradation in SSA is expansion and intensification of agriculture in efforts to feed its growing population. The authors conclude that to mitigate soil degradation, effective solutions must support resilient systems and cut across agricultural, environmental, and socioeconomic objectives.

Two of the contributions [28,29] critically examine the factors causing soil degradation in order to highlight the interconnected nature of social and economic causes of soil degradation. They stress that as the intensity and frequency of both droughts and flooding increase, consumer confidence and the ability of crops to reach important new yield goals are also threatened. Glæsner *et al.* [29] point out that currently no European-scale legislation focuses exclusively on soil conservation. Rather, they found that three soil threats (compaction, salinization and soil sealing) were not even addressed in any of the 19 legislative policies they analyzed. In contrast, erosion, decline in organic matter, and loss of biodiversity and contamination were covered in existing legislation, but only a few directives provided targets for reducing the soil threats.

The final two papers [30,31] focus on strategies that may help prevent further soil degradation or help remediate soil resources that have experienced degradation. Lal (30) summarize strategies to reverse degradation and concludes they are to: (i) reduce soil erosion; (ii) create a positive soil C budget; (iii) improve nutrient availability; (iv) increase soil biodiversity; and (v) enhance rhizosphere processes. However Lal (30) stressed the importance of managing soil organic C as other soil properties are associated with soil organic C.

In the final paper [31], the authors examine how soil biology influences soil health and how biological properties and processes contribute to sustainability of agriculture and ecosystem services. They also critically reviewed what could be done to manipulate soil biology to: (i) increase nutrient availability for production of high yielding, high quality crops; (ii) protect crops from pests, pathogens, weeds; and (iii) manage other factors limiting production, provision of ecosystem services, and resilience to stresses such as drought.

Finally, during the review process associated with this research editorial, it was suggested that too much blame was placed on human-induced soil degradation and that natural, environmental causes were not given sufficient attention. Our initial reaction was that the individual contributions were well balanced, but since the reviewer's concern focused primarily on Andean and other tropical regions of Mesoamerica and the Caribbean Basin, and that region was not addressed in the Special Issue, we decided to provide selected examples of agroforestry projects [32–37] that have addressed soil degradation in that tropical region.

A detailed evaluation of the selected studies and numerous others in the literature is beyond the scope of this editorial, but in general, agroforestry systems have been identified as an effective way to mitigate soil degradation in the humid tropics. Agroforestry systems consisting of various tropical hardwoods, cacao (*Theobroma cacao*), and coffee (*Coffea* spp.), with and without cattle, have been evaluated to quantify their effects on soil erosion, conservation, pesticide requirements, biodiversity, nutrient leaching and other ecosystem services. Many agroforestry systems have been successful because they mimic the natural forest ecosystems that are being lost to agricultural development. However, others [36] have shown no advantage when compared with pasture areas. We suggest that as with all soil and crop management practices, the variability associated with the selected studies simply emphasizes the importance of site specific management and accounting for both anthropogenic and non-anthropogenic causes of soil degradation.

4. Summary and Conclusions

Humankind's history of making inappropriate land use and soil management decisions can be depressing, but we are optimistic that as our knowledge increases through research, sustainable development, and improved education, our collective decision-making processes will also be improved. It is our hope that the Special Issue of *Sustainability* entitled "Enhancing Soil Health to Mitigate Soil Degradation," which clearly establishes that erosion and SOM loss are almost universal indicators of soil degradation, will provide a foundation for studies designed to improve long-term soil and crop management. Implicit in all of the reports is a recommendation for coordinated planning to address physical, chemical, and biological properties and processes that are essential for restoring degraded soils and improving soil health. Addressing soil functions individually has not and will not be successful because of the multi-functionality of soil resources.

Furthermore, our climate is changing and future weather patterns are increasingly uncertain. Therefore, the improved management practices must integrate unique differences in climate and site-specific soil properties. This means that it will be impossible to develop a single, common solution or priority for mitigating soil degradation. Prevention and remediation will require integrated solutions that control all processes governing wind and water erosion, contamination, acidification, salinization, nutrient depletion, and SOM loss. For agricultural soils, management practices including the use of cover crops and/or appropriate crop residue management to reduce raindrop impact, maintain good infiltration rates, and increase soil water retention and release to plants will ultimately increase crop production, increase carbon sequestration and improve soil health. In other areas, agroforestry and planting of windbreaks has been demonstrated to be an effective means for increasing C storage, restoring soil fertility, improving nutrient cycling and availability. The common factor among these various management practices is improved soil carbon management.

Finally, we argue that continuous monitoring of soil resources is needed to document the direction and extent of change in soil resources. This information is needed by land managers and policy makers. Future research efforts should focus on how soil degradation leads to changes in soil ecosystem services, and what land management strategies make systems resilient and thus, more sustainable. Information about soils, particularly degraded soils, must be integrated into climate. This will require cooperation, innovation and communication across many groups, and specifically for soil scientists to become actively involved in trans-disciplinary studies, to broaden their focus, and to publish their results in a language that is accessible to others. One of the most promising endeavors where such interaction is desperately needed is for increased public-private research efforts focused on soil biology. We consider this of utmost importance because of the three indicator regimes (physical, chemical, and biological) influencing soil health/quality, biological relationships are by far the most complex and least understood. Fortunately, many new tools and techniques have been or are being developed to unravel these complex systems. Ultimately, this new knowledge will be used to develop appropriate, site-specific management practices that can restore degraded soils and thus enable humankind to meet rapidly increasing food, feed, fiber, and fuel needs of an expanding global population, while protecting our vital soil resources.

Acknowledgments: The U.S. Department of Agriculture (USDA) prohibits discrimination in all its programs and activities on the basis of race, color, national origin, age, disability, and where applicable, sex, marital status, familial status, parental status, religion, sexual orientation, genetic information, political beliefs, reprisal, or because all or part of an individual's income is derived from any public assistance program. (Not all prohibited bases apply to all programs.) Persons with disabilities who require alternative means for communication of program information (Braille, large print, audiotape, *etc.*) should contact USDA's TARGET Center at (202) 720-2600 (voice and TDD). To file a complaint of discrimination, write to USDA, Director, Office of Civil Rights, 1400 Independence Avenue, S.W., Washington, D.C. 20250-9410, or call (800) 795-3272 (voice) or (202) 720-6382 (TDD). USDA is an equal opportunity provider and employer. Mention of trade names or commercial products in this publication is solely for the purpose of providing specific information and does not imply recommendation or endorsement by the U.S. Department of Agriculture or Kansas State University.

Author Contributions: Both authors contributed equally to the development, writing, review and approval of the final manuscript.

Conflicts of Interest: The authors declare no conflict of interest.

References

1. Sanchez, P.A.; Swaminathan, M.S. Cutting world hunger in half. *Science* **2005**, *307*, 357–359. [CrossRef] [PubMed]
2. Tully, K.; Sullivan, C.; Weil, R.; Sanchez, P. The State of Soil Degradation in Sub-Saharan Africa: Baselines, Trajectories, and Solutions. *Sustainability* **2015**, *7*, 6523–6552. [CrossRef]
3. Hillel, D. *Out of the Earth: Civilization and the Life of the Soil*; University of California Press: Oakland, CA, USA, 1991.
4. Lowdermilk, W.C. *Conquest of the Land through 7000 Years*; USDA Soil Conservation Service; U.S. Department of Agriculture: Washington, DC, USA, 1953.
5. Montgomery, D.R. Soil erosion and agricultural sustainability. *Proc. Natl. Acad. Sci. USA* **2007**, *104*, 13268–13272. [CrossRef] [PubMed]

6. Karlen, D.L.; Peterson, G.A.; Westfall, D.G. Soil and water conservation: Our history and future challenges. *Soil Sci. Soc. Am. J.* **2014**, *78*, 1493–1499. [CrossRef]

7. Janzen, H.H.; Fixen, P.A.; Franzluebbers, A.J.; Hattey, J.; Izaurralde, R.C.; Ketterings, Q.M.; Lobb, D.A.; Schlesinger, W.H. Global Prospects Rooted in Soil Science. *Soil Sci. Soc. Am. J.* **2011**, *75*, 1–8. [CrossRef]

8. Ball, B.C.; Batey, T.; Munkholm, L.J. Field assessment of soil structural quality—A development of the Peerlkamp test. *Soil Use Manag.* **2007**, *23*, 329–337. [CrossRef]

9. Guimarães, R.M.L.; Ball, B.C.; Tormena, C.A. Improvements in the visual evaluation of soil structure. *Soil Use Manag.* **2015**, *27*, 395–403. [CrossRef]

10. Bai, Z.G.; Dent, D.L.; Olsson, L.; Schaepman, M.E. *Global Assessment of Land Degradation and Improvement. 1. Identification by Remote Sensing*; Report 2008/01; ISRIC-World Soil Information: Wageningen, The Netherlands, 2008.

11. Ellis, E.C.; Goldewijk, K.K.; Siebert, S.; Lightman, D.; Ramankutty, N. Anthropogenic Biomes of the World. Version 2: 1700. NASA Socioeconoic Data and Applications Center (SEDAC): Palisades, NY, 2013. Available online: http://sedac.ciesin.columbia.edu/datda/set/anthromes-anthropogenic-biomes-world-v2-1700 (accessed on 9 September 2015).

12. Latham, J.; Cumani, R.; Rosati, I.; Bloise, M. *Global Land Cover SHARE (GLC-SHARE) Database Beta-Release Version 1.0*; FAO: Rome, Italy, 2014.

13. Ramankutty, N.; Foley, J.A.; Olejniczak, N.J. People on the land: Changes in global population and croplands during the 20th century. *AMBIO* **2002**, *31*, 251–257. [CrossRef] [PubMed]

14. Cherubin, M.R.; Department of Soil Science, "Luiz de Queiroz" College of Agriculture, University of São Paulo, 11 Pádua Dias Avenue, Piracicaba, SP 13418-900 Brazil. Personal communication, 2015.

15. Ellis, E.C.; Goldewijk, K.K.; Siebert, S.; Lightman, D.; Ramankutty, N. Anthropogenic Biomes of the World. Version 2: 2000. Palisades, NY; NASA Socioeconoic Data and Applications Center (SEDAC), 2013. Available online: http://sedac.ciesin.columbia.edu/datda/set/anthromes-anthropogenic-biomes-world-v2-2000 (accessed on 9 September 2015).

16. World Resources Institute. *Creating a Sustainable Food Future. Report 2013–2014: Interim Findings*; World Resources Institutute: Washington, DC, USA, 2014.

17. Wall, D.; Six, J. Give soils their due. *Science* **2015**, *347*, 695. [CrossRef]

18. Amundson, R.; Berhe, A.A.; Hopmans, J.W.; Olson, C.; Sztein, A.E.; Sparks, D.L. Soil and human security in the 21st century. *Science* **2015**, *348*. [CrossRef] [PubMed]

19. Baumhardt, R.L.; Stewart, B.A.; Sainju, U.M. North American Soil Degradation: Processes, Practices, and Mitigating Strategies. *Sustainability* **2015**, *7*, 2936–2960. [CrossRef]

20. Wingeyer, A.B.; Amado, T.J.C.; Pérez-Bidegain, M.; Studdert, G.A.; Varela, C.H.P.; Garcia, F.O.; Karlen, D.L. Soil Quality Impacts of Current South American Agricultural Practices. *Sustainability* **2015**, *7*, 2213–2242. [CrossRef]

21. Virto, I.; Imaz, M.J.; Fernández-Ugalde, O.; Gartzia-Bengoetxea, N.; Enrique, A.; Bescansa, P. Soil Degradation and Soil Quality in Western Europe: Current Situation and Future Perspectives. *Sustainability* **2015**, *7*, 313–365. [CrossRef]

22. Günal, H.; Korucu, T.; Birkas, M.; Özgöz, E.; Halbac-Cotoara-Zamfir, R. Threats to Sustainability of Soil Functions in Central and Southeast Europe. *Sustainability* **2015**, *7*, 2161–2188. [CrossRef]

23. Bhattacharyya, R.; Ghosh, B.N.; Mishra, P.K.; Mandal, B.; Rao, C.; Sarkar, D.; Das, K.; Anil, K.S.; Lalitha, M.; Hati, K.M.; *et al.* Soil Degradation in India: Challenges and Potential Solutions. *Sustainability* **2015**, *7*, 3528–3570. [CrossRef]

24. Guo, Z.; Huang, N.; Dong, Z.; van Pelt, R.S.; Zobeck, T.M. Wind Erosion Induced Soil Degradation in Northern China: Status, Measures and Perspective. *Sustainability* **2015**, *6*, 8951–8966. [CrossRef]

25. Chendev, Y.G.; Sauer, T.J.; Ramirez, G.H.; Burras, C.L. History of East European Chernozem Soil Degradation: Protection and Restoration by Tree Windbreaks in the Russian Steppe. *Sustainability* **2015**, *7*, 705–724. [CrossRef]

26. Koch, A.; Chappell, A.; Eyres, M.; Scott, E. Monitor Soil Degradation or Triage for Soil Security: An Australian Challenge. *Sustainability* **2015**, *7*, 4870–4892. [CrossRef]

27. Gelaw, A.M.; Singh, B.R.; Lal, R. Soil Quality Indices for Evaluating Smallholder Agricultural Land Uses in Northern Ethiopia. *Sustainability* **2015**, *7*, 2322–2337. [CrossRef]

28. DeLong, C.; Cruse, R.; Wiener, J. The Soil Degradation Paradox: Compromising Our Resources When We Need Them the Most. *Sustainability* **2015**, *7*, 866–879. [CrossRef]

29. Glæsner, N.; Helming, K.; de Vries, W. Do Current European Policies Prevent Soil Threats and Support Soil Functions? *Sustainability* **2015**, *6*, 9538–9563. [CrossRef]

30. Lal, R. Managing Carbon for Restoring Degraded Soils. *Sustainability* **2015**, *7*, 5875–5895. [CrossRef]

31. Lehman, R.M.; Cambardella, C.A.; Stott, D.E.; Acosta-Martinez, V.; Manter, D.K.; Buyer, J.S.; Maul, J.E.; Smith, J.L.; Collins, H.P.; Halvorson, J.J.; *et al.* Understanding and Enhancing Soil Biological Health: The Solution for Reversing Soil Degradation. *Sustainability* **2015**, *7*, 988–1027. [CrossRef]

32. Beenhouwer, M.; de Aerts, R.; Honnay, O. A global meta-analysis of the biodiversity and ecosystem service benefits of coffee and cacao agroforestry. *Agric. Ecosyst. Environ.* **2013**, *175*, 1–7. [CrossRef]

33. Blanco, R.; Nieuwenhuyse, A. Influence of topographic and edaphic factors on vulnerability to soil degradation due to cattle grazing in humid tropical mountains in northern Honduras. *Catena* **2011**, *86*, 130–137. [CrossRef]

34. Blanco, R.; Aguilar, A. Soil erosion and erosion thresholds in an agroforestry system of coffee (*Coffea Arabica*) and mixed shade trees (*Inga* spp. and *Musa* spp.) in nNorthern Nicaragua. *Agric. Ecosyst. Environ.* **2015**, *210*, 25–35. [CrossRef]

35. Borkhataria, R.; Collazo, J.A.; Groom, M.J.; Jordan-Garcia, A. Shade-grown coffee in Puerto Rico: Opportunities to preserve biodiversity while reinvigorating a struggling agricultural commodity. *Agric. Ecosyst. Environ.* **2012**, *149*, 164–170. [CrossRef]

36. Tornquist, C.G.; Hons, F.M.; Feagley, S.E.; Haggar, J. Agroforestry system effects on soil characteristics of the Sarapiqui region of Costa Rica. *Agric. Ecosyst. Environ.* **1999**, *73*, 19–28. [CrossRef]

37. Tully, K.L.; Lawrence, D.; Scanlon, T.M. More trees less loss: Nitrogen leaching losses decrease with increasing biomass in coffee agroforests. *Agric. Ecosyst. Environ.* **2012**, *161*, 137–144. [CrossRef]

sustainability

MDPI

Article

Restoring Soil Quality to Mitigate Soil Degradation

Rattan Lal

The Ohio State University, Columbus, OH 43210, USA; lal.1@osu.edu; Tel.: +1-614-292-9069

Academic Editor: Marc A. Rosen

Received: 31 March 2015; Accepted: 5 May 2015; Published: 13 May 2015

Abstract: Feeding the world population, 7.3 billion in 2015 and projected to increase to 9.5 billion by 2050, necessitates an increase in agricultural production of ~70% between 2005 and 2050. Soil degradation, characterized by decline in quality and decrease in ecosystem goods and services, is a major constraint to achieving the required increase in agricultural production. Soil is a non-renewable resource on human time scales with its vulnerability to degradation depending on complex interactions between processes, factors and causes occurring at a range of spatial and temporal scales. Among the major soil degradation processes are accelerated erosion, depletion of the soil organic carbon (SOC) pool and loss in biodiversity, loss of soil fertility and elemental imbalance, acidification and salinization. Soil degradation trends can be reversed by conversion to a restorative land use and adoption of recommended management practices. The strategy is to minimize soil erosion, create positive SOC and N budgets, enhance activity and species diversity of soil biota (micro, meso, and macro), and improve structural stability and pore geometry. Improving soil quality (*i.e.*, increasing SOC pool, improving soil structure, enhancing soil fertility) can reduce risks of soil degradation (physical, chemical, biological and ecological) while improving the environment. Increasing the SOC pool to above the critical level (10 to 15 g/kg) is essential to set-in-motion the restorative trends. Site-specific techniques of restoring soil quality include conservation agriculture, integrated nutrient management, continuous vegetative cover such as residue mulch and cover cropping, and controlled grazing at appropriate stocking rates. The strategy is to produce "more from less" by reducing losses and increasing soil, water, and nutrient use efficiency.

Keywords: soil resilience; climate change; soil functions; desertification; soil carbon sequestration

---

1. Introduction

Of the 5.5 billion people living in developing countries in 2014 [1], a large proportion of them depend on agriculture for their livelihood. In fact, one billion of these people are small landholders who cultivate <2 ha of land [2]. With limited resources and poor access to inputs, management of soil quality is essential to strengthen and sustain ecosystem services. Soil degradation is a 21st century global problem that is especially severe in the tropics and sub-tropics. Some estimates indicate degradation decreased soil ecosystem services by 60% between 1950 and 2010 [3]. Accelerated soil degradation has reportedly affected as much as 500 million hectare (Mha) in the tropics [4], and globally 33% of earth's land surface is affected by some type of soil degradation [5]. In addition to negatively impacting agronomic production, soil degradation can also dampen economic growth, especially in countries where agriculture is the engine for economic development [6]. Over and above the environmental and economic impacts, there are also health risks of soil erosion [7] and other degradation processes [8].

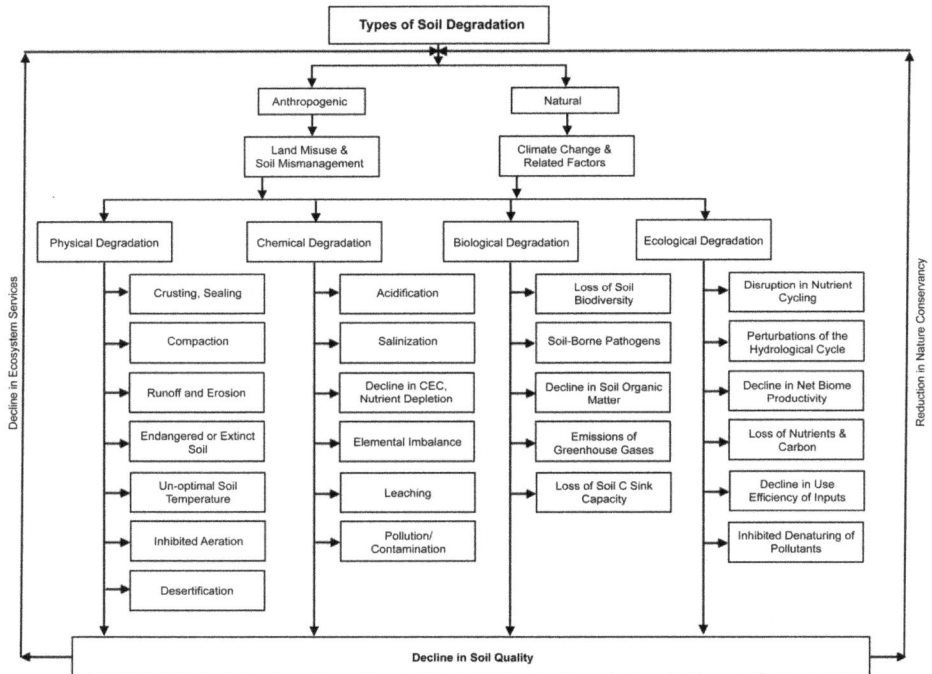

Figure 1. Types of soil degradation.

Soil degradation implies a decline in soil quality [8] with an attendant reduction in ecosystem functions and services. Conceptually, there are four types of soil degradation: (i) physical; (ii) chemical; (iii) biological; and (iv) ecological (Figure 1). Soil physical degradation generally results in a reduction in structural attributes including pore geometry and continuity, thus aggravating a soil's susceptibility to crusting, compaction, reduced water infiltration, increased surface runoff, wind and water erosion, greater soil temperature fluctuations, and an increased propensity for desertification. Soil chemical degradation is characterized by acidification, salinization, nutrient depletion, reduced cation exchange capacity (CEC), increased Al or Mn toxicities, Ca or Mg deficiencies, leaching of NO_3-N or other essential plant nutrients, or contamination by industrial wastes or by-products. Soil biological degradation reflects depletion of the soil organic carbon (SOC) pool, loss in soil biodiversity, a reduction in soil C sink capacity, and increased greenhouse gas (GHG) emissions from soil into the atmosphere. One of the most severe consequences of soil biological degradation is that soil becomes a net source of GHG emissions (*i.e.*, CO_2 and CH_4) rather than a sink. Ecological degradation reflects a combination of other three, and leads to disruption in ecosystem functions such as elemental cycling, water infiltration and purification, perturbations of the hydrological cycle, and a decline in net biome productivity. The overall decline in soil quality, both by natural and anthropogenic factors, has strong positive feedbacks leading to a decline in ecosystem services and reduction in nature conservancy. Once the process of soil degradation is set-in-motion, often by land misuse and soil mismanagement along with the extractive farming, it feeds on itself in an ever-increasing downward spiral (Figure 2).

The objectives of this review are to: (1) deliberate the role of soil resources in provisioning essential ecosystem services; (2) illustrate the impacts of soil degradation on decline in ecosystem services; and (3) identify strategies for improving soil quality to mitigate risks of soil degradation.

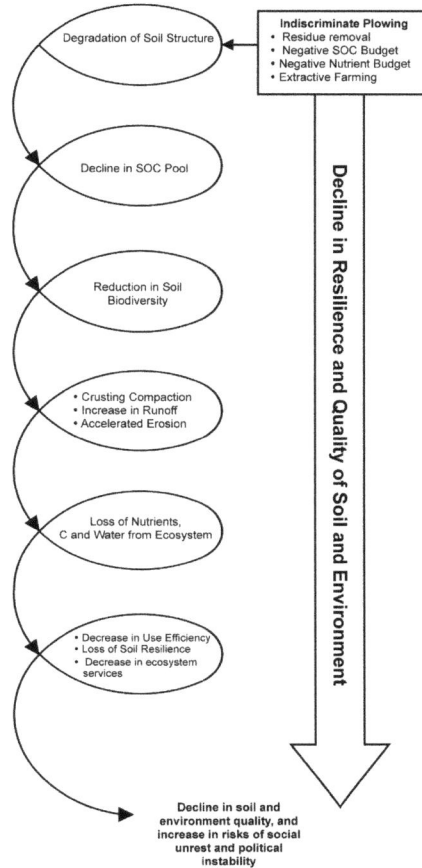

Figure 2. The downward spiral of decline in soil and environment quality exacerbated by indiscriminate plowing, residue removal and extractive farming.

2. Soil and Ecosystem Services

Soil, the most basic of all resources, is the essence of all terrestrial life and a cultural heritage [9]. Yet, soil is finite in extent, prone to degradation by natural and anthropogenic factors, and is non-renewable over the human timescale (decades). Soil quality also has strong implications to human health [8,10], thus illustrating its important role in both society and the environment. Because of numerous ecosystem services provisioned through soils (e.g., food, feed, fiber, climate moderation through C cycling, waste disposal, water filtration and purification, elemental cycling) [11,12], soil quality must be protected or restored to enhance these services. Increased public awareness and a fundamental understanding of basic pedospheric processes (*i.e.*, biology, chemistry, physics, pedology, ecology) are essential both to enhancing long-term productivity and improving the environment [13].

3. Soil Organic Carbon and Its Impact on Soil Quality

The SOC pool, including its quantity and quality, is the defining constituent of soil [14,15]. Indeed, SOC pool is the most reliable indicator of monitoring soil degradation, especially that caused by accelerated erosion [16]. Soil degradation depletes the SOC pool, along with it, plant available N and

13

other essential nutrients such as P and S. Furthermore, as identified repeatedly in this special issue of Sustainability, depletion of SOC pool is a global issue and a principal cause of soil degradation, especially in the European semi-arid Mediterranean regions [17]. Developing strategies to ensure the SOC pool is to increase and preferably maintain above the threshold or critical level of 10 to 15 g/kg (1.0%–1.5%), which is essential for reducing soil degradation risks and reversing degradation trends. Integrated nutrient management (INM) is one strategy that embodies sustainable management of the SOC pool and its dynamics [18]. Adoption of INM or similar management practices that create a positive soil/ecosystem C budget can not only increase productivity but also sequester additional atmospheric CO_2 into the SOC pool. This has been documented for many surface soils within the U.S. Corn Belt which act as C sinks when corn (*Zea mays* L.) is grown using recommended management practices (RMPs) such as conservation agriculture (CA) [19]. There also exists a strong relationship between vegetation cover and the SOC pool, such that excessive reductions in vegetation cover exacerbates risks of soil degradation and SOC depletion. A study conducted in the sub-tropical humid grasslands in South Africa indicated that the decline in grass (vegetative) cover from 100% to 0%–5% reduced the SOC pool by 1.25 kg/m^2 and the soil organic N (SON) pool by 0.074 kg/m^2 [20], There were also attendant declines in the C:N ratio and proportion of SOC and SON in the silt + clay fraction with the decline in aerial grass cover which negatively affected ecosystem functions of the acidic sandy loam soils. Similarly, transformation of a thicket vegetation to an open savanna (dominated by grasses) due to intensive grazing decreased soil quality in the Eastern Cape region of South Africa [21]. Indeed, savanna soils have lower SOC concentration and a greater tendency to crust than thicket soils because of the decreased quantity and stability of structural aggregates.

The widespread prevalence of degraded soils in sub-Saharan Africa (SSA), a classic example of a downward spiral, is attributed to over exploitation, extractive farming, low external inputs, and poor or improper management (Figure 2). Accelerated degradation is shrinking the finite soil resource even more rapidly in these regions of harsh climate and fragile soils. In this context, enhancing the SOC pool is important to sustain soil fertility and agronomic productivity [22]. Simply adding chemical fertilizers or improved varieties, as is often erroneously recommended even by well-intended advocates, is not enough.

The SOC pool of agricultural soils of West Africa, similar to those of croplands in other developing counties (e.g., South Asia), is severely depleted by over-exploitation of natural resources [23]. These soils must therefore be managed to increase both soil C and vegetation [24,25] to restore the degraded agroecosystem services. Changes in aerial vegetative cover could thus be used as an early indicator of shifts in soil ecosystem functions within fragile environments. A shift in vegetative cover may be caused by alterations in land use or climate change. In addition to SOC and SON pools, soil moisture regime is another important indicator of climate change [26]. In conjunction with changes in soil moisture regimes, projected global warming may also influence SOC decomposition rates [27,28] including that of fine woody debris. Field experiments have shown that warming increases mass loss for all vegetative species and size classes by as much as 30%. However, larger debris and that with higher initial lignin content decomposes more slowly than smaller debris and that with lower antecedent lignin content. Indeed, degradation of lignin may not follow the same trend as that of total mass loss [29]. Along with the adverse effects of soil erosion and other degradation processes, the SOC pool is also prone to climate change and associated alterations in temperature and moisture regimes.

The self-reinforcing soil degradation process (Figure 2) is strongly exacerbated by the interaction between processes, factors and causes of soil degradation (Figure 3). Processes include the mechanisms (types) of soil degradation. Factors comprise agents of degradation related to natural or anthropogenic drivers such as climate, physiography, socio-economic or ethnic/cultural parameters. Causes of soil degradation include specific activities which aggravate the adverse effects of processes and factors. Examples of specific causes include activities such as deforestation, land use conversion, extractive farming practices or over-exploitation, excessive grazing, excessive plowing *etc.* (Figure 3). The process-factor- cause nexus is strongly impacted by site-specific conditions. Thus, understanding the nexus or connectivity is critical to restoring soil quality and mitigating degradation.

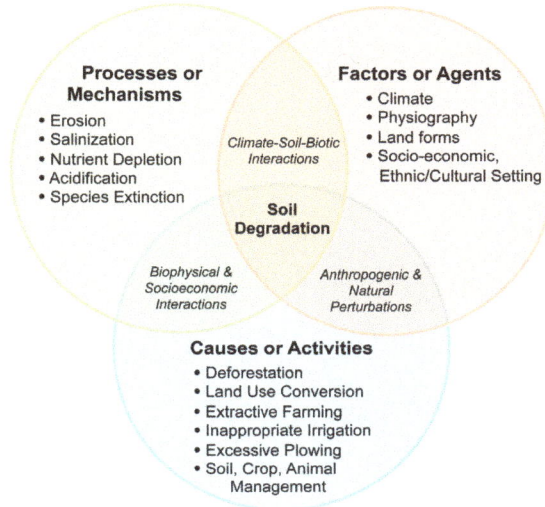

Figure 3. The process-factor-cause nexus as a driver of soil degradation.

4. Soil Quality Index

The SOC pool is a key indicator of soil quality, and an important driver of agricultural sustainability. In addition to its amount, other parameters of SOC include its depth distribution, quality or attributes (physical, chemical, biological), and the turnover rate or the mean residence time (MRT). Relevant indicators of soil physical quality include amount and stability of aggregates; susceptibility to crusting and compaction; porosity comprising of pore geometry and continuity; water transmission (infiltration rate and amount) and retention as plant-available water capacity (PAWC); aeration and gaseous exchange; effective rooting depth; soil heat capacity and the temperature regime. Similarly, appropriate indicators of soil chemical quality include pH, CEC, nutrient availability; and favorable elemental balance and lack of any toxicity or deficiency. Relevant indicators of soil biological quality are microbial biomass C (MBC), activity and diversity of soil fauna and flora, absence of pathogens and pests as indicated by a soil's disease-suppressive attributes. An optimal combination of these properties affects agronomic productivity; use efficiency of water, nutrients and other inputs; and sustainability of management systems. Indicators of soil quality differ among soil types, climates and land uses. For example, there are specific soil quality indicators for the intensively managed soils of the Indo-Gangetic Plains [30] that will differ from those for tropical Alfisols in semi-arid regions [31–34]. A spectral soil quality index based on application of reflectance spectroscopy has also been proposed as a diagnostic tool to assess soil quality [35]. This technique can provide a characterization of physical, chemical and biological attributes that can be merged together to indicate how well a soil is functioning for a specific use [36–39].

5. Conservation Agriculture and Soil Quality

Four basic principles of CA are [40]: (i) retention of crop residue mulch; (ii) incorporation of a cover crop in the rotation cycle; (iii) use of INM involving combination of chemical and bio fertilizers; and (iv) elimination of soil mechanical disturbances. Properly implemented on suitable soil types, CA has numerous co-benefits including reduced fuel consumption and increased soil C sequestration.

15

Mechanical tillage is an energy-intensive process [41] and its reduction or elimination can decrease consumption of fossil fuels. For example, conversion from plow tillage (PT) to CA can reduce diesel consumption by as much as 41 L/ha [42].

In addition, an increase in SOC pool under CA can occur in soils not prone to accelerated erosion, and those which have optimal management strategies. A modeling study in Western Kenya showed that site-specific optimal management strategies can lead to SOC pool of 20 to 40 Mg/ha in 0.1 m depth and corn grain yield of 3.5 to 4.2 Mg/ha [43]. The most desirable tillage systems are those which restore soil quality, minimize risks of soil erosion, improve use efficiency of rain water and fertilizers [44] and minimize risks of SOC and nutrient depletion. Impacts of CA on soil quality restoration, an example of an upward spiral, is outlined in Figure 4.

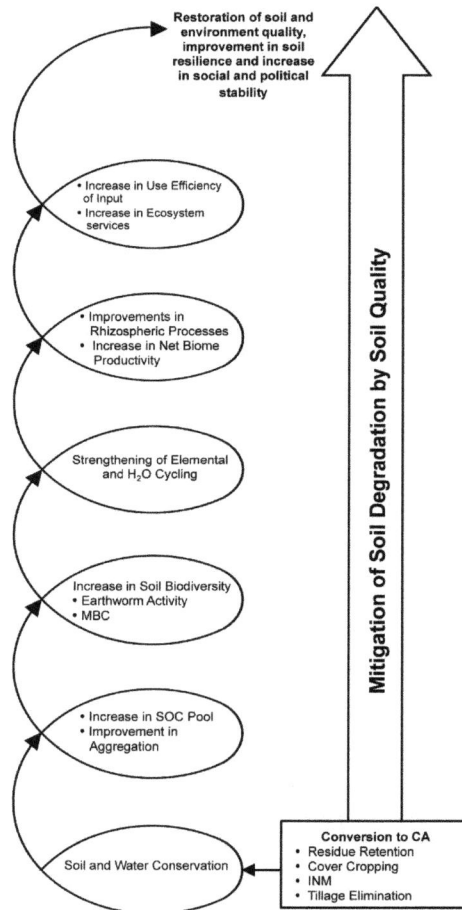

Figure 4. Increase in soil resilience and mitigation of soil degradation by conservation agriculture (CA). Meta-analyses and any other comparisons among unrelated soil management practices can lead to misinterpretation of SOC sink capacity by CA [45] and erroneous inferences on agronomic productivity [46]. The mission is to identify site-specific packages of CA practices to make it functional (INM = integrated nutrient management).

Sustainability **2015**, *7*, 5875–5895

Indiscriminate use of plowing, coupled with excessive removal of crop residues and unbalanced use of chemical fertilizers, can degrade soil quality, deplete SOC pool, and aggravate risks of soil erosion (Figure 2). In contrast, conversion of plow/traditional tillage to CA, especially on sloping lands and those vulnerable to accelerated erosion by water and wind under conventional management, can be conservation-effective, reverse degradation trends, and set-in-motion soil restoration processes with an upward spiral (Figure 4). Retention of crop residue mulch, and incorporation of a cover crop (forages) in the rotation cycle while eliminating bare fallows, can conserve soil and water and improve SOC pool in the surface layer (Figure 4). Increases in soil biodiversity, MBC and activity of earthworms and termites can all improve aggregation and encapsulate C within stable micro-aggregates as outlined in the hierarchy concept [47]. Strengthening elemental cycling, in conjunction with coupled cycling of C and H_2O, can increase the solum's C sink capacity and soil profile depth through increased bioturbation by earthworms or termites, and use of deep-rooted plants such as pigeon pea (*Cajanus cajan*), townsville stylo (*Stylosanthes humilis* (Kunth) Hester), or alfalfa (*Medicago sativa*). Long-term (>10 years) soil quality improvement will increase net biome productivity, improve water and nutrient use efficiencies, and increase above and below-ground biomass-C within the ecosystem. Progressive improvements in rhizospheric processes, driven by biotic mechanisms, would restore soil quality and mitigate degradation (Figure 4). It is important to note, however, that CA is a holistic and system-based approach. Mere elimination of plowing, while removing an excessive amount of crop residues and biomass for other uses (e.g., biofuels, industrial purposes) is not CA, and is rather an extractive farming system with negative impacts on soil and the environment. Thus, comparative analyses of un-related datasets can lead to erroneous assessments of SOC sink capacity associated with a properly implemented CA system [45] and misinterpretation of agronomic yields [46]. Furthermore, improvements in soil quality require more than the input of new varieties and chemical fertilizers [48]. The real issue is improving physical, biological and ecological components of soil quality, and set-in-motion an upward spiral eventually leading to social and political stability and international security (Figure 4).

6. Soil Fertility Management to Restore Soil Quality

Sustainable intensification (SI), producing more from less by reducing losses and increasing the use efficiency, is attainable only through improvement of soil quality including chemical quality or soil fertility. Although not the only way to increase soil fertility, the use of INM is a very effective approach for achieving SI. Nutrient depletion and loss of soil fertility are major causes of low productivity [49] in many developing countries. Use of organic amendments, by recycling organic by-products including urban waste, is a useful strategy to enhance soil fertility [50], and improve structural stability or aggregates [51]. While, nitrogen (N) input is important to improving soil fertility, its improper and/or excessive use can also lead to environmental pollution. China consumes about 30% of the world's N fertilizer [52], and is able to feed ~22% of the world population on just 6.8% of the global cropland area. However, the country has severe environmental problems because of low N use efficiency, leaching of reactive N into surface and groundwater resources, and emission of N (as N_2O) into the atmosphere.

7. Soil Quality and Water Resources

High soil quality or a healthy soil provides the foundation for a healthy economy, environment, and terrestrial biosphere. Thus, there exists a close link between soil quality and water resources in close proximity, such as the health of coastal ecosystems [53]. Changes in land use often affect water quality and pollutant loading [54,55]. Off-site movement of agricultural chemicals is often a significant source of non-point pollution. Many of the major rivers in countries with emerging economies have severe water pollution, contamination and eutrophication problems [56]. Downstream areas are often adversely affected because of adhoc agricultural development activities upstream. Among the most important adverse impacts are river desiccation, ground water depletion, surface and ground water pollution, accelerated erosion, sedimentation, salinization, and nutrient depletion [57].

These problems are especially severe in densely populated regions (e.g., East Asia, South Asia). Rapid urbanization, industrialization and increases in water demand have created severe water quality problems and degradation of the Ganges [58]. Land use changes induced by urbanization have had strong impacts both on soil and water quality in northern Iran [54]. Release of P from agricultural sources to streams and rivers has become a major factor affecting surface water quality not only in China [59], but also in the Great Lakes of North America (e.g., algal bloom of 2014). In Thailand, agriculture [especially rice (*Oryza sativa*) farming] is the key nutrient source within the Thachin River Basin [56]. Recent industrialization and community development have exacerbated water pollution and habitat degradation in the Gulf of Thailand [60]. The problem has been exacerbated by a rapid decrease in mangrove forest, coral reefs, and fishery resources due to misuse and mismanagement.

Irrigated agriculture, an important management strategy for high agronomic productivity in arid and semi-arid regions, is a mixed blessing. Mismanagement of irrigation waters has exacerbated problems with saline-sodic soils which now occupy more than 20% of the irrigated lands [61]. Furthermore, arid wetlands are also prone to contamination by sub-surface agriculture irrigation and drainage as in the Western USA [62]. These areas often experience toxicity problems in fish and wildlife due to drainage of water contaminants. In response, provisions must be made to reduce the amount of contaminants entering wetlands, and to provide for better allocation of freshwater between agriculture and wildlife [62]. Salinity problems are also often confounded by the reuse of untreated waste water (gray/black water) in agriculture [63], especially in urban areas prone to water shortages or those having water resources of marginal quality.

Restoring soil quality within managed ecosystems is critical to improving and sustaining water quality. To accomplish this goal, it is essential to develop strategies for integrated management of soil and water resources because of their strong inter-connectivity or the soil-water-waste nexus. While integrated water management alone is useful [64], the importance of the soil-water nexus cannot be over-emphasized [24]. Management of sediments, especially contaminated ones [65], is another important component of the soil-water nexus that must be critically examined.

8. Strategies for Soil Quality Restoration

Restoring the quality of degraded soils is a challenging task, especially in regions dominated by small, resource-poor landholders. Re-carbonization of the depleted SOC pool, which is essential to numerous functions, requires regular input of biomass-C and essential elements (*i.e.*, N, P, and S) [66]. Thus, restoration of soil quality is a societal, national and international task that necessitates a coordinated approach.

There are three basic strategies of restoring soil quality (Figure 5): (i) minimizing losses from the pedosphere or soil solum; (ii) creating a positive soil C budget, while enhancing biodiversity; and (iii) strengthening water and elemental cycling. There is no silver bullet or panacea to accomplish these basic tasks, and site-specific factors (biophysical, social, economic, cultural) play a significant role. Some examples of site-specific, RMPs are outlined in Table 1. However, each of these technologies has their own tradeoffs, which must be duly considered and minimized.

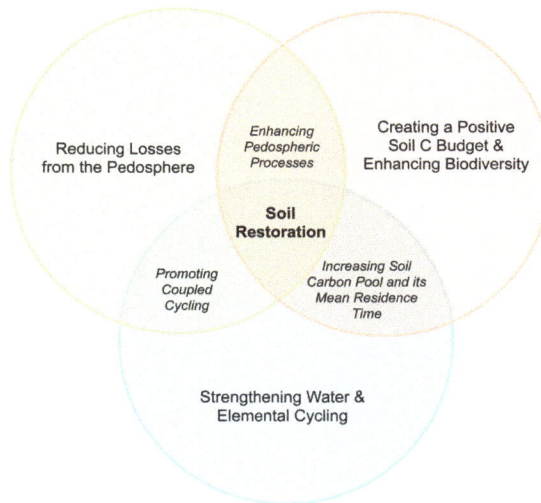

Figure 5. Three strategies of restoring and managing soil quality for mitigating risks of soil degradation.

Table 1. Strategies to improve soil quality.

Strategy	Region	Process	Reference
Litter turnover	Tropics	The rate of organic matter and C supply and nutrient cycling reactivation	[3]
Forestry Plantations	Tropics	Silvo-pastoral system for nutrient cycling	[67]
Woodlot Islets	Degraded drylands	Silvo-pastoral systems in drylands	[68]
Soil Carbon Sequestration	Agroecosystems	Optimal management strategies	[69]
Integrated Nutrient Management	Sub-Saharan Africa	Soil quality management	[17]
Nutrient Management for SOC Sequestration	Sub-Tropical Red Soils (China)	Soil carbon buildup	[70]
Manuring	Indus Plains	Application of farm manure	[71]
Residue Retention as Mulch	Mexican Highlands	Improvement of soil structure	[72]
Regular Organic Inputs	Western Kenya	Nutrient retention and soil structure improvement	[43,73]
Urban Waste	Mediterranean Europe	Enhancing soil fertility	[16,74]
Soil Biological Management	Global soils	Enhance ecosystem services provisioned by SOC pool	[15]
Environmental Awareness	U.S.	Promoting technology adoption	[75]

8.1. Soil Erosion Management

Soil erosion must be curtailed to within the tolerable limits, which is often much less than the presumed value of 12.5 Mg/ha per year. Accelerated erosion also depletes the SOC pool and nutrient reserves. In general, the enrichment ratio of SOC, clay and essential plant nutrients (N, P, S) is >1 (and most often as much as 5 or more) because of the preferential removal of these constituents. Conversion from PT to CA can reduce risks associated with soil erosion and nutrient loss while also providing numerous on- and off-site benefits (Figure 4) [12]. An important strategy is to establish cause-effect relationships, alleviate the causative factors and minimize the risks. Accelerated erosion is a symptom of land misuse and soil mismanagement. Reductions in plant cover caused by over-grazing and the trampling effect can degrade soil structure, reduce water infiltration, increase runoff, aggravate soil erosion, and cause severe economic losses [76]. As an example, experiments conducted in South Africa indicated that plant cover reduction by overgrazing significantly decreased the SOC pool with strong impact on the C cycle [77]. In arid regions, fire-induced depletions of the vegetation cover can also exacerbate the problem, especially after a torrential rainfall because ash left on the soil surface can aggravate hydrophobicity by creating an obtuse contact angle between the solid and liquid phases. When the protective litter cover is burned, the very first rainfall generally results in high surface runoff and aggravates erosion as was the classic situation of wild fire in May–June 2000 in hills which threatened the Los Alamos National Laboratory by runoff and sedimentation. Experiments conducted in Spain indicated high post-fire soil degradation risks and the need for identification of a short-term strategy to conserve soil and water on steep slopes with erodible soils [78]. In addition to the adverse effects on soil quality and productivity, there are also health risks associated with soil erosion [79]. This is especially true for regions prone to wind erosion and dust storms, such as the Harmattan in the Sahel [80].

8.2. Improving Soil/Agro-Biodiversity

Soil biota are important to soil quality and reduce risks of degradation and desertification. Indeed, soil biota comprise a major component of global terrestrial biodiversity and perform critical roles in key ecosystem functions (e.g., biomass decomposition, nutrient cycling, moderating CO_2 in the atmosphere, creating disease suppressive soils, *etc.*). Improving activity and species diversity of soil fauna and flora (micro, meso and macro) is therefore essential to restoring and improving soil quality and reducing risks of soil degradation. Adverse effects of agricultural management on soil microbiological quality is another global concern. As a management tool, either a microbiological quality index [81] or a microbiological degradation index [82] can be useful for decision-making processes [82] Relevant parameters include MBC, respiration, water soluble carbohydrates, enzymatic activities, dehydrogenase activity and activities of other important hydrolases (e.g., urease, protease, phosphatas and β-glucosidase) [82]. There are also marked seasonal changes in biotic and abiotic factors that affect the biological component of soil resources. Vegetative cover, influenced by seasonal changes, has a strong impact on soil microbiological processes. In degraded soils of arid and semi-arid regions, changes in soil moisture regimes can also affect MBC and activity [83].

The importance of macro-organisms (e.g., earthworms, termites) for restoring soil quality has been widely recognized for centuries [84]. Conversion of PT to CA, with crop residue mulch and cover cropping (Figure 4), can increase earthworm activity and also improve structural properties [85,86], but the conversion can also have implications regarding transport of agricultural pollutants into the drainage water [85]. Experiments conducted in central Mexico indicated that conversion of PT to CA improved soil surface aggregation and aggregate stability, increased water infiltration, and enhanced most parameters related to soil quality [87]. Therefore, risks of soil degradation can be mitigated through adoption of land use and management systems which improve soil biological processes, and introduction of beneficial organisms into soils by selective inoculation. For these and other reasons, the presence of earthworms, termites and other soil biota are often identified as important indicators of quality in tropical soils [86,88].

8.3. Soil Restorative Farming/Cropping Systems

Farming/cropping systems (rotations, soil fertility management, erosion control, grazing/stocking rate, water management) affect the type, rate and severity of soil degradation by altering the SOC pool, structural morphology, and other properties. Specifically, crop rotations and grazing can significantly impact SOC pool and the attendant soil properties [89]. Similar to arable lands, managing quality of rangeland soils is also essential for reducing risks of degradation. Sustainable management of rangeland soils is especially challenging because of high variability, harsh environments, and the temptation for over-grazing. A reduction in the proportion of palatable perennials, increases in densification (compaction), and declines in SOC are some of the constraints that need to be alleviated [90–92]. An important strategy to reduce the risks of degrading rangeland soils is to conserve and efficiently manage soil water through an improved understanding of the hydrological attributes [93,94]. Under West African conditions, construction of stone bunds and establishment of contour vegetative hedgerows can be effective for water conservation [95]. Establishment and management of forage trees (*i.e., Acacia fadherbia*) [96] and grass-legume mixtures [97] can also improve the quality of rangeland soils.

9. Soil Resilience

The term soil resilience refers to the ability of the soil to recover its quality in response to any natural or anthropogenic perturbations. Soil resilience is not the same as soil resistance, because resilience refers to "elastic" attributes that enable a soil to regain its quality upon alleviation of any perturbation or destabilizing influence [98,99]. Sound rhizospheric processes are essential for soil resilience against anthropogenic/natural perturbations. Being a dominant site of microbial metabolism, it is pertinent to identify management systems that stimulate soil microbiotic activity and related microbial processes. In this context, an "eco-physiological index" has been proposed to assess the impact of soil resilience [100] on soil processes. Managing the quantity and quality of SOC pool is once again a crucial guiding principle in identifying appropriate management practices that will strengthen resilience and reduce risks for soil degradation [101]. The SOC pool size is strongly related to the quantity of both above and below-ground biomass-C inputs. It is the assured, continuous input of the biomass-C that moderates MBC, provides a reservoir of plant nutrients (e.g., N, P, S), influences nutrient cycling, and improves/stabilizes soil structural morphology and geometry [98,99]. The so-called "sustainable land management (SLM)" concept is based on similar strategies of preserving productivity of the resource base for future generations [101]. There are also some organic management options for reducing soil degradation risk and improving human health [102], that may have site-specific niches. Biochar, a C-rich soil amendment derived from biomass by pyrolysis, can be produced from human sewage [103] and used to improve soil resilience [98] while also mitigating climate change.

In addition to biotic techniques, ancient farmers also developed mechanical/engineering techniques to sustain and improve their soils. Terraced agriculture evolved independently at several locations around the world (e.g., East Asia, the Himalayan region, Yemen, the Andean region). A study of pre-Columbian terraces from the Paca Valley, Peru, indicated that soil depletion from cultivation has compromised soil quality through loss of fine material and SOC. Overall, however, Paca Valley terraces have improved topsoil retention and supported deep profile with a good soil resilience [104].

There are no universally applicable techniques of managing soil resilience, but there are several approaches for ensuring sustainable soil management. Each of these approaches has tradeoffs that must be objectively and critically assessed (Table 1, Figure 6). In view of the heavy demands for agricultural produce to meet the needs of the growing and increasingly affluent population and emerging economies (e.g., India, China, Brazil, Mexico), the role of agricultural practices and their impact on soil, climate, gaseous emission, water resources, biodiversity, along with economic, political, social and ecologic dimensions [105] must be considered more now than in the past. The ideal strategy is to meet increasing global food demands while simultaneously restoring soil quality, improving the environment, and minimizing the tradeoffs.

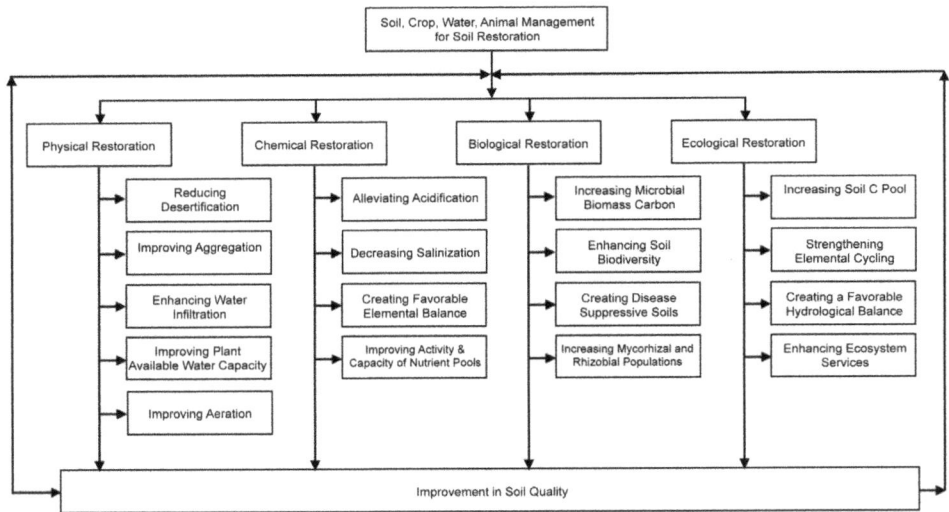

Figure 6. Strategies of restoring soil quality.

In the context of social, economic and cultural issues, it would be a serious omission to ignore poverty, human drudgery and social/gender equity. There is a strong link between poverty and soil quality. When people are poor, desperate and hungry, they pass on their sufferings to the land. Yet, some sociologists have questioned poverty as a major cause of soil/environmental degradation [106]. Stewardship and desperateness are mutually exclusive. Yield gaps, due to lack of adoption of RMPs in SSA and SA, are poverty traps that require a paradigm shift [107]. Since the 1980s, China has bridged the yield gap, alleviated poverty and improved soil quality [108]. Nonetheless, environmental issues remain to be addressed, effectively and immediately, to make China's agricultural revolution a successful venture [109].

10. Peak Soil *vs.* Endangered Soil

Endangered soil [110] and peak soil (losing soil more quickly than it is replaced [111] are concepts that need to be considered philosophically and scientifically because soil resources are finite and non-renewable over the human time frame. They are also geographically disturbed in a non-uniform manner. Excessive plowing, accelerated erosion, and over fertilization are all depleting soil resources, threatening food security, and jeopardizing the environment. Food insecurity is made even more acute by scarcity of soil resources of good quality (prime land) and risks of soil degradation [112]. Soil degradation is affecting 33% of all soils. Indeed, soil is an essentially forgotten resource [113]. In response, a holistic management approach is needed to improve soil quality. Site-specific and appropriate land husbandry practices must be identified to restore physical, chemical, biological, and ecological components of soil quality [114]. The goal should be to increase productivity per unit area, time, and energy input; while restoring soil quality and reducing environmental degradation risks. In terms of soil quality, ecosystem services provisioned by the SOC pool cannot be by-passed by applying commercial chemicals or other technologies [15]. Only by increasing SOC pool can the need for additional inorganic fertilizer N be reduced, especially in degraded soils. Conversion to CA, combined with residue retention and other components, is a sustainable option for several soil-specific conditions (Figure 4). Environmental awareness and stewardship are also important for improving adoption of RMPs and promoting soil restoration.

11. Conclusions

Soil resources are finite in extent, unequally distributed geographically, prone to degradation by land misuse and mismanagement, but essential to all terrestrial life and human wellbeing. Soil degradation can be physical (e.g., decline in structure, crusting, compaction, erosion, anaerobiosis, water imbalance), chemical (e.g., acidification, salinization, elemental imbalance comprising of toxicity or deficiency, nutrient deficiency), biological (depletion of SOC pool, reduction in soil biodiversity, decline in microbial biomass-C), or ecological (e.g., disruption in elemental cycling, decline in C sink capacity). Soil degradation leads to reduction in ecosystem functions and services of interest to human and conservation of nature. The SOC pool, its amount and depth-distribution along with turnover and mean residence time, is a critical component of soil quality and source of numerous ecosystem services. Soil degradation depletes SOC pool, and its restoration to threshold levels of at least 11 to 15 g kg^{-1} (1.1%–1.5% by weight) within the root zone is critical to reducing soil and environmental degradation risks. Important strategies for soil quality restoration and reducing environmental degradation risks are: (i) reducing soil erosion; (ii) creating a positive soil/ecosystem C budget; (iii) improving availability of macro (N, P, S) and micro-nutrients (Zn, Fe, Cu, Mo, Se); (iv) increasing soil biodiversity especially the microbial process; and (v) enhancing rhizospheric processes. The ultimate goal should be to adopt a holistic and integrated approach to soil resource management. The finite nature of soil resources must never be taken for granted—they must be used, improved, and restored.

Conflicts of Interest: There is no conflict of interest in publishing this article by the author.

References

1. Van Pham, L.; Smith, C. Drivers of agricultural sustainability in developing countries: A review. *Environ. Syst. Decis.* **2014**, *34*, 326–341. [CrossRef]
2. IFAD. *The Rural Poverty Report 2011*; International Fund for Agricultural Development: Rome, Italy, 2010.
3. Leon, J.; Osorio, N. Role of Litter Turnover in Soil Quality in Tropical Degraded Lands of Colombia. *Sci. World J.* **2014**, *13*. [CrossRef]
4. Lamb, D.; Erskine, P.; Parrotta, J. Restoration of degraded tropical forest landscapes. *Science* **2005**, *310*, 1628–1632. [CrossRef] [PubMed]
5. Bini, C. Soil: A precious natural resource. In *Conservation of Natural Resources*; Kudrow, N.J., Ed.; Nova Science Publishers: Hauppauge, NY, USA, 2009; pp. 1–48.
6. Scherr, S.J. The future food security and economic consequences of soil degradation in the developing world. In *Response to Land Degradation*; Oxford Press: New Delhi, India, 2001; pp. 155–170.
7. Guerra, A.; Marcal, M.; Polivanov, H.; Lima, N.; Souza, U.; Feitosa, A.; Davies, K.; Fullen, M.A ; Booth, C.A. Environment management and health risks of soil erosion gullies in São Luíz (Brazil) and their potential remediation using palm-leaf geotextiles. In *Environmental Health Risk II*; WIT Press: Southampton, UK, 2005; pp. 459–467.
8. Lal, R. Soil degradation as a reason for inadequate human nutrition. *Food Sec.* **2009**, *1*, 45–57. [CrossRef]
9. Bini, C.; Zilioli, D. Is Soil a Cultural Heritage? Proc. III Conv. Int. Architettura del Passaggio, 20 suppl. DVD. Available online: http://arca.unive.it/handle/10278/22330 (accessed on 12 May 2015).
10. Abrahams, P. Soils: Their implications to human health. *Sci. Total Environ.* **2002**, *291*, 1–32. [CrossRef] [PubMed]
11. Robinson, D.; Emmett, B.; Reynolds, B.; Rowe, E.; Spurgeon, D.; Keith, A.; Lebron, I.; Hockley, N.; Hester, R.; Harrison, R. Soil Natural Capital and Ecosystem Service Delivery in a World of Global Soil Change. *Soils Food Secur.* **2012**, *35*, 41–68.
12. Graber, D.R.; Jones, W.A.; Johnson, J.A. Human and ecosystem health. *J. Agromedicine* **1995**, *2*, 47–64. [CrossRef] [PubMed]
13. Singer, M.J.; Warkentin, B.P. Soil in an environmental context: An American perspective. *Catena* **1996**, *27*, 179–189. [CrossRef]

14. Krupenikov, I.; Boincean, B.; Dent, D.; Krupenikov, I.; Boincean, B.; Dent, D. Humus-Guardian of Fertility and Global Carbon Sink. In *The Black Earth: Ecological Principles for Sustainable Agriculture on Chernozem Soils; International Year of Planet Earth 2011;* Springer: Dordrecht, The Netherlands, 2011; pp. 39–50.

15. Manlay, R.; Feller, C.; Swift, M. Historical evolution of soil organic matter concepts and their relationships with the fertility and sustainability of cropping systems. *Agric. Ecosyst. Environ.* **2007**, *119*, 217–233. [CrossRef]

16. Rajan, K.; Natarajan, A.; Kumar, K.; Badrinath, M.; Gowda, R. Soil organic carbon—the most reliable indicator for monitoring land degradation by soil erosion. *Curr. Sci.* **2010**, *99*, 823–827.

17. Diacono, M.; Montemurro, F. Long-term effects of organic amendments on soil fertility: A review. *Agron. Sustain. Dev.* **2010**, *30*, 401–422. [CrossRef]

18. Vanlauwe, B.; Hester, R.; Harrison, R. Organic Matter Availability and Management in the Context of Integrated Soil Fertility Management in sub-Saharan Africa. *Soils Food Secur.* **2012**, *35*, 135–157.

19. Clay, D.; Chang, J.; Clay, S.; Stone, J.; Gelderman, R.; Carlson, G.; Reitsma, K.; Jones, M.; Janssen, L.; Schumacher, T. Corn Yields and No-Tillage Affects Carbon Sequestration and Carbon Footprints. *Agron. J.* **2012**, *104*, 763–770. [CrossRef]

20. Dlamini, P.; Chivenge, P.; Manson, A.; Chaplot, V. Land degradation impact on soil organic carbon and nitrogen stocks of sub-tropical humid grasslands in South Africa. *Geoderma* **2014**, *235*, 372–381. [CrossRef]

21. Mills, A.; Fey, M. Transformation of thicket to savanna reduces soil quality in the Eastern Cape, South Africa. *Plant Soil* **2004**, *265*, 153–163. [CrossRef]

22. Tiessen, H.; Cuevas, E.; Chacon, P. The role of soil organic-matter in sustaining soil fertility. *Nature* **1994**, *371*, 783–785. [CrossRef]

23. Lal, R. Soil carbon sequestration impacts on global climate change and food security. *Science* **2004**, *304*, 1623–1627. [CrossRef] [PubMed]

24. Lal, R. Soil carbon sequestration in natural and managed tropical forest ecosystems. *J. Sust. For.* **2005**, *21*, 1–30. [CrossRef]

25. Batjes, N. Options for increasing carbon sequestration in West African soils: An exploratory study with special focus on Senegal. *Land Degrad. Dev.* **2001**, *12*, 131–142. [CrossRef]

26. Eaton, W.; Roed, M.; Chassot, O.; Barry, D. Differences in soil moisture, nutrients and the microbial community between forests on the upper Pacific and Caribbean slopes at Monteverde, Cordillera de Tilaran: Implications for responses to climate change. *Trop. Ecol.* **2012**, *53*, 235–240.

27. Melillo, J.M.; Steudler, P.A.; Tian, H.; Butler, S. Fertilizing change: Carbon-nitrogen interactions and carbon storage in land ecosystems. In *Handbook of Climate Change and Agroecosystems: Impact, Adaptation and Mitigation;* Hillel, D., Rosenzweig, C., Eds.; Imperial College Press: London, UK, 2010; pp. 21–36.

28. Melillo, J.; Steudler, P.; Aber, J.; Newkirk, K.; Lux, H.; Bowles, F.; Catricala, C.; Magill, A.; Ahrens, T.; Morrisseau, S. Soil warming and carbon-cycle feedbacks to the climate system. *Science* **2002**, *298*, 2173–2176. [CrossRef] [PubMed]

29. Berbeco, M.; Melillo, J.; Orians, C. Soil warming accelerates decomposition of fine woody debris. *Plant Soil* **2012**, *356*, 405–417. [CrossRef]

30. Ray, S.; Bhattacharyya, T.; Reddy, K.; Pal, D.; Chandran, P.; Tiwary, P.; Mandal, D.; Mandal, C.; Prasad, J.; Sarkar, D.; *et al.* Soil and land quality indicators of the Indo-Gangetic Plains of India. *Curr. Sci.* **2014**, *107*, 1470–1486.

31. Sharma, K.; Grace, J.; Mandal, U.; Gajbhiye, P.; Srinivas, K.; Korwar, G.; Bindu, V.; Ramesh, V.; Ramachandran, K.; Yadav, S. Evaluation of long-term soil management practices using key indicators and soil quality indices in a semi-arid tropical Alfisol. *Aust. J. Soil Res.* **2008**, *46*, 368–377. [CrossRef]

32. Karlen, D.; Mausbach, M.; Doran, J.; Cline, R.; Harris, R.; Schuman, G. Soil quality: A concept, definition, and framework for evaluation. *Soil Sci. Soc. Am. J.* **1997**, *61*, 4–10. [CrossRef]

33. Karlen, D.; Stott, D.; Doran, J.; Coleman, D.; Bezdicek, D.; Stewart, B. A framework for evaluating physical and chemical indicators of soil quality. *Defin. Soil Q. Sustain. Environ.* **1994**, *35*, 53–72.

34. Andrews, S.S.; Carroll, C.R. Designing a soil quality assessment tool for sustainable agroecosystems management. *Ecol. Appl.* **2001**, *11*, 1573–1585. [CrossRef]

35. Paz-Kagan, T.; Shachak, M.; Zaady, E.; Karnieli, A. A spectral soil quality index (SSQI) for characterizing soil function in areas of changed land use. *Geoderma* **2014**, *230*, 171–184. [CrossRef]

36. Lal, R. *Methods and Guidelines for Assessing Sustainable Use of Soil and Water Resources in the Tropics*; USDA/SMSS Bull: 21; US Department of Agriculture: Washington, DC, USA, 1994.

37. Andrews, S.; Karlen, D.; Cambardella, C. The soil management assessment framework: A quantitative soil quality evaluation method. *Soil Sci. Soc. Am. J.* **2004**, *68*, 1945–1962. [CrossRef]

38. Andrews, S.; Karlen, D.; Mitchell, J. A comparison of soil quality indexing methods for vegetable production systems in Northern California. *Agric. Ecosyst. Environ.* **2002**, *90*, 25–45. [CrossRef]

39. Gugino, B.K. *Cornell Soil Health Assessment Training Manual*; New York State Agricultural Experiment Station Cornell University: New York, NY, USA, 2009.

40. Lal, R. On Sequestering Carbon and Increasing Productivity by Conservation Agriculture. *J. Soil Water Conserv.* **2015**, in press.

41. Rodríguez, Y.E.O.; Hernández, D.C. Energetic balance of two farming tools in a Fluvisol for the cultivation of sweet potato (Ipomoea batatas Lam). *Rev. Cienceis Tecinas Agropecu.* **2013**, *22*, 21–25.

42. Labreuche, J.; Lellahi, A.; Malaval, C.; Germon, J. Impact of no-tillage agricultural methods on the energy balance and the greenhouse gas balance of cropping systems. *Cah. Agric.* **2011**, *20*, 204–215.

43. Kimetu, J.; Lehmann, J.; Ngoze, S.; Mugendi, D.; Kinyangi, J.; Riha, S.; Verchot, L.; Recha, J.; Pell, A. Reversibility of soil productivity decline with organic matter of differing quality along a degradation gradient. *Ecosystems* **2008**, *11*, 726–739. [CrossRef]

44. So, H.; Kirchhof, G.; Bakker, R.; Smith, G. Low input tillage/cropping systems for limited resource areas. *Soil Tillage Res.* **2001**, *61*, 109–123. [CrossRef]

45. Powlson, D.; Stirling, C.; Jat, M.; Gerard, B.; Palm, C.; Sanchez, P.; Cassman, K. Limited potential of no-till agriculture for climate change mitigation. *Nat. Clim. Chang.* **2014**, *4*, 678–683. [CrossRef]

46. Pittelkow, C.; Liang, X.; Linquist, B.; van Groenigen, K.; Lee, J.; Lundy, M.; van Gestel, N.; Six, J.; Venterea, R.; van Kessel, C. Productivity limits and potentials of the principles of conservation agriculture. *Nature* **2015**, *517*, U365–U482. [CrossRef]

47. Tisdall, J.; Oades, J. Organic-matter and water-stable aggregates in soils. *J. Soil Sci.* **1982**, *33*, 141–163. [CrossRef]

48. Sanchez, P.A. En route to plentiful food production in Africa. *Nat. Plants* **2015**, *1*, 1–2.

49. Hüttl, R.; Frielinghaus, M. Soil fertility problems—An agriculture and forestry perspective. *Sci. Total Environ.* **1994**, *143*, 63–74. [CrossRef]

50. Abbott, L.K.; Murphy, D.V. What is soil biological fertility? In *Soil Biological Fertility—A Key to Sustainable Land Use in Agriculture*; Springer: Berlin/Heidelberg, Germany, 2007.

51. Abiven, S.; Menasseri, S.; Chenu, C. The effects of organic inputs over time on soil aggregate stability. *Soil Biol. Biochem.* **2008**, *41*, 1–12. [CrossRef]

52. Li, S.; Wang, Z.; Hu, T.; Gao, Y.; Stewart, B.; Sparks, D. Nitrogen in dryland soils of China and its management. *Adv. Agron.* **2009**, *101*, 123–181.

53. Sherman, K. Sustainability, biomass yields, and health of coastal ecosystems: An ecological perspective. *Marine Ecol. Progr. Series* **1994**, *112*, 277–301. [CrossRef]

54. Khaledian, Y.; Kiani, F.; Sohaila, E. The effect of land use change on soil and water quality in northern Iran. *J. Mt. Sci.* **2012**, *9*, 798–816. [CrossRef]

55. Tsatsaros, J.; Brodie, J.; Bohnet, I.; Valentine, P. Water Quality Degradation of Coastal Waterways in the Wet Tropics, Australia. *Water Air Soil Pollut.* **2013**, *224*. [CrossRef]

56. Schaffner, M.; Bader, H.; Scheidegger, R. Modeling the contribution of point sources and non-point sources to Thachin River water pollution. *Sci. Total Environ.* **2009**, *407*, 4902–4915. [CrossRef] [PubMed]

57. Atapattu, S.; Kodituwakku, D. Agriculture in South Asia and its implications on downstream health and sustainability: A review. *Agric. Water Manag.* **2009**, *96*, 361–373. [CrossRef]

58. Trivedi, R. Water quality of the Ganga River—An overview. *Aquat. Ecosyst. Health Manag.* **2010**, *13*, 347–351. [CrossRef]

59. Chen, M.; Chen, J. Phosphorus release from agriculture to surface waters: Past, present and future in China. *Water Sci. Technol.* **2008**, *57*, 1355–1361. [CrossRef] [PubMed]

60. Cheevaporn, V.; Menasveta, P. Water pollution and habitat degradation in the Gulf of Thailand. *Marine Pollut. Bull.* **2003**, *47*, 43–51. [CrossRef]

61. Qadir, M.; Oster, J. Crop and irrigation management strategies for saline-sodic soils and waters aimed at environmentally sustainable agriculture. *Sci. Total Environ.* **2004**, *323*, 1–19. [CrossRef] [PubMed]

62. Lemly, A.; Finger, S.; Nelson, M. Sources and impacts of irrigation drainwater contaminants in arid wetlands. *Environ. Toxicol. Chem.* **1993**, *12*, 2265–2279. [CrossRef]
63. Dakoure, M.; Mermoud, A.; Yacouba, H.; Boivin, P. Impacts of irrigation with industrial treated wastewater on soil properties. *Geoderma* **2013**, *200*, 31–39.
64. Bouwer, H. Integrated water management for the 21st century: Problems and solutions. *J. Irrig. Drain. Eng.-Asce* **2002**, *128*, 193–202. [CrossRef]
65. Apitz, S.E.; Brils, J.; Marcomini, A.; Critto, A.; Agostini, P.; Micheletti, C.; Pippa, R.; Scanferla, P.; Zuin, S.; Lanczos, T.; *et al.* Approaches and frameworks for managing contaminated sediments—A European perspective. In *Assessment and Remediation of Contaminated Sediments*; Springer: Berlin/Heidelberg, Germany, 2006.
66. Lal, R. Societal value of soil carbon. *J. Soil Water Conserv.* **2014**, *69*, 186A–192A. [CrossRef]
67. Kohli, R.V.; Singh, H.P.; Batish, D.R.; Jose, S. Ecological interactions in agroforestry: An overview. In *Ecological Basis of Agroforestry*; Kohli, R.V.S., Batish, D.R., Jose, S., Eds.; CRC Press: Boca Raton, FL, USA, 2008; pp. 3–14.
68. Helman, D.; Lensky, I.; Mussery, A.; Leu, S. Rehabilitating degraded drylands by creating woodland islets: Assessing long-term effects on aboveground productivity and soil fertility. *Agric. For. Meteorol.* **2014**, *195*, 52–60. [CrossRef]
69. Berazneva, J.; Conrad, J.; Guerena, D.; Lehmann, J. Agricultural productivity and soil carbon dynamics: A bio-economic model. In Proceedings of the Agricultural and Applied Economics Association 2014 Annual Meeting, Minneapolis, MN, USA, 27–29 July 2014.
70. Gong, X.; Liu, Y.; Li, Q.; Wei, X.; Guo, X.; Niu, D.; Zhang, W.; Zhang, J.; Zhang, L. Sub-tropic degraded red soil restoration: Is soil organic carbon build-up limited by nutrients supply. *For. Ecol. Manag.* **2013**, *300*, 77–87. [CrossRef]
71. Iqbal, M.; van Es, H.; Anwar-ul-Hassan, R.R.; Schindelbeck, R.; Moebius-Clune, B. Soil Health Indicators as Affected by Long-term Application of Farm Manure and Cropping Patterns under Semi-arid Climates. *Int. J. Agric. Biol.* **2014**, *16*, 242–250.
72. Govaerts, B.; Sayre, K.; Deckers, J. A minimum data set for soil quality assessment of wheat and maize cropping in the highlands of Mexico. *Soil Tillage Res.* **2006**, *87*, 163–174. [CrossRef]
73. Moebius-Clune, B.; van Es, H.; Idowu, O.; Schindelbeck, R.; Kimetu, J.; Ngoze, S.; Lehmann, J.; Kinyangi, J. Long-term soil quality degradation along a cultivation chronosequence in western Kenya. *Agric. Ecosyst. Environ.* **2011**, *141*, 86–99. [CrossRef]
74. Sortino, O.; Montoneri, E.; Patane, C.; Rosato, R.; Tabasso, S.; Ginepro, M. Benefits for agriculture and the environment from urban waste. *Sci. Total Environ.* **2014**, *487*, 443–451. [CrossRef] [PubMed]
75. Baumgart-Getz, A.; Prokopy, L.; Floress, K. Why farmers adopt best management practice in the United States: A meta-analysis of the adoption literature. *J. Environ. Manag.* **2012**, *96*, 17–25. [CrossRef]
76. Pimentel, D.; Harvey, C.; Resosudormo, P.; Sinclair, K.; Kurz, D.; McNair, M.; Crist, S.; Shpritz, L.; Fitton, L.; Saffouri, R.; *et al.* Environmental and economic costs of soil erosion and conservation benefits. *Science* **1995**, *267*, 1117–1123. [CrossRef] [PubMed]
77. Mchunu, C.; Chaplot, V. Land degradation impact on soil carbon losses through water erosion and CO_2 emissions. *Geoderma* **2012**, *177*, 72–79. [CrossRef]
78. Badia, D.; Marti, C. Fire and rainfall energy effects on soil erosion and runoff generation in semi-arid forested lands. *Arid Land Res. Manag.* **2008**, *22*, 93–108. [CrossRef]
79. Du Preez, C.C.; van Huyssteen, C.W.; Mnkeni, P.N.S. Land use and soil organic matter in South Africa 2: A review on the influence of arable crop production. *South African J. Sci.* **2011**, *107*, 35–42.
80. Longeuville, D.; Henry, S.; Ozer, P. Saharan dust pollution: Implications for the Sahel. *Epidemiology* **2009**, *20*. [CrossRef]
81. Moreno, J.; Bastida, F.; Hernandez, T.; Garcia, C. Relationship between the agricultural management of a semi-arid soil and microbiological quality. *Commun. Soil Sci. Plant Anal.* **2008**, *39*, 421–439. [CrossRef]
82. Bastida, F.; Moreno, J.; Hernandez, T.; Garcia, C. Microbiological degradation index of soils in a semiarid climate. *Soil Biol. Biochem.* **2006**, *38*, 3463–3473. [CrossRef]
83. Fterich, A.; Mahdhi, M.; Mars, M. Seasonal Changes of Microbiological Properties in Steppe Soils from Degraded Arid Area in Tunisia. *Arid Land Res. Manag.* **2014**, *28*, 49–58. [CrossRef]
84. Darwin, C.R. *The Formation of Vegetable Mould, through the Action of Worms, with Observations on their Habitats*; John Murray: London, UK, 1881.

85. Edwards, W.M.; Shipitalo, M.J.; Norton, L.D. Contribution of macropososity to infiltration into a continuous corn no-tilled watershed: Implications for contaminant movement. *J. Contam. Hydrol.* **1988**, *3*, 193–205. [CrossRef]

86. Lal, R. *Tropical Ecology and Physical Edaphology*; John Wiley Sons: Chichester, UK, 1987.

87. Castellanos-Navarrete, A.; Rodriguez-Aragones, C.; de Goede, R.; Kooistra, M.; Sayre, K.; Brussaard, L.; Pulleman, M. Earthworm activity and soil structural changes under conservation agriculture in central Mexico. *Soil Tillage Res.* **2012**, *123*, 61–70. [CrossRef]

88. Ayuke, F.; Karanja, N.; Okello, J.; Wachira, P.; Mutua, G.; Lelei, D.; Gachene, C.; Hester, R.; Harrison, R. Agrobiodiversity and Potential Use for Enhancing Soil Health in Tropical Soils of Africa. *Soils Food Secur.* **2012**, *35*, 94–134.

89. Ryan, J.; Masri, S.; Ibrikci, H.; Singh, M.; Pala, M.; Harris, H. Implications of cereal-based crop rotations, nitrogen fertilization, and stubble grazing on soil organic matter in a Mediterranean-type environment. *Turkish J. Agric. For.* **2008**, *32*, 289–297.

90. Teague, W.; Foy, J.; Cross, B.; Dowhower, S. Soil carbon and nitrogen changes following root-plowing of rangeland. *J. Range Manag.* **1999**, *52*, 666–670. [CrossRef]

91. Emmerich, W.; Heitschmidt, R. Drought and grazing: II. Effects on runoff and water quality. *J. Range Manag.* **2002**, *55*, 229–234. [CrossRef]

92. Snyman, H.; du Preez, C. Rangeland degradation in a semi-arid South Africa—II: Influence on soil quality. *J. Arid Environ.* **2005**, *60*, 483–507. [CrossRef]

93. Oesterheld, M.; Loreti, J.; Semmartin, M.; Sala, O. Inter-annual variation in primary production of a semi-arid grassland related to previous-year production. *J. Veg. Sci.* **2001**, *12*, 137–142. [CrossRef]

94. Wiegand, T.; Snyman, H.; Kellner, K.; Paruelo, J. Do grasslands have a memory: Modeling phytomass production of a semiarid South African grassland. *Ecosystems* **2004**, *7*, 243–258. [CrossRef]

95. Thapa, G.; Yila, O. Farmers land management practices and status of agricultural land in the Jos Plateau, Nigeria. *Land Degrad. Dev.* **2012**, *23*, 263–277. [CrossRef]

96. Garrity, D.; Akinnifesi, F.; Ajayi, O.; Weldesemayat, S.; Mowo, J.; Kalinganire, A.; Larwanou, M.; Bayala, J. Evergreen Agriculture: A robust approach to sustainable food security in Africa. *Food Secur.* **2010**, *2*, 197–214. [CrossRef]

97. Muir, J.; Pitman, W.; Foster, J. Sustainable, low-input, warm-season, grass-legume grassland mixtures: Mission (nearly) impossible? *Grass Forage Sci.* **2011**, *66*, 301–315. [CrossRef]

98. Lal, R. Degradation and resilience of soils. *Phil. Trans. R. Soc. Lond. B.* **1997**, *352*, 997–1010. [CrossRef]

99. Greenland, D.J.; Szabolcs, I. (Eds.) *Soil Resilience and Sustainable Land Use*; CAB International: Wallingford, UK, 1994.

100. Lynch, J. Resilience of the rhizosphere to anthropogenic disturbance. *Biodegradation* **2002**, *13*, 21–27. [CrossRef] [PubMed]

101. Syers, J.K. Manging soils for long-term productivity. *Philos. Trans. R. Lond. B* **1997**, *352*, 1011–1021. [CrossRef]

102. Horrigan, L.; Lawrence, R.; Walker, P. How sustainable agriculture can address the environmental and human health harms of industrial agriculture. *Environ. Health Perspect.* **2002**, *110*, 445–456. [CrossRef] [PubMed]

103. Breulmann, M.; van Afferden, M.; Fühner, C. Biochar: Bring on the sewage. *Nature* **2015**, *518*, 483. [CrossRef]

104. Goodman-Elgar, M. Evaluating soil resilience in long-term cultivation: A study of pre-Columbian terraces from the Paca Valley, Peru. *J. Archaeol. Sci.* **2008**, *35*, 3072–3086. [CrossRef]

105. Ogaji, J. Sustainable agriculture in the UK. *Environ. Dev. Sustain.* **2005**, *7*, 253–270. [CrossRef]

106. Ravnborg, H. Poverty and soil management—Relationships from three Honduran watersheds. *Soc. Nat. Resour.* **2002**, *15*, 523–539. [CrossRef]

107. Tittonell, P.; Giller, K. When yield gaps are poverty traps: The paradigm of ecological intensification in African smallholder agriculture. *Field Crops Res.* **2013**, *143*, 76–90. [CrossRef]

108. Rozelle, S.; Huang, J.; Zhang, L. Poverty, population and environmental degradation in China. *Food Policy* **1997**, *22*, 229–251. [CrossRef] [PubMed]

109. Fan, M.; Shen, J.; Yuan, L.; Jiang, R.; Chen, X.; Davies, W.; Zhang, F. Improving crop productivity and resource use efficiency to ensure food security and environmental quality in China. *J. Exp. Bot.* **2012**, *63*, 13–24. [CrossRef] [PubMed]

110. Tennesen, M. Rare Earth. *Science* **2014**, *346*, 692–695. [CrossRef] [PubMed]

Sustainability **2015**, *7*, 5875–5895

111. Reuters. Peak soil threatens global food security. *Reuters*, 17 July 2014.
112. Flora, C. Food security in the context of energy and resource depletion: Sustainable agriculture in developing countries. *Renew. Agric. Food Syst.* **2010**, *25*, 118–128. [CrossRef]
113. Wall, D.; Six, J. Give soils their due. *Science* **2015**, *347*, 695. [CrossRef] [PubMed]
114. Gregory, P.; Hester, R.; Harrison, R. Soils and Food Security: Challenges and Opportunities. *Soils Food Secur.* **2012**, *35*, 1–30.

sustainability

MDPI

Article

Soil Quality Indices for Evaluating Smallholder Agricultural Land Uses in Northern Ethiopia

Aweke M. Gelaw [1,*]**, B. R. Singh** [2] **and R. Lal** [3]

[1] Ethiopian Agricultural Transformation Agency, P.O. Box 708, Off Meskel Flower Road across Commercial Graduates, Addis Ababa, Ethiopia

[2] Norwegian University of Life Sciences, P.O. Box 5003, 1432 Ås, Norway; balram.singh@nmbu.no

[3] Carbon Sequestration and Management Center, The Ohio State University, Columbus, OH 43210, USA; lal.1@osu.edu

* Correspondence: awekegelaw@gmail.com; Tel.: +251-920803761

Academic Editor: Douglas L. Karlen

Received: 12 January 2015; Accepted: 15 February 2015; Published: 27 February 2015

Abstract: Population growth and increasing resource demands in Ethiopia are stressing and degrading agricultural landscapes. Most Ethiopian soils are already exhausted by several decades of over exploitation and mismanagement. Since many agricultural sustainability issues are related to soil quality, its assessment is very important. We determined integrated soil quality indices (SQI) within the surface 0–15 cm depth increment for three agricultural land uses: rain fed cultivation (RF); agroforestry (AF) and irrigated crop production (IR). Each land use was replicated five times within a semi-arid watershed in eastern Tigray, Northern Ethiopia. Using the framework suggested by Karlen and Stott (1994); four soil functions regarding soil's ability to: (1) accommodate water entry (WE); (2) facilitate water movement and availability (WMA); (3) resist degradation (RD); and (4) supply nutrients for plant growth (PNS) were estimated for each land use. The result revealed that AF affected all soil quality functions positively more than the other land uses. Furthermore, the four soil quality functions were integrated into an overall SQI; and the values for the three land uses were in the order: 0.58 (AF) > 0.51 (IR) > 0.47 (RF). The dominant soil properties influencing the integrated SQI values were soil organic carbon (26.4%); water stable aggregation (20.0%); total porosity (16.0%); total nitrogen (11.2%); microbial biomass carbon (6.4%); and cation exchange capacity (6.4%). Collectively, those six indicators accounted for more than 80% of the overall SQI values.

Keywords: soil quality; soil functions; land degradation; land use; Ethiopia

---

1. Introduction

Land degradation and declining soil fertility are critical problems affecting agricultural productivity and human welfare in Sub-Saharan Africa [1]. The main soil-environmental concerns in the region are nutrient depletion, loss of soil organic matter (SOM) and loss of soil functions (*i.e.*, productivity) [1,2]. In Ethiopia, total cultivated land has reached ~12 million hectares in mid-2013, but most of the soils are highly degraded [3]. Further, population growth and agricultural production are not growing *at par*. As a result, expansion to marginal lands and protected areas has become a common practice.

Tigray, the northernmost region in Ethiopia, is most known for its serious land degradation problems. Much of the woodland in Tigray started to disappear in the early 1960s under pressure from the rapidly growing population [4]. Hengsdijk *et al.* [5] wrote their observations as follows: "perhaps nowhere in the world land degradation and soil nutrient depletion are more evident than in the marginal highlands of Tigray". In the region, a short and variable rainy season in combination with degraded soils resulted in low soil productivity and frequent crop failures. As a result, the local population is structurally dependent on food aid [6]. If unattended to, land degradation and soil nutrient depletion would further reduce agricultural productivity and increase pressure on marginal environments, adversely affecting food security and livelihoods of smallholder farmers in the region [6].

Indeed, Tigray is not only known for its severe land degradation, but also for its vast environmental rehabilitation efforts in the last two decades [7]. Among the recent efforts towards enhancing agricultural development in the region, rainwater harvesting has been widely adopted [8] because supplementary irrigation is essential for crop production in arid regions as it increases soil water availability during dry spells [9]. Further, farmers in Tigray have a culture of selectively taking care of trees, which are remnants of the original woodlands. *Acacia albida* Del. (Syn. *Faiderbhia albida* (Del.) A Chev.) trees are among the most selected ones in the region. Nowadays, farmers grow these trees in and around their farmlands in order to improve soil fertility and increase crop yields [10].

Sustainability of agricultural systems is an important issue in Ethiopia. Many of the issues of agricultural sustainability are related to soil quality. Thus, its assessment and the direction of change with time is a primary indicator of whether agriculture is sustainable [11,12]. Soil quality is a combination of soil physical, chemical and biological properties that are able to change readily in response to variations in soil conditions [13]. It may be affected by land use type and agricultural management practices because these may cause alterations in soil's physical, chemical and biological properties, which in turn results in change in land productivity [14,15]. Integrated soil quality indices based on a combination of soil properties provide a better indication of soil quality than individual parameters. Karlen and Stott [16] developed a soil quality index (SQI) based on four soil functions, namely the ability of the soil to: (1) accommodate water entry (WE); (2) facilitate water movement, and absorption (WMA); (3) resist surface degradation (RD); and (4) supply nutrients for plant growth (PNS). Each soil function was explained by a set of indicators. Several authors among them Glover *et al.* [17], Masto *et al.* [12] and Fernandes *et al.* [18] used a similar framework.

A soil quality index (SQI) helps to assess the soil quality of a given site or ecosystem and enables comparisons between conditions at plot, field or watershed level under different land uses and management practices. Several studies were conducted to assess fertility statuses of soils in SSA [1–8]; however, almost all were only based on evaluation of individual soil parameters. Therefore, this study was conducted at a typical semi-arid agricultural watershed in Eastern Tigray, Northern Ethiopia, with the following objectives:

(1)   To evaluate effects of *F. albida* based agroforestry (AF), irrigation based *Psidium guajava* fruit production (IR) and a tree-less row-crop management (RF) (Figure 1) on selected physical, chemical and biological soil quality indicators and,

(2)   To compute an overall integrated soil quality index (SQI) for each land use system and compare among the indices.

Figure 1. The three agricultural land use systems at a semi-arid watershed in Tigray, Northern Ethiopia, with dryland crop production (RF), *F. albida*-based agroforestry (AF) and irrigation-based *P. guajava* fruit production (IR).

The study was conducted to test the hypothesis that land use change from dry land rainfed cultivation (RF) to *F. albida* agroforestry (AF) and irrigation based *P. guajava* fruit production (IR) systems improves physical, chemical, and biological soil quality indicators and the overall integrated soil quality index.

2. Materials and Methods

2.1. Descriptions of the Study Site

Mandae watershed is located in Eastern Tigray, Northern Ethiopia. Geographically, it is located between 15°26′00N to 15°32′00N latitude and 55°00′00E to 55°60′00E longitude, with an area of about 10 km^2, and an elevation of 1960 to 2000 m a.s.l. Average daily air temperature of the area ranges between 15 °C and 30 °C in winter and summer, respectively. Mean annual rainfall of the area is 558 mm, with a large inter-annual variation. Soils are classified as Arenosols, and associations of Arenosols with Regosols according to the World Reference Base for soil resources [19]. These soils are developed from alluvial deposits and Adigrat sandstones. Their textures are dominated by sand, loamy sand and sandy loam fractions [20]. Major land uses of the watershed include *Faidherbia albida* based agroforestry (27.7 ha), rainfed crop production (11.9 ha), open pasture (23.2 ha), and irrigation-based guava (*P. guajava)* fruit production (11.3 ha). Agricultural rotation in the area is usually maize (*Zea mays*)-teff (*Eragrostis tef*)-field beans (*Vicia faba*)-finger millet (*Eleusine coracana*) in the agroforestry and rainfed cultivation land use systems. Fallowing is not practiced in the area due to population pressure and scarcity of farmlands. Use of chemical fertilizers is minimal and land is prepared for cultivation by using a wooden plow with oxen. Crop residues and manures are used for animal feed and household fuel, respectively. No pesticides and other agricultural inputs are used in the area. Irrigation from shallow wells started in the area in late 1990s and currently most of the irrigated areas are covered by guava fruits. Smallholder mixed crop-livestock farming is a typical farming system of the region.

2.2. Soil Sampling and Analysis

Fifteen soil samples were collected in May 2010 from the surface (0–15 cm) layer of five sites randomly chosen at different locations from three agricultural land uses (AF, IR and RF). The summit position of the watershed was excluded to minimize confounding effects of slope and soil erosion. The samples were air-dried, mixed, ground, and passed through a 2-mm sieve for chemical analyses. Core samples were also collected from the same depth using 100 cm^3 volume stainless steel tubes (5-cm diameter and 5.1-cm height). Initial weights of the soil cores were measured in the laboratory immediately after collection. Simultaneously, soil moisture content was determined gravimetrically by oven drying the whole soil at 105 °C for 24 h to compute dry bulk density (ρ_b) [21]. No adjustment was made for rock volume because it was rather minimal. The major parts of the soil analyses were carried out at Mekelle University soil laboratory, Ethiopia. Soil organic carbon (SOC) and total nitrogen (TN) were analyzed at the Carbon Sequestration and Management Center (C-MASC) Laboratory (The Ohio State University, Columbus, OH, USA) using auto CN analyzer (Vario Max CN Macro Elemental Analyser, Elementar Analysensysteme GmbH, Hanau, Germany) by the dry combustion method [22]. Similarly, water stable aggregation (WSA) was measured at C-MASC soil physics laboratory by the wet sieving method [23]. Because soils did not show carbonates when tested with 10% HCl, it was assumed that the total C obtained in the analysis closely estimates soil organic carbon (SOC) concentration. Available P (Olsen) was analyzed using a standard Olsen method [24]. Cation exchangeable capacity (CEC) was estimated titrimetrically by ammonium distillation method [25]. Lastly, total porosity was calculated from particle density of 2.65 g/cm^3.

Microbial Biomass Carbon (MBC)

Another set of nine field-moist soil samples (40 g each) from the surface (0–15 cm) depth were collected in three replications from the three agricultural land uses (AF, IR and RF) in May 2012 for the determination of microbial biomass carbon (MBC). The samples were transported in an icebox to the Norwegian University of Life Sciences soil laboratory, Ås, Norway. The MBC analysis was carried out following the fumigation-extraction method [26,27]. At first, each sample was divided in to three subsamples, and one out of the three (10.0 g) was fumigated with ethanol-free chloroform for 24 h at 25 °C in an evacuated extractor. Afterwards, from the remaining two subsamples, one was used for moisture determination and the other treated as control for each plot. Fumigated and non-fumigated soils were extracted with 40-mL 0.5-mol·L^{-1} K_2SO_4 (1:4 soil:extractant) and shaken for 1-h on a reciprocal shaker. The extracts were filtered using Whatman No. 42 filter paper of 7-cm diameter and stored frozen at -15 °C prior to analysis. Finally, total organic carbon in the extracts was measured using Total Organic Carbon Analyzer (SHIMADZU) at NMBU laboratory, Ås, Norway. Microbial Biomass Carbon (MBC) was calculated as follows:

$$\text{MBC} = \frac{E_C}{KE_C} \tag{1}$$

where E_C = (organic C extracted from fumigated soils) $-$ (organic C extracted from non-fumigated soils) and KE_C = 0.45 [28].

2.3. Soil Quality Assessment

Soil quality assessment tools need to be flexible in terms of selection of soil functions to be assessed and indicators to be measured to ensure that assessments are appropriate for specific management goals [29]. Effects of land use on soil quality were assessed following the framework suggested by Karlen and Stott [16]. We followed this framework because of its flexibility, ease of use and its potential for interactive use. It is the same approach that became the Soil Management Assessment Framework (SMAF) [30]. It uses selected soil functions, which are weighted and integrated according to the following expression:

$$\text{SQI} = \text{WE(wt)} + \text{WMA(wt)} + \text{RD(wt)} + \text{PNS(wt)} \tag{2}$$

where, wt is a numerical weighting for each soil function.

These numerical weights were assigned to each soil function according to their importance in fulfilling the overall goals of maintaining soil quality under specific conditions of this study. According to Karlen and Stott [16], the sum of weights for all soil functions must equal 1.0. Karlen and Stott [16] assigned equal weight to each soil function. However, different weight values of 0.2, 0.2, 0.2 and 0.4 were assigned for this study for WE, WMA, RD, and PNS, respectively (Table 1). For this study, PNS was assigned with more value than other functions, because use of chemical fertilizers was minimal in the area and hence nutrient supply was considered the most important production constraint. Further, sustaining crop production is the major goal of soil management strategies in most developing countries including Ethiopia. The PNS function was further divided into three second-level functions viz. nutrient storage, nutrient cycling and nutrient availability (Table 1).

An ideal soil would fulfill all the functions considered important, and would have an integrated SQI of 1.0 under the proposed framework. However, as a soil fails to meet the ideal criteria, its SQI would decrease, with zero being the lowest rating. Associated with each soil function are soil quality indicators that influence, to varying degrees, the specific soil function. Threshold values for each soil quality indicator were set based on the range of values measured in natural ecosystems (the adjacent grass pasture in our case) and on critical values in the literature (Table 2). Glover *et al.* [17] also used adjacent grass pasture areas to determine critical values for a study conducted in Washington State, USA. After finalizing the thresholds, the soil property values recorded under the three agricultural land use systems were transformed into unit-less scores (between 0 and 1), using the following equation [12]:

$$\text{Non-linear score}(Y) = \frac{1}{(1 + e^{-b(x-A)})} \tag{3}$$

where, x is the soil property value, A the baseline or value of the soil property where the score equals 0.5 and b is the slope of the tangent to the curve at the baseline.

Table 1. Soil quality indexing framework (adapted from Glover *et al.* [17]).

Function	Weight	Indicator Level 1	Weight	Indicator Level 2	Weight	Source for Indicators/Weights
Accommodate Water Entry	0.20	WSA	0.40			[17,31]
		BD	0.20			[17]
		POR	0.20			[12]
		SOC	0.20			[12]
Facilitate Water Movement and Availability	0.20	POR	0.60			[12,17,31]
		SOC	0.40			[17,31]
Resist Surface Degradation	0.20	WSA	0.60			[17,31]
		Microbial Processes	0.40	MBC	0.60	[12,17,31]
				SOC	0.20	[12,17,31]
				TN	0.20	[12,31]
Supply Plant Nutrient	0.40	Nutrient Storage	0.40	CEC	0.40	[12]
				SOC	0.40	[12]
				TN	0.20	[12]
		Nutrient Cycling	0.20	SOC	0.40	[12,31]
				MBC	0.20	[12,31]
				TN	0.40	[31]
		Nutrient Availability	0.40	SOC	0.20	[12]
				pH	0.20	[31]
				TN	0.20	[12]
				AVP	0.20	[12]
				AVK	0.20	[12]

Table 2. Relative importance of the different soil properties used for the soil quality indexing.

Soil Quality Indicator	Weight	Soil Function
Soil organic carbon	0.264	Accommodate water entry Facilitate Water movement and availability Resist Surface structure degradation Supply plant nutrients
Aggregate Stability	0.200	Accommodate water entry Facilitate Water movement and availability Resist surface structure degradation
Bulk density	0.040	Accommodate water entry
Porosity	0.160	Accommodate water entry Facilitate water movement and availability
Microbial biomass carbon	0.064	Resist surface structure degradation Supply plant nutrients
Cation exchange capacity	0.064	Supply plant nutrients
Total Nitrogen	0.112	Supply plant nutrients Resist surface structure degradation
Available phosphorus	0.032	Supply plant nutrients
Available Potassium	0.032	Supply plant nutrients
pH	0.032	Supply plant nutrients
Total	1.00	

The score for each indicator was calculated after establishing the baseline, the lower, and the upper threshold values (Table 3). Threshold values are soil property values where the score equals one (upper threshold) when the measured soil property is at the most favorable level; or equals zero (lower threshold) when the soil property is at an unacceptable level. Baseline values are generally regarded as minimum target values [12]. There are two baselines for "Optimum" curves, lower base line and upper base line, which corresponds to 0.5 score of the growth and death curves, respectively [12].

Table 3. Scoring function values and references used for evaluating the soil quality indices (adapted from Masto *et al.* [12]).

Indicator	Scoring Curve	Depth (cm)	LT	UT	LB	UB	OPT	Slope	Source of Threshold/Baseline Values
				Physical properties					
BD (Mgm^{-3})	Less is better	0–15 cm	1.0	2.0	1.5	-	-	−2.0832	[31]; Adjacent grass pasture
WSA (>0.5 mm)	More is better	0–10 cm	0.0	40.0	20.0	-	-	0.0339	Adjacent grass pasture
TP (V%)	Optimum	0–15 cm	20.0	80.0	40.0	60.0	50.0	0.0644	[12,31]; Adjacent grass pasture
				Chemical Properties					
CEC (cmol (+) kg^{-1})	More is better	0–15 cm	0.0	18.0	9.0	-	-	0.0757	[12]; Adjacent grass pasture
pH (1:2.5)	Optimum	0–15 cm	3.0	9.0	5.0	8.0	7.0	0.5332; −0.496	[18]
TN (kgha^{-1})	More is better	0–15 cm	0.0	2000.0	1000.0	-	-	0.0007	[12]; Adjacent grass pasture
AVP (kgha^{-1})	More is better	0–15 cm	0.0	50.0	25.0	-	-	0.0226	[12]
AVK (kg·ha^{-1})	More is better	0–15 cm	0.0	400.0	200.0	-	-	0.0036	[12]
				Biological Properties					
SOC (gkg^{-1})	More is better	0–15 cm	0.0	10.0	5.0	-	-	0.1341	[12]; Adjacent grass pasture
MBC (mgkg^{-1})	More is better	0–15 cm	0.0	300.0	150.0	-	-	0.0042	[12]; Adjacent grass pasture

Using this non-linear scoring curve equation, three types of standardized scoring functions typically used for soil quality assessments were generated: (1): More is better"; (2) "Less is better"; and (3) "Optimum" as per earlier studies [12,16–18,31,32]. The equation defines a "More is better" scoring curve for positive slopes, a "Less is better" curve for negative slopes, and an "Optimum" curve is defined by the combination of both positive and negative slopes. These scoring curves are presented in detail by many authors [17,18,31–34].

2.4. Statistical Analyses

Effects of different land use systems on soil quality indicators, functions and integrated quality indices were subjected to one-way ANOVA. Excel spreadsheet was used for transforming soil quality indicator values into unit-less scores. Differences between means of parameters were considered significant at the 0.05 level using the Tukey's studentized (HSD) test. The data were analyzed using R version 3.02 software package [35].

3. Results and Discussion

3.1. Soil Physical Quality Indicators

Bulk density ranged from 1.48 Mg·m^{-3} in AF to 1.57 Mg·m^{-3} in both IR and RF land use systems (Table 4). However, there was no significant difference in BD among land uses. Although soils under AF land use contained SOC concentration twice more than that under RF, the detrimental effects of tillage may have offset the beneficial effects of SOC on BD [17,32]. Soils under AF land use also had the highest percentage of water stable aggregates (WSA) of 17.3%, but it was not significantly higher than that under IR and RF land uses. Addition of more organic matter from leaf and root litters from the *F. albida* trees in AF than the other land uses likely explains the improved WSA in AF [36]. Similarly, a study by Gelaw *et al.* [37] at the same site found that soils under natural grazing lands adjacent to cultivated lands were well structured, and contained higher SOC concentrations. Total porosity (TP) ranged from 35.5% in RF to 43.5% and 44.9% in AF and IR land uses, respectively. However, the difference among land uses was not statistically significant. Similarly, the detrimental effects of tillage may have offset the beneficial effects of SOC on TP [17,32,37].

Table 4. Effects of land use systems on selected soil physical, chemical and biological quality indicators at Mandae watershed in eastern Tigray, north Ethiopia.

Soil Quality Indicator	Land Use			
	RF	AF	IR	
Physical				
BD (Mg·m^{-3})	1.57 (0.03)	1.48 (0.05)	1.57 (0.02)	NS
WSA (>0.5 mm)	11.3 (1.8)	17.3 (2.5)	13.6 (3.6)	NS
TP (V%)	35.4 (3.6)	43.5 (2.0)	44.9 (2.7)	NS
Chemical				
CEC (cmol (p+) kg^{-1})	5.4 (1.0) [b]	11.5 (0.8) [a]	4.8 (1.8) [b]	**
pH	6.6 (0.3) [b]	6.4 (0.2) [b]	8.0 (0.03) [a]	***
TN (kg·ha^{-1})	809.7 (134.6) [b]	1568.6 (85.4) [a]	1042.7 (244.6) [a,b]	*
AVP (kg·ha^{-1})	24.4 (10.7)	39.1 (4.3)	39.8 (4.7)	NS
AVK (kg·ha^{-1})	216.5 (56.9) [b]	1019.1 (161.0) [a]	297.7 (71.8) [b]	***
Biological				
SOC (g·kg^{-1})	3.2 (0.7) [b]	6.4 (0.3) [a]	5.9 (1.1) [a,b]	*
MBC (mg·kg^{-1})	75.5 (24.1)	95.9 (10.3)	100.1 (31.3)	NS

RF, Dryland crop production; AF, *Faidherbia albida* based agroforestry; IR, irrigation based fruit production; ± Mean values followed by standard errors in the parentheses; values with different letters are significantly different.* $p < 0.05$; ** $p < 0.01$; NS = not significant (Tukey's test, $p = 0.05$).

3.2. Soil Chemical Quality Indicators

CEC of the soils studied ranged from the highest under AF (11.5 cmol p+ kg^{-1}) to the lowest under IR (4.8 cmol p+ kg^{-1}). It was significantly higher ($p < 0.01$) under AF than that under IR and RF land uses (Table 4). Generally, CEC was low with an exception of some improvements under AF land use. Rabia *et al.* [20] also reported similar results for the same area. Accordingly, up to 90% of soil samples from this area had extremely-low (<5)-to-low (5–15 cmol p+ kg^{-1}) CEC values [20]. EC values of the soils were also much lower than the FAO salinity hazard levels for most crops [20] (Table 4).

In general, Arenosols have neutral pH values [38]. However, soils under IR land use showed a significantly higher ($p < 0.001$) pH value than that under other land uses, and it was slightly alkaline. The source of this slight alkalinity development in the soil under IR land use could be from the supplemental irrigation. Similar results were also reported by Rabia *et al.* [20].

Soils under AF contained the highest total nitrogen (TN) stock (1568.6 kg·ha^{-1}), and it was significantly higher ($p < 0.05$) than that in IR and RF land uses (Table 4). Hadgu *et al.* [10] reported similar results in their study in central Tigray, Northern Ethiopia, which compared TN contents of soils under canopies of *F. albida* and eucalyptus trees with those from tree-less fields. Similarly, available potassium (AVK) was significantly higher ($p < 0.001$) under AF than that under other land uses (Table 4). In contrast, available phosphorus (AVP) contents did not differ among land uses. The higher AVK under AF than that under other land uses could be related to the recycling of nutrients in the aboveground biomass, root biomass or through the recycling of depositions by cattle, which gather for shade under the tree-canopies during sunny days [39]. Sanchez [40] also reported a significant increase both in soil K content and sorghum (*Sorghum bicolar*) yield on soils under the canopy of *F. albida* trees from that on soils 15-m away in two parklands in Burkina Faso. Results presented here are also in accord with reports by Nair [41] that microsite enrichment qualities of trees such as *F. albida* in West Africa and *P. cineraria* in India have long been recognized in many traditional farming systems.

3.3. Soil Biological Quality Indicators

Both SOC and MBC are among principal soil parameters, which affect biological processes and soil quality. The highest SOC concentration was measured in AF (6.4 g·kg^{-1}) followed by that in IR (5.9 g·kg^{-1}), and the lowest was in RF (3.2 g·kg^{-1}) (Table 4). Thus, SOC was significantly higher ($p < 0.05$) in AF than that in RF land use. However, it did not statistically differ between AF and IR, and between IR and RF land uses (Table 4). On the other hand, MBC was slightly higher in soils under IR (100.1 mg·kg^{-1}) than that under AF and RF, but the differences were not statistically significant (Table 4). Higher MBC values under IR than that under AF and RF may be explained by less disturbance of soils under IR than those under the other intensively tilled land uses. The intensity of tillage in IR was less than that under AF and RF land uses. Besides, irrigation farms under guava fruits were not convenient for oxen plowing. Weed control and irrigation in IR land use were also practiced by hand. Soil organic carbon in intensively cultivated soils has less physical protection than that in less cultivated soils because tillage disrupts macroaggregates and exposes previously protected SOM microbial processes [14,37]. Similarly, Franchini *et al.* [42] reported an increase in MBC under no-till (NT) than that under conventional tillage systems (CT) receiving more plant residues in Southern Brazil. The lower MBC regardless of more plant residue addition under CT was due to higher CO_2-emissions, which implies little conversion of carbon from plant residues into MBC [42]. Indeed, parameters associated with soil microbiological activities are sensitive, considered rapid indicators of effects of soil management, and are useful as indicators of soil quality [42].

3.4. Soil Quality Indicators Integration and Assessment

For this study, four soil functions contributed to the overall soil quality index (SQI) (Table 1). They were weighted according to their relative importance in fulfilling the goals of maintaining soil quality in the area. Thus, the major driving soil parameters for the integrated SQI were SOC (26.4%), WSA (20.0%), TP (16.0%), TN (11.2%), MBC (6.4%) and CEC (6.4%). Those six soil quality indicators together contributed for more than 80% of the variability in the overall SQI (Table 2). Further, BD contributed 4.0% followed by AVP, AVK and pH with a contribution of each 3.2% to the overall SQI. Regarding the soil's function for plant nutrient supply, SOC, TN, and CEC contributed 32%, 24%, and 16% of the PNS function, respectively. Available P, AVK and pH each contributed 8% of the soil's function for plant nutrient supply. The soil's MBC contribution to this function was minimal (4%). Overall, SOC alone contributed for more than 25% and 30% of SQI and PNS values, respectively.

Integration of the soil property values into SQI using the framework resulted in a significantly higher ($p < 0.05$) score in AF than in RF land use system for its ability to accommodate water entry (Table 5). The relatively higher WSA, TP and SOC values of the soil under AF land use than those in the soil under RF were largely responsible for the improvement in its ability to accommodate water entry in AF (Table 4). Glover *et al.* [17] also reported higher scores for soil's ability to accommodate water entry because of higher WSA and lower BD under integrated and organic management systems than those under a conventional system in Washington State, USA. Regarding the soil's ability to facilitate water movement and availability, AF also scored significantly higher ($p < 0.05$) value than RF because of the relatively higher TP and SOC values in AF (Table 5). These results indicated that AF land use improved the soil's ability to hold and release water mainly due to its higher SOC content (Table 4). However, land use had no significant effect on soil's resistance to surface degradation (Table 5). This may be a clear indication of the detrimental effects of tillage on soil structure [17,32,37]. In contrast, AF scored significantly higher ($p < 0.05$) value for the soil's ability to supply plant nutrients than RF largely due to higher levels of AVK, CEC, SOC, TN and AVP in the rooting zones of AF land use (Table 5). The score for the soil under IR land use was not significantly different from that under RF

(Table 5). Further, the score for nutrient storage capacity of soils under AF land use was significantly higher ($p < 0.05$) than that under RF, but it was not significantly different from that under IR (Figure 2). However, nutrient cycling was not significantly affected by land use regardless of some improvements in AF. Trees in agroforestry systems can improve nutrient cycling and increase soil chemical fertility through bringing up nutrients from deeper layers and minimizing leaching hazards [41]. In contrast, nutrient availability was affected by land use. Thus, AF scored significantly higher ($p < 0.01$) value for its capacity in nutrient availability than that in other land uses (Figure 2).

Table 5. Soil quality ratings for the different land uses at the watershed.

Soil Function	Land Use			
	RF	AF	IR	
Accommodate Water Entry (0.20)	0.09 (0.00) [b]	0.11 (0.002) [a]	0.10 (0.004) [a,b]	*
Facilitate Water Entry and Availability (0.20)	0.10 (0.004) [b]	0.12 (0.004) [a]	0.11 (0.004) [a,b]	*
Resist Surface Degradation (0.20)	0.09 (0.003)	0.11 (0.002)	0.09 (0.005)	NS
Source of Plant Nutrients (0.40)	0.19 (0.01) [b]	0.24 (0.004) [a]	0.21 (0.015) [a,b]	*
Integrated Soil Quality Index (1.00)	0.47 (0.01) [b]	0.58 (0.01) [a]	0.51 (0.02) [a,b]	**

RF, Dryland crop production; AF, *Faidherbia albida* based agroforestry; IR, irrigation based fruit production; ± Mean values followed by standard errors in the parentheses; values with different letters are significantly different. * $p < 0.05$; ** $p < 0.01$; NS = not significant (Tukey's test, $p = 0.05$).

Finally, the integrated SQI calculated for the land uses using the framework by Karlen and Stott [16] were in the following order: 0.58 (AF) > 0.51 (IR) > 0.47 (RF) (Table 5). Soil quality index differed significantly ($p < 0.01$) between AF and RF land use systems (Table 5). Similarly, Karlen *et al.* [31] reported an improvement in soil quality rating from 0.45 to 0.86 in over ten-year period by retention or addition of crop residues on a no-till (NT) continuous corn in Wisconsin, USA. In another study, Karlen *et al.* [32] reported a significant improvement in SQI ratings from 0.48 and 0.49 under plow and chisel, respectively, to 0.68 under NT using selected physical, chemical and biological soil quality indicators on Rozetta and Palsgrove silt loam soils in Wisconsin, USA. Stott *et al.* [43] in a recent study on Vertisols in Texas using the SMAF model also reported an improvement in overall SQI ranging from 75% to 94% of an optimum when compared with similar soils after 57 years of different agricultural management systems.

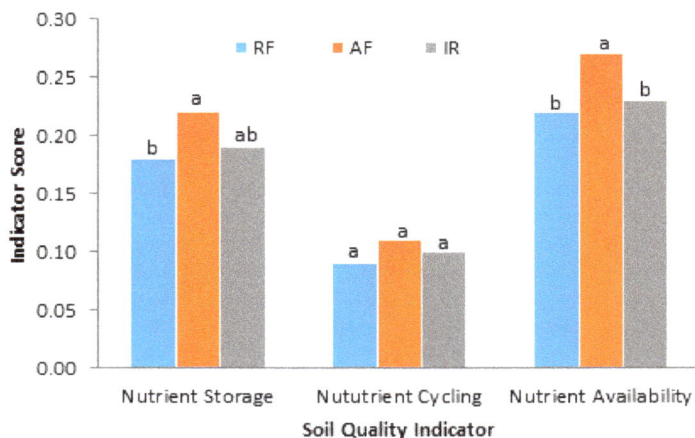

Figure 2. Effects of three agricultural land use systems (RF, AF and IR) on nutrient supplying capacities of soils at the watershed.

Regardless of a significant improvement in AF than that in RF land use, SQI ratings in all the three land use systems were very small compared with an ideal soil (Table 5). This result was in agreement with findings from other authors [5,44] who reported that low organic matter and nutrient stocks are typical characteristics of soils in Tigray, mainly due to nutrient mining because of crop harvests and complete removal of crop residues for feed and fuel. One fundamental principle of sustainability is to return to the soil the nutrients removed through harvests and other loss pathways [45], and one of the main tenets of agroforestry is that trees enhance soil fertility [45,46]. This is supported by observations of higher crop yields near *F. albida* tree canopies in Ethiopia [10,47–49] and elsewhere [50,51], which showed the potentials of agroforestry systems in improving soil quality and productivity of smallholder farms in Ethiopia and the wider region.

4. Conclusions

Relatively higher WSA, TN and SOC concentrations measured in soils under AF land use resulted in improved water entry, movement and availability than those under IR and RF. Soil's ability to supplying plant nutrients was also improved under AF than under RF land use largely due to higher levels of AVK, CEC, SOC, TN and AVP in the rooting zones of AF land use. However, there was no significant improvement in the soil's resistance to surface degradation in all land uses, which may be because of the detrimental effects of tillage. Further, when selected physical, chemical, and biological soil quality indicators were integrated into an overall SQI, AF land use received a higher soil quality rating (0.58) than that of RF (0.47). Thus, the result of this study highlighted the potentials of *F. albida* based AF systems for improving soil quality and productivity of smallholder farms in the area. Further, it demonstrated the effectiveness of the soil quality indexing framework in the study area and beyond to assess soil quality and thus recognized that changes in soil and crop management are needed for a more efficient and sustainable use of soil resources.

Acknowledgments: The International Foundation for Science (IFS) (Grant No. C/4687) provided financial support for the fieldwork. Some laboratory facilities were provided by Mekelle University, Ethiopia. Practical laboratory work for WSA, SOC and TN determination was performed at the laboratory facility of the Carbon Management and Sequestration Center (C-MASC), at The Ohio State University (OSU), Columbus, OH, USA. Financial assistance from the Norwegian State Education Loan Fund for the first author is gratefully acknowledged. The department of Environmental Science (IMV), Norwegian University of Life Sciences (NMBU) is also acknowledged for paying the publishing fee.

Author Contributions: Aweke M. Gelaw: Conception of the idea, designing the experiment, collecting and analyzing the data, and writing the article; Bal Ram Singh: Conception of the idea, Supervision and reviewing the article; Rattan Lal: Conception of the idea, supervision and reviewing the article.

Conflicts of Interest: There is no conflict of interest among the authors.

References

1. Sanchez, P.A.; Shepherd, K.D.; Soule, M.J.; Place, F.M.; Buresh, R.J.; Izac, A.M.N.; Mokwunye, A.U.; Kwesiga, F.R.; Ndiritu, C.G.; Woomer, P.L.; *et al.* Soil Fertility Replenishment in Africa: An Investment in Natural Resource Capital. In *Replenishing Soil Fertility in Africa*; Buresh, R.J., Sanchez, P.A., Calhoun, F., Eds.; SSSA and ICRAF: Madison, WI, USA, 1997; pp. 1–46.
2. Smaling, E. An Agroecological Framework for Integrating Nutrient Management, with Special Reference to Kenya. Ph.D. Thesis, Agricultural University of Wageningen, Wageningen, The Netherlands, 1993; p. 250.
3. Agricultural Transformation Agency (ATA). Status of soil resources in Ethiopia and priorities for sustainable management. In Proceedings of the Global Soil Partnership (GSP) for Eastern and Southern Africa Launching Workshop, Nairobi, Kenya, 25–27 March 2013.
4. Eweg, H.P.A.; van Lammeren, R.; Deurloo, H.; Woldu, Z. Analysing degradation and rehabilitation for sustainable land management in the highlands of Ethiopia. *Land Degrad. Dev.* **1998**, *9*, 529–542. [CrossRef]
5. Hengsdijk, H.; Meijerink, G.W.; Mosugu, M.E. Modeling the effect of three soil and water conservation practices in Tigray, Ethiopia. *Agric. Ecosyst. Environ.* **2005**, *105*, 29–40. [CrossRef]

6. Belete, T. Efforts for Sustainable Land Management in Tigray: The role of Extension. In *Policies for Sustainable Land Management in the Highlands of Tigray, Northern Ethiopia*; Gebremedhin, B., Pender, J., Ehui, S., Haile, M., Eds.; International Food Policy Research Institute (IFPRI): Mekele, Ethiopia, 2002.

7. Girmay, G. Land Use Change Effects in Northern Ethiopia: Runoff, Soil and Nutrient Losses, Soil Quality, and Sediment as Nutrient Sources. PhD Thesis, Norwegian University of Life Sciences, Ås, Norway, 2009; p. 140.

8. Moges, G.; Hengsdijk, H.; Jansen, H.C. Review and quantitative assessment of *ex situ* household rainwater harvesting systems in Ethiopia. *Agric. Water Manag.* **2011**, *98*, 1215–1227. [CrossRef]

9. Feng, Z.; Wang, X.; Feng, Z. Soil N and Salinity leaching after the autumn irrigation and its impact on ground water in Hetao irrigation district, China. *Agric. Water Manag.* **2005**, *71*, 131–143. [CrossRef]

10. Hadgu, K.M.; Kooistra, L.; Rossing, W.A.H.; van Bruggen, A.H.C. Assessing the effect of Faidherbia albida based land use systems on barley yield at field and regional scale in the highlands of Tigray, Northern Ethiopia. *Food Secur.* **2008**, *1*, 337–350. [CrossRef]

11. Karlen, D.L.; Mausbach, J.W.; Doran, J.W.; Cline, R.G.; Harris, R.F.; Schuman, G.E. Soil quality: A concept, definition and framework for evaluation. *Soil Sci. Soc. Am. J.* **1997**, *61*, 4–10. [CrossRef]

12. Masto, R.E.; Chhonkar, P.K.; Singh, D.; Patra, A.K. Soil quality response to long-term nutrient and crop management on a semi-arid Inceptisol. *Agric. Ecosyst. Environ.* **2007**, *118*, 130–142. [CrossRef]

13. Brejda, J.J.; Moorman, T.B.; Karlen, D.L.; Dao, T.H. Identification of regional soil quality factors and indicators: I. Central and southern high plains. *Soil Sci. Soc. Am. J.* **2000**, *64*, 2115–2124. [CrossRef]

14. Islam, K.R.; Weil, R.R. Land use effects on soil quality in a tropical forest ecosystem of Bangladesh. *Agric. Ecosyst. Environ.* **2000**, *79*, 9–16. [CrossRef]

15. Sanchez-Maranon, M.; Soriano, M.; Delgado, G.; Delgado, R. Soil quality in Mediterranean mountain environments: Effects of land use change. *Soil Sci. Soc. Am. J.* **2002**, *66*, 948–958. [CrossRef]

16. Karlen, D.L.; Stott, D.E. A framework for evaluating physical and chemical indicators of soil quality. In *Defining Soil Quality for a Sustainable Environment*; Doran, J.W., Coleman, D.C., Bezdicek, D.F., Stewart, B.A., Eds.; ASA and SSSA: Madison, WI, USA, 1994; pp. 53–72.

17. Glover, J.D.; Reganold, J.P.; Andrews, P.K. Systematic method for rating soil quality of conventional, organic, and integrated apple orchards in Washington State. *Agric. Ecosyst. Environ.* **2000**, *80*, 29–45. [CrossRef]

18. Fernandes, J.C.; Gamero, C.A.; Rodrigues, J.G.L.; Mirás-Avalos, J.M. Determination of the quality index of a Paleudult under sunflower culture and different management systems. *Soil Tillage Res.* **2011**, *112*, 167–174. [CrossRef]

19. WRB; International Union of Soil Science Working Group. *World Soil Resources Reports No. 103*; FAO: Rome, Italy, 2006.

20. Rabia, A.H.; Afifi, R.R.; Gelaw, A.M.; Bianchi, S.; Figueredo, H.; Huong, T.L.; Lopez, A.A.; Mandala, S.D.; Matta, E.; Ronchi, M.; *et al.* Soil mapping and classification: A case study in the Tigray Region, Ethiopia. *JAEID* **2013**, *107*, 73–99.

21. Blake, G.R.; Hartge, K.H. Bulk density. In *Methods of Soil Analysis. Part 1*, 2nd ed.; Klute, A., Ed.; American Society of Agronomy—Soil Science Society of America: Madison, WI, USA, 1986; pp. 363–375.

22. Nelson, D.W.; Sommers, L.E. Total carbon, Organic carbon and Organic matter. In *Methods of Soil Analysis. Part 3*; Book Series 5; Sparks, D.L., Page, A.L., Helmke, P.A., Loeppert, R.H., Soltanpour, P.N., Tabatabai, M.A., Johnston, C.T., Sumner, M.E., Eds.; SSSA: Madison, WI, USA, 1996; pp. 961–1010.

23. Yoder, R.E. A direct method of aggregate analysis of soils and a study of the physical nature of erosion losses. *J. Am. Soc. Agron.* **1936**, *28*, 337–351. [CrossRef]

24. Olsen, S.R.; Cole, C.V.; Watanabe, F.S.; Dean, L.A. *Estimation of Available Phosphorous in Soils by Extraction with Sodium Bicarbonate*; USDA: Washington, DC, USA, 1954.

25. Chapman, H.D. Cation exchange capacity. In *Methods of Soil Analysis*; Black, C.A., Evans, D.D., Ensminger, L.E., White, J.L., Clark, F.E., Eds.; American Society of Agronomy: Madison, WI, USA, 1965; Volume 9, pp. 891–901.

26. Brookes, P.C.; Landman, A.; Pruden, G. Chloroform fumigation and release of soil N: A rapid direct extraction method to measure microbial biomass N in soil. *Soil Biol. Biochem.* **1985**, *17*, 837–842. [CrossRef]

27. Vance, E.D.; Brookes, P.C.; Jenkinson, D.S. An extraction method for measuring soil microbial biomass C. *Soil Biol. Biochem.* **1987**, *19*, 703–707. [CrossRef]

28. Wu, J.; Joergensen, R.G.; Pommerening, B.; Chaussod, R.; Brookes, P.C. Measurement of soil microbial biomass Carbon by fumigation-extraction an automated procedure. *Soil Biol. Biochem.* **1990**, *22*, 1167–1169. [CrossRef]

29. Wienhold, B.J.; Karlen, D.L.; Andrews, S.S.; Stott, D.E. Protocol for Soil Management Assessment Framework (SMAF) soil indicator scoring curve development. Renew. *Agric. Food Syst.* **2009**, *24*, 260–266. [CrossRef]

30. Andrews, S.S.; Karlen, D.L.; Cambardella, C.A. The soil management assessment framework: A quantitative soil quality evaluation method. *Soil Sci. Soc. Am. J.* **2004**, *68*, 1945–1962. [CrossRef]

31. Karlen, D.L.; Wollenhaupt, N.C.; Erbach, D.C.; Berry, E.C.; Swan, J.B.; Eash, N.S.; Jordahl, J.L. Residue effects on soil quality following 10-years of no-till corn. *Soil Tillage Res.* **1994**, *31*, 149–167. [CrossRef]

32. Karlen, D.L.; Wollenhaupt, N.C.; Erbach, D.C.; Berry, E.C.; Swan, J.B.; Eash, N.S.; Jordahl, J.L. Long-term tillage effects on soil quality. *Soil Tillage Res.* **1994**, *32*, 313–327. [CrossRef]

33. Wymore, A.W. Model-Based Systems Engineering. In *An Introduction to the Mathematical Theory of Discrete Systems and to the Tricotyledon Theory of System Design*; CRC: Boca Raton, FL, USA, 1993.

34. Hussain, I.; Olson, K.R.; Wander, M.M.; Karlen, D.L. Adaptation of soil quality indices and application to three tillage systems in southern Illinois. *Soil Tillage Res.* **1999**, *50*, 237–249. [CrossRef]

35. R Core Team. R: A Language and Environment for Statistical Computing. R Foundation for Statistical Computing: Vienna, Austria, 2012. Available online: http://www.R-project.org. (Accessed on 9 March 2014).

36. Tisdale, J.M.; Oades, J.M. Organic matter and water stable aggregates in soils. *J. Soil Sci.* **1982**, *33*, 141–161. [CrossRef]

37. Gelaw, A.M.; Singh, B.R.; Lal, R. Organic carbon and nitrogen associated with soil aggregates and particle sizes under different land uses in Tigray, Northern Ethiopia. *Land Degrad. Dev.* **2013**. [CrossRef]

38. Hartemink, A.E.; Huting, J. Land Cover, Extent, and Properties of Arenosols in Southern Africa. *Arid Land Res. Manag.* **2008**, *22*, 134–147. [CrossRef]

39. Arevalo, L.A.; Alegre, J.C.; Bandy, D.E.; Szott, L.T. The effect of cattle grazing on soil physical and chemical properties in a silvopastoral system in the Peruvian Amazon. *Agrofor. Syst.* **1998**, *40*, 109–124. [CrossRef]

40. Sanchez, P.A. Science in agroforestry. *Agrofor. Syst.* **1995**, *30*, 5–55. [CrossRef]

41. Nair, P.K.R. *An Introduction to Agroforestry*; Kluwer Academic Publishers: London, UK, 1993; p. 489.

42. Franchini, J.C.; Crispino, C.C.; Souza, R.A.; Torres, E.; Hungria, M. Microbiological parameters as indicators of soil quality under various soil management and crop rotation systems in southern Brazil. *Soil Tillage Res.* **2007**, *92*, 18–29. [CrossRef]

43. Stott, D.E.; Cambardella, C.A.; Karlen, D.L.; Harmel, R.D. A Soil Quality and Metabolic Activity Assessment after Fifty-Seven Years of Agricultural Management. *Soil Sci. Soc. Am. J.* **2013**, *77*, 903–913. [CrossRef]

44. Girmay, G.; Singh, B.R.; Mitiku, H.; Borresen, T.; Lal, R. Carbon Stocks in Ethiopian Soils in Relation to Land Use and Soil Management. *Land Degrad. Dev.* **2008**, *19*, 351–367. [CrossRef]

45. Sanchez, P.A. Tropical soil fertility research, towards the second paradigm. In *Transactions 15th World Congress of Soil Science*; International Society of Soil Science and Mexican Society of Soil Science: Acapulco, Mexico, 1994; Volume 1, pp. 65–88.

46. Palm, C.A. Contribution of agroforestry trees to nutrient requirements of intercropped plants. *Agrofor. Syst.* **1995**, *30*, 105–124. [CrossRef]

47. Poschen, P. An evaluation of the Acacia albida-based agroforestry practices in the Hararghe highlands of Eastern Ethiopia. *Agrofor. Syst.* **1986**, *4*, 129–143. [CrossRef]

48. Kamara, C.S.; Haque, I. Faidherbia albida and its effects on Ethiopian highland Vertisols. *Agrofor. Syst.* **1992**, *18*, 17–29. [CrossRef]

49. Asfaw, Z.; Ågren, G.I. Farmers' local knowledge and topsoil properties of agroforestry practices in Sidama, Southern Ethiopia. *Agrofor. Syst.* **2007**, *71*, 35–48. [CrossRef]

50. Kwesiga, F.; Coe, R. The effect of short-rotation Sesbania sesban planted fallows on maize yields. *For. Ecol. Manag.* **1994**, *64*, 199–208. [CrossRef]

51. Sanchez, P.A.; Palm, C.A. Nutrient cycling and agroforestry in Africa. *Unasylva* **1996**, *47*, 24–28.

![sustainability logo] *sustainability*

MDPI

Article

Threats to Sustainability of Soil Functions in Central and Southeast Europe

Hikmet Günal [1,†,*], **Tayfun Korucu** [2,†], **Marta Birkas** [3,†], **Engin Özgöz** [4,†] and **Rares Halbac-Cotoara-Zamfir** [5,†]

1   Department of Soil Science, Gaziosmanpasa University, Tokat 60240, Turkey; hikmet.gunal@gop.edu.tr
2   Department of Biosystem Engineering, Kahramanmaraş Sutcu Imam University, Kahramanmaras 46100, Turkey; tkorucu@hotmail.com
3   Institute of Crop Production, Szent István University, H-2103 Gödöllő, Hungary; Birkas.Marta@mkk.szie.hu
4   Department of Biosystem Engineering, Gaziosmanpasa University, Tokat 60240, Turkey; engin.ozgoz@gop.edu.tr
5   Department of Hydrotechnics, Politehnica University of Timisoara, Timisoara 300006, Romania; raresh_81@yahoo.com
*   Correspondence: hikmet.gunal@gop.edu.tr; Tel.: +90-533-738-4759; Fax: +90-356-252-1488
†   These authors contributed equally to this work.

Academic Editor: Marc A. Rosen
Received: 14 December 2014; Accepted: 12 February 2015; Published: 16 February 2015

Abstract: A diverse topography along with deforestation, changing climatic conditions, long-term human settlement, overuse of agricultural lands without sustainable planning, cultural difficulties in accepting conservative land management practices, and wrong political decisions have increased the vulnerability of many soils to degradation and resulted in a serious decline in their functional capacity. A progressive reduction in the capacity of soils to support plant productivity is not only a threat in the African continent and its large desert zone, but also in several parts of Central and Southeastern Europe (CASEE). The loss of soil functions throughout CASEE is mainly related to the human activities that have profound influence on soil dynamic characteristics. Improper management of soils has made them more vulnerable to degradation through water and wind erosion, organic matter depletion, salinity, acidification, crusting and sealing, and compaction. Unmitigated degradation has substantial implications for long term sustainability of the soils' capability to support human communities and resist desertification. If sustainable agricultural and land management practices are not identified, well understood and implemented, the decline in soil quality will continue and probably accelerate. The lack of uniform criteria for the assessment and evaluation of soil quality in CASEE countries prevents scientific assessments to determine if existing management practices are leading to soil quality improvement, or if not, what management practices should be recommended to mitigate and reverse the loss of soil health.

Keywords: soil health; degradation; land management; erosion; Central and Southeast Europe

---

1. Introduction

Rapid human population growth, along with the spread of technology and culture have significantly increased the rate of natural degradation processes in the pedosphere. Several parts of Central and Southeastern Europe (CASEE) are characterized by severe soil degradation due to accelerated water and wind erosion, nutrient imbalance, depletion of soil organic matter, waterlogging, salinization, contamination, acidification, landslides, soil sealing and compaction by both farm machinery and grazing. Many of these processes cause land abandonment which, in turn, may accelerate degradation processes due to desertification. Abandoned poor agricultural land in Poland constitutes at least 1/3 of all waste land (almost 0.5 million ha) [1]. Both policies and planning instruments for agriculture in many CASEE countries were missing prior to establishing the European Union. Therefore, migration from rural to urban areas and the lack of rural infrastructure development led to an increase in negative anthropogenic influence on soils [2].

The decline in soil quality or degradation of soil, as a consequence of intensive or improper land use, is a problem with ancient roots. Degradation impairs soil quality by partially or entirely influencing one or more of its functions [3]. Although some destructive processes occur naturally, human activity can accelerate the rate of destruction, initially causing a decline in functioning capacity of soils and finally resulting in a loss of the biological production capacity. Therefore, a desert condition is often associated with long-term human habitation in a region [4]. For sustainable development, soils (or soil functions) need to be protected from degradation [5]. Turkey was once the breadbasket for civilization and food production within the region. It has been inhabited since the Paleolithic era, including various Ancient Anatolian civilizations and ancient Thracians [6]. Many of the fertile lands located in semi-arid to semi-humid regions that provided the most favourable sites for the early development of human culture and were once used by archaic civilizations are now buried in debris, because of destructive treatment of the land [4]. Tillage-based agricultural production during those ancient times led to soil degradation resulting in reduced human carrying capacity of the land. Tillage accelerates the destruction of soil organic matter, diminishes microorganism populations, weakens the strength of soil aggregates, and impacts many of the soil-mediated ecosystem functions that ensure, adjust and conserve environmental services. Montgomery [7] concludes that tillage influences soil stability, resilience and quality. He states that the notion of soil quality is referring to the soil's capacity to perform three main functions: economic productivity, environmental regulation and aesthetic or cultural value.

Soil quality has been defined by Doran and Parkin [8] as "the capacity of a soil to function, within ecosystem and land-use boundaries, to sustain biological productivity, maintain environmental quality, and promote plant and animal health." Maintaining soil quality is essential to meet growing human needs for sustainable food and fiber production. Unlike air and water, soil is a limited, non-renewable resource that is not readily movable and does not recover from damage as easily as those resources [9]. Accumulation of salts and in particular sodium in soils when irrigation water is applied to land with inadequate or inefficient drainage will result in deterioration of soil physical structure that can restrict crop establishment and growth [10].

As a multifunctional part of the environment, soil is a conditionally renewable natural resource. It is the most important medium for multipurpose biomass production; the integrator and reactor of other natural resources; a natural repository of water, heat and plant nutrients; a substance with a huge buffering and detoxifying capacity for natural and human-induced stresses; a habitat for soil-dependent organisms; and a mediator of biodiversity. Soil resources may be used and conserved at the same time, but the preconditions of soil resilience must be ensured: constant attention and special care are needed to preserve the unique ability of soil resources [11]. Modification of soil physical, chemical and biological properties through tillage has negative impacts on the functioning capacity of soils. The alteration of soil conditions caused by tillage might seem useful in lowering bulk density while increasing porosity and infiltration. However, in the long-term, tillage causes a decline in soil quality [12] that can eventually threaten the sustainability of food and fiber production in agricultural lands. This will result in poverty of rural areas, force people to migrate from rural to urban areas, and increase urban sprawl onto fertile agricultural lands. Urban sprawl is one of the most prominent threats to agricultural lands surrounding industrialized cities. In order to prevent further degradation of ever widening bands of current agricultural land surrounding large cities, farmers have to be convinced that agriculture can be profitable and sustainable if they are willing to adopt conservation tillage and other management practices that are being developed and demonstrated by researchers and the Extension Service.

2. Historical Changes Contributing to Soil Degradation

Economic and social situations in the CASEE countries were quite diverse at the end of the 1980s. Political system changes in the former 'socialist' countries had great effects on agricultural production through land privatization, new farm establishments, market liberalization, and attraction of foreign capital. However, there were unexpected socio-economic consequences including a decrease in population, migration of people from rural to urban areas, and an increase in the amount of uncultivated land. Changes in economic and political circumstances, poverty, shortage of production inputs and population growth in CASEE countries are the main causes for a decline in soil quality in agricultural fields [1]. Some of the negative effects of the post-communist land reform in Romania were the excessive fragmentation of farming lands, emergence of a large number of individual farms practicing subsistence agriculture and poor services for agriculture (*i.e.*, support for irrigation, fertilization, and mechanization). All have contributed to severe degradation of soil quality [13]. Secondary but no less important causes of soil quality decline include an aging population, agricultural industrialization and climate change. Furthermore, in countries like Romania, improper agricultural water management through intensive land reclamation works (*i.e.*, irrigation and drainage) without considering climate change forecasts or the links between land reclamation and climate change have all contributed to severe degradation [14].

The conventional primary cultivation practices that led to loss of soil functions prevailed until the end of the 1970s and in some CASEE countries until the end of the 1990s. Unfortunately, in countries like Turkey and Romania, conventional tillage is still the main practice applied for crop production. Conventional tillage in many of CASEE countries consists of ploughing in autumn to a depth of 18–30 cm to control weeds and bury plant residues, and a secondary tillage operation to create a seed bed. Özgöz *et al.* [15], used the Soil Management Assessment Framework (SMAF) to quantitatively evaluate farmland and pasture management on soil quality of fine, smectitic, active Typic Haplustolls in Turkey. The pasture had never been cultivated, whereas conventional tillage was used on the farmland for approximately 50 years. Quality assessment indicated that soils within farmland were functioning at 71 and 70 percent of their full potential at the 0–15 cm and 15–30 cm depth increments, whereas pasture soils were functioning at 73 and 69 percent, respectively. The lowest indicator scores were obtained for total organic carbon (TOC) and bulk density (BD) at both depths, presumably due to conventional tillage, intensive grazing and compaction. Overgrazing by sheep and cattle actually resulted in higher bulk density in pasture and a lower overall soil quality index (SQI) than cultivated areas. A significant reduction in TOC score indicated substantial loss of organic carbon in farmland soils where soil organic matter was inherently high before conversion.

Traditionally, the importance of creating a good seedbed for plants, including the improvement of soil fertility, has been emphasized [16] to producers. From a physical perspective, tillage was regarded as playing a very important role in creating a favorable seedbed. Consequently, a period of several centuries was dominated by this approach and is referred to as the era of crop oriented tillage. Over-estimation of the importance of tillage for crop production resulted in damage to soils that ultimately led to an era of "soil oriented" tillage starting in the mid-1960s. By using soil-preserving tillage practices, soil quality could be protected and all crop requirements could be met by keeping the soil in a good physical and biological condition. In addition to causing less damage, soil oriented tillage also reduced costs of production. Following the recent recognition of increasing climate change effects, new trends are emerging recognizing that tillage also has a climate effect and must also be managed with the aim of reducing greenhouse gas (GHG) emissions through improved soil quality.

Strategies for Overcoming Historical Degradation

Arable land use systems can be classified into various categories depending on their impacts on the soil, environment, and farmers practices: early low intensity (~1000s–1800s), conventional (~1800–1960s, from the first year of deeper tillage), early intensive (~1960–1980), integrated (~1980–), modern intensive (~1990–), modern low intensity (~1990–), and ecological/organic (~1980–). The factors taken into account in their review and appraisal are yield, productivity, crop species, manure application, chemical load, weed control, energy input, equipment level, required expertise, tillage and environmental damage [17]. Soil degradation is considered to be a permanent threat and originated from the first land use systems (Table 1).

Table 1. Assessment of the land use systems within the CASEE region [17].

Land use pattern	Positive to soil attributes	Negative to soil attributes	Long-term consequences (±)
Early low intensity (~1000s–1800s)	No chemical soil contamination; Moderated soil diseases; Moderated deepening of the tilled layer	Soil compaction; Moderated decreasing OM resources; Water and wind erosion	Extending arable area at the expense of forests, swamps, *etc.*; Arable area exposing to climate threats
Conventional (~1800–1960s)	Slight chemical contamination of soils	OM loss due to multi-ploughing systems; Soil physical deterioration	Decreasing humus content of soils; Extending water and wind eroded area
Early intensive (~1960–1980)	Recognizing the threats of reduction in soil fertility	Declining soil biological activity due to higher chemical and physical load	Increasing intensity of soil physical deterioration and expose soil to different danger
Integrated (~1980–)	Harmony between soil physical, chemical and biological factors	More soil disturbance to limit new pests, diseases and weeds	Improvement of soil biological and physical characters
Modern intensive (~1990–)	Moderated chemical load, site specific physical intervention	Expose to climate phenomena (silting, crusting *etc.*)	Higher input requires to maintain soil production ability
Modern low intensity (~1990–)	Soil condition may be improved in longer period	Soil condition may deteriorate during *non-hoped* wet seasons	Soil productivity affecting by climate, site and technology level
Organic (~1980–)	Favorable soil biological activity, great number of earthworms	More intervention in soil state requires organic matter compensation	Ploughing is used as crop protection method—soil structure deterioration seems a real threat

Tillage has been an important factor of the land use pattern for centuries.Conventional systems include primary and secondary tillage operations used in preparation of a seedbed for a given crop and area. On the other hand, conservation systems combine tillage and planting operations striving to maintain at least 30% surface cover after planting. Erosion is reduced by at least 50% in conservation tillage compared to bare soils [18,19]. Conservation tillage systems can reduce erosion due to the crop residue left on the soil surface and improve soil conditions for crop growth, while at the same time conserving energy and lowering the cost of farming. Crop residue left on the soil surface is especially effective in reducing evaporation rate, providing plants with nutrients, increasing organic matter levels in the soil, and increasing soil water content by decreasing evaporation and increasing infiltration rate and thus can enhance crop growth [19,20]. Conservation tillage can restore soil structure and improve overall soil drainage, allowing more rapid infiltration of water into soil [21,22]. More recently, providing soil surface protection with residue cover has been more important during the summer, because of intense rainfall and periodic droughts.

In Turkey, conventional tillage methods are dominant as conservation tillage has not yet become a standard practice. However, scientific studies, relevant extension activities and governmental incentives to adopt conservative management practices have increased in the last two decades. The result has been an increased use of reduced tillage in some regions [23].Conservation tillage, however, is mostly practiced at the research level while the government has been employing policies to promote the use of direct planters to benefit from conservation tillage system [24]. Considering the disadvantages of intensive farming and the related costs, direct seeding seems more plausible for farmers with fewer plant production problems for Turkish farmers [25].

Agricultural tillage practices in Romania have changed over decades. Conservation tillage, characterized by leaving residues on the soil surface and reduced- or no-till practices have become more popular [26], but according to Mihovsky and Pachev [27], it has also increased the possibility of soil compaction (already a problem in Romania) when compared to conventionally tilled soil. They argue that compaction due to the use of conservation tillage can also increase the possibility of flooding or occurrence of poor drainage in vulnerable areas.

In most CASEE countries, agricultural-induced environmental loading has remained low. In fact, the primary problem is not over-fertilization but rather (based on country reports) poor plant nutrition management.Prior to the years of political change, the predominant fertilization strategy was based on the crop and/or soil manuring with an attempt to maintain a positive nutrient balance in the soil. Fertilizer consumption in the CASEE countries declined markedly between 1990 and 2010, but to meet future agricultural production demand, fertilizer use is expected to increase substantially.

3. Soil Quality Degradation Symptoms in CASEE Countries

The primary threat to soil quality in CASEE countries is related to human activities. Within the EU, the main symptoms of reduced soil function have been identified as: (1) decline in OM; (2) erosion; (3) compaction; (4) salinization; (5) floods; (6) contamination; and (7) sealing [3,5]. Some of the symptoms are obviously related to each other and one problem can often create and accelerate another (*i.e.*, compaction can cause and in some cases accelerate soil erosion [28]).

3.1. Decline in Organic Matter

Soil organic matter (OM) content is often identified as the most important indicator of soil quality because of its effect on water entry, retention and release, aggregation, wind erosion and nutrient cycling. It protects soil from the erosive forces of wind and raindrop, retards water runoff, provides channels for water to penetrate, increases the water holding capacity of soil and the crop yield [29,30], and provides important buffering and filtration capacities, a rich habitat for soil organisms, and an enhanced sink for atmospheric carbon dioxide. Unfortunately, soil OM is not yet rationally managed for its agronomic, environmental or ecological functions.

A decline in OM began centuries ago and has shown higher levels of decline in three periods: the era of multi-ploughing (in the 1800s), the 1960s, and the 1990s. These three periods are associated with the start of the deeper ploughing, the decade of the early intensive land use, and the first years after land privatization. Loss of OM was greatest in certain soil types, especially dystric Cambisols, Luvisols, Stagnosols and Gleysols. Liming, irrigation, drainage, deep ploughing and removing plant residues by either burning stubble prior to ploughing or gathering the material (e.g., cotton and bushy pasture plants) for firewood also contributed to OM decline. Although there is no data to quantify the original levels of soil organic matter (soil chemical analysis did not begin until the end of the 1880s), recognition of soil deterioration and possible contributing factors can be investigated through classic publications. Symptoms of the soil exhaustion have become apparent since the beginning of the 19th century. This phenomenon may account for the perceptible reduction in soil organic matter within arable soils. A 2003 survey of SOC throughout the EU [31] indicated that compared to virgin soil, the decline in organic matter due to long term tillage ranged from 10%–50% for most soils but was even higher for others (Figure 1) based on data from the European Conservation Agriculture Federation [32].

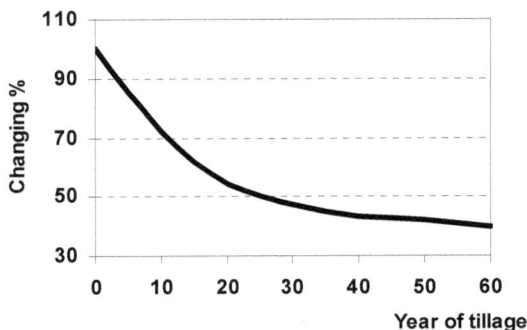

Figure 1. Impact of the years of tillage on changing in organic matter content [32].

The greatest impact on SOM is due to erosion and tillage induced structural degradation. In recent decades, intensive land use and avoiding organic matter recycling (e.g., FYM, stubble residues) both had unfavorable influence on soil processes. As the soil degraded, biomass production was reduced so less was returned to soil and thus the OM content was further depleted [33]. One of the laws of sustainable soil management proposed by Lal [34] states that the rate of restoration of the soil organic matter pool is extremely slow, while its depletion is often very rapid.

Pasture area in Turkey decreased by approximately 47% between 1938 and 1991 falling from 41–21.8 million hectare (Mha), while cultivated agricultural fields increased by about 80% from 13.3–24 Mha [35]. Compared to adjacent pasture, conversion to arable lands with the restricted soil depth resulted in significant loss of OM (up to 49%), decreased in stability of aggregates, reduced mean weight diameter and decreased hydraulic conductivity [36]. Twelve years of continued cultivation of former pasture land in Turkey caused 61% and 64% decreases in mean weight diameter for the 0–10 cm depth and 52% and 62% decreases for the 10–20 cm depth, respectively, when compared to forest and pasture soils. Degradation of soil physical properties due to the loss of OM through cultivation also made soils more vulnerable to erosion [36], because of a decrease in the water infiltration rate that led to increased run-off and soil loss. Montanarella *et al.* [37] stated that loss of OM, particularly in arid and semi-arid areas, is closely linked to the process of soil erosion. Erosion reduces the organic matter content by washing away fertile topsoil that is vitally important for sustaining soil functions. In Romania, the reduction of organic matter and macro-nutrients content affects more than 3.3 million hectares representing 14.1% of total country. According to Bireescu *et al.* [38], the soil organic matter losses, which are caused by the removal of the topsoil, range between 45% and 90% of the total organic matter pool in the soil. At the country level, SOM losses are estimated at 500,000 tons per year.

Stubble that was maintained on soil surface, especially in the long-term, increases soil organic matter content, enhances aggregate stability of soils, and provides soil and water conservation [39]. Monoculture farming and stubble burning are thought to be the major causes of low organic matter content of soils in Turkey. Stubble burning increased with double-cropping that increased rapidly following the introduction of machinery into farming operations in the 1960s. Burning stubble is perhaps the most controversial crop residue management option. There are advantages to burning, but some of the perceived advantages are not as great as some believed [40]. Stubble burning is important in assisting normal tillage operations reducing or removing the vegetative cover from the soil surface. Burning has also been used as a substitute for herbicides and pesticides in the control of weeds, pests and diseases [39,41]. In minimum or zero tillage systems, burning is often used as means of land clearing to prepare the field for seeding [42]. However, maintaining plant residue on the soil surface is favorable for protecting the soil against wind and water erosion. Burning of stubble removes the entire beneficial plant residue and leaves the soil surface bare and consequently unprotected from raindrop impact and an increased erosion risk.

3.2. Soil Erosion

Soil erosion is one of the major and most widespread threats on soil quality. Inappropriate soil management practices led to physical degradation of soils and are major causes for water and wind erosion [43]. Physical degradation of soil may be recognized as the loss of soil structural stability and ability to resist the destructive impacts of wind and water. Structural degradation can be observed both on the surface, where thin crusts may occur, and below the surface when compacted zones form in or below the ploughed layer. The absence of aggregate resistance to disintegration reduces water infiltration and increases runoff and erosion rates [44].Wind or water erosion occurs and is a problem in agricultural lands within almost every CASEE country; even those with flat topography such as Lithuania or the other Baltic states [1]. Soil erosion occurs in vast areas of Ukraine, with 41% (17 million ha) of agricultural land having been characterized as being subjected to water and wind erosion in 1996 [9]. In the Balkan Peninsula, particularly in Bulgaria and Romania, around 40% of land is affected by soil erosion [45]. According to Debicki [1], in some of these countries (e.g., Bulgaria, Romania, Albania, Slovenia, FYR of Macedonia, and Georgia), water erosion is very severe and may ultimately lead to desertification.

Soil erosion, mainly due to water and to a lesser extent wind, is still the most important degradation process in most CASEE countries. It has resulted in shallow soil depths, loss of most fertile topsoil and organic matter from eroded surfaces, and irreversible loss of natural farmland over time-scales of tens to hundreds of years [46]. Even where soil is deep and loss of the topsoil is often not apparent, the effects are nevertheless potentially very damaging to sustainability. The rate of erosion is sensitive to climate and management practices, as well as to conservation practices applied at the farm level. Sauerborn *et al.* [47] indicated that detrimental impacts of soil erosion are anticipated to increase, since climate change is expected to influence the characteristics of rainfall in ways that might increase the intensity of water erosion in central Europe.

In addition to inherent soil properties (e.g., slope and texture), unsustainable agricultural management practices such as forming large fields with no anti-erosion protection before the change in political systems, growing wide-row crops (e.g., maize or sunflower) on sloped fields, and overgrazing are also major causes of erosion. Deforestation and farming in uplands and mountains, overgrazing, use of heavy machinery, excessive irrigation of vulnerable agricultural fields, and poverty also result in severe damage to both land and permanent plant cover. Converting natural vegetation cover to field crops requires mechanical soil cultivation that intensifies the erosion [1].

In Romania, more than 40% of the total agricultural area is situated on the slopes higher than 5%. Because of their soil characteristics, the main problem Romanian agriculture faces in the hilly areas is soil erosion. Almost 5.3 million hectares of agricultural land are vulnerable to surface and depth erosion as well as to landslides. Water erosion is considerable on about 3.5 million hectares of this area. Approximately 55% of the 4.8 million hectares of pasture and meadow have been negatively affected by erosion and landslides due to inappropriate management. The areas affected by water erosion, which includes agricultural lands, forests and the unproductive areas on slopes, are as follows: slight erosion—46.3%; moderate and high erosion—41.5% and severe to excessive erosion—12.2% [48,49]. In southern Romania, soil erosion has increased because forest belts have been destroyed, droughts are becoming more frequent, and crop growth is often poor because there are very few irrigation systems.

Deforestation, conventional tillage practices and improper irrigation management have led to increasing rates of soil erosion for a long time in Turkey [50]. Continuing loss of soil functions is now threatening some of the country's most fertile agricultural fields. Although not located within a desert belt *per se*, improper agricultural practices led to degradation of agricultural fields and put many regions of Turkey at the risk of desertification. Furthermore, with 46% of the land area having slopes of 40% or more, many agricultural practices are complicated and erosion is easily increased.

Overall, 59% of the agricultural land, 64% of rangeland and 54% of forestland are subjected to erosion in Turkey, and approximately 180 million tons of sediments are transported to seas and lakes every year. Aykas *et al.* [51] indicated that sediment lost by erosion is equal to losing 25 cm of soil from 400 thousand hectares land. This soil loss is an even greater problem considering the political instabilities of neighboring countries such as Syria and Iraq that have been continuing for nearly four decades. Turkey now has to feed almost two million refugees in addition to the 77 million residents of the country. Therefore, any decline in soil quality due to severe erosion and consequent decline in agricultural productivity increases food security risks for Turkey.

Wind erosion is common in the plains of the arid and semi-arid climatic regions, as well as on sandy and silty soils within other CASEE countries. Wind erosion occurs throughout the year on bare lands, and mainly in the spring and summer months on the overgrazed rangelands and over-cultivated/tilled soils. Avci *et al.* [52] reported severe wind erosion in winter months, especially during the Lodos-south wind in Central Anatolia of Turkey. Lack of plant cover on rangelands and low organic matter content of arable lands are the major causes of wind erosion. In Turkey, wind erosion has not been considered to be as important as water erosion since it is generally confined to special areas such as Karapinar-Konya, Incesu-Kayseri [52], Aralik-Igdir, and coastal regions in Mediterranean and Aegean Sea (Figure 2).

Figure 2. Severe wind erosion, sediments filled the irrigation channel in Aralik-Igdir/Turkey.

Wind erosion is site-specific, especially for the southern part of Romania. The absence of irrigation and the uncontrolled deforestation of protection belts accelerated the northward extension of desertification-affected surfaces and movement of sand dunes. It has also been conducive to depletion of arable-land productivity and, in time, abandonment of those lands [13].

The functions of soils, mainly biomass production, crop yields due to removal of nutrients for plant growth, and soil filtering capacity due to disturbance of the hydrological cycle (from precipitation to runoff) are decreased or totally lost by removal of soil [9]. The estimated area (Table 2) damaged by water erosion calls attention to the need for prevention and alleviation. In the past, loss of productivity was compensated for by installing modern irrigation systems and applying additional mineral fertilizers, so that the impact of the erosion did not appear on time [9]. Nowadays, however, anti-erosion measures must be included in land management plans. Increasing areas of permanent grasslands in hilly and mountainous regions also represent a positive trend toward reducing soil loss.

3.3. Soil Compaction

Compaction is one of the most common forms of soil physical degradation which can cause a serious reduction in water penetration and seedling emergence. Soil compaction has been described as one of the five threats to sustained soil quality by the EU Soil Framework Directive. Where crop production has been severely affected by compaction of arable soils in many of CASEE countries, a primary cause has been an increase in field traffic.Globally, more than 68 million ha of land are classified as compacted, with 4% being associated with anthropogenic soil degradation [53]. In Europe alone, compaction accounts for about 17% of the total degraded area [54].

Soil compaction represents damage of several soil physical properties, including the breakdown of soil structure, decreased loosening, limited water transport and consequently higher risk for water erosion and drought stress. Soil compaction also negatively impacts other soil processes and can have a range of negative consequences depending upon the inherent soil properties, bearing capacity, moisture condition, relief, field patterns, and applied technologies such as irrigation, fertilizer application, and many other factors [1]. Allen [55] reported that compaction decreases macro porosity and hydraulic conductivity of soils which increases the susceptibility of soils to erosion. Soil compaction results in high mechanical resistance to root growth in a compact dry soil and poor aeration in a compact wet soil (Figure 3) [56]. Both natural and induced compaction can occur because of a high content of fine clay- and silt-sized particles and may be caused by either drying out or being covered by water. Traffic induced compaction is caused by heavy machinery and frequent loading on wet soils and is most common between cultivated and undisturbed layers. Under the less favorable economic conditions of CASEE countries, tillage-induced soil compaction (plough or disk pan) occurs more frequently than the traffic-induced variant. Overgrazing can also induce crust formation by surface compaction of wet or moist soils and mechanical destruction of the surface soil aggregates [57]. In Turkey, low organic matter and high clay content often leads to formation of a dense plow pans. Çarman [58] noted that annual yield losses due to soil compaction in Turkey were over one billion U.S. dollars ($). Furthermore, compaction can also accelerate other threats such as wind and water erosion [28].

Subsurface soil compaction occurs as a result of the forces applied when agricultural machineries are used on the field [56]. Deep soils with less than 25% clay content are the most sensitive to subsoil compaction [59]. In contrast to the compaction of surface soils, subsurface compaction cannot easily be reversed and may last longer until broken by a ripper [60]. In many CASEE countries, compaction became a very serious degradation agent as the size and weight of farm machinery increased. For example, since 1955 the mass of tractors and tillage implements have increased by 68% and 200%, respectively, in the Czech and Slovak Republics [1].

Figure 3. Subsoil compaction severely constrains root growth.

Soil compaction and crusting are most prevalent in the plains region of southern and western Romania, where use of heavy machinery is widespread. Unfortunately, farmers in many CASEEcountries are not aware of the seriousness of subsoil compaction. Restoring drainage, increasing plant nutrition, and improving irrigation systems can sometimes mask the detrimental effects of subsoil compaction on crop production, but those temporary solutions for preventing yield reduction due to compaction often increase expenses for farmers and contribute to environmental problems due to increased use of water and nutrients. Sustainable agricultural production with reasonable management practices requires no subsoil compaction [61].

3.4. Salinization

Soilsalinization occurs in areas with saline soils such as the solontsak, solonetz, salinemeadow, and saline chernozems. Itis also a local risk if temporary water logging occurs and brings excess salt from deeper layers to the surface.One cause for salinization is irrigation which is vital for agricultural production in arid lands, but with improper management (*i.e.*, lack of drainage)arid land irrigation can negatively impact soil quality through salinization and alkalinization [57]. Salt-induced land degradation is a major drawback to optimal functioning of soils in arid and semiarid regions of CASEE countries. For example, in Romania 4% of the total agricultural land was affected by salinization in 2002. Inappropriate water regulation, land use changes (conversion to from pasture to arable) deep ploughing, disturbance of deeper soil layers, and irrigation (without proper drainage systems) can exacerbate the salinization problems. Low amounts of precipitation, dry conditions and very high temperatures during summer seasons, topographic properties and parent material, along with wrong (mainly water) management practices, are the major causes.

Salinization and waterlogging are serious problems, especially in areas where large but poorly managed irrigation systems were constructed. Applying excess irrigation water in the absence of a well maintained drainage system may cause the water table to rise and can also result in an increase in secondary soil salinity. To avoid worse effects, strict land use (e.g., avoid deep ploughing) and water management (e.g., no irrigation, using good quality irrigation water, and keeping the water table down) practices must be followed. Intensive tillage of saline soils in arid regions will destroy aggaegates, increase the capillary rise of salts to soil surface and eventually leave soils vulnerable to wind erosion (Figure 4).

November 5, 2010 Nigde/Turkey

Figure 4. Intensive tillage of saline soils in arid soils increases capillary rise of salts to soil surface.

Unsustainable irrigation practices and inappropriate water management at the farm-level stimulates raising of groundwater and contributes to salt accumulation, particularly in irrigated fields of arid and semi-arid regions [62], and eventually causes salt-induced land degradation. The potential of economically irrigable agricultural lands in Turkey is about 8.5 million ha, of which 4.9 million ha are now irrigated. The government has initiated many projects to expand the irrigated area, strengthen the economy, and meet a growing food demand based on agriculture [62]. The

Southeastern Anatolia Project, commonly referred to by its Turkish acronym "GAP", is about to be completed as a large integrated water resources development project in the semi-arid Southeastern region of Turkey. The project includes 22 dams in the upper Euphrates-Tigris Basin, and aims to provide irrigation for 1.7 million hectares of land. However, salinity is a major problem in the GAP region, and in particular in the Harran Plain which was first opened for irrigation in 1995. Prior to excessive and uncontrolled irrigation, an insufficient and uncared for drainage system, and an increase in the groundwater level caused by the improper irrigation management practices [63] have all contributed to the problem.

Tillage can be used to improve soil permeability in saline soils, but if it is not properly practiced a compacted plough layer might form and salts will be accumulated above it and potentially bring salts even closer to the soil surface [64]. Timely and convenient monitoring and assesment of soil quality will help guide adoption of corrective mesures to control the salinity that might otherwise threaten the sustainability of crop production. Minimum soil disturbance at a shallow depth is recommended for seedbed preparation in saline soils. Salts leached to the lower part of the soil profile by winter snow melts can be returned to the surface by deep spring tillage. Thus, deep tillage operations on saline land can unnecessarily increase surface salt concentrations [65], and should not be used unless they are needed to ameliorate subsurface compaction.

In Romania, salinization is primarily a natural process, but some poorly applied, intensive land improvement works, such as embankment, drainage and irrigation, have aggravated the problem. Currently, salinization and sodification problems affect less than 600,000 ha of land, and occur mainly in the eastern part of the Romanian Danube Plain and in the Western Plain. In 1989, Romania had more than 3 million hectares of irrigated land and another 3 million hectares of land with adequate drainage systems. By the end of the 1990s, all these projects had been severely degraded, leaving many areas without any cover against extreme drought or intensive precipitation. By 2006, Romania irrigated only 3% of the overall managed agricultural land and had suffered a decrease in cereal grain output of 35%–40%.

3.5. Flooding

Localized flooding is a serious problem, associated with extreme rainfall and unpredictable rainy periods. The frequency of floods seems to have increased over the last decade in all CASEE countries, maybe in response to global climate change.

In 2005, the surface exposed to flood danger in natural regime of flow was up to 30,000 km^2 (3500 km^2 representing agricultural areas), representing about 13% of the Romanian territory. Romania was severely affected by several floods in recent years that resulted in some important arable areas being no longer suitable for agriculture. An important factor that increased the severity and impact of the floods was that several pumping stations associated with land drainage systems were no longer functioning properly. The drainage canals which were not properly maintained together with an underestimation of pumping stations discharge capacities contributed to long-term stagnation of water from floods and implicitly to land degradation. Several other factors have also contributed to increased flooding problems. These include: (1) a reduced capacity in minor flow paths which in Romania are exceeded about 30%–50% of the time; (2) various construction projects that divert overflow in meadows; (3) abundant rainfall which often exceeds the 20–40 ratio between maximum flood discharge and average discharge; (4) increased hillside runoff; (5) inadequate maintenance of flow paths; (6) improper bridge construction and maintenance that includes obtrusion of bridge sections with floats, clogging of canals, inadequate maintenance of gutters in most villages, under dimensioning of bank requirements and cutting large areas of forests; and (7) an increase in natural maximum flood discharge due to long banking without measures concerning to take over these effects [66].

As most high fertility soils are located in floodplains, they are also affected by the floods, through compaction, alluvial deposits, and under certain circumstances, heavy metal pollution. Experts from academia and land reclamation have issued a disastrous scenario for agriculture in the western part of Romania: without rapid intervention inland improvements, floods will cause Romania to lose more than one million hectares of arable land in the Western Plain. Flash floods, which are specific to hilly areas and have been a main factor causing massive deforestations during recent years, can also cause significant land degradation, especially when they are coupled with other phenomena such as landslides, even though they usually affect a relatively small area.

An indirect factor of land degradation due to flooding is represented by political involvement in flood management. The current land reclamation system in Romania, which is based on embankments, drainage and floods, has in recent years passed through a series of perpetual reorganizations making it very unclear what purpose these actions are intended to have. Currently, the National Administration of Land Reclamation is reorganizing and has resulted in massive layoffs with a severe negative impact on maintaining and operating the existing flood management infrastructure. There are numerous examples in which interventions to restore flood defense works were limited to simply recovering the affected works and not according to the physical condition and their continued degradation. Insufficient staff and funds made impossible for maintenance and repair of embankments, dams, channels, and culverts.

3.6. Soil Contamination

Soil contamination is an important threat in the CASEE contries due to rapid industrialization and urbanization which often results in soilsbeing used for the disposal of waste products. Fortunately, pollution with heavy metals and radioactive nuclides has a very local character, so large territories of the CASEE countries are suitable for producing environmentally clean products. High competition of soil and land for different uses often causes contamination problems to be particularly high in densely populated areas [9]. The results of 40,000 soil analyses from all over Poland indicated that 79% of the soils had heavy metal concentrations at background levels while 18% had concentrations that were slightly over threshold levels. Less than 1% of the soils (2.3% of agricultural land) were polluted with heavy metals. Agricultural production that does not directly affect the food chain should be performed on these lands [67]. Muranyi [68] reported that results of monitoring heavy metal pollution in Hungary were similar to those reported for Poland. Heavy metal pollution was encountered in only certain hot spots, but soil acidification and soil erosion problems were noted throughout the country. The case for pollution of soils with heavy metals in Czech Republic was also similar to other CASEE countries with high heavy metal concentrations generally being associated with long term industrial emissions [69]. Soil degradation and soil pollution show a manageable rate in some countries (e.g., Czech Republic, Hungary, Slovak Republic), but cautious monitoring and control of the threatening factors need to be continued. The buffering, filtering and transforming functions of soils are mostly affected by local and diffuse contaminates. Soils can absorb toxic metals without harm up to a critical point, but loadings over the buffering capacity can result in a release of the substance back to the environment [9].

Acidification, both natural and anthropogenic, is the most widespread type of soil contamination in some Western and Central European countries, with especially large areas having been identified in Poland and Ukraine. Under acidic conditions, exchangeable base cations (Ca, Mg, K, and Na) are highly mobilized and leached from soil profile. This leads to a depletion in buffering capacity of soils. The pH of soils will start to decline and with increasing acidity, ions such as aluminum are mobilized. Higher concentrations of aluminum are toxic to most plants. Acidification also mobilizes heavy metals that were accumulated and bound in the soil under higher pH conditions [1].

3.7. Soil Sealing

Sealing means an irreversible loss of soil multi functionality as a consequence of the competition between their use for infrastructure development *versus* biomass production. Reviewing the CASEE country reports it seems that the area affected by surface sealing has increased at an average annual rate of about 6000–10,000 ha/year for the last two decades. Since approximately 2010, the total area under agriculture has remained relatively stable because the new economic situation has resulted in less investment in industry and only moderate deforestation. Soil degradation due to urbanization and industrial development in most of the Central and Eastern European countries has significantly increased, due to recent population shifts and a more extensive urban pattern [9]. Sealing is particularly apparent in coastal zones, where urban and recreation areas, agriculture, industry, commercial activities and tourism are all concentrated, and in many cases, in competition for the same land area. Soil sealing in Romania between 1989 and 1994 increased almost 19% [70]. Based on a 2012 report published by the Institute of Regional Development Planning at the University of Stuttgart, 45% of the land area in Hungary was sealed. In contrast, only 20% and 26% of the urban land was sealed in Romania and Estonia [70].

Sealing of soils through urban and industrial development, such as the construction of roads, houses, industrial premises, and sporting facilities in Turkey began in the 1950s and accelerated through the 1960s due to the lack of legal enforcement to prevent conversion of agricultural lands, unplanned industrial sprawl upon agricultural and natural areas, and increased population growth [50,72]. The most striking change in demographic structure of Turkey from 1927, when the first consensus of the Turkish Republic was made, to 2011 was the high population increase in urban regions and the decrease in rural areas. Urbanization caused particularly intensive use of arable land around cities. According to the 1927 census, 75.8% of Turkey's population were living in rural areas and 24.2% in urban areas. In contrast, the 2011 census showed 23.2% of the population living in rural areas and 76.8% living in urban areas.

The severity of sealing is even more complex than the simple loss of land, since the growth of many urban regions often affects high quality soils (Figure 5). Fertile soils around rivers and river deltas are mainly occupied by cities, and valuable land is therefore lost to food production [71]. The occupation of agricultural lands by settlements and commercial and industrial facilities has reduced the productivity of the agricultural sector, while at the same time, increasing the likelihood of floods. Occupation of high quality land for settlements and industrial purposes in Turkey has almost reached 172,000 ha [72].

Figure 5. Open dumping of municipal wastes at the edge of agricultural fields in Tokat Province of Turkey.

Table 2. Percentage of agricultural land affected by natural and human-induced soil degradation in selected countries.

Selected countries	Agricultural land area (1000 ha)	Soil compaction	Temporary drought/water-logging effects	Wind erosion	Water erosion	Decline in OM %
		As a percentage of agricultural or total (t) land area				
Albania	699	36	?/18.5	no data	50	0–35
Bulgaria	5123	47	40/35	29	72	10–40
Croatia	3220	25–35	35/25	10	35	F
Czech Republic	3101	28–34	31/27	33	14	F
Hungary	5585	30–35	27/23	24	39	F
Poland	18,512	20–25	16/24	28	28.5	F
Romania	14,714	5.6 (t)	48/26	1.6 (t)	26.4(t)	14 (t)
Serbia	5109	F	35/30	13	33	10–37
Slovakia	2466	28	F	6.5	43.3	F
Slovenia	480	F	39 (t)/23	23 (t)	44 (t)	F
Turkey	28,050	F	F	0.65	87.9	F

Note: (t): from total area; F: it was found, but not determined.

4. Other Human Induced Soil Threats in the CASEE Countries

Managing Crop Residues

Crop residues are very important for soil conservation. From the 1800s until the 1970s, crop residues were managed primarily by using tillage to create suitable soil conditions, often with fine structure, for plant germination, emergence and growth. As discussed previously, overestimation of crop requirements for tillage have likely contributed to the deterioration in soil quality [17].

Recently, attention has again been focused on crop residues as a potential source of "bio-energy" [34]. The challenge is balancing this new demand for crop residue with the traditional uses that are important for soil conservation. Currently, farmers either burn the stubble and residue or leave it on the field surface and plant through it using minimum or no-till practices to incorporate some or all of it into the soil [40]. Surface residue provides protection during the summer and is indisputably important for water conservation in Eastern, Central, and Southern Europe. Crop residues are also being recognized for their ability to buffer climate-induced damage that is being observed more frequently throughout the region during and outside the growing season. Furthermore, although the amount of summer rainfall has been decreasing, the rain storms have become more frequent and devastating. Soil resources are being degraded by the kinetic energy associated with those storms. For example, a 25.4 mm (1 inch) rainfall event applied uniformly across a 0.4 ha (1 acre) surface delivers a force of 2.7 MPa. This amount of energy hitting bare soil in the fields without any plant residue will cause a breakdown of soil aggregates, increase the potential for sealing and make the fields more susceptible to water and wind erosion. Having crop residues on the surface to absorb this energy will prevent degradation of the soil aggregates [51].

Soils deprived of their protective straw are also increasingly exposed to summer climate stress. Crop residues are thus needed to not only keep the soil in place but also to alleviate heat stress and reduce evaporative water loss [74,75]. In addition to crop residues *per se*, green manure mulch, chemically treated weeds, and volunteer weeds can also provide protective surface for soils. Where the crop residue is left on the soil surface, the level of protection is first affected by the ratio of the cover, and later by the mode and quality of stubble tillage. Table 3 shows that the amount of soil removed from field was significantly reduced as the amount of crop residue on the soil surface increased [76]. Similarly, Kalmár *et al.* [77] cited Schertz [78] who stated that soil conserving tillage is characterized by having at least 30% cover ratio after sowing.

Figure 6 shows surface soil that has no residue cover in a typical fallow-wheat system in Central Anatolia. This widely used system is incompatible with the conservation agriculture concept due to frequent tillage operations for weed control and seedbed preparation during the 16 months of fallow. Even after planting the crop (wheat, barley or rye), the fields have no residue cover to protect them from wind and water erosion during the initial part of the growing period [52].

Table 3. The relationship between crop residue on soil surface and soil loss [76].

Crop Residue tons/ha	Runoff %	Infiltration %	Soil Loss tons/ha
0.00	45.0	54	13.00
0.63	40.0	60	7.50
1.25	25.0	74	2.50
2.50	0.5	99	0.75
5.00	0.1	99	0.00
10.00	0.0	100	0.00

Figure 6. Fallow-wheat system applied in Central Anatolia.

5. Conservation Farming for Achieving Sustainable Soil Systems

Soil deterioration has occurred for centuries primarily due to "conventional" soil management. When the processes causing deterioration of soil quality are traced and controlled, both soil-use and soil quality are sustainable [5]. The concept of sustainability has environmental, economic and social aspects as well as an institutional dimension. Therefore, processes affecting all of these issues should be taken into account to maintain soil health.

Beginning as early as the 1860s, the practice of ploughing to depths exceeding 25 cm was increasingly adopted in response to encouragement for increased sugar beet production. For beets, tillage depth was significantly relevant to water and soil conservation, because as soil porosity increased, more water was retained in tilled layer. However, the increased porosity and available water also stimulated the mineralization of soil organic matter. Burying crop stubble to a depth of 12 cm also enhanced decomposition such that nearly all the residue was mineralized during the 18 month fallow period, whereas only 33% of residue was decomposed when left on the soil surface [79].

Tillage has been an integral part of agriculture for several hundred years, although the standards by which the process was evaluated declined following the two world wars and as a consequence of land redistribution and privatization. Farmers failed to recognize the importance soil tillage research but they were quick to respond to changes in economic conditions.

Excessive tillage cannot be directly linked to any particular time period [17,80] or economic condition, although economizing under the force of necessity has always been a typical human response to periods of economic difficulty. Therefore, farmer attitudes with respect to rationalizing tillage could, in retrospect, be explained by shortage of capital. Often it was easier to follow traditional methods than to invest in new equipment, fertilizers, or pesticides. Simple tradition may also be an explanation for farmer aversion to new production methods even though soil deterioration symptoms that originated due to long-term traditional tillage [81] are becoming increasingly abundant.

Other authors have stated that without remedying the condition of the soils, it will be nearly impossible to achieve adoption of new techniques introduced into the CASSE region. This conclusion is supported by soil protection research which has been a key subject for decades. Only now are the results achieved so far being taken into account for development and application of new cultivation practices [82].

Table 4 lists several different methods of soil protection with the first of them coming from work by North-American researchers [83]. The different methods of soil protection have been developed and are being conducted in parallel with no-till experiments within areas exposed to water and/or wind erosion [84]. Currently, there is growing interest in other soil conservation techniques (e.g., till-plant, mulch-till, and strip-till), to some extent perhaps as a consequence of increasing climate threats.

One of the new management strategies, known as conservation agriculture (CA) requires adoption of supplemental agricultural practices that minimize degradation of the soil organic matter and soil structure, protect against soil erosion and degradation, and preserve soil biodiversity [85]. The first step toward adoption of CA involves recognition of the risks (*i.e.*, wrong practices, bad habits, poor soil quality, extreme climate phenomena) and the desire for improvement, while the second step involves improvement or conservation of soil quality in harmony with ecological conditions, mechanization and the farm management conditions.

Table 4. Soil tillage trends, objectives and realization in the CASEE region.

Trends	Time and place of developing	Aims of the system	In the CASEE region	
			appearance	realization
Minimum tillage	1950s (USA)	cutting tillage depth, passes and costs	mid-1970s	reduced constraint e.g., disk tillage
Reduced tillage	1960s (USA)	cutting tillage passes and costs	mid/end-1970s	tool/element combination
Conservation tillage	1960s (USA)	effectual soil preserving by surface cover ($\geq$30%) after sowing	end-1980s, first years of the 2000s	surface cover after stubble tillage and after some types of primary tillage
No-till	1950s (USA)	soil and water preserving by minimized soil disturbance	from the 1960s	problems in the first years limited the interests
Mulch-till	1980s (USA)	soil and water preserving by whole surface disturbance and by fair surface cover	mid-1980s, first years of the 2000s	good: by tine, by loosening, risky: by disking
Ridge-till	1980s (USA)	soil and water preserving in sloped fields	1990s	in experiments only
Strip-till—1st	1970s (USA)	clean sowing strips, covered inter rows—reducing tillage intervention and costs; improved by satellite guidance and automatic positioning	1990s	tepid interest
Strip-till—2nd	2000s (USA)		2010s	field trials with hope of the extending
Climate mitigating	mid-1990s (Europe)	all systems are adaptable to site and climate conditions	first years of the 2000s	step by step, however time presses

Twelve factors have been selected to outline the fundamental requirements of sustainable soil tillage [86,87]. These are:

(1) Avoiding farming- and tillage-induced soil damage, including occurrence and extension of soil compaction, degradation of soil structure, water and wind erosion, high CO_2 emission, and loss of organic material.

(2) Maintaining soil moisture transport by improving infiltration and storage during wet periods and decreasing moisture loss during dry and average seasons. As indicated by Avci [52], crop residue

cover is very important for meeting this requirement because it reduces wind speed, prevents sunlight from penetrating the soil surface and increasing evaporation, and if standing traps snow for extra soil moisture.

(3) Preserving soil organic matter to increase water-holding capacity, structure stability, loading capacity, and workability while decreasing soil compactibility and vulnerability. Again, leaving plant residue on soil surface will provide erosion control and lead to an increase in soil organic matter content.

(4) Managing stubble residues by applying harvest and tillage techniques that leave mulch cover. Covering the surface for as long as possible after harvest will help preserve soil structure, moisture and mitigate heat and rainfall stresses outside the growing season. Surface cover, particularly with small-stemmed crop residues such as wheat or barley, creates a friction with the wind and effectively reduce the erosion [52].

(5) Recycling stubble residues to increase soil organic matter, promote favorable biological activity, and improve workability through the mellowing processes.

(6) Optimizing machinery (*i.e.*, tractor selection, tool mass, running gear, working speed) and arable site factors to reduce energy consumption and decrease environmental load.

(7) Minimizing soil loading from stubble to sowing phases.

(8) Applying optimal crop sequences to reduce fertilizer needs and improve soil biological activity.

(9) Maintaining infiltration, storage capacity, and aggregation on irrigated soils.

(10) Applying tools without creating tillage pans, particularly in wet soils.

(11) Assessing possible risks prior to establishment of new tillage and sowing systems. Soil condition assessment will have greater importance before tillage interventions, in crop stands and after sowing.

(12) Selecting the most adaptable soil conservation methods that conform to site and crop production requirements.

Table 5 provides a summary of tillage and sowing methods that appear to be adaptable to CASEE soil conditions. Overall, *mulch-till with subsoiling* or *tine* tillage appear to be indispensable for maintaining stability and reliability of cropping in extreme seasons. *Tine tillage* is also recommended for gently mixing the upper (0–30 cm) layer of soil after three to four years of *strip-till*. *Mulch-till with disking* should only be applied if deeper soil layers are in good condition, and the soil is dry. *Composting tillage* shows similar advantages and risks.

Table 5. Experiences in soil conservation solutions in the CASEE region.

System/method	Method	Main advantages	Main considerations	First adoption
Mulch-till	Subsoiling	Deep rooting, less climate dependence due to improved water transport	Weed infestation in the first years	-mid 1980s -from the 2000s
Mulch-till	Tine	Soil structure preserving and improvement, less dependence on soil water content	Same diseases, weed infestation in the first years	-mid 1980s -from the 2000s
Mulch-till	Disking	Saving time and energy	Shallow loosened layer, higher climate dependence	-from the 1980s
Till-plant	Shallow (2–5 cm) or deeper (810–15 cm)	Saving time and energy	State of the root zone	-from the 2010s
No-till	Continuous or short term	Saving time and energy	Continuous: long-term soil conversion; occasional: soil water content	-1960s, 1990s -2010s

Table 5. *Cont.*

System/method	Method	Main advantages	Main considerations	First adoption
Strip-till	Depth is varied for crops	Loosened soil to the created depth, saving time and energy	Uncrushed maize stalks (good habitat to E. corn borer)	-from the 2010s
	Twin-row sowing	Deep rooting in subsoiled variant	Endeavors to optimizing crop root development and placement	-Kolbai, 1956, Hungary -2010s (USA)
Composting tillage	All crops	Soil structure preserving and improvement	Depth of the loosened layer	-from the 2010 (Slovenia)

Strip-till is suitable for mid-technology farming because it can create a good soil state with only a small amount of tillage. *No-till* is a special cropping method that minimizes soil disturbance but requires modern machinery, sound experience, frequent technology updates, and adjustments to the site, year and crop being grown. The risks associated with conservation farming can be minimized through planning and progressive management. Therefore, considering the disadvantages of intensive farming and related costs, direct seeding becomes more viable and may have fewer plant production problems for Turkish farmers [25,52,88].

6. Conclusions

Soil characteristics are directly or indirectly altered and their capacity to function is either limited or enhanced by humans. Since degradation of lands cannot solely be accounted for by physical or technical causes, social and political dynamics along with any activities that threaten soil quality in CASEE countries should be controlled by laws and regulations. Environmental awareness and improved socio-economic status of people in rural areas are also important incentives for encouraging farmer adoption of new conservation and agricultural practices. The negative effects of unfavorable agricultural management on the environment generally originate at the single farm level. Therefore adaptation of tools for improving soil quality is needed at this level. The main limitations and uncertainty regarding soil sustainability can often be traced back the economic situation which causes fluctuations in agricultural activities, including soil remediation.

Most soils have the potential to resist degradation processes to some extent. Therefore, the rate of degradation can be efficiently decreased and their unfavorable consequences can be at least moderated by maintaining and continuing to use appropriate land management and water conservation practices.

Finally, we conclude that low productivity soils in CASEE countries and elsewhere around the world can be eliminated by implementing site-specific tillage and intensive crop production systems that improve low organic matter soils by minimizing conventional tillage, residue removal, soil compaction, water and wind erosion.

Acknowledgments: The authors would like to thank Douglas L. Karlen for the useful comments and the English language review of the Manuscript.

Author Contributions: All authors contributed equally to this work. All authors read and approved the final manuscript.

Conflicts of Interest: The authors declare no conflict of interest.

References

1. Debicki, R. State of the Land Degradation in Central and Eastern Europe: Proceedings of the Workshop on Land Degradation/Desertification in Central and Eastern Europe in the Context of the UNCCD. Available online: http://www.unccd.int/Lists/SiteDocumentLibrary/Regions/CEE/meetings/regional/brussels05_2000/proceedings.pdf (accessed on 8 May 2000).

2. Andronikov, S. The present status of the soil environment in Russia. In *Soil Quality, Sustainable Agriculture and Environmental Security in Central and Eastern Europe*; Springer Netherlands: Heidelberg, Germany, 2000; pp. 87–95.

3. Blum, W.E.H. Characterisation of soil degradation risk: An overview. In *Threats to Soil Quality in Europe*; Tóth, G., Montanarella, L., Rusco, E., Eds.; EUR 23438 EN; Publications Office: Luxembourg, 2008; pp. 5–10.

4. Ozturk, M.; Ozcelik, H.; Sakcali, M.S.; Guvensen, A. Land degradation problems in the Euphrates basin, Turkey. *Environews* **2004**, *10*, 7–9.

5. Tóth, G. Soil quality in the European Union. In *Threats to soil quality in Europe*; Tóth, G., Montanarella, L., Rusco, E., Eds.; EUR 23438 EN; Publications Office: Luxembourg, 2008; pp. 11–20.

6. Lloyd, S. *Ancient Turkey: A Traveller's History of Anatolia*; University of California Press: Oakland, CA, USA, 1989.

7. Montgomery, D. *Dirt: The Erosion of Civilizations*; University California Press: Los Angeles, CA, USA, 2007.

8. Doran, J.W.; Parkin, T.B. Defining and assessing soil quality. In *Defining Soil Quality for a Sustainable Environment*; Doran, J.W., Coleman, D.C., Bezdicek, D.F., Sterwart, B.A., Eds.; SSSA Special Publication; Soil Science Society of Amer Madison: Madison, WI, USA, 1994; pp. 3–21.

9. Gentile, A.R. Soil degradation in Europe. In *Soil Degradation in Central and Eastern Europe: The Assessment of the Status of Human-induced Degradation*; United Nations Environment Programme (UNEP), and ISRIC—World Soil Information: Wageningen, The Netherland, 2000; pp. 68–89.

10. Corwin, D.L.; Kaffka, S.R.; Hopmans, J.W.; Mori, Y.; van Groenigen, J.W.; van Kessel, C.; Oster, J.D. Assessment and field-scale mapping of soil quality properties of a saline-sodic soil. *Geoderma* **2003**, *114*, 231–259. [CrossRef]

11. Várallyay, G. Soil degradation processes and extreme soil moisture regime as environmental problems in the Carpathian Basin. *Agrokémia és Talajtan* **2006**, *55*, 9–18. [CrossRef]

12. Gajri, P.R.; Arora, V.K.; Prihar, S.S. *Tillage for Sustainable Cropping*; Food Products Press: New York, NY, USA, 2002.

13. Balteanu, D.; Dragota, C.S.; Popovici, A.; Dumitrascu, M.; Kucsicsa, G.; Grigorescu, I. Land use and crop dynamics related to climate change signals during the post-communist period in the south Oltenia, Romania. *Proc. Rom. Acad.* **2013**, *15*, 265–278.

14. Halbac, C.Z.R. Changes in agricultural water demands for western Romania. In Proceedings of the 42nd International Symposium on Agricultural Engineering, Opatija, Croatia, 25–28 February 2014; pp. 35–46.

15. Özgöz, E.; Gunal, H.; Acir, N.; Gokmen, F.; Birol, M.; Budak, M. Soil quality and spatial variability assessment of land use effects in a typic haplustoll. *Land Degred. Dev.* **2013**, *24*, 277–286. [CrossRef]

16. Birkás, M.; Antal, J.; Dorogi, I. Conventional and reduced tillage in Hungary—A review. *Soil Tillage Res.* **1989**, *13*, 233–252. [CrossRef]

17. Birkás, M.; Antos, G.; Neményi, M.; Szemők, A. Environmentally-sound adaptable tillage. *Acta Agron. Hung.* **2008**, *57*, 191–194.

18. McCarthy, J.R.; Pfost, D.L.; Currence, H.D. *Direct Planting and Residue Management to Reduce Soil Erosion*; Agricultural Publication G1650: Columbia, MO, USA, 1993.

19. Fallahi, F.; Raoufat, M.H. Row-crop planter attachments in a direct planting system: A comparative study. *Soil Tillage Res.* **2008**, *98*, 27–34. [CrossRef]

20. Chastin, T.G.; Ward, J.K.; Wysocki, D.J. Stand establishment response of soft winter wheat to seed bed residue and seed size. *Crop Sci.* **1995**, *35*, 213–218. [CrossRef]

21. Derpsch, R.; Florentin, M.; Moriya, K. The laws of diminishing yields in the tropics. In Proceedings of the 17th ISTRO Conference, Kiel, Germany, 28 August–3 September 2006; International Soil Tillage Research Organization: Kiel, Germany, 2006.

22. Gus, P. The influence of Soil Tillage on yield and on some soil characteristics. In Proceeding of the Alternatives in Soil Tillage Symposium, Cluj-Napoca, Romania, 9–10 October 1997; pp. 151–155.

23. Korucu, T.; Arslan, S. Effects of direct and conventional planting on soil properties and yield characteristics of second crop maize. *Agric. Sci.* **2009**, *15*, 157–165.

24. Arslan, S.; Korucu, T. Conventional and conservation tillage systems performance on two different soils in Turkey. In Proceeding of the 10th International Agricultural Engineering Conference, Bangkok, Thailand, 7–10 December 2009; Salokhe, V.M., Soni, P., Eds.; Asian Association for Agricultural Engineering: Klong Luang, Thailand, 2009.

25. Yalçın, H.; Çakır, E. Tillage effects and energy efficiencies of subsoiling and direct seeding in light soil and yield of second crop corn for tillage in western Turkey. *Soil Tillage Res.* **2006**, *90*, 250–255. [CrossRef]

26. Topa, D.; Ailincai, C.; Raus, L.; Cara, M.; Jitareanu, G. Tillage effects on soil structure and grain yield of maize. *Lucrări Ştiinţifice* **2012**, *55*, 237–240.

27. Mihovsky, T.; Pachev, I. Reduced tillage practices. Available online: http://www.bjbabe.ro/reduced-tillage-practices-3/ (accessed on 10 December 2014).

28. Houšková, B.; Montanarella, L. The natural susceptibility of European soils to compaction. In *Threats to Soil Quality in Europe EUR*; Tóth, G., Montanarella, L., Rusco, E., Eds.; EUR 23438 EN; Publications Office: Luxembourg, 2008; pp. 23–35.

29. Wysocki, D. Measuring residue cover. Available online: http://pnwsteep.wsu.edu/tillagehandbook/chapter3/030988.htm (accessed on 10 August 2008).

30. Mc Kenney, D.J.; Wang, S.W.; Drury, C.F.; Findlay, I. Denitrification and mineralization in soil emended with legume, grass and corn residues. *Soil Sci. Soc. Am. J.* **1993**, *57*, 1013–1020. [CrossRef]

31. Jones, R.J.; Hiederer, R.; Rusco, E.; Loveland, P.J.; Montanarella, L. *The Map of Organic Carbon in Topsoils in Europe*; Office for Official Publications of the European Communitie: Luxembourg, 2003.

32. European Conservation Agricultural Federation. *Conservation Agriculture in Europe: Environmental, Economic and EU Policy Perspectives*; European Conservation Agricultural Federation: Brussels, Belgium, 1999.

33. Lal, R. Carbon sequestration in dryland ecosystems of West Asia and North Africa. *Land Degrad. Dev.* **2002**, *13*, 45–59. [CrossRef]

34. Lal, R. Soil quality impacts of residue removal for bioethanol production. *Soil Tillage Res.* **2009**, *102*, 233–241. [CrossRef]

35. TurkStat. *Statistical Yearbook of Turkey, 2001*; Turkish Statistical Institute: Ankara, Turkey, 2009.

36. Çelik, I. Land-use effects on organic matter and physical properties of soil in a southern Mediterranean highland of Turkey. *Soil Tillage Res.* **2005**, *83*, 270–277. [CrossRef]

37. Montanarella, L.; Olazabal, C.; Selvaradjou, S.K. *Reports of the Technical Working Groups Established under the Thematic Strategy for Soil Protection*; Office for Official Publications of the European Communities: Luxembourg, 2004.

38. Bireescu, G.; Ailincai, C.; Raus, L.; Bireescu, L. Studding the impacts of technological measures on the biological activity of pluvial eroded soils. In *Land Degradation and Desertification: Assessment, Mitigation and Remediation*; Zdruli, P., Pagliai, M., Kapur, S., Cano, A.F., Eds.; Springer Netherlands: Heidelberg, Germany, 2010; pp. 529–547.

39. Bescansa, P.; Imaz, M.J.; Virto, I.; Enrique, A.; Hoogmoed, W.B. Soil water retention as affected by tillage and residue management in semiarid Spain. *Soil Tillage Res.* **2006**, *87*, 19–27. [CrossRef]

40. Korucu, T.; Arslan, S.; Günal, H.; Şahin, M. Spatial and temporal variation of soil moisture content and penetration resistance as affected by post harvest period and stubble burning of wheat. *Fresenius Environ. Bull.* **2009**, *18*, 1736–1747.

41. Valzano, F.P.; Greene, R.S.B.; Murphy, B.W. Direct effects of stubble burning on soil hydraulic and physical properties in direct drill tillage. *Soil Tillage Res.* **1997**, *42*, 209–219. [CrossRef]

42. Are, K.S.; Oluwatosin, G.A.; Adeyolanu, O.D.; Oke, A.O. Slash and burn effect on soil quality of an Alfisol: Soil physical properties. *Soil Tillage Res.* **2008**, *103*, 4–10. [CrossRef]

43. Food and Agriculture Organization of the United Nations. *Manual on Integrated Soil Management and Conservation Practices (FAO Land and Water Bulletin)*; Food and Agriculture Organization of the United Nations: Rome, Italy, 2000.

44. Cabeda, M.S.V. Degradação física e erosão. In *I Simpósio de Manejo do solo e Plantio Direito no sul do Brasil e III Simpósio de Conservação de Solos do Planalto*; Passo Fundo: Univerdidade de Passo, Fundo, Brasil, 1984.

45. IRENA indicator 23—Soil erosion. Available online: http://epp.eurostat.ec.europa.eu/statistics_explained/index.php/Agri-environmental_indicator_-_soil_erosion (accessed on 25 September 2014).

46. Blum, W.E. The challenge of soil protection in Europe. *Environ. Conserv.* **1990**, *17*, 72–74. [CrossRef]

47. Sauerborn, P.; Klein, A.; Botschek, J.; Skowronek, A. Future rainfall erosivity derived from large-scale climate models—Methods and scenarios for a humid region. *Geoderma* **1999**, *93*, 269–276. [CrossRef]

48. Motoc, M.; Ionita, I.; Nistor, D.; Vatau, A. *Soil Erosion Control in Romania*; Regional Environmental Centre: Budapest, Hungary, 1992.

49. Mircea, S.; Petrescu, N.; Musat, M.; Radu, A.; Sarbu, N. Soil erosion and conservation in Romania—Some figures, facts and its impact on environment. *Ann. Food Sci. Technol.* **2010**, *11*, 105–110.

50. Kapur, S.; Akça, E.; Kapur, B.; Öztürk, A. Migration: An irreversible impact of land degradation in Turkey. In *Desertification in the Mediterranean Region*; Springer Netherlands: Heidelberg, Germany, 2006; pp. 291–301.

51. Aykas, E.; Çakır, E.; Yalçın, H.; Çelik, A.; Okur, B.; Nemli, Y. Koruyucu toprak işleme, doğrudan ekim ve türkiye'deki uygulamalari. In Proceedings of the Ziraat Mühendisliği VII Teknik Kongresi, 11–15 Ocak 2010; pp. 269–292.

52. Avci, M. Conservation tillage in Turkish dryland research. In *Sustainable Agriculture*; Springer Netherlands: Heidelberg, Germany, 2011; pp. 351–361.

53. Oldeman, L.R.; Hakkeling, R.T.A.; Sombroeck, W.G. *World Map of the Status of Human-Induced Soil Degradation: An Explanatory Note*; United Nations Environment Programme and ISRIC—World Soil Information: Wageningen, The Netherlands, 1991.

54. Khan, T.O. *Soil Degradation, Conservation and Remediation*; Springer Netherlands: Heidelberg, Germany, 2014; p. 237.

55. Allen, J.C. Soil response to forest clearing in the United States and the tropics: Geological and biological factors. *Biotropica* **1985**, *17*, 15–27. [CrossRef]

56. Håkansson, I. Subsoil compaction caused by heavy vehicles: A long-term threat to soil productivity. *Soil Tillage Res.* **1994**, *29*, 105–110. [CrossRef]

57. Fabiola, N.; Giarola, B.; da Silva, A.P.; Imhoff, S.; Dexter, A.R. Contribution of natural soil compaction on hardsetting behavior. *Geoderma* **2003**, *113*, 95–108. [CrossRef]

58. Çarman, K. Compaction characteristics of towed wheels on clay loam in a soil bin. *Soil Tillage Res.* **2002**, *65*, 37–43. [CrossRef]

59. Hebert, J. About the problems of structure in relation to soil degradation. In *Soil Degradation: Proceedings of the Land Use Seminar on Soil Degradation Wageningen 13–15 October 1980*; Boels, D., Davies, D.B., Johnston, A.E., Eds.; CRC Press: Boca Raton, FL, USA, 1982.

60. European Environment Agency. *Europe's Environment: The Dobris Assessment*; Publications of the European Communities: Luxembourg, 1995.

61. Van den Akker, J.J.H.; Schjonning, P. Subsoil compaction and ways to prevent it. In *Managing Soil Quality: Challenges in Modern Agriculture*; Schjonning, P., Elmholt, S., Christensen, B.T., Eds.; CABI: Wallingford, UK, 2004.

62. Çullu, M.A.; Aydemir, S.; Qadir, M.; Almaca, A.; Öztürkmen, A.R.; Bilgic, A.; Ağca, N. Implication of groundwater fluctuation on the seasonal salt dynamic in the Harran Plain, south-eastern Turkey. *Irrig. Drain.* **2010**, *59*, 465–476. [CrossRef]

63. Kendirli, B.; Cakmak, B.; Ucar, Y. Salinity in the Southeastern Anatolia Project (GAP), Turkey. Issues and Options. *Irrig. Drain.* **2005**, *54*, 115–122. [CrossRef]

64. Abrol, I.P.; Yadav, J.S.P.; Massoud, F.I. *Salt-Affected Soils and Their Management (No. 39)*; Food and Agriculture Organization of the United Nations: Rome, Italy, 1988.

65. Frizen, D. Managing saline soils in North Dakota. NDSU Extension Service, Fargo, ND 58105. Available online: http://www.ag.ndsu.edu/pubs/plantsci/soilfert/sf1087.pdf (accessed on 13 January 2015).

66. Halbac, C.Z.R. *Romania's Western Part (Timis County) Facing Climatic Changes*; Lambert Academic Publishing: Saarbrücken, Germany, 2010.

67. Terelak, H.; Stuczynski, T.; Piotrowska, M. Heavy metals in agricultural soils in Poland. *Pol. J. Soil Sci.* **1997**, *2*, 35–42.

68. Murányi, A. Quality and contamination of agricultural soils in Hungary as indicated by environmental monitoring and risk assessment. In *Soil Quality, Sustainable Agriculture and Environmental Security in Central and Eastern Europe*; Springer Netherlands: Heidelberg, Germany, 2000; pp. 61–77.

69. Podlešáková, E.; Nemecek, J. Contamination and degradation of soils in the Czech Republic—Contemporary and future state. In *Soil Quality, Sustainable Agriculture and Environmental Security in Central and Eastern Europe*; Springer Netherlands: Heidelberg, Germany, 2000; pp. 79–86.

70. *Ministry of Waters, Forests and Environmental Protection of Romania (2003)*; Environment Protection Strategy: Romania, Bucharest, 2003; p. 178.

71. Soil Sealing. Available online: http://globalsoilweek.org/wp-content/uploads/2013/10/GSW_factsheet_Sealing_en.pdf (accessed on 3 November 2014).

Sustainability **2015**, *7*, 2161–2188

72. Dinç, U.; Akca, E.; Dinç, D.M.; Özden, D.M.; Tekinsoy, P.; Alagöz, H.A.; Kapur, S. Soil sealing: The permanent loss of soil and its impacts on land use. In *First MEDRAP Workshop on Sustainable Management of Soil and Water Resources—Greece/European Union Concerted Action to Support the Northern Mediterranean RAP*; University of Sassari: Athens, Greece, 2001; pp. 18–19.

73. Haktanir, K.; Karaca, A.; Omar, S.M. The prospects of the impact of desertification on Turkey, Lebanon, Syria and Iraq. In *Environmental Challenges in the Mediterranean 2000–2050*; Springer Netherlands: Heidelberg, Germany, 2004; pp. 139–154.

74. Birkás, M. Tillage, impacts on soil and environment. In *Encyclopedia of Agrophysics*; Glinski, J., Horabik, J., Lipiec, J., Eds.; Springer: Dordrecht, The Netherlands, 2011; pp. 903–906.

75. Turk, A.; Mihelič, R. Wheat straw decomposition, N-mineralization and microbial biomass after 5 years of conservation tillage in Gleysol field. *Acta Agric. Slov.* **2013**, *101*, 69–75.

76. Korucu, T.; Kirişci, V.; Görücü, S. Korumalı toprak işleme ve Türkiyedeki uygulamaları. In Proceedings of the Tarımsal Mekanizasyon 18, Ulusal Kongresi, Tekirdağ,Türkiye, 17–18 Eylül 1998. (In Turkish)

77. Kalmár, T.; Pósa, B.; Sallai, A.; Csorba, S.; Birkás, M. Soil quality problems induced byextreme climate conditions. *Növénytermelés* **2013**, *62*, 209–212.

78. Schertz, D.L. Conservation tillage: An analyis of acreage projections in the United States. *J. Soil Water Conserv.* **1988**, *43*, 256–258.

79. Özbek, H.; Dinç, U.; Güzel, N.; Kapur, S. *Çukurova Bölgesinde anız Yakmanın Toprağın Fiziksel ve Kimyasal Özellikleri Üzerine Etkisi*; Toag No: 182; Tubitak: Ankara, Turkey, 1976; pp. 3–5. (In Turkish)

80. Jug, D.; Birkás, M.; Šeremešić, S.; Stipešević, B.; Jug, I.; Žugeć, I.; Djalović, I. Status and perspective of soil tillage in South-East Europe. In Proceedings of the 1st International Scientific Symposium on Soil Tillage—Open Approach, Osijek, Croatia, 9–11 September 2010; Jug, I., Vukadinović, V., Eds.; CROSTRO: Osijek, Croatia; pp. 50–64.

81. Kovačević, D.; Lazić, B. Modern trends in the development of agriculture and demands on plant breeding and soil management. *Genetika* **2012**, *44*, 201–216. [CrossRef]

82. Spoljar, A.; Kisic, I.; Birkás, M.; Gunjaca, J.; Kvaternjak, I. Influence of crop rotation, liming and green manuring on soil properties and yields. *J. Environ. Prot. Ecol.* **2011**, *12*, 54–69.

83. Allen, R.R.; Fenster, C.R. Stubble-mulch equipment for soil and water conservation in the Great Plains. *J. Soil Water Conserv.* **1986**, *41*, 11–16.

84. Kisić, I.; Bašić, F.; Nestroy, O.; Mesić, M.; Butorac, A. Soil erosion under different tillage methods in central Croatia. *Die Bodenkultur* **2003**, *53*, 197–204.

85. Agri-environmental indicator-tillage practices. Available online: http://epp.eurostat.ec.europa.eu/statistics_explained/index.php/Agri-environmental_indicator_-_tillage_practices (accessed on 02 December 2014).

86. Bašić, F. Land degradation in Croatia. In *Land Degradation*; Jones, R.J.A., Montanarella, L., Eds.; Publications of the European Communities: Luxembourg, 2003; pp. 165–176.

87. Birkás, M.; Mesić, M. Impact of tillage and fertilization on probable climate threats in Hungary and Croatia, soil vulnerability and protection. In *Hungarian—Croatian Intergovernmental S&T Cooperation 2010–2011*; Szent István Egyetemi Kiadó: Gödöllő, Magyarország, 2012; p. 186.

88. Rasmussen, K.J. Impact of ploughless soil tillage on yield and soil quality: A Scandinavian review. *Soil Tillage Res.* **1999**, *53*, 3–14. [CrossRef]

sustainability

MDPI

Article

Understanding and Enhancing Soil Biological Health: The Solution for Reversing Soil Degradation

R. Michael Lehman [1,*], Cynthia A. Cambardella [2], Diane E. Stott [3], Veronica Acosta-Martinez [4], Daniel K. Manter [5], Jeffrey S. Buyer [6], Jude E. Maul [6], Jeffrey L. Smith [7,†], Harold P. Collins [8], Jonathan J. Halvorson [9], Robert J. Kremer [10,‡], Jonathan G. Lundgren [1], Tom F. Ducey [11], Virginia L. Jin [12] and Douglas L. Karlen [2]

[1]   USDA-ARS North Central Agricultural Research Laboratory, 2923 Medary Ave., Brookings, SD 57006, USA; jonathan.lundgren@ars.usda.gov
[2]   USDA-ARS National Laboratory for Agriculture and the Environment, 2110 University Blvd, Ames, IA 5001, USA; cindy.cambardella@ars.usda.gov (C.A.C.); doug.karlen@ars.usda.gov (D.L.K.)
[3]   USDA-ARS National Soil Erosion Research Laboratory, 275 S. Russell St., West Lafayette, IN 47907, USA; diane.stott@ars.usda.gov
[4]   USDA-ARS Wind Erosion and Water Conservation Research Laboratory, 3810 4th Street, Lubbock, TX 79417, USA; veronica.acosta-martinez@ars.usda.gov
[5]   USDA-ARS Soil Plant Nutrient Research Unit, Natural Resources Research Center, 2150 Centre Ave., Bldg. D, Suite 100, Fort Collins, CO 80526-8119, USA; daniel.manter@ars.usda.gov
[6]   USDA-ARS Sustainable Agricultural Systems Laboratory, Room 124, 10300 Baltimore Ave., Bldg. 001, BARC-WEST, Beltsville, MD 20705-2350, USA; jeffrey.buyer@ars.usda.gov (J.S.B.); jude.maul@ars.usda.gov (J.E.M.)
[7]   USDA-ARS Land Management and Water Conservation Research, 215 Johnson Hall, Washington State University, Pullman, WA 99164, USA; jeffrey.smith@ars.usda.gov
[8]   USDA-ARS Grassland Soil and Water Research Laboratory, Temple, TX 76502, USA; hal.collins@ars.usda.gov
[9]   USDA-ARS Northern Great Plains Research Laboratory, 1701 10th Ave. SW, PO Box 459, Mandan, ND 58554, USA; jonathan.halvorson@ars.usda.gov
[10]   USDA-ARS Cropping Systems and Water Quality Research Laboratory, 269 Agricultural Engineering Bldg., University of Missouri, Columbia, MO 65211, USA; kremerr@missouri.edu
[11]   USDA-ARS Coastal Plain Soil, Water and Plant Conservation Research Center, 2611 W. Lucas St, Florence, SC 29501, USA; thomas.ducey@ars.usda.gov
[12]   USDA-ARS Agroecosystem Management Research Unit, 137 Kiem Hall, University of Nebraska, Lincoln, NE 68583, USA; virginia.jin@ars.usda.gov
*   Correspondence: michael.lehman@ars.usda.gov; Tel.: +605-693-5205; Fax: +605-693-5240
†   This author is deceased.
‡   This author has retired.

Academic Editor: Marc A. Rosen

Received: 13 November 2014; Accepted: 12 January 2015; Published: 19 January 2015

Abstract: Our objective is to provide an optimistic strategy for reversing soil degradation by increasing public and private research efforts to understand the role of soil biology, particularly microbiology, on the health of our world's soils. We begin by defining soil quality/soil health (which we consider to be interchangeable terms), characterizing healthy soil resources, and relating the significance of soil health to agroecosystems and their functions. We examine how soil biology influences soil health and how biological properties and processes contribute to sustainability of agriculture and ecosystem services. We continue by examining what can be done to manipulate soil biology to: (i) increase nutrient availability for production of high yielding, high quality crops; (ii) protect crops from pests, pathogens, weeds; and (iii) manage other factors limiting production, provision of ecosystem services, and resilience to stresses like droughts. Next we look to the future by asking what needs to be known about soil biology that is not currently recognized or fully understood and how these needs could be addressed using emerging research tools. We conclude, based on our perceptions of how new knowledge regarding soil biology will help make agriculture more sustainable and productive, by recommending research emphases that should receive first priority through enhanced public and private research in order to reverse the trajectory toward global soil degradation.

Keywords: soil biology; sustainable agriculture; soil health; soil management; soil organic matter (SOM)

1. Introduction

One of the most unexplored frontiers associated with understanding the dynamics of soil resources and their subsequent health or quality is that of soil biology. We suggest this reflects the challenges associated with understanding biological properties and processes when compared to soil physical and chemical manipulations that can be used to influence soil quality/health. As a result, multiple post-World War II developments leading to agriculture as we know it today [1] placed a greater emphasis on physical and chemical manipulation than on soil biology [2]. These developments included: (i) increased availability and use of synthetic fertilizers, herbicides, and pesticides; (ii) an improved understanding of plant nutrition and an infrastructure for delivering fertilizers to farmers; (iii) improved tillage, planting, harvesting equipment; (iv) cost-effective subsurface drainage; (v) increased efficiencies for both animal and crop production systems; and (vi) development of global markets. Unfortunately, soil biological responses to these developments were often overlooked or not recognized, so the rapid changes also resulted in unintended consequences, especially with regard to soil health and long-term agricultural sustainability.

Optimistically recognizing the challenges associated with stopping and even reversing soil degradation, our objectives are to identify critical soil biological questions and to suggest various strategies for answering them through enhanced public and private research efforts focused on the concept of soil health. In order to identify knowledge gaps, we review previous literature on soil health and the role of soil biology, and frame future prospects in terms of emerging analytical capabilities.

2. What Constitutes a Healthy Soil?

2.1. Definition of Soil Health

Soil is a dynamic, living, natural body that is vital to the function of terrestrial ecosystems [3]. Farmers intuitively recognize the importance of healthy soils and have used qualitative terms (*i.e.*, color, taste, touch and smell) to describe soil condition and performance for crop production since the dawn of agriculture about 10,000 years ago [4]. At the beginning of the 20th Century, qualitative descriptions were gradually replaced by analytical procedures to assess and evaluate soil almost exclusively from the perspective of inorganic nutrients and crop yield [5].

Warkentin and Fletcher [6] were among the first to introduce the soil quality concept as an approach to improve the process of land use planning. The soil quality concept evolved rapidly during the 1990's, an outcome of increased emphasis on sustainable land use and a growing consensus that soil quality in agriculture should no longer be limited to productivity goals [3,7–12]. As the soil quality concept evolved, methods and tools for soil quality assessment were developed to facilitate comparisons between soil management systems and to document changes in soil properties and processes that occurred in response to land-use or soil management decisions [7,9,13–16]. There was agreement that the design of any generalized soil quality assessment tool must be flexible enough to capture multiple soil functions in various combinations [14] while respecting the broader goals of sustaining plant and animal productivity, erosion control, maintaining or enhancing water and air quality, and supporting human health and habitation [9,15,17].

Soil quality is most simply defined as "the capacity of the soil to function" [9]. Important soil functions include: water flow and retention, solute transport and retention, physical stability and support; retention and cycling of nutrients; buffering and filtering of potentially toxic materials; and maintenance of biodiversity and habitat [18]. A broader, ecologically-based approach was presented by Doran *et al.* [3], where they defined soil health as "the continued capacity of soil to function as a vital living system, within ecosystem and land-use boundaries, to sustain biological productivity, maintain the quality of air and water environments, and promote plant, animal, and human health." The terms soil quality and soil health are often used interchangeably, although farmers and some members of the research community favor the term soil health [19] because it more clearly conveys the idea that soil is a living dynamic system [3]. Most soil scientists, however, reluctantly prefer the term soil quality because of its focus on quantitative soil properties and the quantitative linkages between those properties and various soil functions [19].

Soil taxonomy (the set of innate soil characteristics conveyed by the classification) is the foundation for the soil quality/health framework. Each specific soil has inherent soil quality characteristics that are determined by the interaction of climate, topography, living organisms (vegetation, microorganism, humans) and parent material over long periods of time [20]. The term "dynamic soil quality" refers to the effects of human use and management on soil function [21,22], reflecting changes associated with current or past land use and crop and soil management decisions. Dynamic soil quality can be measured and used to compare different practices on similar soils or temporal trends on the same soil. The inherent properties of different soils may limit the extent of changes due to dynamic processes and need to be accounted for within management strategies to producer healthier soils.

2.2. Existing Soil Quality/Health Assessment

Assessment of soil quality is usually accomplished through direct measurement of a suite of soil biological, chemical, and physical properties and processes that have the greatest sensitivity to changes in soil function [14]. Soil quality indicators should correlate well with ecosystem processes, integrate soil properties and processes, be accessible to many users, sensitive to management and climate, and, whenever possible, be components of existing databases [23]. Selected groups of soil indicators, also referred to as minimum datasets (MDSs), that are used to indirectly measure soil function must also be sufficiently diverse to represent chemical, biological, and physical properties and processes of complex systems [23,24]. Researchers have given particular attention to soil indicators that can serve as early and sensitive indicators of longer-term changes in soil ecosystem function [25]. Frequently recommended soil quality indicators include: soil organic matter (SOM), particulate organic matter (POM), microbial biomass carbon (MBC), potentially mineralizable nitrogen (PMN), macroaggregate stability, electrical conductivity (EC), sodium absorption ratio (SAR), pH, inorganic N, P, potassium (K), and magnesium (Mg), available water-holding capacity (AWC), bulk density (BD), topsoil depth, and infiltration rate [9,23,26]. Soil enzyme activity, specifically β-glucosidase activity which is involved in plant residue degradation, and water-filled pore-space were recently added to the recommended list of important soil quality indicators because of their association with soil biological properties and processes [27].

The issue of spatial and temporal scale affects both the sensitivity of assessment and the choice of indicators that are evaluated. Both scales vary depending upon the type of soil management questions that are being asked or the purpose for which soil quality is being evaluated [28]. In general, soil quality evaluations at the farm, watershed, county, state, regional, or national scales are more general and less precise than those made at the point or plot scale [29]. Large-scale assessments often rely on databases, simulation models, and remote sensing in conjunction with statistically representative point sampling to verify the projections [21]. For instance, Potter *et al.* [30] used a combination of model simulations and data point measurements across the U.S. to assess soil organic carbon and identify areas most at risk for soil quality/health degradation and loss of soil function.

2.3. The Significance of Soil Health to Agroecosystems and Soil Restoration

The single most important soil quality indicator for nearly all soils throughout the world is SOM. It is also one of the most common deficiencies identified in degraded soils because of the numerous chemical, physical, and biological properties and processes it influences. Soil organic matter is generally measured based on the concentration of soil organic carbon (SOC), because about 50% of the SOM is accounted for by SOC. Increases in SOM, particularly in biologically-available forms, are intimately linked to changes in the size, activity and composition of the soil microbial community, enhanced cycling and retention of nutrients, improved aggregate stability, and increased water-holding capacity.

Effective SOM management involves balancing two ecological processes: mineralization of carbon (C) and nitrogen (N) in SOM for short-term crop uptake, and sequestering C and N in SOM pools for long-term maintenance of soil quality, including structure and fertility. Agricultural land management options recommended to increase SOM and improve soil quality nearly always include some reduction in tillage intensity and implementation of integrated, multifunctional cropping rotations that include forage legumes, and/or small grains.

Integrated, extended crop rotations that include small grains and forage legumes have been shown to increase SOC compared to mono- or bi-crop rotations [31–34] with positive impacts being especially evident in the biologically active fractions of SOM [35–37]. Cover crops increase the complexity of rotations and extend the duration of photosynthetic capture in annual crop rotations, thus increasing organic C inputs to the soil and the potential for soil C sequestration—a critical process for restoring degraded soils and addressing increasing concerns regarding global climate changes.

Cover crops can also provide important ecosystem services when planted within corn (*Zea mays* L.) soybean [*Glycine max* (L.) Merr.] systems and extended cropping rotations. Environmental benefits such as decreased soil erosion [10,38] and decreased nitrate leaching [39–41] have been consistently demonstrated in cover crop studies. In general, leguminous cover crops provide the greatest potential for improving yields, but cereal crops generally result in higher levels of SOM, greater weed suppression, and more soil N immobilization, which can reduce nitrate leaching during winter months [42]. Planting small grains and N-fixing cover crops together may be an effective management strategy to simultaneously increase soil C and optimize soil N cycling processes, and thereby reduce both leaching and gaseous emission losses of N.

Conservation tillage increases surface SOC content compared to plow tillage [43–45], but some studies indicate subsoil C content is higher under plowing [46–48]. There is evidence that changes in tillage management alter C cycling processes, resulting in greater retention of corn-derived C in no-till (NT) compared to plowed systems. Type and intensity of tillage directly controls substrate availability to soil organisms and rate of decomposition of substrates by affecting the quantity and distribution of plant residues and roots [49,50]. Tillage factors can also exert indirect control on residue decomposition processes by influencing soil aeration, water content, soil temperature, and especially soil aggregate properties.

Soil management practices that increase SOM and enhance soil health create expanded habitat and greater niche diversity for soil biological communities. It is the inputs of organic matter from plant residues and exudates that provide carbon and energy sources for soil organisms. Net increases in SOM improve soil aeration, temperature, moisture, and aggregate stability, and provide a resilient resource base for a wide variety of soil organisms through the maintenance of a rich and varied source of OM and the efficient supply of nutrients. Improving the quality and health of the soil is important not only for those that manage the land, but for anyone who enjoys a cup of clear water or access to a plentiful and consistent food supply.

3. How Does Soil Biology Influence Soil Health, or What's Missing in a Degraded Soil?

3.1. Soil Biology Overview

Soil biology encompasses the collective biomass and activities of soil-dwelling organisms from an array of trophic levels that are present in staggering quantities, even though individuals may not be visible to the unaided eye. For example, it is estimated that there are at least one billion bacterial cells per gram of soil distributed among thousands to millions of individual species [51]. It has been calculated that the microbial biomass existing underground may approach the sum of all living biomass on the earth's surface [52]. Viewing the tree-of-life (based on genetic relatedness), one begins to understand the diversity of the unseen microbial world, especially since only the three branches at the top right (Animalia, Fungi, and Plantae) contain individual organisms that can be seen with the unaided eye (Figure 1).

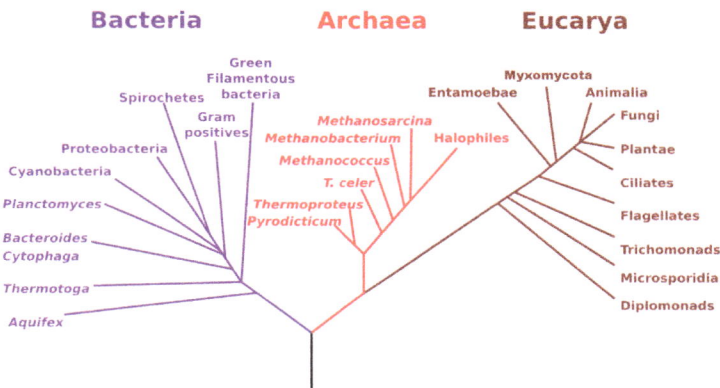

Figure 1. Tree of life based on genetic relatedness using the ribosomal RNA gene sequence.

One of the three domains (domain is the highest taxonomic level of life), the Archaea, was first described in 1977. Archaea appear morphologically similar to bacteria; however, they possess fundamental biochemical similarities with Eucarya and fundamental biochemical distinctions from Bacteria. In short, Archaea are genetically and phylogenetically as different from Bacteria as they are from any of the members of the Eukaryotic domain. The discovery of major taxonomic groups containing microscopic life continues at a rapid pace with a significant modification of the Archaeal domain now becoming apparent—all within in the last 25 years. In 1987 the domain Bacteria contained just 12 phyla (phyla is highest taxonomic group within a domain); today over 70 bacterial phyla are recognized or under consideration for recognition [53].

Each of the primary "Tree-of-Life" branches represents numerous species such that a detailed view would show each branch giving way to successively smaller branches, which are further studded with bushes. Within the two prokaryotic domains (Archaea and Bacteria), even the lowest taxonomic level of species often contains considerable diversity (*i.e.*, microdiversity) that manifests itself in strains, ecotypes, biotypes, serotypes, *etc*. For example, all *Escherichia coli* are considered the same species, but there are numerous strains that are distinctive not only genetically but functionally as well. This means that one strain can be a deadly pathogen while other strains are either benign or even beneficial partners co-existing with plants and animals. This is just one example of the subtle differences associated with soil biology and why it can be difficult to identify exactly what's wrong in degraded soils that simply are not performing as expected.

The variety of physiological capabilities, tolerances, and energy sources of soil microorganisms are extraordinary, and new discoveries are common. A useful tool to comprehend the physiological diversity is the electron tower which displays standard electrode potentials of redox couples (Figure 2). Plants can photosynthesize by fixing CO_2 using water. Animals and plants respire organic compounds at the expense of oxygen as an electron acceptor. Microbes, on the other hand, can use all of these compounds (and more) as either an energy source or an electron acceptor so that energy can be gained from hydrogen gas (H_2) and inorganic molecules in their reduced form (e.g., nitrogen, sulfur, iron, manganese, *etc.*), while CO_2 and those same molecules (e.g., nitrogen, sulfur, iron, manganese, *etc.*) in their oxidized forms can be used as electron acceptors. Bacteria that can oxidize ammonium using nitrite as an electron acceptor have been described in the 21st century and have been found to play key roles in wastewater treatment. In addition to heterotrophic metabolism using exogeneous electron acceptors, microbes can ferment organic compounds, reducing one part while oxidizing the other. Microbes can fix CO_2 by not only standard photosynthetic processes, but also by anoxigenic photosynthesis using other compounds (e.g., sulfur) as electron donors in lieu of water, plus three other pathways not found in eukaryotic organisms [54]. As recently as 2000, it was discovered that some bacteria contained a protein, bacteriorhodopsin, which creates energy from light allowing photo-heterotrophic growth [55]. Similar bacteriorhodopsin molecules had previously only been detected in extremely halophilic archaea. This previously overlooked bacterial metabolism has since been found to be performed by a significant fraction of the world's marine bacterioplankton.

Microbes can even partner with others to perform metabolic processes thought to be energetically unfavorable such as anaerobic methanotrophy which couples methane oxidation with sulfate reduction. Anaerobic oxidation of methane has been described largely in the last decade, and new details such as the use of nitrate (in lieu of sulfate) in this reaction are still emerging.

The physiological tolerances of bacteria far exceed that of eukaryotes. Biological activity of microorganisms can proceed at environmental extremes including temperatures below freezing and above boiling, at pH approaching acid and alkaline endpoints, under very low water tensions, at very high ionic strength, in high radiation fields, and in the presence of high concentrations of toxic compounds. Viable bacteria have been retrieved from 2 miles below the earth's surface [56]; in fact, the existence of a sterile location on earth is difficult to prove.

Several recent findings highlight the on-going transformation in understanding of soil organisms and their processes. In 2006, it was determined that members of the Archaea were actually responsible for most of the nitrification occurring in many soils [57]. This completely changed what was "known" for decades—that nitrification was performed strictly by a very limited number of Bacterial genera. Bacteria belonging to the phylum Acidobacteria are now thought to be the numerically-dominant organism in many soils, but were first described in 1991 and virtually unheard of 15 years ago. Unfortunately, due to their resistance to laboratory culturing, there is yet insufficient information to establish their functional roles. Clearly, the basic understanding of the microbial world remains incomplete, and therefore represents an impediment to assessing and promoting soil health. The continuing exponential increase in soil biological knowledge is also why we are optimistic that the pathway for mitigation and even reclamation of degraded soils is through an increased emphasis on research and education.

Frequent gene exchange, even between members of different domains [58], by multiple mechanisms, further emphasizes the genetic and functional fluidity of the unseen world that exists in soil. From a scientific standpoint, gene exchange among unrelated organisms greatly complicates attempts to classify them, to study their ecological relationships, and to develop useful models that will lead to predictive power necessary for applications. Yet, these difficulties do not diminish the potential value of understanding and influencing the power of the soil biota.

In addition to the prokaryotic organisms, there are enormous numbers of microscopic eukaryotes living in the soil. The net result is that in one gram of soil, there may be a million fungi comprised of hundreds of different species that can produce over 100 m of mycelial filaments. Add to the mix some thousands to millions of algae (classified as Plants), and millions of Protozoa belonging to several different phyla, and several dozen microscopic nematodes. Beyond the microbiota, soil supports a great diversity of invertebrates, ranging across many Phyla and Classes of organisms that are frequently larger and termed meso- and macro-biota. Phyla include Annelida, Nematoda, Nematomorpha, and Arthropoda, of which the last is by far the best studied group. At least five Classes of Arthropods reside within soil food webs: Arachnida (spiders, mites, *etc.*), Chilopoda (centipedes), Diplopoda (millipedes), Crustacea (isopods), and Hexapoda (insects, collembolans, diplurans, *etc.*). Constraints of modern taxonomic tools notwithstanding, scientists regard insects, spiders, and mites as the most diverse macro-taxa within soil food webs, and their numbers are overwhelming. Within conventional agroecosystems, density estimates reveal 100,000–160,000 insects and spiders per ha near the soil surface in soybean, and 340,000–680,000 per ha within the soil column in corn. Healthy soil arthropod communities within agroecosystems are composed of hundreds of species, each with a distinct function and biology. Altogether, soil inhabitants form a food web (Figure 3) that extends above-ground to plants and all other living organisms. Some of the more notable roles for soil fauna in contributing to healthy functioning soils are recognized in the next section; however, the remainder of this review is largely confined to the consideration of microbiota living in the soil. For literature reviews on the contribution of soil fauna to soil health, please refer to Lavelle *et al.* [59] and Blouin *et al.* [60].

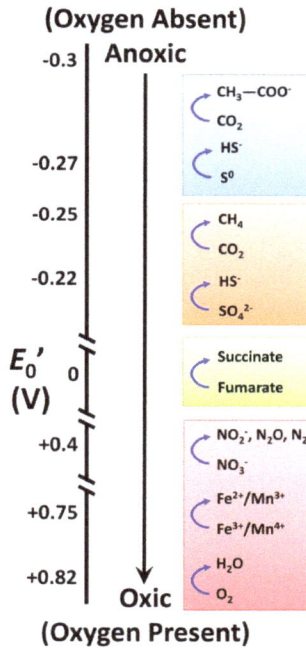

Figure 2. Standard electrode potentials of selected redox couples.

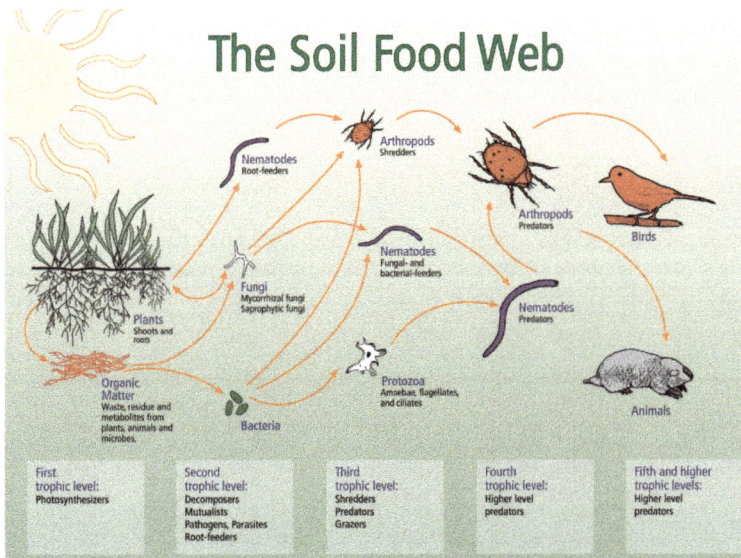

Figure 3. Soil food web (Reprinted with permission from: Soil Biology Primer. 2000. Soil and Water Conservation Society, Ankeny, IA, USA). Please note that the graphic does not represent all the important groups of soil organisms such as enchytraeids and earthworms.

3.2. Relationships between Soil Biology, Soil Quality and Restoration Strategies

Soil was characterized by Doran and Parkin [7] as having good "quality" if it could:

(1)   Accept, hold and release nutrients and other chemical constituents.
(2)   Accept, hold and release water to plants, streams and groundwater.
(3)   Promote and sustain root growth.
(4)   Maintain suitable soil biotic habitat.
(5)   Respond to management.
(6)   Resist degradation.

All of these attributes of soil quality are largely a function of soil biology and why we continue to emphasize that the most optimistic solution for reversing soil degradation is to enhance soil biology. It is widely recognized that soil microorganisms enable other forms of life to exist on Earth [61,62]. By catalyzing redox reactions, soil microorganisms directly mediate the biogeochemical cycling of carbon, nutrients and trace elements. These activities moderate atmospheric composition, water chemistry, and the bioavailability of elements in soil. Soil fertility and other properties of soil (e.g., texture, aeration, available moisture, *etc.*) that support agricultural production are directly dependent on the biomass, metabolites, and activities of microorganisms. Specific populations of microbes are known to exert largely beneficial effects on plants (*i.e.*, symbiotic nitrogen fixing bacteria, mycorrhizal fungi) while others may exert deleterious effects (*i.e.*, pathogens). These microbes can be endophytic (living inside the plant) or free-living soil organisms living adhered to the root surface (*i.e.*, the rhizoplane), in close proximity to roots (*i.e.*, the rhizosphere) or further away in the bulk soil. At their most basic level, microbes and soil invertebrates are an important source of carbon and other nutrients. Soil invertebrates alter the structural components of the soil, increasing soil porosity, changing aggregate structure, and redistributing nutrients throughout the soil column and across the landscape. Invertebrates return nutrients and organic matter to the soil, either directly by breaking down plant material, or indirectly by consuming animal waste (e.g., dung beetles) or saprophytes like fungi. It follows that a well-poised and active soil biological community will be responsive to management and resist degradation.

3.3. How does Soil Biology Influence Ecosystem Services that Are Crucial for Well-Functioning Soils?

Soil biota are integral providers of fundamental ecosystem services such as those listed in Table 1. These are also among the most critical functions that need to be restored in degraded soil resources. Using a meta-analysis of published studies, Benayas *et al.* [63] documented the positive linkage between total biodiversity and provision of ecosystem services in terrestrial ecosystems. The multifaceted contributions of soil macroorganisms to ecosystem services has been well-described by Lavelle *et al.* [59]. The overall economic benefit of soil biodiversity to ecosystem services, and thus well-functioning soil resources, was estimated to be 1.5 quadrillion U.S. dollars [64]. In recent publications, biodiversity was also shown to influence global C [65] and greenhouse gas budgets [66], enhance water quality [67], moderate soil organic matter decomposition [68,69], regulate nutrient retention and availability [69], and determine the susceptibility of soil to invasion by a pathogen [70]. Synthesis papers by Kremen [71] and Hooper *et al.* [72] have summarized the established linkage between biological communities and ecosystems services, while emphasizing the need to understand biological complexity to properly manage the systems, particularly in agroecosystems.

Table 1. Ecosystem services provided by soil biota [†].

Ecosystem Services Provided by Soil Biota
Regulation of biogeochemical cycles
Retention and delivery of nutrients to primary producers
Maintenance of soil structure and fertility
Bioremediation of pollutants
Provision of clean drinking water
Mitigation of floods and droughts
Erosion control
Regulation of atmospheric trace gases
Pest and pathogen control
Regulation of plant production via non-nutrient biochemicals

[†] Modified from [73].

3.4. The Significance of Soil Biology to Sustaining Agriculture and Restoring Soil Health

Numerous examples of failed societies can be linked to degradation of soils by agricultural practices [74]; which by definition, must be considered examples of unsustainable practices. The characteristics of sustainable farming practices which maintain and/or restore soil resources are those that can be used over the long-term to produce adequate yields without severe degradation of soil, water and air resources that would limit agricultural production, cause human morbidity and mortality, and otherwise incur off-site economic costs. In practice, this means the soil and crop management practices must: (1) maintain soil carbon; (2) control erosion; (3) maintain soil structure; (4) maintain soil fertility; (5) increase nutrient cycling efficiency; (6) reduce export of nutrients and thus the need for increased inputs; and (7) reduce pesticide input requirements and potential export of either the materials or their residuals [73]. Once again, these are all attributes of a well-functioning soil and thus our premise that to restore degraded soils, the first step must be to enhance and maintain soil biological properties and processes.

The mechanism to achieve all of these goals is take advantage of inherent biological services to the greatest extent possible. Obviously, in entirely undisturbed grasslands, there is no human management to achieve the seven sustainability goals listed above, but on cultivated and range lands, soil and crop management practices can have positive or negative effects and thus influence the potential for soil degradation or enhancement. Sustainable agricultural management systems strive to integrate complexity into the management approach to include cover crops, filter strips, and non-crop landscapes such as grasslands and forest areas that provide vital habitats for beneficial organisms and serve as nutrient sinks to capture soluble nutrients and trap contaminants before these impact aquatic ecosystems [75].

The biomass of soil organisms nominally accounts for 2% of the SOC, but contribute to a much larger proportion of the actively cycled carbon fraction. At the decomposer level and higher, soil organisms represent the transformers of all fixed soil carbon and determine its fate. Soil microorganisms are well-documented to promote soil aggregation by their biomass and by their secretions. Microcolonies of bacteria and thin coatings of bacteria known as biofilms are held together and attached to their substrata by extracellular secretions largely composed of polysaccharides. Arbuscular mycorrhizal (AM) fungi have been shown to produce a glycoprotein, glomalin, which is responsible for aggregating soil particles [76]. Filamentous microbes, largely fungi, are particularly effective in mechanical binding of soil particles with their thread-like morphology. Plant roots, proliferating throughout the upper soil profile, support microbial communities actively involved in soil aggregation by providing organic carbon through rhizodeposition and thus helping stabilize soil structure and abate potential erosion [15]. Macro-invertebrates promote soil aggregation and create structures at a larger scale by tunneling, ingesting and depositing organic matter, producing secretions, and transforming organic residues [77]. The activities of ants [78] and earthworms [60] are widely recognized for promoting soil structure. Naturally, the degree of soil aggregation is directly related the soil's resistance to degradation and erosion by wind and water. Soil structure promoted by soil organisms is also central to soil water dynamics, increasing water infiltration and holding capacity.

Soil microorganisms are responsible for mineralizing organic compounds, including potential contaminant molecules such as pesticides. Half-lives of agrichemicals are based on the biodegradative abilities of the soil microbial community, as well as the local environmental conditions. In mineralizing organic compounds (native or added), microbial communities release combined elements (e.g., N, P) in their chemically-reduced forms, generally increasing their availability to plants. Soil microbes also perform direct redox transformations of many inorganic elements using them as electron donors or acceptors in energy-yielding metabolic processes. In short, microorganisms moderate the abundance, speciation, and plant bioavailability of nutrients in the soil. Nitrogen-fixing bacteria exist in symbiotic and associative relationships with plants and as free-living communities in the soil to provide N to plants. Symbioses of N-fixing bacteria with soil invertebrates have been shown to be particularly important to the N cycling in some soils [79]. Nitrogen-transforming microorganisms (e.g., nitrifiers, denitrifiers) also moderate the speciation and therefore mobility of soil N affecting its propensity to stay or leave the system. Phosphate-solubilizing bacteria and fungi produce organic acids that either complex P or change microsite local pH to increase plant-available P. General activities of soil microbes result in the release of extracellular phosphatase enzymes which mineralize organic P, some of which becomes available to plants. Obligate plant symbiotic fungi, AM fungi, use a variety of mechanisms to uptake and translocate immobile nutrients (*i.e.*, P, Zn, Cu) and water to their host plants in exchange for fixed carbon [80]. A healthy soil food web with a diversity of macroinvertebrates has been shown to increase the release of P via the activities of grazers and predators [81]. The activity of tunneling organisms such as earthworms redistributes carbon and nutrients in the soil profile [60].

Phosphorus is a major nutrient with dwindling global supplies and rising prices. At the same time, only a small amount of P applied (20%) to crops is taken up by plants in the year of application [82,83]. The remaining P becomes sequestered in the soil, with limited availability to plants, or is lost by erosion and leaching (including tile drainage) to the watershed where it impacts other downstream populations and water quality by eutrophication which may culminate in the formation of marine dead zones. Similarly, only about one-quarter of annually applied N is taken up by crops in the year of application; some of the remaining N enters the watershed by leaching through the soil profile, tile drainage, or by overland flow processes to cause eutrophication and water treatment costs at downstream sites.

Nutrient-use efficiency is often defined based on the amount of N or P accumulated by a crop in comparison to the amount applied through manures or inorganic fertilizers. However, a portion of the P and N in the crop has originated from within the soil, where it was already present and probably in a stable organic form resistant to export. Therefore, traditional nutrient use efficiency calculations often overestimate the efficiency of fertilizer application and fail to reflect added nutrients that were lost from the soil by leaching and/or erosion. A more reasonable goal would be to export fewer nutrients and consider how much of the added nutrient remains in the soil [84]. This should mean that inputs are reduced, while increasing the amount being provided by the soil through biologically-fixed N, or mineralization of P and N from organic matter at just the right time. In the case of P, there are large amounts of P already in the soil, unavailable to plants without the appropriate microorganisms and proper levels of activity. By considering the nutrient balance of the entire system, agricultural soils could be managed to stabilize at lower soil nutrient levels that make more efficient use of resources [85,86]. Some P exported with the crop will have to be replenished from external sources, but there is great room for improvement in promoting organic P cycling in soils and biological mobilization of "occluded" P already present in the soil.

There is a long history of using bacteria and fungi as control agents for a variety of insect pests [87]. One example is the use of the entomophagous fungi to control insect pests such as aphids [88]. Contemporary use of proteins native to *Bacillus thuringiensis*, as whole cells, protein extracts, or expressed by genetically-engineered plants to control insect pests is widespread. Among the many potential benefits that AM fungi have been shown to confer to their plant hosts is pest- and pathogen-resistance [89,90]. Predatory insects and spiders within the soil readily attack soil-dwelling pests, often maintaining these pests at low levels [91]. Invertebrates are also important herbivores of weeds, and reduce weed seed density and emergence by consuming many of the weed seeds that fall to the soil surface [92–94]. Microbes also affect weed seed banks, either directly by degradation [95] or indirectly as symbionts within insects, influencing their consumption of weed seeds [96]. Soils which inherently reduce weed seed germination are known as "weed-suppressive soils". Although the exact biological qualities that contribute to control of weeds are not well known, one of the mechanisms is the production of allelochemicals that reduce weed germination [97]. Similarly, some soils are considered to be "disease suppressive" wherein often poorly-defined components of a diverse soil microbial community confer disease resistance to plants [70,98,99]. Use of inherent (or perhaps added) organisms to manage pest, disease, and weeds in agroecosystems would provide opportunities for lowered use of biocidal agrichemical use, export, and residuals.

Soil bacteria that produce a positive effect on plant growth and vigor have been termed "plant growth promoting bacteria" (PGPB), or if they are located in the rhizosphere, rhizoplane, or inside the root (endophytic), they are termed "plant-growth promoting rhizobacteria" (PGPR) [100–102]. Sometimes the endophytes are considered separate from other PGPR [103]. There are, of course, also fungi that are endophytic like AM fungi and some *Trichoderma* sp. [104] which are often considered beneficial to the plant host. Soil organisms belonging to these groups have been identified to specific strains (*i.e.*, *Enterobacter* sp. 638 [105]) or have been more generally categorized (*i.e.*, fluorescent pseudomonads [106]). The functional contributions of the PGPR/B include repression of pests and diseases, and so there is overlap with the phenomena of disease-suppressive soils and pest protection discussed above. The putative mechanisms for pest and pathogen resistance include the production of antibiotics and siderophores, the physical (preventive) colonization of root tissue, interspecific-competition for resources, biodegradation of biogenic toxic substances, and the production of chemical signals (e.g., salicylic acid) that induce systemic resistance by the plants [102,107–109]. There are also PGPR/B that contribute to plant nutrient acquisition like the well-know symbiotic nitrogen fixers, *Rhizobium*, but also free-living N_2 fixers such as *Azospirillum* and *Azotobacter* [110]. Some PGPR assist in mobilizing P for plant uptake using mechanisms such as production of acidity, organic ligands including siderophores, and extracellular phosphatases [110]. Other PGBR assist plants by degrading toxic organic compounds in the soil or immobilizing toxic metals [102].

Another distinct soil microbial function is the production of growth factors and metabolic products that positively influence plant metabolism in ways not directly associated with pest or pathogen resistance. For instance, the enzyme 1-aminocyclopropane-1-carboxylase (ACC) produced by soil bacteria degrades an ethylene precursor that, in turn, depresses the plant's stress response to a variety of biotic and abiotic stress factors [111]. Soil microbes also can stimulate plant growth via the production of plant hormones such as auxins and cytokines. The auxin, indole acetic acid (IAA), is a phytohormone produced by soil bacteria which influences plant physiology, often resulting in enhanced root growth [112]. Naturally, microbial metabolites that positively influence plant vigor also impact plant resistance to pests and pathogens.

Biological production and reception of chemical signals are a common feature of the integrated biome present in agricultural soils and the net outcome of these interactions on crop production may be positive as described above, or negative [113]. In opposition to PGPR/B is a loosely-defined group of microorganisms termed "deleterious rhizosphere bacteria" (DRB) [114,115]. These soil bacteria have been determined to have negative consequences for plant growth and vigor via mechanisms that include phytotoxin and phytohormone production, nutrient competition, and inhibition of AM fungi [115]. The DRB are usually not considered to be plant pathogens, but this is not always the case. Interestingly, groups such as the fluorescent pseudomonads have been identified as DRB [116], even though other studies have identified them as PGPR/B [106]. One view is that a single organism can be a DRB under one set of environmental conditions and a PGPR/B under a different set of conditions [115].

4. How Can Soil Biology Be Used More Effectively to Mitigate Soil Degradation?

4.1. Strategies to Manipulate Soil Biology Focusing on Soil Microorganisms

The benefits of a healthy soil and the role of the biological community in soil health have been covered in the previous two sections. Therefore, we now shift our focus to examine the potential for influencing soil biological communities to (i) increase nutrient availability for production of high yielding, high quality crops; (ii) protect crops from pests, pathogens, and weeds; and (iii) manage other factors that limit or threaten the stability of production and ecosystem services. As with any management decision, the process or tools selected to manipulate soil biological communities will be defined by the desired goals and objectives. With this in mind, we envision two strategies for management of soil microbial communities to obtain beneficial functions: (i) specific approaches; or (ii) general approaches. Specific approaches will require knowing the service that specific microbes are providing (*i.e.*, nutrient acquisition, disease suppression) so they can be targeted to provide immediate relief for problems or degraded soil conditions identified within a specific field, farm or other location. Typical options for this approach include selection of disease resistant plants and/or cultivars with desired exudates. The specific approach is hindered by the lack of reliable information on the specific role(s) of more than a handful of the diverse taxa in soil. In contrast, the general approach seeks to provide a suitable environment to enrich the abundance and/or diversity of the entire microbiome through management practices. However, as with the specific approach, this will require knowledge of the current plant-soil-microbiome status in order to focus on any missing or limiting conditions for establishment of a robust and diverse soil microbial community.

4.2. Specific Approach: Plant Selection and Microbial Amendments

Plant root exudates include a variety of sugars, amino acids, flavonoids, proteins, and fatty acids [117], that can serve as growth substrates, signal molecules for suitable microbial partners, or growth deterrents for microbes [118]. The composition of plant root exudates can vary by plant species, and even cultivars within a species [119–121], resulting in concomitant changes in the composition of the soil microbial community [122–124].

Despite a general knowledge of the growth requirements for microbes in culture (which may or may not translate to the field), knowledge of the relative importance of various root exudates with regard to shaping soil microbial communities or restoring degraded soils is lacking. Can selective effects be explained by a small number of high-impact compounds? How important are the diversity, quantity, or consistency of exudation to host plant-selective effects? The impact of particular aspects of root exudation on soil microbes has begun to be addressed for model plant species [117,125] through the use of ABC transporter mutants to alter root exudates; however, this should be made a priority for agriculturally relevant species as well.

With emerging extreme climatic changes, another critical question is whether cropping system sustainability can be increased by using plants that can interact with a variety of PGPR/B that are capable of increasing photosynthetic capacity [126,127], conferring drought and salt tolerance [126,128–130], and improving the effectiveness of the plant's own iron acquisition mechanisms [129]. A variety of companies have begun to offer new products that consist of PGPR/B inoculants (soil and/or seed treatments), or chemicals aimed at increasing root exudation to help foster PGPR/B establishment. However, field studies with PGPR/B inoculants often result in limited PGPR/B establishment and colonization, highlighting the need to better understand the factors involved in successful PGPR/B establishment.

The use of amendments, either as live organisms or solutions applied in small amounts that are promoted to stimulate microorganisms, is increasing. The use of seed-applied, symbiotic N_2-fixing bacteria to enhance the performance of legumes has a long, successful history. However, there are few other well-documented success stories to report. The use of AM fungi inoculants has been rising, but few refereed publications exist to support the benefits of this practice in production agriculture. In a three-year field study, the application of PGPR and AM fungal amendments was reported to positively affect plant nutrient uptake and conservation in corn plots [131]. However, for many commercial live biological amendments, there is little data beyond yield comparisons from company-sponsored field trials to evaluate these products. It is impossible to determine the potential benefits or risks of amendments without an increased basic understanding of soil microbial functional groups, their distributions, and their ecology (e.g., dispersal, survival).

Agricultural chemicals applied to the foliage of crops or in-furrow can also impact soil organisms. Biostimulants (e.g., products containing plant hormones and other organic and inorganic compounds) and liquid fertilizers affect soil microorganisms by providing additional nutrients or growth factors that alter soil and plant metabolic activities for improving crop growth and productivity [132–134]. Biostimulants applied at extremely low dosages affect rates of organic matter decomposition, nutrient mineralization, and soil microbial activity [132,134]. Depending on the product, amendments could be classified as either a specific or general strategy. Developing a more complete understanding of how biostimulants and other formulations could be used to help restore degraded soils also provides a strong argument for increased public-private partnerships designed to address these complex and "wicked" [135] problems. Such partnerships could be very effective for overcoming current barriers to understanding appropriate uses and modes of action for the various amendments created by the proprietary nature of product formulations.

4.3. A General Approach: Modify the Whole Soil Community

The future of any soil microbial community is determined by the capacity of its individual members to adapt or modify to "negative" soil characteristics [136], and to challenges such as climate change. Any potential manipulation of soil microbial communities must consider that ambient soil characteristics (e.g., water potential, aggregation, salinity, legacy of past management, pH, texture, SOM content/quality) influence the existing community and will consequently influence attempts at manipulation. Some of these soil factors can be positively manipulated through management (within local limitations), with considerable feedback from the soil biota (e.g., SOM, aggregation), while other soil factors are more resistant to modification (e.g., pH, texture). For example, one of the most influential factors on the microbial community is soil pH as different strains exhibit optimum pH in which they can function. Soil pH not only affects the cell functioning (*i.e.*, enzymes), but also reactions altering the availability of nutrients and metals. Studies on several soils have observed a positive correlation between bacterial diversity and soil pH within a range of 4 to 7 [137,138]. In terms of the response between different groups, the fungal community composition appears to be less strongly affected by pH than the bacterial community composition, and thus, wider pH ranges are observed for optimal growth of the fungal community [139].

Different microbial communities can be expected under different soil types due to variation in soil physical properties (*i.e.*, texture, bulk density, water infiltration), chemical properties (*i.e.*, mineralogy, SOM, nutrient availability, pH) and other factors (*i.e.*, soil genesis and morphology, climatic conditions). The challenge of selecting approaches to manipulate the microbial communities is, therefore, site-specific. For example, soils with higher SOM and clay content will show higher microbial community size and activities than a sandy soil, but it is still not clear whether a soil that is higher in organic matter and clay content is more resistant to manipulation. Even within a given soil profile, distributions of organisms and activities will vary according to heterogeneity in key soil properties. Further, the plant-microbe interaction is difficult to separate from the influence of soil characteristics on the microbial diversity as there are many examples of shifts in microbial community composition without changes in the SOM as affected by vegetation. Vegetation also introduces heterogeneity to the soil habitat. As studies are designed to determine how to most effectively remediate degraded soils, they will have to recognize that each set of soil characteristics and environmental boundaries will be an important determinant influencing the response of microbial communities for that soil.

There is no doubt that agricultural management practices can influence soil biological populations and processes and thus have a positive or negative effect on soil health. Agricultural management effects on soil health, in turn, influence the type and magnitude of ecosystem services provided by the soil biota. One measurement of soil health is biodiversity, which has been shown to influence global C [65] and greenhouse gas budgets [66], water quality [67], SOM decomposition [68,69], nutrient retention and availability [69], and the susceptibility of soil to invasion by a pathogen [70]. Many assessments of soil health based on measurements of soil microorganisms have relied on estimates of total biomass and activity. The following agricultural practices have been observed to modify the whole soil biological community (biomass, numbers, diversity, activity) in a generally positive manner: no till or conservation tillage, cover cropping, elimination of fallow, incorporation of perennial crops, retention of crop residues, diverse crop rotation, use of organic fertilizer sources, and implementation of integrated pest management practices (Figure 4) [75,140–142]. Many of these same practices have been shown to increase PGPR/B and reduce DRB [100,101].

Tillage represents a disturbance of the soil habitat and can mechanically disrupt filamentous organisms, decrease soil structure, temporarily increase organic matter decomposition, and alter water and nutrient content and distribution [143,144]. Tillage-induced disturbance often has a negative impact on soil biota and the services that they supply [144–146]. Tillage most noticeably impacts large soil biota like earthworms [144] and filamentous organisms like fungi, particularly AM fungi [147]. Reductions in tillage are frequently linked to increased fungal biomass, and therefore have been suggested as strategies to increase microbial C use efficiency and soil C sequestration potential [148]. Reduced tillage is generally thought to increase microbial biomass in the long term [149] and has been associated with reductions in DRB in wheat cropped fields [150]. The combination of reduced or no-tillage with crop rotation or incorporation of perennial crops for integrated livestock and cropping systems promote AM fungi which enhance plant uptake of phosphorus and water, and disease resistance potential [151]. Although conservation tillage has been reported to impact beneficial microbial communities in certain scenarios (e.g., soils in humid regions), Acosta-Martinez *et al.* [152] reported that semiarid soils under different cropping systems showed no differences in microbial community size or structure when no-tillage and conventional tillage systems were compared after five years.

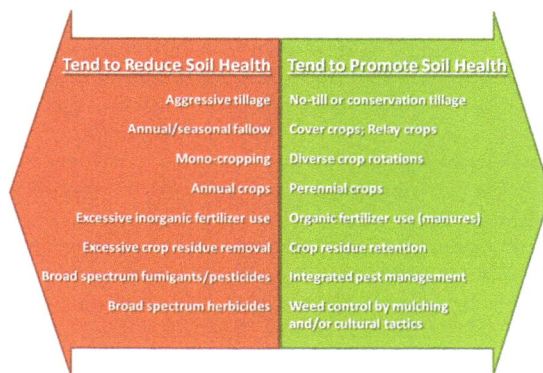

Figure 4. Generalized Effects of Agricultural Management Practices on Soil Health (information compiled from: [75,140–142]).

Cover crops were originally defined as crops grown to protect the soil from erosion and nutrient losses [153]. However, it has become clear that cover crops have a wide array of benefits that depend on local soil-climatic conditions [154]. By reducing seasonal fallow, cover crops have enormous influence on soil biology by increasing the quantity and variety of C entering the soil through plant biomass, exudates, and residues. Additionally, cover crops increase N in the soil by stimulating the free-living N fixing bacteria and symbiotic N fixers when leguminous cover crops are planted. The inclusion of cover crops in a variety of corn production systems has been shown to significantly increase native AM fungal numbers and diversity [155–157] and P availability [158]. Recent research suggests that the benefits of cover crops include many additional factors, such as weed suppression and pest management that are likely connected to the larger soil biological community [159]. A study comparing four different cover crops in potato systems of the San Luis Valley show that they can support a disease-suppressive microbiome (Manter, unpublished data). In particular, the soil community under Sudan grass 79 is enriched for siderophore microbes that can not only provide disease suppression against fungal pathogens but also increase nutrients available to the subsequent crop species.

Conversion of lands for biofuel feedstocks using either perennial vegetation such as switchgrass (*Panicum virgatum* L.) or rotations using corn or sorghum (*Sorghum bicolor* L.) may help meet increasing national energy demands, but require careful evaluation of impacts on overall ecosystem functioning. Despite the potential negative impact of excessive corn stover removal on SOM dynamics in the Midwest [160], other studies have shown that conversion of marginal lands to rotations involving high-residue crops (e.g., cotton (*Gossypium* spp.)) to high-yielding sorghum on low SOM soils can increase microbial biomass and metabolic capacity related to biochemical cycling [161]. In experimental cellulosic ethanol production systems where corn stover was harvested, no-till and addition of cover crops limited extensive changes in soil microbial communities [162]. Additional studies quantifying biofuel-cropping system effects on soil microbial communities are also needed to be sure such practices are not detrimental to biological soil quality.

Crop rotation has long been noted for disrupting pest cycles and adding N fertility with legume crops [85,86,163]. In a study of five long-term diversified cropping systems, crop diversity (rotation) increased soil microbial biomass and activity and was associated with positive changes in soil C and nutrient dynamics [164]. Crop rotations have been specifically noted for increasing soil fungal biomass, which in turn aids in soil aggregation and C sequestration [148]. Rotating corn with other crops increased soil microbial biomass, C availability [165] and numbers of AM fungi [166]. In comparison to continuous corn, rotating corn with canola resulted in greater microbial biomass, activity, and functional diversity [167]. On the other hand, continuously cultivated crops are most commonly associated with increased incidence of DRB which impair plant growth through numerous modes of action [115,168]. The absence of a crop (fallow) is an obvious factor in decreased soil health as there is no plant host for obligate symbionts, no exudates for the rhizosphere community, and no residues for the bulk soil community. Fallow is associated with poor nutrient conservation [169], lowered AM fungi populations [140,170] and other impacts to soil health that affect crop production [171]. While crop rotation is known to benefit crop production via modification of the soil microbial communities, many details are still unknown [142]. Specific crop sequences have been shown to be particularly effective for controlling weeds, but often the mechanism remains unexplained, and probably involves modification to the soil biota [172].

Organic amendments enhance the physical environment for nutrient retention and bioavailability causing alterations to the existing microbial community. Depending upon their composition and nutrient content, they can also cause significant shifts in the existing microbial community of soil by introducing another diverse microbial pool plus their metabolites into the soil. However, some researchers argue that organic amendments seem to have less prolonged effects on soil microbial communities than seasonal variations or other anthropogenic factors such as the mechanical management of the soil [173]. Recent studies using molecular techniques have identified detectable changes within *Proteobacteria*, *Acidobacteria* and *Bacteroidetes* with the use of organic amendments [174,175]. Another recent study suggested that compost effects were mainly caused by physicochemical characteristics of the compost matrix rather than by compost-borne microorganisms and that there was no resilience of microbial characteristics during the study (6–12 months) after applying a high amount of compost [176]. However, a comprehensive meta-analysis showed that organic amendments routinely increased soil microbial biomass in agricultural soils [177]. And, while excessive inorganic P fertilization is known to suppress AM fungi [80,140,178], meta-analysis results show equivocal effects of mineral N fertilizers on aspects of the entire soil microbial community [179]. Thus, more information is needed on actual comparisons of the microbial communities within different organic amendments and the extent of alteration and resilience of the inherent soil microbial community over time.

While the effects of fumigants are relatively easy to predict—they are used as a soil biocide—the non-target effects of other agrichemicals such as insecticides, herbicides, fungicides on soil biota are less clear. Determining agrichemical effects on soil biota is complicated by different modes of delivery (seed applied, foliar, soil drench, *etc.*), the concentration, mixtures, the specificity of the target(s), and the mechanism(s) of action. Most agrichemicals represent a C and nutrient source for some soil microorganisms. The most widely-applied herbicide, glyphosate, is relatively non-toxic to most soil biota in laboratory bioassays [180]. Largely negligible impacts on soil biota have been observed in field or greenhouse studies of potential glyphosate treatment effects [181–185]. On the other hand, extensive research has indicated negative impacts of glyphosate application on symbiotic N-fixing bacteria when applied to glyphosate-resistant soybean [186]. In the absence of any additional stressor, the inhibition of these symbiotic N-fixers is transient, and not expected to affect yields [186]. Some recent reports indicate the potential for indirect effects of glyphosate via its complexation with trace nutrients resulting in increases in pathogenic soil microorganisms, perhaps due to stressed plants [187,188]. However, there is a lack of consensus in the literature on the potential for glyphosate to select for soil pathogens [189]. One possible outcome that is not well-documented is that large areas that are devoid of vegetation due to glyphosate application will have lower soil microbial biomass and activities simply due to the lack plant hosts, exudates, and residues. The lack of weeds has been shown to negatively influence the diversity of some insects and birds in agroecosystems [190]. Some agricultural pesticides have been indirectly linked to increased DRB numbers [114,116,191,192], and it was recently concluded that the fungicide carbendazim inhibits AM fungal colonization of pepper plants [193]. It is difficult to generalize non-target effects of agrichemicals (herbicides, fungicides, and insecticides) on beneficial soil biota because the experimental conditions and results of individual studies are variable.

5. What Are the Primary Knowledge Gaps Limiting Manipulation of Soil Biological Communities and Mitigation of Degraded Soils?

Despite the amount of research already conducted, we do not know how soil microbial communities are controlled. One model proposes that the control is balanced between the soil (texture), the plant (maize or Arabidopsis), and the particular microorganism (an actinomycete or *Pseudomonas* sp.) [99] (Figure 5); however, the actual situation is most assuredly more complicated. If we are to manipulate soil biology in order to optimize ecosystem services and restore degraded soil resources, we need to understand what controls soil microbial community structure, function, and biomass under a given set of conditions, how much it varies according to conditions, and distinguish these effects from seasonal influences. Further, the duration of effects due to changes in management or crop is an unresolved question with conflicting research findings.

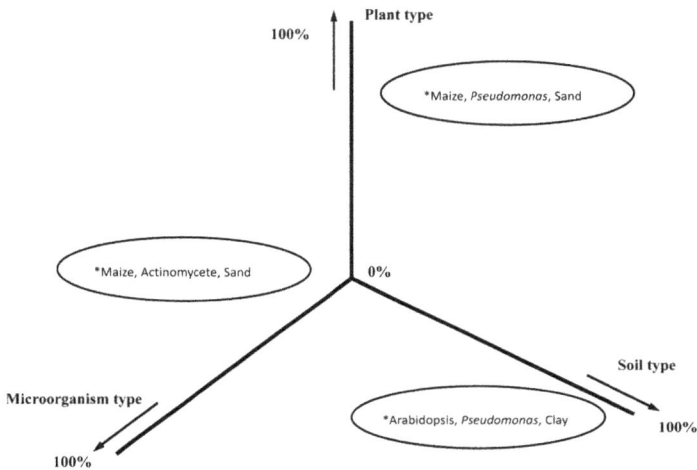

Figure 5. Conceptual model of the relative strengths of forces shaping microbial communities in the soil (from Garbeva *et al.* [99], with permission of the publisher).

While microbial community function should theoretically vary with community structure, it is not known how common this linkage actually is within soils. If functional redundancy is very high across different phylogenetic groups, large changes in microbial community structure could occur without corresponding changes in soil function or possibly resiliency. If functional redundancy occurs across ecotypes, changes in soil conditions could occur yet function might remain unchanged. The distribution of soil microbial populations and functions at local and larger scales (*i.e.*, biogeography) and their colonization abilities are largely unknown for most taxa. The extent of microbial species endemism and functional redundancy are central to measurement of soil health and resilience, particularly in relation to biodiversity [194].

We do not know how soil microbial community function is related to microbial biomass. As biomass increases, the potential for function should increase due to a higher number of organisms carrying out that function, but other factors may limit gene expression or enzyme activity and therefore function. While soil microbial community function should be related to the number of copies of that functional gene in the community and the degree of expression of that gene, in many cases we do not know how gene expression is controlled or the factors controlling enzyme activity in soil microbial communities.

A predictive model that combined all of these factors to explain how soil microbial community structure, relative abundances, and function were controlled would permit us to maximize soil health and optimize ecosystem services on a site and management-specific basis. In part our understanding has been limited by available methodology. Only recently has a method (next- generation sequencing) been developed which has the potential to identify changes in soil microbial communities at the species or genus level. Before next-generation sequencing, scientists could detect alterations in microbial community structure but could not determine which genera changed, or were restricted to the very small proportion of soil microbes that could be cultured in the laboratory. Similarly, functional genes and gene expression of the entire soil metagenome can now be measured using microarrays and next-generation sequencing.

6. What Are the Highest Priority Research Needs to Improve Soil Health and Reverse Soil Degradation?

6.1. Framing High Priority Research Questions

Soil microbial communities can be manipulated to enhance ecosystem services and improve crop productivity, but this requires an understanding of the genetic potential of the soil microbiome [195]. Given this enormous amount of functional diversity, substantial research is needed to link microbial species, or assemblages, with key function(s) in the soil, and in particular how they are influenced by management [152,196–198]. Furthermore, addressing emerging challenges such as climate change and land use will be reliant upon the identification of microbial species and/or assemblages that enhance soil structure, nutrient and water uptake by plants, and protection from pathogens, pests, and weeds. Our goals are to understand these interactions and apply that understanding to increase agroecosystem productivity, to document suitable indicators of soil health, and to provide guidelines for restoring and then maintaining the health of degraded soil resources.

6.2. Fundamental Information Lacking Regarding the Identities, Distributions, Ecology, and Functionality of Soil Biota

Fundamental information is required to answer simple questions like "What organisms are there?" and "What are they doing?" Projects such as TerraGenome (www.terragenome.org) are an important step in our efforts to better understand the true diversity of genes and functions residing in the soil. When a sufficient amount of the census information exists, the next questions that require more complete answers are: "How are they distributed?" and "What do we know about their ecology?" Determining the extent of a biogeography for individual taxa or functional capabilities is key to understanding how factors influence communities and their function, and what management practices will inhibit degradation of soil health. For instance, if certain AM fungi with specific functional abilities or host preferences are endemic to a given soil-climatic region, and they are eliminated by soil-degrading practices, then appropriate management will be required for re-establishment. The required management will depend on the ecological characteristics of the AM fungi such as life-cycle and dispersal abilities. Management options could be creating better conditions (*i.e.*, cover cropping, crop rotation, avoiding fallow) or inoculation with non-native, commercial AM fungal inocula, or on-site amplification of native AM fungal inocula [199], depending on what information is available for local conditions.

Basic descriptive information is required for taxa associated with soil biological functions that are fairly cosmopolitan (e.g., denitrification), but also for more specialized functional guilds (e.g., symbiotic N-fixers). The current information void concerning soil organisms and consortia that are known, or suspected to be particularly influential to plant development (e.g., AM fungi, PGPR/B and DRB), limits opportunities to exploit these organisms to improve soil health and function.

The nature of interactions of plants and rhizosphere microbial communities deserve special attention. Future research should investigate plant characteristics that are related to aspects of microbiome diversity, *i.e.*, the richness and evenness of species composing the community. For instance, how important are adaptation or long-term association in maintaining evenness among rhizosphere microbes? Over long time scales, does rhizosphere microbial evenness increase as many microbial community members undergo adaptation or niche differentiation in the context of a stable assemblage of interacting organisms? Does increasing exudate diversity sustain greater microbial richness in the rhizosphere? Or, can simple exudates be transformed by microbial activity into sufficiently diverse metabolites to allow for niche differentiation of many microbes? If so, simply increasing exudate quantity may be as effective in maintaining a rich microbiome as increasing exudate diversity. The relative importance of carbon source identity *vs.* diversity has begun to be explored through simple studies of resource amendment using defined compounds [200] but much more work of this sort is needed.

Research should consider the role of soil microbial richness and evenness on plant performance and address whether a greater functional gene diversity and/or functional redundancy associated with increased taxonomic diversity leads to a more resilient and consistent functioning of the soil microbiome across changing environments [201]. Furthermore, while only a portion of the soil microbial pool is metabolically active (at different rates) at any given time [202], a more diverse community should increase the metabolically active pool of microbes, but also provide the genetic diversity to function under changing environmental conditions. A more abundant and diverse community would also maximize microbial competition and/or niche saturation rendering the soil more resistant to new invasion. For example, soils with higher microbial biomass and/or diversity have been found to be more disease-suppressive [203–205] and resilient to invasive organisms [98]. The role of community evenness has received less attention than richness or diversity; however, evidence supports an important role for evenness in community functioning and plant productivity, particularly under stresses or perturbations [206]. The mechanistic basis behind these benefits still need to be explored; however, like community richness and diversity, may be associated with a more complete resource utilization that reduces niche space available for invaders. In particular, community evenness has been shown to be important to limit invasive plants [207] or insect pests [208].

6.3. Defining Relationships among Climate, Edaphic Factors, and management with Respect to Soil Biota

Overall, there remains insufficient information to quantify effects of agricultural management practices on key soil biological functions under a range of soil-climatic conditions. It is also essential to incorporate the temporal element as the timing of disturbances (managed or natural) could determine their significance, and length of time needed to recover critical soil functions.

While numerous studies have found the effect of one or more specific factors (edaphic, management) on soil microbial communities to be significant, very few researchers have integrated a wide range of factors into one study, and interactions were generally not identified. In one of the very few attempts to look at this problem on a broad scale, a study on bulk soil from field plots in California found that microbial communities were affected by the following variables, in order of decreasing importance: soil type > time > specific farming operation > management system > spatial variation [209]. Studies like this need to be repeated with modern methods across a wide range of soil types and climate. New metagenomic tools including high-throughput sequencing and functional gene arrays now make it possible to directly address this question. This is a critical question which must be answered in order to build a model that uses agricultural management and environmental factors to predict soil biological health and ecosystem services. A step in this direction has been made with the inclusion of AM fungi in modeling the services provided by cover crops [154].

Both short term and long term disturbances affect the soil biological community and its function. An example of a short term or acute change would be the transient change in overall biomass or activity due to a single event such as tillage or fertilizer application. These changes may or may not be significant depending on the stage of crop growth and its current requirements. Long term changes in the soil biological community are those occurring in response to persistently applied management approaches such as tillage regime, crop rotation, or cover cropping. These changes result in alterations within the soil communities as some members are lost while others become dominant. If an organism that is lost has limited dispersal mechanisms, such as AM fungi, then recovery of these populations may take a lot of time or require intervention by inoculation.

6.4. Development of Improved Indicators of Soil Health

Another challenge is to identify those soil biological functions or variables that are sensitive and have short-term biological relevance but also integrate management history. Measuring such functions could then be used to inform management decisions. The natural temporal or spatial scales of some soil functions will likely not correspond to the scale of management. Highly variable, but biologically important, soil parameters such as soil moisture, temperature, mineralization rates, and pools of labile C and N may be most useful for understanding short term, localized patterns of soil functions but their relatively high spatial and temporal heterogeneity hamper meaningful measurement and limit their use for determining prescriptive management activities at the field scale [210]. Moreover, parameters with variable tendencies may not adequately detect baseline shifts in key soil biological activities without a robust temporal and spatial historical dataset. Conversely, relatively large scale soil parameters that impact soil biological functions may not be manageable (soil texture) or change slowly (soil organic matter), making them less useful for modifying management plans in the short term. A truly defensible measure of *in situ* biological function remains a challenge, as the act of measuring or sampling will influence the target measurement. Improved relevance of functional measurements is imperative for understanding the dynamic processes occurring in soils.

Identification of optimal soil functions and a suitable set of representative soil variables must be specific enough to be useful at the local scale but also capture information that will allow meaningful comparisons across geographic gradients or over time. Coordinated research, using standardized methodology and development of appropriate methods for normalizing soil biological functions may be one means for such comparisons. A related but more difficult task is the development of forward-looking information to accommodate anticipated, but uncertain changes to soil-climate linkages in the future. Future changes are expected to manifest as shifts in the overall trends of major environmental factors such as temperature and precipitation but accompanying these may be increasing variability and thus risk.

One challenge to understanding the relationships between management and soil function, whether under different management options, combinations of soil climate, or scenarios of change, is to move beyond descriptive soil biology towards mechanistic characterizations of community composition and activities that are directly related to productivity or sustainability and are amenable to management [210]. Productivity is relatively easy to measure but sustainability is more complicated given our imprecise understanding of how the communities of soil biota link to ecosystem functions and how they can respond to change, whether planned or stochastic. It has proven difficult to comprehensively define "ambient" or "optimum" levels of soil health in part because these are context dependent terms that depend on intended land use. Thus further work is needed to provide suitable baseline criteria about manageable, functionally-related soil traits in order to compare among various management approaches such as conventional *versus* low input *versus* integrated approaches that combine livestock and crop production. Such comparisons may be at the local scale (e.g., nutrient cycling, pathology, or aggregate stability) or have broader consequences (e.g., water quality, C sequestration, greenhouse gas formation erosion).

One area that demands a comprehensive level of effort is the role of soil biology in improving nutrient use efficiency by plants. Current nutrient recommendations are primarily based on a single, point-in-time measurement of soluble and easily-exchangeable soil nutrients. However, the chemical speciation of nutrients changes frequently, often catalyzed by biological processes. However, nutrient recommendations are commonly developed under standard test conditions, usually similar to conventional farming practices, where soil biological contributions to soil fertility are likely to be minimized. Consequently, while nutrient recommendations do predict the average crop response, they do not reliably predict plant response and soil fertility under many site-specific conditions, particularly where soil biology has been enhanced by management practices [211–213]. Typical calculations of nutrient use efficiency contribute to excessive nutrient application because they fail to account for loss of nutrients from the system [84]. Improved nutrient use indexes that account for nutrient loss from the system implicitly include the extensive effects of soil biota on nutrient dynamics. Plant nutrition models fail to capture many biological rhizosphere processes, particularly the kinetic aspects, and enhancement of root-rhizosphere processes is the most probable path for ecologically-sustainable intensification of agriculture [214]. Managing the nutrient balance of the entire soil system allows the system to stabilize at lower nutrient levels that take advantage of biological means of nutrient retention and makes the most efficient use of resources [85,86].

Simple and effective indicators of soil quality/health which have meaning to land managers remain inadequate for assessing the sustainability of management. Indices and models are needed to link changes in microbial community composition and activities to a change in metabolic functions (*i.e.*, C cycling, and nitrous oxide (N_2O) and methane (CH_4) fluxes) for different soils and crop scenarios. A recent report by a group of scientists for the American Academy of Microbiology (AAM) stressed the importance of incorporating microbial processes into climate models [215]. Currently, no index includes the microbial portion of soil, which poses another challenge to assess the success of benefits to soil health provided by conversion of cropland to conservation programs (*i.e.*, The Conservation Reserve Program). Quality and quantity of SOM is coupled with composition and functioning of the microbial community and therefore, SOM quality assessments must also be a component of future research/indices. Perhaps, soil microbial community characteristics (e.g., size, composition and specific activities) and changes occurring with management can be assigned a ranking number to guide management decisions and policy.

The proposed introduction of several new organic amendments or 21st Century by-products (e.g., biochars or nanomaterials) that can last longer in the environment than traditional amendments, create another level of complexity. Critical assessments are needed to quantify the impacts of these products on resident microbial communities and their associated—but largely unknown—activities. Similarly, the use of microbial amendments and stimulants are difficult to justify without better understanding of the baseline contributions of soil biota and suitable indicators to evaluate if modification of the soil biota results in significant improvements to soil health and function.

7. Soil Biology Research Investments Needed to Ensure Our Future by Promoting Soil Health and Mitigate Soil Degradation

The challenge for agriculture in the 21st century is to implement more sustainable farming systems that are economically viable and accommodate changing technologies and climate. The production of food and fiber continues to increase agriculture's C footprint through the increased use of fuel and fertilizer, and contributes to widespread soil and water quality degradation, and loss of habitat diversity and biodiversity. To decrease this footprint, nutrient management in sustainable systems must be a top priority [216]. Soil biology is the foundation for soil health and the biological processes which moderate nutrient availability to plants, in addition to buffering plants from changes in water availability and pest, pathogen, and weed pressures. The health of the soil biota is strongly linked to the resistance of soils to erosion. Soil biological diversity is positively linked to ecosystem level processes such as C and nutrient dynamics [69] and has a central role in agroecosystems that are operated in an environmental- and economically-sustainable manner [217,218]. Soil biology is the key to ensuring the ability to "Feed the World" [219] and reversing the degradation of soils that support crop production.

As farming systems constantly change due to economic and technical drivers, soil biological functions need to be continually re-evaluated [220]. Synthesis papers by Kremen [71] and Hooper *et al.* [72] detail the linkages between biological communities and ecosystems services; understanding soil biological complexity is essential to properly manage agroecosystems. Recent advances in DNA and biochemical methods in characterizing biological activity and biodiversity will help better understand the complex nature of life in soil, provide new insights into functional mechanisms of soil microbial communities, and thus be useful for restoring degraded soil resources. This new knowledge will also greatly aid and drive development of innovative agricultural production systems that are economically and environmentally sustainable [220].

Climate change models suggest that modified cropping systems will be needed for optimal production under extreme weather events, such as the recent drought facing much of the U.S. The resiliency and resistance of agroecosystems depends, in part, on the functioning of the microbial community. Changes in cropping systems resulting from an earlier growing season, emerging plant pathogens and lower yields, and cropping sequence disruption due to drought cycles in certain regions will challenge land and water resources to maintain food, fiber, and feed production for the growing population. For example, frequent drought cycles in the U.S. Southern Plains have resulted in transition from irrigated to dryland production with possible total crop abandonment and/or interruptions in production cycles [152,221]. Identification of key soil microbial assemblages and the soil management practices that support these key microbial assemblages may assist the recovery of soils from major disturbances. Climate change may result in even more soil degradation through greater wind erosion and increased use of fallow periods to compensate for periodic droughts in some rotations. Greater knowledge of microbes and their roles in essential soil processes will aid in quickly adapting to these climate changes and other factors contributing to soil degradation. As cropping systems evolve with changing technologies, producer views and environmental constraints, specific bacterial-fungal assemblages that foster efficient nutrient and water uptake under modified or new cropping systems will need to be identified.

Research investment is required to significantly advance basic knowledge of soil biology and to properly assess soil biological responses in agricultural systems. Research should be designed with particular agricultural applications in mind and sites need to accommodate regionally- different soil-climatic regimes and agricultural practices. Long-term, multi-location, multidisciplinary team research with shared goals and protocols is required to thoroughly and productively advance this area of research.

Significant progress toward enabling predictable application of soil biology manipulation in agricultural systems could be made using currently-available analytical tools provided a critical mass of effort is assembled. A hierarchical set of analyses should be applied, such as that proposed by Kowalchuk *et al.* [222] to assess the effects of GM plants on soil microorganisms (Figure 6). These analyses would include basic measures of the size of the microbial community (e.g., biomass and numbers), bulk activities (respiration, enzyme activities), community composition (PLFA, molecular profiles) as well as quantification of subsets of microbes and their activity potentials using molecular probes and soil metagenomic approaches.

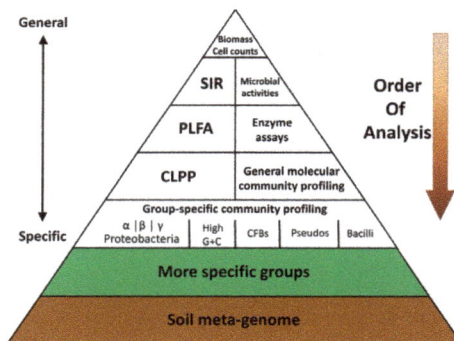

Figure 6. Hierarchy of soil microbial analyses to characterize soil microbial communities. Modified with permission from Kowalchuk *et al.* [222]).

Sustainability **2015**, *7*, 988–1027

8. Summary and Conclusions

After reviewing what's known and unknown regarding soil microbial communities and their relationships to soil health, we remain optimistic that one of the most promising strategies for mitigating and even reversing soil degradation around the world is to significantly increase public-private research efforts focused on soil biology. Of the three indicator regimes (physical, chemical, and biological) influencing soil health/quality at all scales, biological relationships are by far the most complex with large deficiencies in basic understanding. Many new tools and techniques have been or are being developed, thus making it more feasible to unravel these complex systems. Ultimately, this new knowledge will be used for informing management to restore the degraded soils that humankind desperately needs to meet the rapidly increasing food, feed, fiber, and fuel needs of an expanding global population.

Acknowledgments: The USDA is an equal opportunity provider and employer. The use of trade, firm, or corporation names in this publication is for the information and convenience of the reader. Such use does not constitute an official endorsement or approval by the United States Department of Agriculture or the Agricultural Research Service of any product or service to the exclusion of others that may be suitable.

Author Contributions: All authors participated in the literature review and writing of this manuscript. All authors read and approved the final manuscript. We dedicate this contribution to the late Jeffrey L. Smith (USDA-ARS) who made significant contributions not only to research planning documents used for this manuscript but also to the overall development of the entire soil quality/health concept, its application to alternative agricultural management practices, and for his foresight and dedication to mitigating and reversing soil degradation around the world.

Conflicts of Interest: The authors declare no conflict of interest.

References

1. Karlen, D.L.; Andrews, S.S.; Weinhold, B.J.; Zobeck, T.M. Soil quality assessment: Past, present, and future. *Electron. J. Integr. Biosci.* **2008**, *6*, 3–14.
2. Karlen, D.L. Soil health: The concept, its role, and strategies for monitoring. In *Soil Ecology and Ecosystem Services*; Wall, D.H., Bardgett, R.D., Behan-Pelletier, V., Herrick, J.E., Jones, H., Ritz, K., Six, J., Strong, D.R., van der Putten, W.M., Eds.; Oxford University Press: New York, NY, USA, 2012; pp. 331–336.
3. Doran, J.W.; Sarrantonio, M.; Liebig, M.A. Soil health and sustainability. *Adv. Agron.* **1996**, *56*, 1–54.
4. Magdoff, F.; van Es, H. *Building Soils for Better Crops*; Sustainable Agriculture Network Publications: Burlington, VT, USA, 2000; p. 241.
5. Harris, R.F.; Bezdicek, D.F. Descriptive aspects of soil quality/health. In *Defining Soil Quality for a Sustainable Environment*; Doran, J.W., Coleman, D.C., Bezdicek, D.F., Stewart, B.A., Eds.; Soil Science Society of America: Madison, WI, USA, 1994; pp. 23–35.
6. Warkentin, B.P.; Fletcher, H.F. Soil quality for intensive agriculture. In Intensive Agriculture Society of Science, Soil and Manure. Proceedings of the International Seminar on Soil Environment and Fertilizer Management; National Institute of Agricultural Science: Tokyo, Japan, 1977; pp. 594–598.
7. Doran, J.W.; Parkin, T.B. Defining and assessing soil quality. In *Defining Soil Quality for a Sustainable Environment*; Doran, J.W., Coleman, D.C., Bezdicek, D.F., Stewart, B.A., Eds.; Soil Science Society of America: Madison, WI, USA, 1994; pp. 3–21.
8. Karlen, D.L.; Erbach, D.C.; Kaspar, T.C.; Colvin, T.S.; Berry, E.C.; Timmons, D.R. Soil tilth: A review of past perceptions and future needs. *Soil Sci. Soc. Am. J.* **1990**, *54*, 153–161. [CrossRef]
9. Karlen, D.L.; Mausbach, M.J.; Doran, J.W.; Cline, R.G.; Harris, R.F.; Schuman, G.E. Soil quality: A concept, definition, and framework for evaluation. *Soil Sci. Soc. Am. J.* **1997**, *61*, 4–10. [CrossRef]
10. Kaspar, T.C.; Radke, J.K.; Laflen, J.M. Small grain cover crops and wheel traffic effects on infiltration, runoff, and erosion. *J. Soil Water Conserv.* **2001**, *56*, 160–164.
11. Larson, W.E.; Pierce, F.J. Conservation and enhancement of soil quality. In Evaluation for Sustainable Land Management in the Developing World, Proceedings of the International Workshop, Chiang Rai, Thailand, 15–21 September 1991; Dumanski, J., Pushparajah, E., Larson, M., Myers, R., Eds.; Int. Board for Soil Res. and Management: Bangkok, Thailand, 1991; Volume 2, pp. 175–203.

12. National Research Council (NRC). *Soil and Water Quality: An Agenda for Agriculture*; National Academy Press: Washington, DC, USA, 1993.

13. Doran, J.W.; Jones, A.J. *Methods for Assessing Soil Quality*; Soil Science Society of America: Madison, WI, USA, 1996.

14. Andrews, S.S.; Karlen, D.L.; Cambardella, C.A. The soil management assessment framework: A quantitative soil quality evaluation method. *Soil Sci. Soc. Am. J.* **2004**, *68*, 1945–1962. [CrossRef]

15. Karlen, D.L.; Stott, D.E. A framework for evaluating physical and chemical indicators of soil quality. In *Defining Soil Quality for a Sustainable Environment*; Doran, J.W., Coleman, D.C., Bezdicek, D.F., Stewart, B.A., Eds.; Soil Science Society of America: Madison, WI, USA, 1994; pp. 53–72.

16. Smith, J.L.; Halvorson, J.J.; Papendick, R.I. Using multiple-variable indicator kriging for evaluating soil quality. *Soil Sci. Soc. Am. J.* **1993**, *57*, 743–749. [CrossRef]

17. Parr, J.; Papendick, R.; Hornick, S.; Meyer, R. Soil quality: Attributes and relationship to alternative and sustainable agriculture. *Am. J. Altern. Agric.* **1992**, *7*, 5–11. [CrossRef]

18. Daily, G.C.; Matson, P.A.; Vitousek, P.M. Ecosystem services supplied by soil. In *Nature's Services Societal Dependence on Natural Ecosystems*; Daily, G.C., Ed.; Island Press: Washington, DC, USA, 1997; pp. 365–374.

19. Romig, D.E.; Garlynd, M.J.; Harris, R.F.; McSweeney, K. How farmers assess soil health and quality. *J. Soil Water Conserv.* **1995**, *50*, 229–236.

20. Jenny, H. *Factors of Soil Formation*; McGraw-Hill Book Co.: New York, NY, USA, 1941.

21. Seybold, C.A.; Mausbach, M.J.; Karlen, D.L.; Rogers, H.H. Quantification of soil quality. In *Soil Processes and the Carbon Cycle*; Lal, R., Kimble, J.M., Follett, R.F., Stewart, B.A., Eds.; CRC Press Inc.: Boca Raton, FL, USA, 1998; pp. 387–404.

22. Halvorson, J.J.; Smith, J.L.; Papendick, R.I. Issues of scale for evaluating soil quality. *J. Soil Water Conserv.* **1997**, *52*, 26–30.

23. Doran, J.W.; Parkin, T.B. Quantitative indicators of soil quality: A minimum data set. In *Methods for Assessing Soil Quality*; Doran, J.W., Jones, A.D., Eds.; Soil Science Society of America: Madison, WI, USA, 1996; pp. 25–37.

24. Gregorich, E.G.; Carter, M.R.; Angers, D.A.; Monreal, C.M.; Ellert, B.H. Towards a minimum data set to assess soil organic matter quality in agricultural soils. *Can. J. Soil Sci.* **1994**, *74*, 367–385. [CrossRef]

25. Weil, R.R.; Magdoff, F. Significance of soil organic matter to soil quality and health. In *Soil Organic Matter in Sustainable Agriculture*; Magdoff, F., Weil, R.R., Eds.; CRC Press: Boca Raton, FL, USA, 2004; pp. 1–43.

26. Wienhold, B.J.; Karlen, D.L.; Andrews, S.S.; Stott, D.E. Protocol for soil management assessment framework (SMAF) soil indicator scoring curve development. *Renew. Agric. Food Syst.* **2009**, *24*, 260–266. [CrossRef]

27. Stott, D.E.; Andrews, S.S.; Liebig, M.A.; Wienhold, B.J.; Karlen, D.L. Evaluation of β-glucosidase activity as a soil quality indicator for the soil management assessment framework (SMAF). *Soil Sci. Soc. Am. J.* **2010**, *74*, 107–119. [CrossRef]

28. Karlen, D.L.; Andrews, S.S.; Doran, J.W. Soil quality: Current concepts and applications. *Adv. Agron.* **2001**, *74*, 1–40.

29. Karlen, D.L.; Gardner, J.C.; Rosek, M.J. A soil quality framework for evaluating the impact of CRP. *J. Prod. Agric.* **1998**, *11*, 56–60. [CrossRef]

30. Potter, S.R.; Andrews, S.S.; Atwood, J.D.; Kellogg, R.L.; Lemunyon, J.; Norfleet, M.L.; Oman, D. *Model Simulation of Soil Loss, Nutrient Loss, and Change in Soil Organic Carbon Associated with Crop Production*; USDA-NRCS, Ed.; USDA Natural Resources Conservation Service: Washington, DC, USA, 2006.

31. Marriott, E.E.; Wander, M. Qualitative and quantitative differences in particulate organic matter fractions in organic and conventional farming systems. *Soil Biol. Biochem.* **2006**, *38*, 1527–1536. [CrossRef]

32. Pikul, J.L., Jr.; Johnson, J.M.F.; Schumacher, T.E.; Vigil, M.; Riedell, W.E. Change in surface soil carbon under rotated corn in eastern south Dakota. *Soil Sci. Soc. Am. J.* **2008**, *72*, 1738–1744. [CrossRef]

33. Singer, J.W.; Kohler, K.A.; Liebman, M.; Richard, T.L.; Cambardella, C.A.; Buhler, D.D. Tillage and compost affect yield of corn, soybean, and wheat and soil fertility. *Agron. J.* **2004**, *96*, 531–537. [CrossRef]

34. Teasdale, J.R. Strategies for soil conservation in no-tillage and organic farming systems. *J. Soil Water Conserv.* **2007**, *62*, 144A–147A.

35. Marriott, E.E.; Wander, M.M. Total and labile soil organic matter in organic and conventional farming systems. *Soil Sci. Soc. Am. J.* **2006**, *70*, 950–959. [CrossRef]

36. Fließbach, A.; Oberholzer, H.R.; Gunst, L.; Mader, P. Soil organic matter and biological soil quality indicators after 21 years of organic and conventional farming. *Agric. Ecosyst. Environ.* **2007**, *118*, 273–284. [CrossRef]
37. Tu, C.; Louws, F.J.; Creamer, N.G.; Mueller, J.P.; Brownie, C.; Fager, K.; Bell, M.; Hu, S.J. Responses of soil microbial biomass and N availability to transition strategies from conventional to organic farming systems. *Agric. Ecosyst. Environ.* **2006**, *113*, 206–215. [CrossRef]
38. Langdale, G.W.; Blevins, R.L.; Karlen, D.L.; McCool, D.K.; Nearing, M.A.; Skidmore, E.L.; Thomas, A.W.; Tyler, D.D.; Williams, J.R. Cover crop effects on soil erosion by wind and water. In *Cover Crops for Clean Water*; Hargrove, W.L., Ed.; Soil and Water Conservation Society: Ankeny, IA, USA, 1991; pp. 15–22.
39. Kaspar, T.C.; Jaynes, D.B.; Parkin, T.B.; Moorman, T.B. Rye cover crop and gamagrass strip effects on NO_3 concentration and load in tile drainage. *J. Environ. Qual.* **2007**, *36*, 1503–1511. [CrossRef] [PubMed]
40. Kladivko, E.J.; Frankenberger, J.R.; Jaynes, D.B.; Meek, D.W.; Jenkinson, B.J.; Fausey, N.R. Nitrate leaching to subsurface drains as affected by drain spacing and changes in crop production system. *J. Environ. Qual.* **2004**, *33*, 1803–1813. [CrossRef] [PubMed]
41. Strock, J.S.; Porter, P.M.; Russelle, M.P. Cover cropping to reduce nitrate loss through subsurface drainage in the northern US corn belt. *J. Environ. Qual.* **2004**, *33*, 1010–1016. [CrossRef] [PubMed]
42. Snapp, S.S.; Swinton, S.M.; Labarta, R.; Mutch, D.; Black, J.R.; Leep, R.; Nyiraneza, J.; O'Neil, K. Evaluating cover crops for benefits, costs and performance within cropping system niches. *Agron. J.* **2005**, *97*, 322–332.
43. Cambardella, C.A.; Johnson, J.M.F.; Varvel, G.E. Soil carbon sequestration in central U.S. Agroecosystems. In *Managing Agricultural Greenhouse Gases: Coordinated Agricultural Research through Gracenet to Address Our Changing Climate*; Liebig, M.A., Franzleubbers, A.J., Follet, R.F., Eds.; Academic Press, Elsevier: San Diego, CA, USA, 2012; pp. 41–58.
44. Franzluebbers, A.J. Achieving soil organic carbon sequestration with conservation agricultural systems in the southeastern United States. *Soil Sci. Soc. Am. J.* **2010**, *74*, 347–357. [CrossRef]
45. Johnson, D.W.; Verburg, P.S.J.; Amone, J.A. Soil extraction, ion exchange resin, and ion exchange membrane measures of soil mineral nitrogen during incubation of a tallgrass prairie soil. *Soil Sci. Soc. Am. J.* **2005**, *69*, 260–265. [CrossRef]
46. Follett, R.F.; Vogel, K.P.; Varvel, G.E.; Mitchell, R.B.; Kimble, J. Soil carbon sequestration by switchgrass and no-till maize grown for bioenergy. *BioEnergy Res.* **2012**, *5*, 866–875. [CrossRef]
47. Gál, A.; Vyn, T.J.; Michéli, E.; Kladivko, E.J.; McFee, W.W. Soil carbon and nitrogen accumulation with long-term no-till *versus* moldboard plowing overestimated with tilled-zone sampling depths. *Soil Tillage Res.* **2007**, *96*, 42–51. [CrossRef]
48. Yang, X.; Drury, C.; Wander, M.; Kay, B. Evaluating the effect of tillage on carbon sequestration using the minimum detectable difference concept. *Pedosphere* **2008**, *18*, 421–430. [CrossRef]
49. Clapp, C.E.; Allmaras, R.R.; Layese, M.F.; Linden, D.R.; Dowdy, R.H. Soil organic carbon and C-13 abundance as related to tillage, crop residue, and nitrogen fertilization under continuous corn management in minnesota. *Soil Tillage Res.* **2000**, *55*, 127–142. [CrossRef]
50. Huggins, D.R.; Allmaras, R.R.; Clapp, C.E.; Lamb, J.A.; Randall, G.W. Corn-soybean sequence and tillage effects on soil carbon dynamics and storage. *Soil Sci. Soc. Am. J.* **2007**, *71*, 145–154. [CrossRef]
51. Gans, J.; Wolinsky, M.; Dunbar, J. Computational improvements reveal great bacterial diversity and high metal toxicity in soil. *Science* **2005**, *309*, 1387–1390. [CrossRef] [PubMed]
52. Gold, T. The deep, hot biosphere. *Proc. Natl. Acad. Sci. USA* **1992**, *89*, 6045–6049. [CrossRef] [PubMed]
53. Pace, N.R. Mapping the tree of life: Progress and prospects. *Microbiol. Mol. Biol. Rev.* **2009**, *73*, 565–576. [CrossRef] [PubMed]
54. Thauer, R.K. A fifth pathway of carbon fixation. *Science* **2007**, *318*, 1732–1733. [CrossRef] [PubMed]
55. Beja, O.; Aravind, L.; Koonin, E.V.; Suzuki, M.T.; Hadd, A.; Nguyen, L.P.; Jovanovich, S.B.; Gates, C.M.; Feldman, R.A.; Spudich, J.L.; *et al.* Bacterial rhodospin: Evidence for a new type of phototrophy in the sea. *Science* **2000**, *289*, 1902–1906. [CrossRef] [PubMed]
56. Boone, D.; Liu, Y.; Zhao, Z.; Balkwill, D.; Drake, G.; Stevens, T.; Aldrich, H. *Bacillus infernus* sp. Nov., an Fe(iii)- and Mn(iv)-reducing anaerobe from the deep terrestrial subsurface. *Int. J. Syst. Bacteriol.* **1995**, *45*, 441–448. [CrossRef] [PubMed]
57. Leininger, S.; Urich, T.; Schloter, M.; Schwark, L.; Qi, J.; Nicol, G.; Prosser, J.; Schuster, S.; Schleper, C. Archaea predominate among ammonia-oxidizing prokaryotes in soils. *Nature* **2006**, *442*, 806–809. [CrossRef] [PubMed]

58. Nelson, K.E.; Clayton, R.A.; Gill, S.R.; Gwinn, M.L.; Dodson, R.J.; Haft, D.H.; Hickey, E.K.; Peterson, J.D.; Nelson, W.C.; Ketchum, K.A.; *et al.* Evidence for lateral gene transfer between archaea and bacteria from genome sequence of *Thermotoga maritima*. *Nature* **1999**, *399*, 323–329. [CrossRef] [PubMed]
59. Lavelle, P.; Decaëns, T.; Aubert, M.; Barot, S.; Blouin, M.; Bureau, F.; Margerie, P.; Mora, P.; Rossi, J.P. Soil invertebrates and ecosystem services. *Eur. J. Soil Biol.* **2006**, *42*, S3–S15. [CrossRef]
60. Blouin, M.; Hodson, M.E.; Delgado, E.A.; Baker, G.; Brussaard, L.; Butt, K.R.; Dai, J.; Dendooven, L.; Pérès, G.; Tondoh, J. A review of earthworm impact on soil function and ecosystem services. *Eur. J. Soil Sci.* **2013**, *64*, 161–182. [CrossRef]
61. Falkowski, P.G.; Fenchel, T.; DeLong, E.F. The microbial engines that drive earth's biogeochemical cycles. *Science* **2008**, *320*, 1034–1039. [CrossRef] [PubMed]
62. Kowalchuk, G.A.; Jones, S.E.; Blackall, L.L. Microbes orchestrate life on earth. *ISME J.* **2008**, *2*, 795–796. [CrossRef] [PubMed]
63. Benayas, J.M.; Newton, A.C.; Diaz, A.; Bullock, J.M. Enhancement of biodiversity and ecosystem services by ecological restoration: A meta-analysis. *Science* **2009**, *325*, 1121–1124. [CrossRef] [PubMed]
64. Pimental, D.; Wilson, C.; McCullum, C.; Huang, R.; Dwen, P.; Flack, J.; Tran, Q.; Saltman, T.; Cliff, B. Economic and environmental benefits of biodiversity. *Bioscience* **1997**, *47*, 747–757. [CrossRef]
65. Nielsen, U.N.; Ayres, E.; Wall, D.H.; Bardgett, R.D. Soil biodiversity and carbon cycling: A review and synthesis of studies examining diversity-function relationships. *Eur. J. Soil Sci.* **2010**, *62*, 105–116. [CrossRef]
66. Pritchard, S.G. Soil organisms and global climate change. *Plant Pathol.* **2011**, *60*, 82–89. [CrossRef]
67. Cardinale, B.J. Biodiversity improves water quality through niche partitioning. *Nature* **2011**, *472*, 86–89. [CrossRef] [PubMed]
68. Gessner, M.O.; Swan, C.M.; Dang, C.K.; McKie, B.G.; Bardgett, R.D.; Wall, D.H.; Hattenschwiler, S. Diversity meets decomposition. *Trends Ecol. Evol.* **2010**, *25*, 372–380. [CrossRef] [PubMed]
69. Wagg, C.; Bender, S.F.; Widmer, F.; van der Heijden, M.G.A. Soil biodiversity and soil community composition determine ecosystem multifunctionality. *Proc. Natl. Acad. Sci. USA* **2014**, *111*, 5266–5270. [CrossRef] [PubMed]
70. Van Elsas, J.D.; Chiurazzi, M.; Mallon, C.A.; Elhottova, D.; Kristufek, V.; Salles, J.F. Microbial diversity determines the invasion of soil by a bacterial pathogen. *Proc. Natl. Acad. Sci. USA* **2012**, *109*, 1159–1164. [CrossRef] [PubMed]
71. Kremen, C. Managing ecosystem services, what do we need to know about their ecology? *Ecol. Lett.* **2005**, *8*, 468–479. [CrossRef] [PubMed]
72. Hooper, D.U.; Chapin, F.S.; Ewel, J.J.; Hector, A.; Inchausti, P.; Lavorel, S.; Lawton, J.H.; Lodge, D.M.; Loreau, M.; Naeem, S.; *et al.* Effects of biodiversity on ecosystem functioning: A consensus of current knowledge. *Ecol. Monogr.* **2005**, *75*, 3–35. [CrossRef]
73. Wall, D.H.; Bardgett, R.D.; Covich, A.P.; Snelgrove, P.V.R. The need for understanding how biodiversity and ecosystem functioning affect ecosystem service in soils and sediments. In *Sustaining Biodiversity and Ecosystem Services in Soils and Sediments*; Wall, D.H., Ed.; Island Press: Washington, DC, USA, 2004; Volume SCOPE 64, pp. 1–12.
74. Montgomery, D.R. *Dirt: The Erosion of Civilizations*; Univ of California Press: Oakland, CA, USA, 2012.
75. Altieri, M.A. The ecological role of biodiversity in agroecosystems. *Agric. Ecosyst. Environ.* **1999**, *74*, 19–31. [CrossRef]
76. Wright, S.; Upadhyaya, A. A survey of soils for aggregate stability and glomalin, a glycoprotein produced by hyphae of arbuscular mycorrhizal fungi. *Plant Soil* **1998**, *198*, 97–107. [CrossRef]
77. Six, J.; Bossuyt, H.; Degryze, S.; Denef, K. A history of research on the link between (micro)aggregates, soil biota, and soil organic matter dynamics. *Soil Tillage Res.* **2004**, *79*, 7–31. [CrossRef]
78. Peck, S.L.; McQuaid, B.; Campbell, C.L. Using ant species (Hymenoptera: Formicidae) as a biological indicator of agroecosystem condition. *Environ. Entomol.* **1998**, *27*, 1102–1110. [CrossRef]
79. Pinto-Tomás, A.A.; Anderson, M.A.; Suen, G.; Stevenson, D.M.; Chu, F.S.T.; Cleland, W.W.; Weimer, P.J.; Currie, C.R. Symbiotic nitrogen fixation in the fungus gardens of leaf-cutter ants. *Science* **2009**, *326*, 1120–1123. [CrossRef] [PubMed]
80. Smith, S.E.; Read, D.J. *Mycorrhizal Symbiosis*, 3rd ed.; Academic Press: London, UK, 2008.
81. Oberson, A.; Joner, E.J. Microbial turnover of phosphorus in soil. In *Organic Phosphorus in the Environment*; Turner, B.L., Frossard, E., Baldwin, D.S., Eds.; CABI International: Wallingford, CT, USA, 2005; pp. 133–165.

82. Richardson, A.E.; Lynch, J.P.; Ryan, P.R.; Delhaize, E.; Smith, F.A.; Smith, S.E.; Harvey, P.R.; Ryan, M.H.; Veneklaas, E.J.; Lambers, H.; *et al.* Plant and microbial strategies to improve the phosphorus efficiency of agriculture. *Plant Soil* **2011**, *349*, 121–156. [CrossRef]

83. McLaughlin, M.J.; McBeath, T.M.; Smernik, R.; Stacey, S.P.; Ajiboye, B.; Guppy, C. The chemical nature of P accumulation in agricultural soils—Implications for fertiliser management and design: An Australian perspective. *Plant Soil* **2011**, *349*, 69–87. [CrossRef]

84. Simpson, R.J.; Oberson, A.; Culvenor, R.A.; Ryan, M.H.; Veneklaas, E.J.; Lambers, H.; Lynch, J.P.; Ryan, P.R.; Delhaize, E.; Smith, F.A.; *et al.* Strategies and agronomic interventions to improve the phosphorus-use efficiency of farming systems. *Plant Soil* **2011**, *349*, 89–120. [CrossRef]

85. Drinkwater, L.E.; Snapp, S.S. Nutrients in agroecosystems: Rethinking the management paradigm. *Adv. Agron.* **2007**, *92*, 163–186.

86. Dawson, J.C.; Huggins, D.R.; Jones, S.J. Characterizing nitrogen use efficiency in natural and agricultural ecosystems to improved the performance of cereal crops in low-input and organic agricultural systems. *Field Crops Res.* **2008**, *107*, 89–101. [CrossRef]

87. Angus, T.A. Symposium on microbial insecticides. I. Bacterial pathogens of insects as microbial insecticides. *Bacteriol. Rev.* **1965**, *29*, 364–372. [PubMed]

88. Vandenberg, J.D. Standardized bioassay and screening of *Beauveria bassiana* and *Paecilomyces fumosoroseus* against the Russian wheat aphid (Homoptera: Aphididae). *J. Econ. Entomol.* **1996**, *89*, 1418–1423. [CrossRef]

89. Rillig, M.C. Arbuscular mycorrhizae and terrestrial ecosystems processes. *Ecol. Lett.* **2004**, *7*, 740–754. [CrossRef]

90. Gianinazzi, S.; Gollotte, A.; Binet, M.N.; van Tuinen, D.; Redecker, D.; Wipf, D. Agroecology: The key role of arbuscular mycorrhizas in ecosystem services. *Mycorrhiza* **2010**, *20*, 519–530. [CrossRef] [PubMed]

91. Lundgren, J.G.; Fergen, J.K. Predator community structure and trophic linkage strength to a focal prey. *Mol. Ecol.* **2014**, *23*, 3790–3798. [CrossRef] [PubMed]

92. White, S.S.; Renner, K.A.; Menalled, F.D.; Landis, D.A. Feeding preferences of weed seed predators and effect on weed emergence. *Weed Sci.* **2007**, *55*, 606–612. [CrossRef]

93. Shearin, A.F.; Reberg-Horton, S.C.; Gallandt, E.R. Direct effects of tillage on the activity density of ground beetle (Coleoptera: Carabidae) weed seed predators. *Environ. Entomol.* **2007**, *36*, 1140–1146. [CrossRef] [PubMed]

94. Lundgren, J.G.; Shaw, J.T.; Zaborski, E.R.; Eastman, C.E. The influence of organic transition systems on beneficial ground-dwelling arthropods and predation of insects and weed seeds. *Renew. Agric. Food Syst.* **2006**, *21*, 227–237. [CrossRef]

95. Chee-Sanford, J.C.; Williams II, M.M.; Davis, A.S.; Sims, G.K. Do microorganisms influence seed-bank dynamics? *Weed Sci.* **2009**, *54*, 575–587. [CrossRef]

96. Lundgren, J.G.; Lehman, R.M. Bacterial gut symbionts contribute to seed digestion in an omnivorous beetle. *PLoS One* **2010**, *5*, e10831. [CrossRef] [PubMed]

97. Kremer, R.J.; Li, J. Developing weed-suppressive soils through improved soil quality management. *Soil Tillage Res.* **2003**, *72*, 193–202. [CrossRef]

98. Van Elsas, J.D.; Garbeva, P.; Salles, J.F. Effects of agronomic measures on the microbial diversity of soils as related to the suppression of soil-borne pathogens. *Biodegradation* **2002**, *13*, 29–40. [CrossRef] [PubMed]

99. Garbeva, P.; van Veen, J.A.; van Elsas, J.D. Microbial diversity in soil: Selection of microbial populations by plant and soil type and implications for disease suppressiveness. *Ann. Rev. Phytopathol.* **2004**, *42*, 243–270. [CrossRef]

100. Zahir, A.M.; Frankenberger, W.T. Plant growth promoting rhizobacteria: Applications and perspectives in agriculture. *Adv. Agron.* **2004**, *81*, 97–168.

101. Kloepper, J.W.; Zablotowicz, R.M.; Tipping, E.M.; Lifshitz, R. Plant growth promotion mediated by bacterial rhizospheres colonizers. In *The Rhizosphere and Plant Growth*; Kleister, D.L., Cregan, P.B., Eds.; Kluwer: Dordrecht, The Netherland, 1991; pp. 315–326.

102. Lutenberg, B.; Kamilova, F. Plant growth promoting rhizobacteria. *Ann. Rev. Microbiol.* **2009**, *63*, 541–556. [CrossRef]

103. Sturz, A.V.; Christie, B.R.; Nowak, J. Bacterial endophytes: Potential role in developing sustainable systems of crop production. *Crit. Rev. Plant Sci.* **2000**, *19*, 1–30. [CrossRef]

104. Harman, G.E.; Howell, C.R.; Viterbo, A.; Chet, I.; Lorito, M. Trichoderma species—Opportunistic, avirulent plant symbionts. *Nat. Rev. Microbiol.* **2004**, *2*, 43–56. [CrossRef] [PubMed]

105. Taghavi, S.; van der Lelie, D.; Hoffman, A.; Zhang, Y.-B.; Walla, M.D.; Vangronsveld, J.; Newman, L.; Monchy, S. Genome sequence of the plant growth promoting endophytic bacterium Enterobacter sp. 638. *PLoS Genet.* **2010**, *6*, e1000943. [CrossRef] [PubMed]

106. Haas, D.; Défago, G. Biological control of soil-borne pathogens by fluorescent pseudomonads. *Nat. Rev. Microbiol.* **2005**, *3*, 307–319. [CrossRef] [PubMed]

107. Vallad, G.E.; Goodman, R.M. Systemic acquired resistance and induced systemic resistance in conventional agriculture. *Crop Sci.* **2004**, *44*, 1920–1934. [CrossRef]

108. Compant, S.; Duffy, B.; Nowak, J.; Clément, C.; Barka, E.A. Use of plant growth-promoting bacteria for biocontrol of plant diseases: Principles, mechanisms of action, and future prospects. *Appl. Environ. Microbiol.* **2005**, *71*, 4951–4959. [CrossRef] [PubMed]

109. Van Loon, L.C.; Bakker, P.A.H.M.; Pieterse, C.M.J. Systemic resistance induced by rhizosphere bacteria. *Ann. Rev. Phytopathol.* **1998**, *36*, 453–483. [CrossRef]

110. Vessey, J.K. Plant growth promoting rhizobacteria as biofertilizers. *Plant Soil* **2003**, *255*, 571–586. [CrossRef]

111. Glick, B.R.; Todorovic, B.; Czarny, J.; Cheng, Z.; Duan, J.; McConkey, B. Promotion of plant growth by bacterial ACC deaminase. *Crit. Rev. Plant Sci.* **2007**, *26*, 227–242. [CrossRef]

112. Kim, Y.C.; Leveau, J.; Gardener, B.B.M.; Pierson, E.A.; Pierson, L.S.; Ryu, C.-M. The multifactorial basis for plant health promotion by plant-associated bacteria. *Appl. Environ. Microbiol.* **2011**, *77*, 1548–1555. [CrossRef] [PubMed]

113. Anaya, A.L. Allelopathy as a tool in the management of biotic resources in agroecosystems. *Crit. Rev. Plant Sci.* **1999**, *18*, 697–739. [CrossRef]

114. Kloepper, J.W.; Hu, C.-H.; Burkett-Cadena, M.; Liu, K.; Xu, J.; McInroy, J. Increased populations of deleterious fluorescent pseudomonads colonizing rhizomes of leatherleaf fern (*Rumohra adiantiformis*) and expression of symptoms of fern distortion syndrome after application of benlate systemic fungicide. *Appl. Soil Ecol.* **2012**, *61*, 236–246. [CrossRef]

115. Nehl, D.B.; Allen, S.J.; Brown, J.F. Deleterious rhizosphere bacteria: An integrating perspective. *Appl. Soil Ecol.* **1996**, *5*, 1–20. [CrossRef]

116. Kremer, R.J. Deleterious rhizobacteria. In *Plant-Associated Bacteria*; Gnanamanickam, S.S., Ed.; Springer: Dordrecht, The Netherland, 2006; pp. 335–357.

117. Badri, D.V.; Vivanco, J.M. Regulation and function of root exudates. *Plant Cell Environ.* **2009**, *32*, 666–681. [CrossRef] [PubMed]

118. Bais, H.P.; Weir, T.L.; Perry, L.G.; Gilroy, S.; Vivanco, J.M. The role of root exudates in rhizosphere interactions with plants and other organisms. *Ann. Rev. Plant Biol.* **2006**, *57*, 233–266. [CrossRef]

119. Kowalchuk, G.A.; Buma, D.S.; de Boer, W.; Klinkhamer, P.G.; van Veen, J.A. Effects of above-ground plant species composition and diversity on the diversity of soil-borne microorganisms. *Antonie Leeuwenhoek* **2002**, *81*, 509–520. [CrossRef] [PubMed]

120. Högberg, M.N.; Högberg, P.; Myrold, D.D. Is microbial community composition in boreal forest soils determined by pH, C-to-N ratio, the trees, or all three? *Oecologia* **2007**, *150*, 590–601. [CrossRef] [PubMed]

121. Micallef, S.A.; Shiaris, M.P.; Colón-Carmona, A. Influence of arabidopsis thaliana accessions on rhizobacterial communities and natural variation in root exudates. *J. Exp. Bot.* **2009**, *60*, 1729–1742. [CrossRef] [PubMed]

122. Grayston, S.J.; Wang, S.; Campbell, C.D.; Edwards, A.C. Selective influence of plant species on microbial diversity in the rhizosphere. *Soil Biol. Biochem.* **1998**, *30*, 369–378. [CrossRef]

123. Kuklinsky-Sobral, J.; Araújo, W.L.; Mendes, R.; Geraldi, I.O.; Pizzirani-Kleiner, A.A.; Azevedo, J.L. Isolation and characterization of soybean—Associated bacteria and their potential for plant growth promotion. *Environ. Microbiol.* **2004**, *6*, 1244–1251. [CrossRef] [PubMed]

124. Salles, J.F.; van Veen, J.A.; van Elsas, J.D. Multivariate analyses of Burkholderia species in soil: Effect of crop and land use history. *Appl. Environ. Microbiol.* **2004**, *70*, 4012–4020. [CrossRef] [PubMed]

125. Badri, D.V.; Loyola-Vargas, V.M.; Broeckling, C.D.; De-la-Peña, C.; Jasinski, M.; Santelia, D.; Martinoia, E.; Sumner, L.W.; Banta, L.M.; Stermitz, F. Altered profile of secondary metabolites in the root exudates of Arabidopsis ATP-binding cassette transporter mutants. *Plant Physiol.* **2008**, *146*, 762–771. [CrossRef] [PubMed]

126. Xie, X.; Zhang, H.; Paré, P.W. Sustained growth promotion in arabidopsis with long-term exposure to the beneficial soil bacterium Bacillus subtilis (gb03). *Plant Signal. Behav.* **2009**, *4*, 948–953. [CrossRef] [PubMed]

127. Zhang, H.; Xie, X.; Kim, M.S.; Kornyeyev, D.A.; Holaday, S.; Pare, P.W. Soil bacteria augment Arabidopsis photosynthesis by decreasing glucose sensing and abscisic acid levels in planta. *Plant J.* **2008**, *56*, 264–273. [CrossRef] [PubMed]

128. Zhang, H.; Kim, M.-S.; Sun, Y.; Dowd, S.E.; Shi, H.; Paré, P.W. Soil bacteria confer plant salt tolerance by tissue-specific regulation of the sodium transporter HKT1. *Mol. Plant-Microbe Interact.* **2008**, *21*, 737–744. [CrossRef] [PubMed]

129. Zhang, H.; Sun, Y.; Xie, X.; Kim, M.-S.; Dowd, S.E.; Paré, P.W. A soil bacterium regulates plant acquisition of iron via deficiency-inducible mechanisms. *Plant J.* **2009**, *58*, 568–577. [CrossRef] [PubMed]

130. Zhang, H.; Murzello, C.; Sun, Y.; Kim, M.-S.; Xie, X.; Jeter, R.M.; Zak, J.C.; Dowd, S.E.; Paré, P.W. Choline and osmotic-stress tolerance induced in Arabidopsis by the soil microbe Bacillus subtilis (GB03). *Mol. Plant-Microbe Interact.* **2010**, *23*, 1097–1104. [CrossRef] [PubMed]

131. Adesemoye, A.; Torbert, H.; Kloepper, J. Enhanced plant nutrient use efficiency with PGPR and AMF in an integrated nutrient management system. *Can. J. Microbiol.* **2008**, *54*, 876–886. [CrossRef] [PubMed]

132. Chen, S.; Subler, S.; Edwards, C.A. Effects of agricultural biostimulants on soil microbial activity and nitrogen dynamics. *Appl. Soil Ecol.* **2002**, *19*, 249–259. [CrossRef]

133. Kinnersley, A.M. The role of phytochelates in plant growth and productivity. *Plant Growth Regul.* **1993**, *12*, 207–218. [CrossRef]

134. Subler, S.; Dominguez, J.; Edwards, C.A. Assessing biological activity of agricultural biostimulants: Bioassays for plant growth regulators in three soil additives. *Commun. Soil Sci. Plant Anal.* **1998**, *29*, 859–866. [CrossRef]

135. Batie, S.; Nowak, P.; Schnepf, M. *Taking Conservation Seriously as a Wicked Problem. Managing Agricultural Landscapes for Environmental Quality II. Achieving More Effective Conservation*; Soil and Water Conservation Society: Ankeny, IA, USA, 2010; pp. 143–155.

136. Tate, R., III. *Soil Microbiology*; John Wiley & Sons: New York, NY, USA, 2000.

137. Fierer, N.; Jackson, R.B. The diversity and biogeography of soil bacterial communities. *Proc. Natl. Acad. Sci. USA* **2006**, *103*, 626–631. [CrossRef] [PubMed]

138. Lauber, C.L.; Hamady, M.; Knight, R.; Fierer, N. Pyrosequencing-based assessment of soil pH as a predictor of soil bacterial community structure at the continental scale. *Appl. Environ. Microbiol.* **2009**, *75*, 5111–5120. [CrossRef] [PubMed]

139. Rousk, J.; Bååth, E.; Brookes, P.C.; Lauber, C.L.; Lozupone, C.; Caporaso, J.G.; Knight, R.; Fierer, N. Soil bacterial and fungal communities across a pH gradient in an arable soil. *ISME J.* **2010**, *4*, 1340–1351. [CrossRef] [PubMed]

140. Jansa, J.; Wiemken, A.; Frossard, E. *The Effects of Agricultural Practices on Arbuscular Mycorrhizal Fungi*; Geological Society of London: London, UK, 2006.

141. Moonen, A.; Barberi, P. Functional biodiversity: An agroecosystem approach. *Agric. Ecosyst. Environ.* **2008**, *127*, 7–21. [CrossRef]

142. Dias, T.; Dukes, A.; Antunes, P.M. Accounting for soil biotic effects on soil health and crop productivity in the design of crop rotations. *J. Sci. Food Agric.* **2014**. [CrossRef]

143. Young, I.; Ritz, K. Tillage, habitat space and function of soil microbes. *Soil Tillage Res.* **2000**, *53*, 201–213. [CrossRef]

144. Kladivko, E.J. Tillage systems and soil ecology. *Soil Tillage Res.* **2001**, *61*, 61–76. [CrossRef]

145. Köhl, L.; Oehl, F.; van der Heijden, M.G.A. Agricultural practices indirectly influence plant productivity and ecosystem services through effects on soil biota. *Ecol. Appl.* **2014**, *24*, 1842–1853. [CrossRef]

146. Triplett, G.B.; Dick, W.A. No-tillage crop production: A revolution in agriculture! *Agron. J.* **2008**, *100*, S153–S165. [CrossRef]

147. Helgason, T.; Daniell, T.J.; Husband, R.; Fitter, A.H.; Young, J.P.W. Ploughing up the wood-wide web? *Nature* **1998**. [CrossRef]

148. Bailey, V.L.; Smith, J.L.; Bolton, H. Fungal-to-bacterial ratios in soil investigated for enhanced C sequestration. *Soil Biol. Biochem.* **2002**, *34*, 997–1007. [CrossRef]

149. Helgason, B.; Walley, F.; Germida, J. No-till soil management increases microbial biomass and alters community profiles in soil aggregates. *Appl. Soil Ecol.* **2010**, *46*, 390–397. [CrossRef]

150. Schippers, B.; Bakker, A.W.; Bakker, P.A. Interactions of deleterious and beneficial rhizosphere microorganisms and the effect of cropping practices. *Ann. Rev. Phytopathol.* **1987**, *25*, 339–358. [CrossRef]

151. Davinic, M.; Moore-Kucera, J.; Acosta-Martinez, V.; Zak, J.; Allen, V. Soil fungal distribution and functionality as affected by grazing and vegetation components of integrated crop–livestock agroecosystems. *Appl. Soil Ecol.* **2013**, *66*, 61–70. [CrossRef]

152. Acosta-Martinez, V.; Lascano, R.; Calderon, F.; Booker, J.D.; Zobeck, T.M.; Upchurch, D.R. Dryland cropping systems influence the microbial biomass and enzyme activities in a semiarid sandy soil. *Biol. Fertil. Soils* **2011**, *47*, 655–667. [CrossRef]

153. Reeves, D.W. The role of soil organic matter in maintaining soil quality in continuous cropping systems. *Soil Tillage Res.* **1997**, *43*, 131–167. [CrossRef]

154. Schipanski, M.E.; Barbercheck, M.; Douglas, M.R.; Finney, D.M.; Haider, K.; Kaye, J.P.; Kemanian, A.R.; Mortensen, D.A.; Ryan, M.R.; Tooker, J.; *et al.* A framework for evaluating ecosystem services provided by cover crops in agroecosystems. *Agric. Syst.* **2014**, *125*, 12–22. [CrossRef]

155. Boswell, E.P.; Koide, R.T.; Shumway, D.L.; Addy, H.D. Winter wheat cover cropping, VA mycorrhizal fungi and maize growth and yield. *Agric. Ecosyst. Environ.* **1998**, *67*, 55–65. [CrossRef]

156. White, C.M.; Weil, R.R. Forage radish and cereal rye cover crop effects on mycorrhizal fungus colonization of maize roots. *Plant Soil* **2010**, *328*, 507–521. [CrossRef]

157. Lehman, R.M.; Taheri, W.I.; Osborne, S.L.; Buyer, J.S.; Douds, D.D., Jr. Fall cover cropping can increase arbuscular mycorrhizae in soils supporting intensive agricultural production. *Appl. Soil Ecol.* **2012**, *61*, 300–304. [CrossRef]

158. Horst, W.J.; Kamh, M.; Jibrin, J.M.; Chude, V.O. Agronomic measures for increasing P availability to crops. *Plant Soil* **2001**, *237*, 211–223. [CrossRef]

159. Dabney, S.M.; Delgado, J.A.; Reeves, D.W. Using winter cover crops to improve soil and water quality. *Commun. Soil Sci. Plant Anal.* **2001**, *32*, 1221–1250. [CrossRef]

160. Wilhelm, W.W.; Johnson, J.M.; Karlen, D.L.; Lightle, D.T. Corn stover to sustain soil organic carbon further constrains biomass supply. *Agron. J.* **2007**, *99*, 1665–1667. [CrossRef]

161. Cotton, J.; Acosta-Martínez, V.; Moore-Kucera, J.; Burow, G. Early changes due to sorghum biofuel cropping systems in soil microbial communities and metabolic functioning. *Biol. Fertil. Soils* **2012**, *49*, 403–413. [CrossRef]

162. Lehman, R.M.; Ducey, T.F.; Jin, V.L.; Acosta-Martinez, V.; Ahlschwede, C.M.; Jeske, E.S.; Drijber, R.A.; Cantrell, K.B.; Frederick, J.R.; Fink, D.M. Soil microbial community response to corn stover harvesting under rain-fed, no-till conditions at multiple US locations. *BioEnergy Res.* **2014**, *7*, 540–550. [CrossRef]

163. Bullock, D.G. Crop rotation. *Crit. Rev. Plant Sci.* **1992**, *11*, 309–326. [CrossRef]

164. McDaniel, M.D.; Grandy, A.S.; Tiemann, L.K.; Weintraub, M.N. Crop rotation complexity regulates the decomposition of high and low quality residues. *Soil Biol. Biochem.* **2014**, *78*, 243–254. [CrossRef]

165. Bunemann, E.K.; Bossio, D.A.; Smithson, P.C.; Frossard, E.; Oberson, A. Microbial community composition and substrate use in a highly weathered soil as affected by crop rotation and P fertilization. *Soil Biol. Biochem.* **2004**, *36*, 889–901. [CrossRef]

166. Oehl, F.; Sieverding, E.; Ineichen, K.; Mäder, P.; Boller, T.; Wiemken, A. Impact of land use intensity on the species diversity of arbuscular mycorrhizal fungi in agroecosystems of central Europe. *Appl. Environ. Microbiol.* **2003**, *69*, 2816–2824. [CrossRef] [PubMed]

167. Lupwaya, N.Z.; Blackshaw, R.E. Soil microbial properties in Bt (Bacillus thuringiensis) corn cropping systems. *Appl. Soil Ecol.* **2013**, *63*, 127–133. [CrossRef]

168. Barazani, O.; Friedman, J. Allelopathic bacteria and their impact on higher plants. *Crit. Rev. Microbiol.* **2001**, *27*, 41–45. [CrossRef] [PubMed]

169. Bunemann, E.; Smithson, P.C.; Jama, B.; Frossard, E.; Oberson, A. Maize productivity and nutrient dynamics in maize-fallow rotations in western Kenya. *Plant Soil* **2004**, *264*, 195–208. [CrossRef]

170. Rosendahl, S.; Matzen, H.B. Genetic structure of arbuscular mycorrhizal populations in fallow and cultivated soils. *New Phytol.* **2008**, *179*, 1154–1161. [CrossRef] [PubMed]

171. Wetterauer, D.; Killorn, R. Fallow-and flooded-soil syndromes: Effects on crop production. *J. Prod. Agric.* **1996**, *9*, 39–41. [CrossRef]

172. Anderson, R.L. Possible causes of dry pea synergy to corn. *Weed Technol.* **2012**, *26*, 438–442. [CrossRef]

173. Calbrix, R.; Barray, S.; Chabrerie, O.; Fourrie, L.; Laval, K. Impact of organic amendments on the dynamics of soil microbial biomass and bacterial communities in cultivated land. *Appl. Soil Ecol.* **2007**, *35*, 511–522. [CrossRef]

174. Chaudhry, V.; Rehman, A.; Mishra, A.; Chauhan, P.S.; Nautiyal, C.S. Changes in bacterial community structure of agricultural land due to long-term organic and chemical amendments. *Microb. Ecol.* **2012**, *64*, 450–460. [CrossRef] [PubMed]

175. Sun, H.Y.; Deng, S.P.; Raun, W.R. Bacterial community structure and diversity in a century-old manure-treated agroecosystem. *Appl. Environ.Microbiol.* **2004**, *70*, 5868–5874. [CrossRef] [PubMed]

176. Saison, C.; Degrange, V.; Oliver, R.; Millard, P.; Commeaux, C.; Montange, D.; le Roux, X. Alteration and resilience of the soil microbial community following compost amendment: Effects of compost level and compost-borne microbial community. *Environ. Microbiol.* **2006**, *8*, 247–257. [CrossRef] [PubMed]

177. Kallenbach, C.; Grandy, A.S. Controls over soil microbial biomass responses to carbon amendments in agricultural systems: A meta-analysis. *Agric. Ecosyst. Environ.* **2011**, *144*, 241–252. [CrossRef]

178. Cheng, Y.; Ishimoto, K.; Kuriyama, Y.; Osaki, M.; Ezawa, T. Ninety-year-, but not single, application of phosphorus fertilizer has a major impact on arbuscular mycorrhizal fungi communities. *Plant Soil* **2012**, *365*, 397–407. [CrossRef]

179. Geisseler, D.; Scow, K.M. Long-term effects of mineral fertilizers on soil microorganisms: A review. *Soil Biol. Biochem.* **2014**, *75*, 54–63. [CrossRef]

180. Carlisle, S.; Trevors, J. Glyphosate in the environment. *Water Air Soil Pollut.* **1988**, *39*, 409–420.

181. Means, N.E.; Kremer, R.J.; Ramsier, C. Effects of glyphosate and foliar amendments on activity of microorganisms in the soybean rhizosphere. *J. Environ. Sci. Health Part B* **2007**, *42*, 125–132. [CrossRef]

182. Wardle, D.; Parkinson, D. Influence of the herbicide glyphosate on soil microbial community structure. *Plant Soil* **1990**, *122*, 29–37. [CrossRef]

183. Haney, R.; Senseman, S.; Hons, F.; Zuberer, D. Effect of glyphosate on soil microbial activity and biomass. *Weed Sci.* **2009**, *48*, 89–93.

184. Liphadzi, K.B.; Al-Khatib, K.; Bensch, C.N.; Stahlman, P.W.; Dille, J.A.; Todd, T.; Rice, C.W.; Horak, M.J.; Head, G. Soil microbial and nematode communities as affected by glyphosate and tillage practices in a glyphosate-resistant cropping system. *Weed Sci.* **2005**, *53*, 536–545. [CrossRef]

185. Mijangos, I.; Becerril, J.M.; Albizu, I.; Epelde, L.; Garbisu, C. Effects of glyphosate on rhizosphere soil microbial communities under two different plant compositions using cultivation-dependent and -independent methodologies. *Soil Biol. Biochem.* **2009**, *41*, 505–513. [CrossRef]

186. Zablotowicz, R.M.; Reddy, K.N. Impact of glyphosate on the *Bradyrhizobium japonicum* symbiosis with glyphosate-resistant trangenic soybean: A minireview. *J. Environ. Qual.* **2004**, *33*, 825–831. [CrossRef] [PubMed]

187. Johal, G.; Huber, D. Glyphosate effects on diseases of plants. *Eur. J. Agron.* **2009**, *31*, 144–152. [CrossRef]

188. Kremer, R.J.; Means, N.E. Glyphosate and glyphosate-resistant crop interactions with rhizosphere microorganisms. *Eur. J. Agron.* **2009**, *31*, 153–161. [CrossRef]

189. Duke, S.O.; Lydon, J.; Koskinen, W.C.; Moorman, T.B.; Chaney, R.L.; Hammerschmidt, R. Glyphosate effects on mineral nutrition, crop rhizosphere microbiota, and plant disease in glyphosate-resistant crops. *J. Agric. Food Chem.* **2012**, *60*, 10375–10397. [CrossRef] [PubMed]

190. Marshall, E.; Brown, V.; Boatman, N.; Lutman, P.; Squire, G.; Ward, L. The role of weeds in supporting biological diversity within crop fields. *Weed Res.* **2003**, *43*, 77–89. [CrossRef]

191. Greaves, M.P.; Sargent, J.A. Herbicide-induced microbial invasion of plant roots. *Weed Sci.* **1986**, *34*, 50–53.

192. Kuklinsky-Sobral, J.; Welingon, L.A.; Mendes, R.; Pizzirani-Kleiner, A.A.; Azavedo, J.L. Isolation and characterization of endophytic bacteria from soybean (*Glycine max*) grown in soil treated with glyphosate herbicide. *Plant Soil* **2005**, *273*, 91–99. [CrossRef]

193. Ipsilantis, I.; Samourelis, C.; Karpouzas, D.G. The impact of biological pesticides on arbuscular mycorrhizal fungi. *Soil Biol. Biochem.* **2012**, *45*, 147–155. [CrossRef]

194. Griffiths, B.S.; Philippot, L. Insights into the resistance and resilience of the soil microbial community. *FEMS Microbiol. Rev.* **2013**, *37*, 112–129. [PubMed]

195. Morales, S.E.; Holben, W.E. Linking bacterial identities and ecosystem processes: Can "omic" analyses be more than the sum of their parts? *FEMS Microbiol. Ecol.* **2011**, *75*, 2–16. [CrossRef] [PubMed]

196. Acosta-Martinez, V.; Dowd, S.E.; Sun, Y.; Allen, V.G. Tag-encoded pyrosequencing analysis of bacterial diversity in a single soil type as affected by management and land use. *Soil Biol. Biochem.* **2008**, *40*, 2762–2770. [CrossRef]

197. Acosta-Martinez, V.; Dowd, S.E.; Sun, Y.; Wester, D.; Allen, V.G. Pyrosequencing analysis for characterization of soil bacterial populations as affected by an integrated livestock-cotton production system. *Appl. Soil Ecol.* **2010**, *45*, 13–25. [CrossRef]

198. Sugiyama, A.; Vivanco, J.M.; Jayanty, S.S.; Manter, D.K. Pyrosequencing assessment of soil microbial communities in organic and conventional potato farms. *Plant Discuss.* **2010**. [CrossRef]

199. Douds, D., Jr.; Nagahashi, G.; Pfeffer, P.; Reider, C.; Kayser, W. On-farm production of AM fungus inoculum in mixtures of compost and vermiculite. *Bioresour. Technol.* **2006**, *97*, 809–818. [CrossRef] [PubMed]

200. Orwin, K.H.; Wardle, D.A.; Greenfield, L.G. Ecological consequences of carbon substrate identity and diversity in a laboratory study. *Ecology* **2006**, *87*, 580–593. [CrossRef] [PubMed]

201. Loreau, M.; Naeem, S.; Inchausti, P.; Bengtsson, J.; Grime, J.; Hector, A.; Hooper, D.; Huston, M.; Raffaelli, D.; Schmid, B. Biodiversity and ecosystem functioning: Current knowledge and future challenges. *Science* **2001**, *294*, 804–808. [CrossRef] [PubMed]

202. Pennanen, T.; Caul, S.; Daniell, T.; Griffiths, B.; Ritz, K.; Wheatley, R. Community-level responses of metabolically-active soil microorganisms to the quantity and quality of substrate inputs. *Soil Biol. Biochem.* **2004**, *36*, 841–848. [CrossRef]

203. Larkin, R.P.; Honeycutt, C.W. Effects of different 3-year cropping systems on soil microbial communities and *Rhizoctonia* diseases of potato. *Phytopathology* **2006**, *96*, 68–79. [CrossRef] [PubMed]

204. Ochiai, N.; Powelson, M.L.; Crowe, F.J.; Dick, R.P. Green manure effects on soil quality in relation to suppression of *Verticillium* wilt of potatoes. *Biol. Fertil. Soils* **2008**, *44*, 1013–1023. [CrossRef]

205. Postma, J.; Schilder, M.T.; Bloem, J.; van Leeuwen-Haagsma, W.K. Soil suppressivenss and functional diversity of the soil microflora in organic farming systems. *Soil Biol. Biochem.* **2008**, *40*, 2394–2406. [CrossRef]

206. Wittebolle, L.; Marzorati, M.; Clement, L.; Balloi, A.; Daffonchio, D.; Heylen, K.; de Vos, P.; Verstraete, W.; Boon, N. Initial community evenness favours functionality under selective stress. *Nature* **2009**, *458*, 623–626. [CrossRef] [PubMed]

207. Tracy, B.F.; Sanderson, M.A. Forage productivity, species evenness, and weed invasion in pasture communities. *Agric. Ecosyst. Environ.* **2004**, *102*, 175–183. [CrossRef]

208. Crowder, D.W.; Northfield, T.D.; Strand, M.R.; Snyder, W.E. Organic agriculture promotes evenness and natural pest control. *Nature* **2010**, *466*, 109–112. [CrossRef] [PubMed]

209. Bossio, D.A.; Scow, K.M.; Gunapala, N.; Graham, K.J. Determinants of soil microbial communities: Effects of agricultural management, season, and soil type on phospholipid fatty acid profiles. *Microb. Ecol.* **1998**, *36*, 1–12. [CrossRef] [PubMed]

210. Smith, J.L. Soil quality: The role of microorganisms. In *Encyclopedia of Environmental Microbiology*; Bitton, G., Ed.; John Wiley and Sons: New York, NY, USA, 2002; pp. 2944–2957.

211. Kuchenbuch, R.O.; Buczko, U. Re-visiting potassium- and phosphate-fertilizer responses in field experiments and soil-test interpretations by means of data mining. *J. Plant Nutr. Soil Sci.* **2011**, *174*, 171–185. [CrossRef]

212. Frossard, E.; Condron, L.M.; Oberson, A.; Sinaj, S.; Fardeau, J.C. Processes governing phosphorus availability in temperate soils. *J. Environ. Qual.* **2000**, *29*, 15–23. [CrossRef]

213. Condron, L.M.; Newman, S. Revisiting the fundamentals of phosphorus fractionation of sediments and soils. *J. Soils Sediments* **2011**, *11*, 830–840. [CrossRef]

214. Hinsinger, P.; Brauman, A.; Devau, N.; Gerard, F.; Jourdan, C.; Laclau, J.-P.; le Cadre, E.; Jaillard, B.; Plassard, C. Acquisition of phosphorus and other poorly mobile nutrients by roots. Where do plant nutrition models fail? *Plant Soil* **2011**, *348*, 29–61. [CrossRef]

215. American Academy of Microbiology (AAM). *Incorporating Microbial Processes into Climate Change Models*; AAM: Washington, DC, USA, 2011.

216. Smith, J.L.; Collins, H.P. Managing soil microorganisms and their processes. In *Soil Microbiology, Ecology and Biochemistry*; Paul, E.A., Ed.; Academic Press: Burlington, MA, USA, 2007; pp. 471–500.

217. Smith, R.G.; Gross, K.L.; Robertson, G.P. Effects of crop diversity on agroecosystem function: Crop yield response. *Ecosystems* **2008**, *11*, 355–366. [CrossRef]

218. Cook, R.J. Toward cropping systems that enhance productivity and sustainability. *Proc. Natl. Acad. Sci. USA* **2006**, *103*, 18389–19384. [CrossRef] [PubMed]

219. American Academy of Microbiology (AAM). *How Microbes Can Feed the World*; AAM: Washington, DC, USA, 2013; p. 33.

220. Gupta, V.V.S.R.; Rovira, A.D.; Roger, D.K. Principles and management of soil biological factors for sustainable rainfed farming systems. In *Rainfed Farming Systems*; Tow, P., Cooper, I., Partridge, I., Birch, C., Eds.; Springer: Dordrecht, The Netherland, 2011; pp. 149–184.

221. Liebig, M.; Carpenter-Boggs, L.; Johnson, J.; Wright, S.; Barbour, N. Cropping system effects on soil biological characteristics in the Great Plains. *Renew. Agric. Food Syst.* **2006**, *21*, 36–48. [CrossRef]

222. Kowalchuk, G.A.; Bruinsma, M.; van Veen, J.A. Assessing responses of soil microorganisms to GM plants. *Trends Ecol. Evol.* **2003**, *18*, 403–410. [CrossRef]

sustainability

MDPI

Article

History of East European Chernozem Soil Degradation; Protection and Restoration by Tree Windbreaks in the Russian Steppe

Yury G. Chendev [1], **Thomas J. Sauer** [2,*], **Guillermo Hernandez Ramirez** [3] **and Charles Lee Burras** [4]

[1] Department of Natural Resources Management and Land Cadastre, Belgorod State University, Belgorod 308015, Russia; chendev@bsu.edu.ru
[2] USDA-ARS, National Laboratory for Agriculture and the Environment, Ames, IA 50011, USA
[3] Department of Renewable Resources, University of Alberta, Edmonton, AB T6G 2R3, Canada; ghernand@ualberta.ca
[4] Department of Agronomy, Iowa State University, Ames, IA 50011, USA; lburras@iastate.edu
[*] Correspondence: tom.sauer@ars.usda.gov; Tel.: +1-515-294-3416; Fax: +1-515-294-8125

Academic Editor: Marc A. Rosen
Received: 30 October 2014; Accepted: 29 December 2014; Published: 8 January 2015

Abstract: The physiographic region of the Central Russian Upland, situated in the Central part of Eastern Europe, is characterized by very fertile grassland soils—Chernozems (Mollisols in the USDA taxonomy). However, over the last several centuries this region has experienced intense land-use conversion. The most widespread and significant land-use change is the extensive cultivation of these soils. As a result, Chernozems of the region that were some of the most naturally fertile soils in the world with thick A horizons had become, by the second half of the 19th century, weakly productive, with decreased stocks of organic matter. When not protected by plant cover, water and wind erosion degraded the open fields. The investigation of methods for rehabilitation and restoration of Chernozems resulted in the practice of afforestation of agricultural lands (mainly by windbreak planting). Preferences of agroforestry practices were initially connected with protection of cropland from wind and water erosion, improvement of microclimate for crop growth, and providing new refugia for wild animal and plant habitats. During the last several decades, tree windbreaks have begun to be viewed as ecosystems with great potential for atmospheric carbon sequestration, which plays a positive role in climate change mitigation. For the evaluation of windbreak influence on Chernozem soils, a study was developed with three field study areas across a climatic gradient from cool and wet in the north of the region to warm and dry in the south. Windbreak age ranged from 55–57 years. At each site, soil pits were prepared within the windbreak, the adjacent crop fields of 150 years of cultivation, and nearby undisturbed grassland. Profile descriptions were completed to a depth of 1.5 m. A linear relationship was detected between the difference in organic-rich surface layer (A + AB horizon) thickness of soils beneath windbreaks and undisturbed grasslands and a climate index, the hydrothermal coefficient (HTC). These results indicate that windbreaks under relatively cooler and wetter climate conditions are more favorable for organic matter accumulation in the surface soil. For the 0–100 cm layer of the Chernozems beneath windbreaks, an increase in organic C stocks comparable with undisturbed grassland soils (15–63 Mg·ha^{-1}) was detected. Significant growth of soil organic matter stocks was identified not only for the upper 30 cm, but also for the deeper layer (30–100 cm) of afforested Chernozems. These findings illustrate that, in the central part of Eastern Europe, tree windbreaks improve soil quality by enhancing soil organic matter while providing a sink for atmospheric carbon in tree biomass and soil organic matter.

Keywords: Russian Chernozems; soil organic carbon; degradation of soils; restoration of soils; afforestation

1. Introduction

One of the main natural resources of Russia are its Chernozem soils or Chernozems. The history of soil science evolved from studies of these soils. The founder of soil science, Vasily Dokuchaev called Chernozems the "Tsar of soils" [1]. As for any other science, the foundation of soil science was connected with "demands of practice". In 1876, the Free Economic Society of Russia, with support of the Russian government, developed a project to study the causes that led to agricultural depletion of Chernozems. The research program included detailed study of Chernozems, their properties, formation, and patterns of spatial distribution within Eastern Europe. A young Quaternary Geologist, V.V. Dokuchaev, was selected as research leader of this project. During the several years following the execution of the project, he had proposed and argued the main foundations of pedology or soil science using the example of Chernozem soils [2]:

- The soil is an important separate component of the environment with its own history of development, connected by properties and processes with other components of the environment;
- The soil is a product of influence of soil forming factors: parent material, relief, climate, plants and animals, and age of the territory;
- The soil consists of a few genetically interrelated layers or horizons with individual properties, reflecting the history of soil formation;
- The soil is a mirror of the environment, the focus of all natural processes which form landscapes and biosphere in the whole.

Further (after 1883) development of soil science in Russia was associated with expeditions by Dokuchaev and his students with different goals: soil cartography and descriptions of soils in a series of administrative Russian units (provinces), study of soil zones (elevation zones in mountain systems), and the founding of special scientific stations for long term instrumental observations of soil properties and soil regimes in different climatic zones [3]. However, one of the most important activities of Dokuchaev at the end of the 19th century was organization of the so-called "Special Expedition" to Kamennaya Steppe of the Voronezh province to study the effects of a drought that affected the southern provinces of European Russia in 1891–1892. Detailed studies of all components of the environment in the Kamennaya Steppe led Dokuchaev to the scientific evaluation of the idea for the necessity of agroforestry within steppe landscapes, crop fields and grasslands—to improve the soil resistance to adverse climatic changes and to improve their productivity [4]. Since this time, Russia began to systematically introduce agroforestry as an effective way to protect soils from erosion and improve soil and climate conditions for increasing crops yields.

Until the end of the 1940s, many agroforestry activities within steppe agricultural lands were concentrated in experimental scientific stations such as Kamennaya Steppe. Only in the 1950–1960s did afforestation become public agricultural policy in the Soviet Union [5]. The "Plan for the Transformation of Nature"—a comprehensive program of scientific management of nature in the Soviet Union was implemented in the late 1940s. The plan was adopted on the initiative of Stalin and put into effect the decision of the Council of Ministers of the USSR on 20 October 1948 as "On the plan of shelterbelts, the introduction of grass crop rotations, construction of ponds and reservoirs to ensure high crop yields in the steppe and forest-steppe regions of the European part of the USSR". In print, the document is called "Stalin's Plan for the Transformation of Nature". The plan was unprecedented in the global experience in terms of scale. The program included a plan to plant windbreaks to block the hot, dry winds and together with ponds and reservoirs to increase water storage and ameliorate climate change in an area of more than 120 million hectares, which is equal to the total territory of England, France, Italy, Belgium and the Netherlands (Figure 1). The centerpiece of the plan was field-protective afforestation and irrigation. In the course of the project, afforestation of agricultural lands covered four major watershed basins of the rivers Dnieper, Don, Volga, and Ural in the European South of Russia. The first designed windbreak was to stretch from the Ural Mountains to the shores of the Caspian Sea, with a length of more than a thousand kilometers. The most intensive planting of windbreaks took place in 1949–1953 when 2.1 million hectares of windbreaks were planted. In the 1950–1960s, afforestation of the Southern part of Eastern Europe continued, but not as intensively as from 1949–1953. The main result of this State plan for agricultural lands afforestation was an ecological situation of stabilization as well as in increase of crop yields [5].

Given this background, what were the reasons for degradation of Chernozem soils that were the impetus for the emergence of soil science, and then the introduction of windbreak planting? The answer to this question can be found in the history of natural and soil resource management. As a model region, we can observe one subject of the Russian Federation, Belgorod Oblast, located on the southern slopes of the Central Russian Upland. Of the 27,134 km^2 area of Belgorod Oblast, 77% has Chernozem soils [6]. Economic development of the Belgorod Oblast (as in many regions of the South of European Russia) began in the end of 16th–early 17th centuries. Notable anthropogenic changes of the steppe vegetation began in the 16th century with the beginning of the construction of Tatar roads [7], although some lands remained unsettled for a long time. Substantial transformations of steppe vegetation during this time were caused by destruction of natural steppe vegetation on Tatar roads, and fires on the steppes.

Figure 1. Poster showing the plan for the steppe zone transformation in the European part of the USSR through afforestation according to plan of 1948 (photo from the Museum of Kamennaya Steppe).

Transformation of forests in relation to the steppe transformation occurred quickly. First, forests were affected by construction of settlements and fortifications. At the end of 17th to the first half of the 18th century, forests were reduced significantly to build the Azov-Black Sea fleet [8]. In the 18th century, the average annual consumption of wood had increased by at least five times as compared with the 17th century. Figure 2 provides a visual representation of the spatial and temporal variation of the natural vegetation changes within Belgorod Oblast over the past 400 years. Areas of forest and virgin steppe declined, being replaced by settlements, arable lands, and pastures. Calculations on maps show that in the 1780s virgin steppes occupied about 22% of the total area, and forest only 16%. By the middle of 19th century, forests covered 13% of the area and the area of untilled steppes dramatically decreased to just a few percent. In the 1880s, the last virgin steppe areas were plowed except for accidentally surviving islets of steppes in the eastern and northern parts of Belgorod Oblast, which now have the status of natural reserves. The last large-scale deforestation ended in 1917, after which the area of forests varied only slightly, accounting for about 8% of the area. Currently, forest harvests continue to be carried out with selective felling of trees, so the age of trees is young (the vast majority of trees less than 80 years). Artificial planting of Scots pine (*Pinus sylvestris*) on sandy river terraces of the study area [9] can be considered a positive anthropogenic effect of man's influence to restore lost elements of flora.

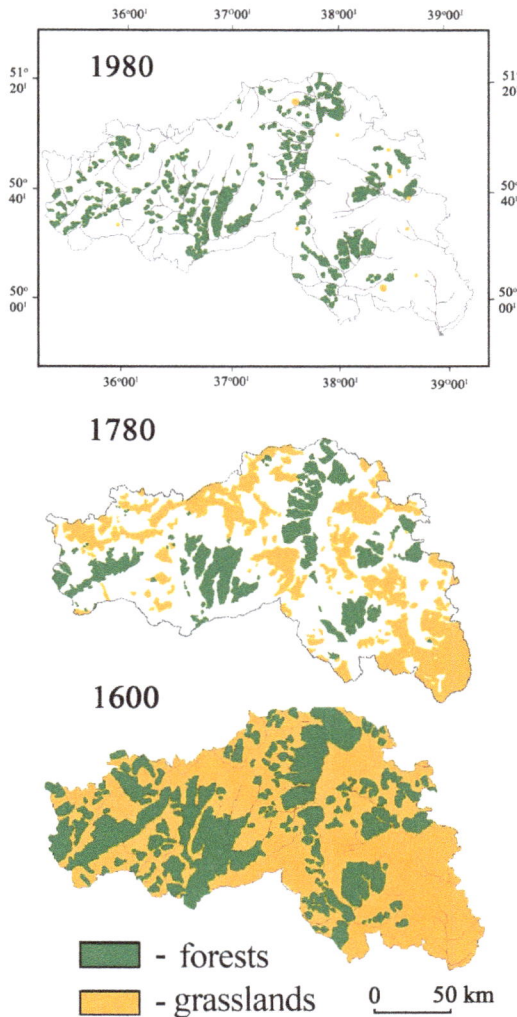

Figure 2. Distribution of forests and grasslands in the Belgorod Oblast in different historical periods (compiled from archival sources and literary and cartographic materials).

It is well known that forests have an important impact on the environment—they play a role in regulating surface runoff as well as form more wet/humid climatic conditions. Therefore, deforestation of Belgorod Oblast impacted the climate (which become drier), which may have also contributed to the further deterioration of soil quality and reduction of crop yields.

Aside from deforestation, significant degradation of Chernozem soils has occurred due to long-term cultivation without or with sporadic replenishment of nutrients with fertilizers. According to Chendev and Petin [10], the mean duration of Chernozem soil cultivation in the Belgorod Oblast is approximately 240 years. Thus, at the beginning of the research by Dokuchaev for 1877–1879 [1], the soils had been already under agricultural management no less than 100 years. The lack of organic fertilizers (manure) in a large area of arable soils at the end of the 19th century (Figure 3) certainly influenced the appearance of an ecological crisis due to Chernozem soil nutrient depletion.

A   B

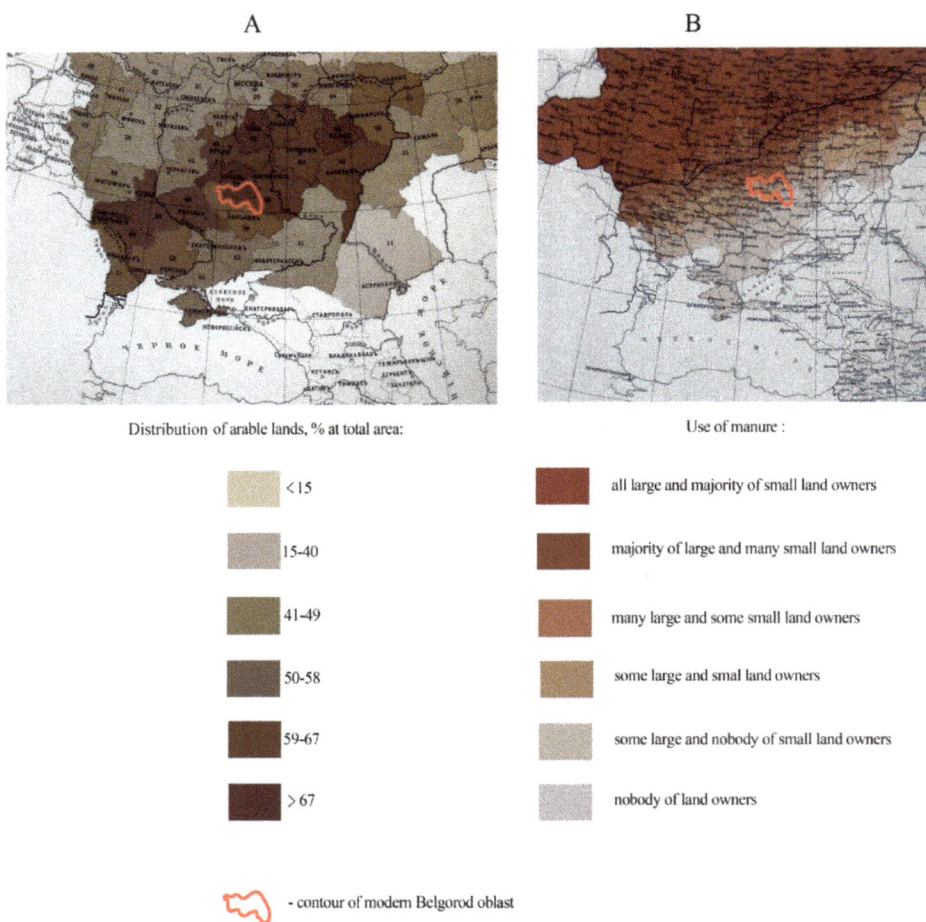

Distribution of arable lands, % at total area: Use of manure :

▢ <15	▢ all large and majority of small land owners
▢ 15-40	▢ majority of large and many small land owners
▢ 41-49	▢ many large and some small land owners
▢ 50-58	▢ some large and smal land owners
▢ 59-67	▢ some large and nobody of small land owners
▢ >67	▢ nobody of land owners

- contour of modern Belgorod oblast

Figure 3. Distribution of arable lands (**A**) and use of manure (**B**) in the Southern part of European Russia at the end of the 19th century (1887) [11].

The unfavorable state of agricultural land fertility was exacerbated by the natural relief with a large proportion of relatively steep slopes and by the practice of plowing along, not across, slopes.

From year to year, plow furrows along slopes led to gully network formation. Realistic documentation of the 19th century cultivation practices were captured by artist's paintings demonstrating these phenomena (Figures 4 and 5).

Figure 4. Painting of I. Levitan "The evening on arable land" (second half of the 19th century) (in the painting plowing along the slope is visible).

Figure 5. Painting of N. Sergeyev "Summer Landscape" (second half of the 19th century). In the enlarged fragment, plow furrows are visible along the slope, with the formation of a new, expanding gully downslope.

The sheet and rill water erosion of arable lands along with the practice of plowing along slopes led to the increase in the number of fields intersected by gullies (Figure 6).

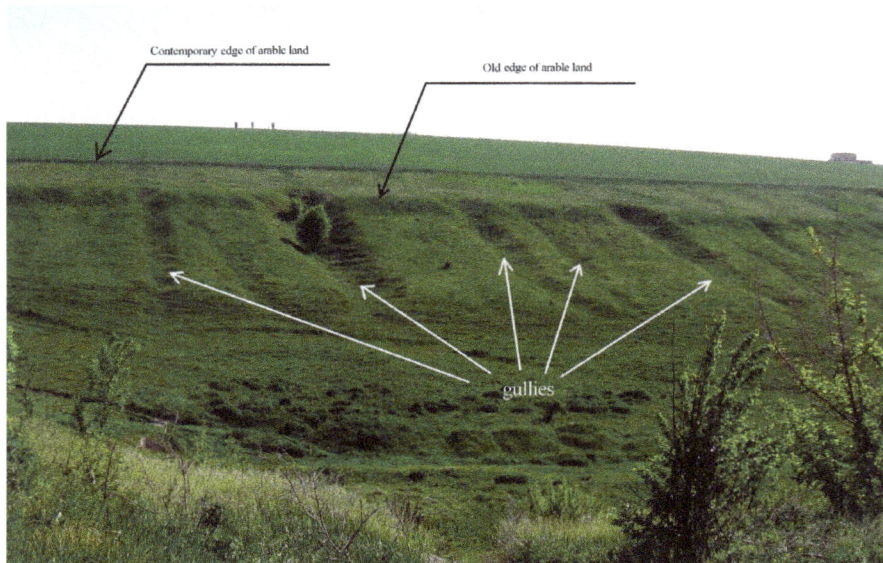

Figure 6. Slope of natural ravine with secondary forms of relief—earthen plowed rampart along old (beginning of 20th century) edge of arable land, and gullies associated with this edge. (Photo in Korocha district of Belgorod Oblast).

It is estimated that, for the history of agricultural management in the Belgorod Oblast, the average rate of gully network growth was equal to 56.5 km·year^{-1} and the annual loss of land by new gully formation was approximately 1.1 km^2. The total volume of soil and rock eroded by gully formation over the history of land management in the study region was estimated to be 2–2.5 km^3, which is equal to a 7-cm layer of soil with bulk density 1.15 g·cm^{-3} distributed over the entire area (27,134 km^2) of Belgorod Oblast [10].

So, we see that many factors contributed to the depletion of Chernozem soils in the 19th–early 20th centuries. Separate consideration is given to historical changes in tillage technologies and their impact on soil fertility. According to an analysis of land use history in the study territory, soils before the 1930s were tilled with wooden and iron plows to a depth of 10–20 cm. Mechanical plowing with tractors led to a gradually increasing depth of the plow layer. At present, primary tillage penetrates to a depth of 30–37 cm. The wide using of mineral and organic fertilizers began only after World War II. However, volumes of fertilization were relatively low. The annual input of farmyard manure has never been higher than 4 Mg·ha^{-1}—at least until the end of the 20th century. Another factor is that, since the 1960s, more sugar beet and other industrial crops have been cultivated in addition to the cereal crops. These crops require higher use of farm machinery on fields. This led to further deterioration of arable Chernozems' physical properties through compaction and destruction of soil structure [12,13].

Therefore, we conclude that cultivation practices for almost all of the history of agriculture did not favorably affect soil quality and soil fertility. An exception to this pattern has occurred only during the last few years in the Belgorod Oblast. Agricultural development with new land owners and agricultural holdings has begun to use advanced technologies of cultivation such as minimum and zero tillage. They use scientifically recommended rates of soil mineral and organic fertilizers. To protect soils on sloping lands from soil erosion, some cultivated fields have been sown to forage grasses. Soil conservation farming systems are being increasingly introduced in the practice of modern agriculture.

Tillage practices and other agricultural technologies during the historical changes of Chernozems in the Belgorod Oblast—like other regions of the Southern part of European Russia—which did not serve to improve soil quality. Many current practices add to this long (several centuries) agricultural management of these soils with an accumulation of negative effects on the current properties of these soils.

As has been shown above, V.V. Dokuchaev [2] founded a new science about soils, which today is known as soil science or pedology, based on the study of Russian Chernozems. The initial reason for their study by Dokuchaev was the Russian government's concern over the decrease in fertility of these soils. The initial focus of soil science after 1883 was to find ways to decrease the rate of Chernozem degradation as well as improve their condition and fertility. Again, Dokuchaev was the first scientist who drew attention to afforestation of agricultural lands within Chernozem area as the most effective measure for their protection and restoration [4]. Further agricultural land management practices in the Soviet Union (since the beginning of 1950s) through afforestation brought positive results until the end of the 20th century. However, afforestation of agricultural lands during the last two decades (since 1991) declined due to changes in the political and economic status of Russia. Many older windbreaks have been removed [5]. In spite of the introduction of new crop production technologies and tillage practices in recent years in Russia, significant rehabilitation of windbreaks has not occurred. There is a need to re-focus attention on soil rehabilitation, and special attention should be given to agroforestry practices as one of the most effective ways to improve, protect and rehabilitate Russian Chernozems.

Afforestation of steppe agricultural lands is a well-known practice and it has a long history of practice in cropland management not only in Russia, but within other regions of Northern Hemisphere [5,14,15]. In the 19th and 20th centuries, preferences for this practice were connected with protection of croplands from wind and water erosion, improvement of microclimate for crop growth, and creation of new refugia for wild animal and plant habitat. At the end of the 20th to the beginning of the 21st centuries, windbreaks were often viewed from the ecosystem perspective for atmospheric carbon sequestration, having positive influence on the climatic balance of CO_2 [16–19]. Therefore, the current objectives of this study are the collection and analysis of new data, reflecting opportunities for windbreaks to sequester soil organic carbon and rehabilitate degraded soils. One such perspective region, naturally presented by Chernozems, is the Central Russian Upland, situated within the central part of Eastern Europe.

2. Materials and Methods

This study focuses on Chernozems of the Central Russian Upland under different types of vegetation—virgin grasslands, crops, and windbreaks. For selection of the study key areas, the authors were guided by the following requirements: key areas must be situated in the forest-steppe zone (a transitional zone between forest and steppe); topographically they must be located on flat summits; and all key areas must be in close proximity to each other. Sub-areas under virgin vegetation, arable land, and a windbreak planted on formerly cultivated land were identified within the same kind of soil having homogeneous parent material. The location of the study key areas is shown in Figure 7. Historical stages of agricultural and agroforestry management of the study soils is illustrated in Figure 8.

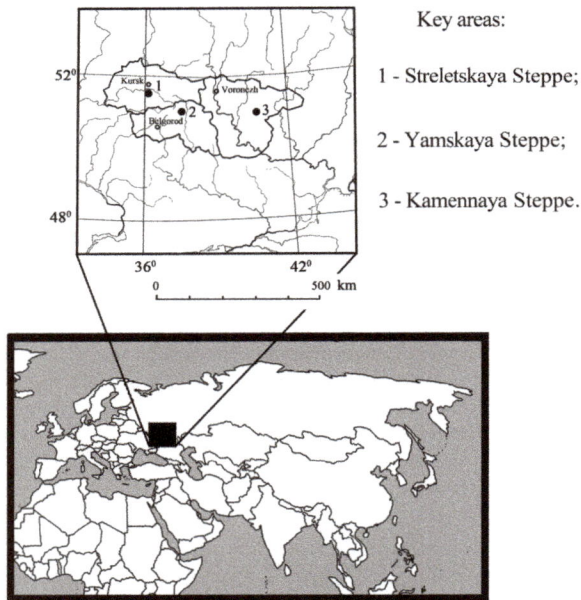

Figure 7. Locations of the study areas.

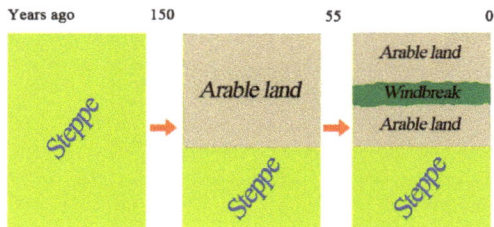

Figure 8. Historical changes of land management within the study areas in the Central Russian Upland.

Prior to undertaking field studies it was necessary to carefully select the key areas. This work included analysis of maps of different ages, selection and study of remote sensing data, and consultations with experts in the fields of geography, geobotany, and agroforestry. Reconnaissance visits to prospective areas of planned field research were also completed. The Central Russian Upland is an agricultural region (arable lands currently occupy about 60% of the total area); thus, it was challenging to find native grasslands in many places within the study area. Therefore, as a base, we selected sites located in the immediate vicinity of three State steppe preserves: near the Central Chernozem Preserve (key area Streletskaya Steppe, Kursk Oblast), in vicinity and on the territory of preserve "Belogorye" (key area Yamskaya Steppe, Belgorod Oblast), and in the preserve "Kamennaya Steppe" (key area Kamennaya Steppe, Voronezh Oblast). Information about elevation, parent materials, and classification status of the studied soils and some climatic indicators of the key areas are given in Table 1. An overview of the virgin Chernozem profiles, in particular thickness of their humus horizons, are shown in Figure 9.

Table 1. Some natural indicators of the key areas studied in the Central Russian Upland.

Topographic Structure	Name of Key Area	Elevation above Sea Level, m	Parent Materials	Soils	Annual Precipitation, mm	Annual Mean Temperature, °C	HTC
Central Russian Upland	Streletskaya Steppe	240	1	Leached Chernozems	580	+5.3	1.23
	Yamskaya Steppe	230	1	Typical Chernozems	530	+5.6	1.1
	Kamennaya Steppe	190	2	Ordinary Chernozems, transitional to typical	480	+5.8	1.0

Parent materials: 1—loess carbonated loams; 2—loess carbonated clays. HTC—Hydrothermal Coefficient, $10R/\Sigma t$, where R—precipitation in millimeters for the period with temperatures above 10 °C, Σt—the sum of temperatures in degrees Celsius for the same period [20].

Croplands of the study areas are either located in research sites (in Yamskaya Steppe and Kamennaya Steppe) or they are in farm use (Streletskaya Steppe). Age of cultivation after transfer of virgin grassland to arable land at all sites is approximately the same and estimated at 150 years. All key areas are using crop rotations with predominance of cereals, but in more cool and wet climatic conditions, (in areas Streletskaya Steppe and Yamskaya Steppe) crop rotations have increased proportions of fodder and sugar beet, and in the more drier climatic conditions (Kamennaya Steppe), increased proportions of cereals and sunflower.

Figure 9. The profiles of virgin Chernozems of the study areas within the Central Russian Upland: 1—Streletskaya Steppe; 2—Yamskaya Steppe; 3—Kamennaya Steppe.

In all key areas, the windbreaks have multi-rows and a full/dense design. Their width varies from 20 (Streletskaya Steppe) to 35 m (Yamskaya Steppe) with intermediate width at Kamennaya Steppe of 25 m. Shelterbelts were created in the middle to second half of the 1950s, and are 55–57 years old. The dominant trees are at Streletskaya Steppe—Black poplar (*Populus nigra*) and Silver birch (*Betula pendula*); at Yamskaya Steppe—Box elder (*Acer negundo*); and at Kamennaya Steppe—English oak (*Quercus robur*) and Balsam poplar (*Populus balsamifera*).

In spite of the desire to find homogeneous soil cover within each key area, we have not completely escaped its natural heterogeneity. Therefore, the study sub-areas in every key area are characterized by certain differences of soil, and therefore could not be considered as a single statistical sample. However, detection of general trends in soil properties (in particular, the thickness of humus profiles, content and stocks of organic carbon) under the windbreaks compared to arable lands on different sites suggests a certain direction of windbreak effects on soil organic matter.

Studies in all key areas were carried out using consistent methodology. Soil samples were collected along three equally spaced (5 m) transects across the windbreak including arable lands on both sides at spacings of 4–5 m. Soil was sampled to 30 cm at every point in triplicate followed by compositing into one sample. The total number of points along each transect was equal to 18, six of which were under the windbreak, and four from both sides of the neighboring arable lands. With three transects within every key area, the total numbers of samples were: 18—under windbreak, and on 12 in every sub-area of arable land. Based on the variation of windbreak width (20–35 m), total area of sampling under tree plantations varied from 200–350 m^2 (band with 10-m width), and in arable lands on both sides of plantations—120–150 m^2 (also in form of a band with 10-m width).

This technique of soil sampling with subsequent statistical analysis of results is widely used in the study of soils in agroforestry areas in the United States [18].

Additionally, in each sub-area, a large soil pit was prepared by hand to a depth of 1.5 m. All soil horizon boundaries and profile descriptions were prepared from observations of three exposed soil faces. Horizon boundaries were measured at five locations on each exposed face. Sampling of soils in every soil pit was executed to a depth of 1 m in two opposite soil faces with following meaning of laboratory analysis results. Soil pits were located in central parts of the windbreaks, and in the fields on both sides of the windbreaks adjacent to of the soil sampling transects at the greatest distance from the edge of windbreak. Distance between soil pits on arable lands and under windbreaks varied depending of windbreak width from 25–35 m. Soils of virgin sub-areas were studied in large pits and with a series of auxiliary deep soil cores.

Thus, within every key area 48 soil samples to depth 0–30 cm and double selections from four soil pits in every 10 cm to depth 1 m were collected for a total of 80 soil samples. After drying, preparation of soil samples for laboratory analyses was executed according to standard methods. Soil samples were analyzed to determine soil organic carbon (SOC) content and bulk density (for the following recalculation of SOC content into soil carbon stocks). Sampling for bulk density in pits at selected points in the profile were collected with steel rings.

The SOC content was determined by dry combustion analysis NA 15000 Fison (ThermoQuest Corp., Austin, TX, USA) at the National Laboratory for Agriculture and the Environment of USDA (Ames, IA, USA). Visible roots were removed and a subsample passed through a 2 mm sieve, air dried, and roller-milled before SOC content analysis. Results are presented for the means of pit and point samples.

In statistical analysis of data in Tables 2 and 3, a 95% confidence level ($p = 0.05$) was used.

3. Results

An important morphogenetic indicator characterizing the direction of change with time of the studied soils transition from "grassland—arable land—windbreak", are the thickness of their humus horizons (A) and humus profiles (A + AB horizons) (Table 2). For the three study key areas, humus horizon thickness of virgin Chernozems varied from 44.7–52.9 cm, and their humus profiles—from 70.3–75.5 cm. After 55–57 years of soil formation of Chernozems under windbreaks, the thickness of their humus horizon relevant to background (virgin) analogues (at Streletskaya Steppe), were significantly higher than background (at Yamskaya Steppe—by 7 cm, and at Kamennaya Steppe—by 5 cm). In Chernozems under windbreaks, humus profile thickness is significantly higher than in virgin grasslands (in Streletskaya Steppe—by 11 cm, in Yamskaya Steppe—by 5 cm). This index is significantly lower in the area Kamennaya Steppe—9 cm, by comparison with the windbreak's Chernozem. At each location, there were significantly greater thicknesses of the A + AB horizons in soils beneath tree plantings compared to the adjacent cultivated soils. The difference in A + AB thickness between tree and crop soils was 18.3, 10.9, and 5.6 cm for the Streletskaya Steppe, Yamskaya Steppe, and Kamennaya Steppe locations, respectively (Table 2). It is likely that these differences in thickness of the A + AB horizons are due to both continued SOC loss from cropping practices, especially tillage, and SOC accumulation beneath the trees where there is both greater biomass input and limited soil disturbance.

Table 2. Statistics of humus horizons and humus profiles thickness of soils within three study areas in the Central Russian Upland.

Sub-Area	Horizon/s	n	Min–Max, cm	$\bar{X} \pm \delta_{\bar{X}}$, cm	δ, cm	V, %
Streletskaya Steppe, Kursk Oblast						
Grassland	A	15	40–54	44.7 ± 0.9	3.30	7.4
	A + AB	15	60–77	70.3 ± 1.0	3.87	5.5
Windbreak	A	15	35–50	44.6 ± 1.3	5.19	11.6
	A + AB	15	75–85	81.4 ± 1.0	3.80	4.7
Cultivated land	Ap + A	30	33–48	37.3 ± 0.6	3.48	9.3
	Ap + A + AB	30	47–80	63.1 ± 2.0	10.83	17.2
Yamskaya Steppe, Belgorod Oblast						
Grassland	A	15	47–58	52.9 ± 0.9	3.45	6.5
	A + AB	15	69–80	75.5 ± 0.9	3.36	4.5
Windbreak	A	15	46–64	60.1 ± 1.9	7.31	12.2
	A + AB	15	71–92	80.1 ± 1.6	6.29	7.9
Cultivated land	Ap + A	30	39–56	45.3 ± 0.9	4.65	10.3
	Ap + A + AB	30	54–82	69.2 ± 1.3	6.87	9.9
Kamennaya Steppe, Voronezh Oblast						
Grassland	A	15	41–55	44.7 ± 0.9	3.56	8.0
	A + AB	15	69–88	75.5 ± 1.3	4.91	6.5
Windbreak	A	15	46–55	49.6 ± 0.8	3.09	6.2
	A + AB	15	57–70	64.1 ± 1.0	3.80	5.9
Cultivated land	Ap + A	30	37–47	41.0 ± 0.4	2.37	5.8
	Ap + A + AB	30	49–64	58.5 ± 0.6	3.52	6.0

A plot of the difference of A + AB horizon thickness of tree and grass soils *vs.* HTC for the three studied sites exhibits a linear dependence (Figure 10).

In accordance with the observed trend (Figure 10), the humus profile thickness in soils under artificial forest plantations increased by 10 cm for every 0.1 unit increase in HTC. For the area of the Central Russian Upland, this corresponds to an advance in a northwesterly direction of a distance of 30–70 (average 50) km.

Statistical analysis of SOC stock distribution in the 0–30 cm of study soils (Table 3) shows an increase of this index in native grasslands from the cool-wet part of Central Russian Upland to its warm-dry locations. In leached Chernozems of Streletskaya Steppe they are 126.2 ± 2.3 Mg·ha^{-1}, in typical Chernozems of Yamskaya Steppe—138.0 ± 3.9 Mg·ha^{-1}, and in ordinary Chernozems of Kamennaya Steppe—152.5 ± 4.7 Mg·ha^{-1}. Spatial variation of SOC stocks in virgin soils in this direction also increase with coefficients of variation of 4.4, 6.8 and 7.5, respectively (Table 3). In the areas Streletskaya Steppe and Yamskaya Steppe, no significant differences between virgin Chernozems and their windbreaks analogues on SOC stocks were determined. In these sites, arable soils were significantly lower on this index than soils of grasslands or windbreaks. At Kamennaya Steppe, SOC stocks in soils under the windbreak were about the same as in arable lands—125.0 ± 3.3 Mg·ha^{-1} (on arable lands—123.6 ± 2.2 Mg·ha^{-1}) and significantly lower than in virgin Chernozems—152.5 ± 4.7 Mg·ha^{-1} (Table 3).

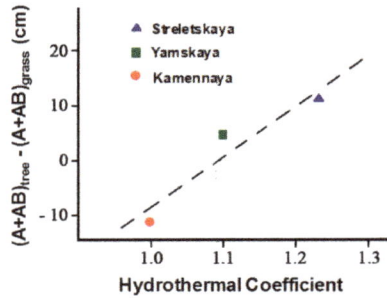

Figure 10. Difference of humus profile thickness in soils under tree plantings and uncultivated grassland soils *vs.* hydrothermal coefficient (HTC) within the Central Russian Upland.

Thus, we conclude that under windbreaks in more cooler and humid climatic conditions of the northern and central forest-steppe of the Central Russian Upland (Streletskaya Steppe and Yamskaya Steppe), an increase of humus reserve by decay of ground litter and roots under windbreaks is obvious in the 0–30 cm layer of Chernozems, while in more dry and warm south-east conditions of the study region, this process is manifested more weakly.

Characteristics of individual soil profiles on SOC stocks to a depth of 100 cm, which have been studied in three sub-areas within every key site, are presented in Table 4. As in the 0–30 cm layer (Table 3), in the layers of 0–50, 50–100 and 0–100 cm, SOC stocks of virgin Chernozems regularly decrease from cool-wet to warm-dry forest-steppe of the Central Russian Upland (Table 4). The 0–50 cm layer of virgin Chernozems contains 66%–72% of the SOC in the surface meter of the soils. In Chernozems under windbreaks, there has been some vertical redistribution of SOC. In the 0–50 cm layer, SOC stocks have decreased to 59%–63% of its total stocks in the meter layer of soils due to replenishment of the organic carbon pool in the 50–100 cm layer. In the adjacent croplands, SOC stocks also decreased in the 0–50 cm layer to 60%–67% of the total stocks in the 0–100 cm layer (Table 4). However, the cause of this decline is likely associated with more intensive decomposition of organic matter in the upper soil layers as a result of continued cultivation.

Table 3. Statistic indexes of spatial distribution of SOC stocks ($Mg \cdot ha^{-1}$) in study soils, layer 0–30 cm (Central Russian Upland).

Sub-Area	n	Min–Max	$\bar{X} \pm \delta_X$	δ	V, %
		Streletskaya Steppe			
Grassland	6	119.9–135.4	126.2 ± 2.3	5.55	4.4
Windbreak	18	109.9–241.1	126.4 ± 7.0	29.74	23.5
Cultivated land	24	91.5–126.7	109.3 ± 2.1	10.24	9.4
		Windbreak—Grassland = +0.2; LSD_{05} = 15.2			
		Windbreak—Cultivated land = +17.1; LSD_{05} = 14.7			
		Yamskaya Steppe			
Grassland	6	131.2–155.9	138.0 ± 3.9	9.44	6.8
Windbreak	18	119.3–163.2	142.1 ± 3.0	12.81	9.0
Cultivated land	24	111.2–156.8	127.2 ± 2.4	11.80	9.3
		Windbreak—Grassland = +4.1; LSD_{05} = 10.1			
		Windbreak—Cultivated land = +14.9; LSD_{05} = 7.7			
		Kamennaya Steppe			
Grassland	6	135.9–170.0	152.5 ± 4.7	11.40	7.5
Windbreak	18	104.4–156.5	125.0 ± 3.3	14.10	11.3
Cultivated land	24	100.0–140.7	123.6 ± 2.2	10.98	8.9
		Windbreak—Grassland = −27.5; LSD_{05} = 11.8			
		Windbreak—Cultivated land = +1.4; LSD_{05} = 8.1			

Note: LSD_{05}—Least significant difference at 95% confidence level.

Table 4. SOC stocks in study soil profiles under virgin grasslands, windbreaks, and adjacent to the windbreaks crop fields within the Central Russian Upland.

Layer, cm	Key Site					
	Streletskaya Steppe		Yamskaya Steppe		Kamennaya Steppe	
	SOC Stocks, $Mg \cdot ha^{-1}$	% of Stocks in Layer 0–100 cm	SOC Stocks, $Mg \cdot ha^{-1}$	% of Stocks in Layer 0–100 cm	SOC Stocks, $Mg \cdot ha^{-1}$	% of Stocks in Layer 0–100 cm
			Grasslands			
0–50	175.0	72	208.2	66	227.9	67
50–100	68.5	28	105.7	34	113.6	33
0–100	243.5	100	313.9	100	341.5	100
			Windbreaks			
0–50	178.6	63	237.3	63	208.6	59
50–100	104.6	37	139.7	37	147.3	41
0–100	283.2	100	377.0	100	355.9	100
			Cultivated lands near windbreaks (mean on 2 profiles)			
0–50	188.5	67	186.6	60	189.9	67
50–100	93.8	33	126.2	40	95.3	33
0–100	282.4	100	312.8	100	285.2	100

The vertical redistribution of SOC in soils beneath windbreaks cannot be explained by these soils having such profiles from their arable period before afforestation since the growth of relative stocks of SOC in the 50–100 cm layer beneath the tree plantations has been identified together with an excess of their absolute values in comparison with the same layers of the virgin Chernozems (Table 4, Figure 11).

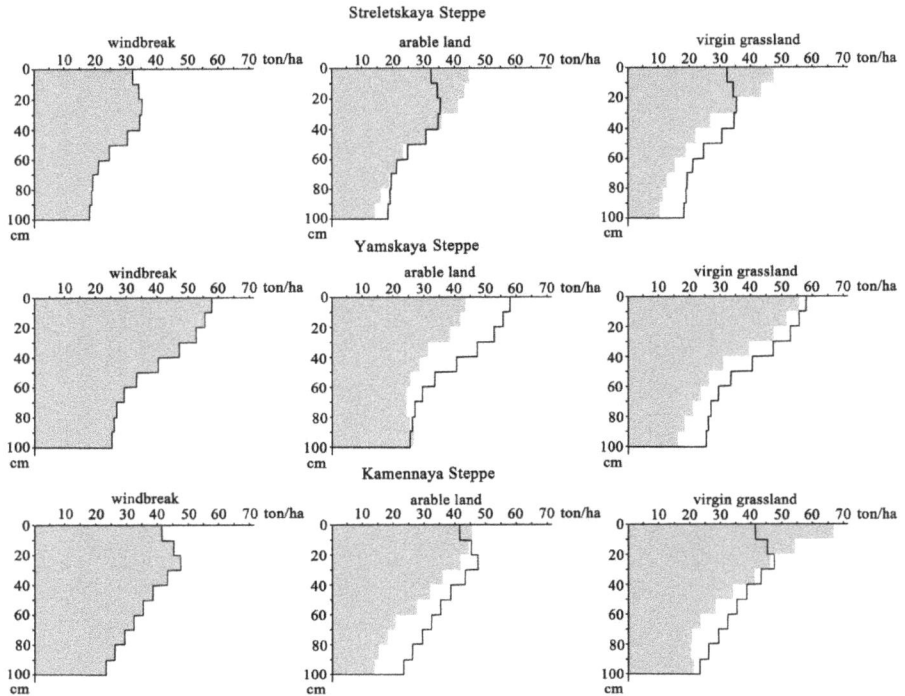

Figure 11. Vertical distribution of SOC stocks in profiles of study soils within the Central Russian Upland. In diagrams, reflecting arable and virgin Chernozems, the profile for soils beneath windbreaks is shown for comparison. Reproduced from [21].

Also, we identified reliable differences in SOC stocks between windbreaks and grasslands, and between windbreaks and cultivated lands within the 50–100 cm layer as average indexes for all key plots (Table 5). This may serve as further evidence of SOC stock accumulation in Chernozems under windbreaks, the result of windbreak influence on soils.

Thus, according to our study, we confirm that 55 year-old windbreaks within the Central Russian Upland forest-Steppe lead to improvement in the quality of Chernozem soils by addition of SOC. This finding is supported by similar studies in other geographic regions, however, not all afforestation projects yield such positive results. Studies from past plantings may also have produced different results than plantings today may produce in the future under potentially different climate conditions. At present, opinions of different authors vary: some of them detected increases of SOC under windbreaks of Europe and the USA [16,18,22,23]; others found changes in comparison of these soils with the initial SOC state of virgin grasslands soils [4,24,25]. A further complication is the change in SOC, and other properties under windbreaks vary with time. For example, a study of windbreaks of different ages in the state of Iowa in the USA found varying responses in Mollisols to sequestration of organic carbon with a tendency of this process slowing 30 years after windbreak planting [26].

Table 5. Average SOC stocks ($Mg \cdot ha^{-1}$) in soils under virgin grasslands, windbreaks, and crop fields adjacent to the windbreaks on three study areas (Streletskaya, Yamskaya, Kamennaya Steppe), Central Russian Upland.

Sub-Area	Layer, cm		
	0–50	50–100	0–100
Windbreaks, $n = 6$	208 ± 11	131 ± 8	339 ± 18
Grasslands, $n = 6$	204 ± 10	96 ± 9	300 ± 18
Cultivated lands, $n = 6$	188 ± 1	105 ± 7	293 ± 6
Windbreaks—Grasslands	+4	+35	+39
LSD_{05}	33	30	57
Windbreaks—Cultivated lands	+20	+26	+46
LSD_{05}	25	24	45

4. Conclusions

Chernozems of the Central Russian Upland respond rapidly to anthropogenic influences in the form of plowing and afforestation of agricultural lands. Tree cover improves soil quality by increasing A and A + AB horizon thickness and total SOC stocks. Maximum effect of this favorable soil development was found in relatively cooler and wetter conditions of central and north-western forest-steppe (Streletskaya Steppe and Yamskaya Steppe).

Depth of the SOC-enriched surface horizon was strongly correlated with a climate index, the HTC. A 10 cm growth of humus profile thickness in soils under windbreaks was observed for every HTC increase by 0.1. For the Central Russian Upland, this corresponds to an advance in a northwesterly direction of 50 km. In every key site, an increase in soil organic carbon stocks in the 0–100 cm layer of afforested (beneath windbreaks) Chernozems was greater than in the virgin grassland Chernozems. More significant change was especially evident in the 30–100 cm layer: at Streletskaya Steppe—40 $Mg \cdot ha^{-1}$, at Yamskaya Steppe—59 $Mg \cdot ha^{-1}$, and at Kamennaya Steppe—42 $Mg \cdot ha^{-1}$.

Afforestation in the form of field windbreaks is an effective means to increase C storage, restore soil fertility by improving nutrient cycling and availability, protect Chernozem soils from water and wind erosion, and produce high and stable crop yields in southern European Russia. Regarding the problem of global climate change, sequestration of atmospheric carbon into tree phytomass and organic matter of soils of the windbreak ecosystems can be seen as a positive effect in the biosphere. Agroforestry management, which had a high significance in the Soviet Union (period from 1948–1960s, during rapid economic growth and industrial reconstruction), needs to continue and expand across agricultural lands today and in the future.

Acknowledgments: The authors are grateful for the assistance of Aleksandr Petin, Larisa Novykh, Eugeny Zazdravnykh (Belgorod State University), Andry Dolgikh (Institute of Geography, Russian Academy of Science, Moscow), Aleksandr Shapovalov (Reserve "Belogorye"), Yury Cheverdin and Valerian Tischenko (Reserve "Kamennaya Steppe").

Author Contributions: Yury G. Chendev supervised all preliminary and field studies in the key sites, analyzed results of field and laboratory data reported in this paper, performed literary review and drafted the paper's text. Thomas J. Sauer participated in field studies in all key sites and supervised laboratory analyses of soil properties reported herein. Both Yury G. Chendev and Thomas J. Sauer have been Principal Investigators of a CRDF project RUG1-7024-BL-11, thematically connected to this paper. Thomas J. Sauer served as secondary author, providing significant input on all drafts. Guillermo Hernandez Ramirez, and Charles Lee Burras helped in analysis of all collected field data, took significant participation in discussions of the results, performed literary review and prepared some parts of the manuscript.

Conflicts of Interest: The authors declare no conflict of interest.

References

1. Dokychaev, V.V. *Russian Chernozem*; Printing house of Dekleron and Evdokimov: Saint Petersburg, Russian, 1883; p. 376. (In Russian)
2. Gennadiev, A.N.; Glazovskaya, M.A. *Geography of Soils with Foundations of Soil Science*; Moscow State University Press: Moscow, Russian, 2005; p. 461. (In Russian)
3. Ivanov, I.V. *History of Russian Pedology*; Nauka: Moscow, Russian, 2003; p. 397. (In Russian)
4. Mil'kov, F.N.; Nesterov, A.I.; Petrov, P.G.; Skachkov, B.I. *Kamennaya Steppe: Agroforestry Landscapes*; Voronezh State University Press: Voronezh, Russian, 1992; p. 224. (In Russian)
5. Erusalimskii, V.I. State protective windbreaks as a part of the protective forest plantations. In *Variety of Soils in the Kamennaya Steppe*; Soil Institute of V.V. Dokuchaev: Moscow, Russian, 2009; pp. 387–405. (In Russian)
6. Akhtyrtsev, B.P.; Solovichenko, V.D. *Soil Cover of Belgorod Oblast: Structure, Zoning, and Rational Management*; Voronezh State University Press: Voronezh, Russian, 1984; p. 224. (In Russian)
7. Zagorovsky, V.P. *The History of the Central Chernozem Region Joining to the Russian State in the XVI Century*; Voronezh State University: Voronezh, Russian, 1991; p. 270. (In Russian)
8. Wrangel, V. *The History of Forest Legislation of the Russian Empire with Assay of the Russian Ship Timbers*; Printing house of Fisher: Saint Petersburg, Russian, 1844; p. 153. (In Russian)
9. Myl'kov, F.N. (Ed.) *Central Russian "Belogorye"*; Voronezh State University Press: Voronezh, Russian, 1985; p. 238. (In Russian)
10. Chendev, Y.G.; Petin, A.N. *Natural Changes and Anthropogenic Transformation of the Environmental Components in Regions with Old History of Mastery (for the Example of Belgorod Oblast)*; Moscow State University Press: Moscow, Russian, 2006; p. 124. (In Russian)
11. Slobodchikov, D.Y. (Ed.) *Agricultural Activity in Russia*; Publication of the Department of Agriculture: Petrograd, Russia, 1914; p. 252. (In Russian)
12. Akulov, P.G.; Azarov, B.F.; Shelganov, I.I.; Yavtushenko, V.E. *Soil-Agrochemical Bases of Crop Farming Stability in the Central Chernozem Zone*; Agropromizdat: Moscow, Russian, 1991; p. 142. (In Russian)
13. Chendev, Y.G.; Aleksandrovskii, A.L.; Khohlova, O.S.; Smirnova, L.G.; Novykh, L.L.; Dolgikh, A.V. Anthropogenic evolution of dark gray forest soils in the Southern Part of the Central Russian Upland. *Eurasian Soil Sci.* **2011**, *1*, 3–15.
14. Brandle, J.R.; Hodges, L.; Zhou, X.H. Windbreaks in North American agricultural systems. *Agrofor. Syst.* **2004**, *61*, 65–78.
15. Read, R.A. *The Great Plains Shelterbelt in 1954 (A Re-Evaluation of Field Windbreaks Planted between 1935 and 1942 and a Suggested Research Program)*; Great Plains Agricultural Council Publication. No. 16; Nebraska Agricultural Experiment Station: Lincoln, NE, USA, 1958; p. 125.
16. Hernandez-Ramirez, G.; Sauer, T.J.; Cambardella, C.; Brandle, J.R.; James, D.E. Carbon sources and dynamics in afforested and cultivated Corn Belt soils. *Soil Sci. Soc. Am. J.* **2011**, *75*, 216–225. [CrossRef]
17. Kort, J.; Turnock, R. Carbon Reservoir and biomass in Canadian prairie shelterbelts. *Agrofor. Syst.* **1999**, *44*, 175–186. [CrossRef]
18. Sauer, T.J.; Cambardella, C.A.; Brandle, R.B. Soil carbon and tree litter dynamics in a red cedar-scotch pine shelterbelt. *Agrofor. Syst.* **2007**, *71*, 163–174. [CrossRef]
19. Sauer, T.J.; Nelson, M.P. Science, ethics, and the historical roots of our ecological crisis: Was white right? In *Sustaining Soil Productivity in Response to Global Climate Change: Science, Policy, and Ethics*; Villey-Blackwell: Chichester, UK, 2011; pp. 3–16.
20. Selyaninov, G.T. On agricultural climate valuation. *Proc. Agric. Meteorol.* **1928**, *20*, 165–177.
21. Chendev, Y.G.; Sauer, T.J.; Hall, R.B.; Petin, A.N.; Novykh, L.L.; Zazdravnykh, E.A.; Cheverdin, Y.I.; Tischenko, V.V.; Filatov, K.I. Stock assessment and balance of organic carbon in the Eastern European forest-steppe ecosystems tree windbreaks. *Reg. Environ. Issues* **2013**, *4*, 7–14. (In Russian)
22. Al'benskiy, A.B. Afforestation and environmental change. *Gerald Agric. Sci.* **1961**, *2*, 96–101. (In Russian)
23. Tumin, G.M. *Impact of Shelterbelts to Soil in Kamennaya Steppe*; Kommuna: Voronezh, Russian, 1930; p. 40. (In Russian)
24. Kaganov, V.V. Change in carbon stocks of ecosystems under afforestation within steppe and semi-desert zones of the European Russia. *Reg. Environ. Issues.* **2012**, *4*, 7–12. (In Russian)

25. Solovyov, P.E. *Influence of Windbreaks to Soil Formation and Fertility of Steppe Soils*; Moscow State University Press: Moscow, Russian, 1967; p. 200. (In Russian)

26. Sauer, T.J.; James, D.E.; Cambardella, C.A.; Hernandez-Ramirez, G. Soil properties following restoration or afforestation of marginal cropland. *Plant Soil.* **2012**, *360*, 1–2. [CrossRef]

sustainability

MDPI

Article

Do Current European Policies Prevent Soil Threats and Support Soil Functions?

Nadia Glæsner [1,2,*], Katharina Helming [1] and Wim de Vries [3]

1   Directorate, Leibniz Centre for Agricultural Landscape Research (ZALF), Müncheberg 15374, Germany;
     helming@zalf.de
2   Department of Plant and Environmental Sciences, University of Copenhagen, Thorvaldsensvej 40,
     Frederiksberg 1871, Denmark
3   Alterra, Wageningen University and Research Centre (WUR), P.O. Box 47, Wageningen 6700 AA,
     The Netherlands; wim.devries@wur.nl
*   Correspondence: nadia.glaesner@gmail.com; Tel.: +45-2614-7726

External Editors: Marc A. Rosen and Douglas L. Karlen

Received: 30 October 2014; in revised form: 8 December 2014; Accepted: 12 December 2014; Published:
22 December 2014

Abstract: There is currently no legislation at the European level that focuses exclusively on soil conservation. A cross-policy analysis was carried out to identify gaps and overlaps in existing EU legislation that is related to soil threats and functions. We found that three soil threats, namely compaction, salinization and soil sealing, were not addressed in any of the 19 legislative policies that were analyzed. Other soil threats, such as erosion, decline in organic matter, loss of biodiversity and contamination, were covered in existing legislation, but only a few directives provided targets for reducing the soil threats. Existing legislation addresses the reduction of the seven soil functions that were analyzed, but there are very few directives for improving soil functions. Because soil degradation is ongoing in Europe, it raises the question whether existing legislation is sufficient for maintaining soil resources. Addressing soil functions individually in various directives fails to account for the multifunctionality of soil. This paper suggests that a European Soil Framework Directive would increase the effectiveness of conserving soil functions in the EU.

Keywords: European Union; soil policy; soil degradation; soil conservation; soil threats; soil functions; grand societal challenges; DPSIR

1. Introduction

Despite growing pressures on European soils and the danger that these pressures pose to the services that healthy soils provide, there is no common EU policy on soil protection. In 2002, the Commission presented its approach to soil protection in a Communication that was titled "Towards a Thematic Strategy on Soil Protection" [1]. The main threats that lead to soil degradation were identified as erosion, decline in organic matter, contamination, sealing, compaction, loss of biodiversity, salinization and floods and landslides. Floods and landslides were later addressed in a separate Directive on flood risk management prevention (2007/60/EC) and have therefore been excluded from the Thematic Strategy on Soil Protection. The Commission stressed the importance of integrating soil aspects into other directives, but also indicated the need for legislation that focuses exclusively on soil. To fill the gap in European environmental legislation and to provide a more holistic approach to soil protection in the EU, the European Commission presented a new policy in 2006 that was titled "Thematic Strategy for Soil Protection" [2]. This followed a comprehensive stakeholder consultation and included a proposal for a Soil Framework Directive [3]. However, the proposal was not adopted. Germany, France, The Netherlands, the United Kingdom and Austria opposed the proposal [4] on the grounds of the subsidiarity and proportionality principles, expected costs and the administrative burden. They also questioned the value that the new policy added to existing Union law [5]. The proposal had been pending since 2006, but was finally withdrawn in May 2014 [6], because the Soil Framework Directive had been pending for eight years during which time no effective action had been taken [7].

The Impact Assessment (IA) [8] that supplemented the proposed Soil Framework Directive was focused on the costs of soil degradation, which were divided into different soil threats. Impact assessment (IA) has been an obligatory EU tool for achieving evidence-based policymaking since 2002 for all new directives to address the three pillars of sustainability, *i.e.*, social, environmental and economic impact. However, the IA could not justify the activity at the European level, because it provided little evidence of the impacts of soil threats. Estimates of the costs of soil degradation and soil threats at the European level were speculative, because the impacts of soil degradation on societal challenges, such as food production, could not clearly be shown, as the range of economic evaluation for each threat was very large and the estimations were not precise enough. Ongoing activities under the EU Soil Thematic Strategy are therefore currently narrowed to raising awareness, conducting research and integrating policies [9].

In addition to identifying soil threats, the 2006 proposal for a Soil Framework Directive introduced the functions that soil provides for humankind, but the impacts on those functions of measures to ameliorate the threats were not mentioned, even in a qualitative sense. Within research, focus has shifted from soil degradation (soil threats) to soil functions in the last decade [10–12]. This is reflected in the international conferences that were titled the *"Wageningen Conference on Applied Soil Science"* that was held in The Netherlands in 2011 [13] and *"Protection of soil functions—challenges for the future"* that was held in Pulawy, Poland in October, 2013. [14]. The concept of soil functions originates from a descriptive analysis by Blum [15]. Bouma [16] further elaborated on soil functions as a fundamental concept for linking soil science to policy and decision support. Indeed, the concept of soil functions can be seen as an early embodiment of the concept of ecosystem functions and services. The ecosystem service concept was developed to express the value of nature to human societies [17]. This concept is used to formulate policy recommendations or, more generally, to support decision-making and is relevant at the interface of science and policy, where it can play two roles: it can translate the link between ecological processes and human wellbeing in a way that is understood by decision makers, and it can also communicate the scientific knowledge that is relevant to decision-making [18,19]. Considering the importance of the services provided by terrestrial ecosystems, Dunbar *et al.* [20], for example, evaluated the impact of various EU policies (Common Agricultural Policy (CAP), Biodiversity Strategy 2020, Habitat Directive, Bird Directive, Soil Thematic Strategy) on those services. The ecosystem service concept distinguishes between "functions", which are defined as the "capacity of ecosystem components and processes to provide goods and services that satisfy human needs" [21] and "services", which are defined as the actual "benefits people derive from ecosystems" [17]. The concept of soil functions seems to include both [12,16].

Current grand societal challenges have been identified at the EU level in "Horizon 2020", which is the Common EU Framework for Research and Innovation (2014–2020). "Horizon 2020" is intended to secure Europe's global competitiveness. The main soil-related challenges to competitiveness are food security, energy security and resource-use efficiency. Food security is becoming increasingly important in light of the growing worldwide food demand that results from an expanding and wealthier world population. Increased resource-use efficiency ("doing more with less") is crucial to increasing the production of food, feed and energy crops while reducing the use of resources, such as energy, water, land and nutrients, and reducing environmental impacts. The maintenance of soil resources plays a vital role in meeting these grand societal challenges and underpins the actuality of the topic of soil protection.

Current EU strategies and communications that are related to soil challenges include the 7th Environmental Action Programme (7 EAP) and the Resource Efficiency Roadmap (COM/2011/571), which are leading to a revival of the political discussion of the importance of soil protection in Europe.

This was reflected in the communication from the European Commission (EC) on the implementation of the soil thematic strategy [9], which argues that the status of soil degradation remains alarming despite considerable efforts to raise awareness, conduct research and integrate policies. Information about the continued soil degradation throughout Europe is gathered in the European Soil Data Centre (ESDAC) [22]. This is done with the aim of not only presenting information relevant for soil policies, but also for other policies, such as CAP, climate change policy, the EU forest action plan, rural development and water management, further illustrated in a recent report by the Joint Research Centre [23]. The report estimates that 20% of European soils are being eroded by water and wind. The 7 EAP sets this at 25%. Within the 34 European countries in Europe (EU 28 plus Norway and Balkan States), moderate and high levels of land susceptibility to wind erosion are further predicted, corresponding to 5.3 and 2.9% of total area [24]. Spatial maps of European soils have been created for estimating organic matter content [25] and soils subject to salinization [26], indicating that 45% of European soils have a low organic matter content (defined as having less than 2% organic carbon) and that 3.8 million ha of soil are subject to salinization [23]. European subsoils have been classified into very high (9%), high (28%), moderate (44%) and low (20%) susceptibility to compaction [27]. This implies that more than a third of the European subsoils are classified as having a high or very high susceptibility to compaction. Potential contaminated sites are estimated to more than 2.5 million, and identified contaminated sites are around 342 thousand [28]. Further, at least 275 ha of soil per day is lost to permanent soil sealing [23]. Finally, soil biodiversity is reported to be declining throughout Europe, mainly because of the abovementioned soil degradation processes. These trends appear to indicate that current legislative action is not adequate, especially considering that safeguarding important soil functions at the European level is a necessary precondition for meeting the upcoming grand societal challenges. The proposal to withdraw the Soil Framework Directive stated that it "opens the way for an alternative initiative in the next mandate" [7].

The general aim of the present study is to analyze the need for such a common EU soil protection framework in view of existing soil-related policies. Addressing grand societal challenges justifies action on soil conservation at the European level (European added value). We therefore analyze the contribution of soil to society and the ways in which existing policies address soil. We identify gaps and overlaps that exist between those policies and whether there is a need for a new soil directive to replace the one that is currently withdrawn.

2. Materials and Methods

2.1. Analytical Framework

A policy analysis was carried out to assess the need for a separate soil directive at the European level. The analysis addressed the state of existing soil-related policies in terms of soil threats and soil functions and identified gaps and overlaps in soil protection in existing policies. The paradigm shift from soil degradation to the societal value of soil (which are both mentioned in the proposal for a Soil Framework Directive [3]) is the rationale behind using the concept of soil functions in combination with soil threats. Protection of soil resources plays a vital role in meeting the grand societal challenges. Therefore, the shift of focus from soil degradation (soil threats) to soil functions is relevant because of their relationship to the grand societal challenges at a European level (Figure 1).

Figure 1 shows how the DPSIR approach [29] is used to illustrate the links among existing policies (drivers of change) of sectors that affect soils (pressures) and soil threats (states), soil functions and grand societal challenges (impacts), as a way to address the need for new policy targets (separate legislation on soil conservation) (responses). Note that a DPSIR structure of policy evaluation had been mentioned in previous studies that were related to the implementation of a soil protection strategy [30–32], but those studies focused on a source-pathway-receptor approach, such as a health risk assessment, to support effective country-specific regulatory decisions for managing contaminated land [31]. We did not make this type of assessment in the present study. Rather, we evaluated the need for separate legislation on soil conservation by assessing whether existing policies adequately prevent or reduce soil threats and prevent the reduction of or improve soil functions in view of grand societal challenges.

The concept of soil functions connects physical, chemical and biological processes with the benefits of soil to society in environmental, economic and social terms. Similar concepts are "ecosystem services", which include provisioning, regulating, supporting and cultural services [17], and "landscape services", which was introduced as a bridge between landscape ecology and sustainable landscape development [33]. The term "soil functions" has been limited to agricultural soils by Schulte *et al.* [34], who distinguished five agricultural soil functions: (i) biomass production; (ii) water purification; (iii) carbon sequestration; (iv) biodiversity habitat; and (v) recycling of nutrients and agro-chemicals. We have chosen to address all soils, not just those that are related to agriculture, so that we could use the soil functions concept that is mentioned in the proposal for a Soil Framework Directive [3].

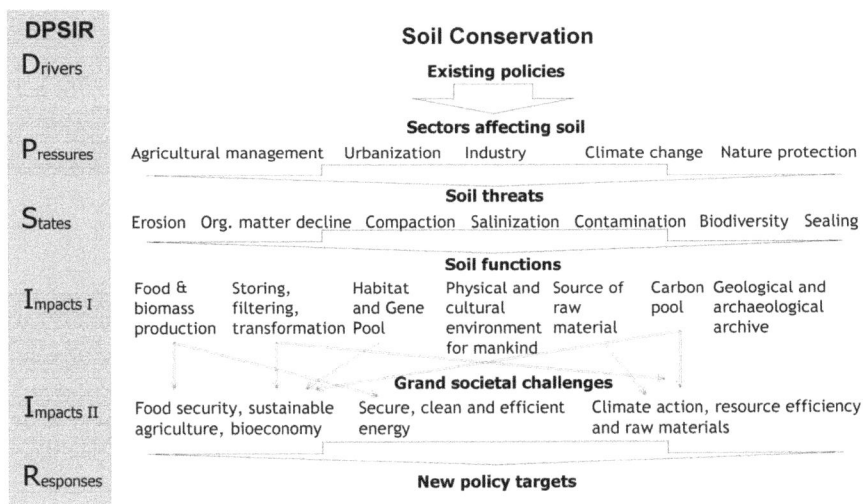

Figure 1. The analytical DPSIR framework that links policies to soil threats [1], soil functions [2,16] and grand societal challenges (Horizon 2020).

2.2. Cross-Policy Analysis (Gap and Overlap Analysis)

A sector's having a direct relationship with soil constituted the criterion for policy selection. The main sectors that lead to soil degradation (pressures) fall into four main categories (Figure 1). The agricultural management category is related to the decline in soil organic matter, salinization, erosion and compaction as a result of biomass production for food, feed and energy. The industry category is related to the contamination that is associated with industrial sites; the urbanization category is related to soil sealing and land take for urban structures and infrastructure and for tourism; and the climate change category is related to greenhouse gas emissions and carbon pool changes. We therefore selected policies from these four sectors, as well as nature conservation policies for the analysis. Nature conservation policies were included in the analysis because of their role in preserving environment, including soil. A total of 19 legislative policies and two recent EC communications that were related to soil were analyzed (Table 1). Based on this analysis, gaps and overlaps in soil threats and soil functions that were addressed in existing policies were identified (Table 2). The criterion for inclusion was directly addressing a specific soil threat or soil function. Directives that may have indirect effects, such as livestock grazing, which results in greater soil compaction, were not included. A distinction was made between directives that "prevent acceleration" and those that "reduce" soil threats and similarly between those that "prevent reduction" or "improve" soil functions (Table 3). The analysis was related to the counterfactual issue that no policy (directive) was in place. The analysis was carried out at the EU level, therefore, national policies were not included.

Table 1. Policies analyzed in the study.

Policies	Number	Title
Agricultural policies		
Common Agricultural Policy (CAP)	1305/2013	European Agricultural Fund for Rural Development (EAFRD)
	1306/2013	European financing, management and monitoring of the common agricultural policy
	1307/2013	Common rules for direct support schemes
	1308/2013	European common organization of the markets in agricultural products
Plant Protection Products Directive	91/414/EEC	Concerning the placing of plant protection products on the market
Nitrates Directive	1991/676/EEC	Concerning the protection of waters against pollution caused by nitrates from agricultural sources
GMO Directive	2001/18/EC	Deliberate release into the environment of genetically-modified organisms
Pesticide Use Directive	2009/128/EC	Action to achieve the sustainable use of pesticides
Industrial policies		
Industrial Emissions Directive	2010/75/EU	Industrial emissions (integrated pollution prevention and control)
Landfill Directive	1999/31/EC	Landfill of waste
Mining Waste Directive	2006/21/EC	Management of waste from extractive industries
Biocidal Products Regulation	(EU) 528/2012	Concerning making available on the market and use of biocidal products
Waste Directive	2008/98/EC	Waste
Urban policies		
Sewage Sludge Directive	86/278/EEC	Protection of the environment and, in particular, of the soil, when sewage sludge is used in agriculture
Urban Waste Water Directive	91/271/EEC	Concerning urban waste water treatment
Climate policies		
Carbon Storage Directive	2009/31/EC	Geological storage of carbon dioxide
Renewable Energy Directive	2009/28/EC	Promotion of the use of energy from renewable sources
Nature conservation policies		
Habitat Directive	92/43/EEC	Conservation of natural habitats and of wild fauna and flora
Water Framework Directive	2000/60/EC	Establishing a framework for community action in the field of water policy
Air quality Framework Directive	2004/107/EC	Relating to arsenic, cadmium, mercury, nickel and polycyclic aromatic hydrocarbons in ambient air
Environmental Liability Directive	2004/35/CE	Environmental liability with regard to the prevention and remedying of environmental damage
Floods Directive	2007/60/EC	Assessment and management of flood risks
Non-binding EC communications		
Resource Efficiency Roadmap	COM/2011/571	Roadmap to a Resource Efficient Europe
7 EAP		7th Environmental Action Programme to 2020 "Living well, within the limits of our planet"

Table 2. Existing policies that directly address soil threats and soil functions.

Policies		Agricultural				Industrial					Urban		Climate		Nature conservation			Air quality Framework	Environmental liability	Non-binding EC Communications	
	CAP	Plant protection products	Nitrates	Pesticide use	GMO	Industrial emissions	Biocidal products	Waste	Landfill	Mining waste	Urban waste water	Sewage sludge	Carbon storage	Renewable energy	Habitat	Water Framework	Floods			Resource efficiency roadmap	7 EAP
Year	2013	1991	1991	2009	2001	2010	2012	2008	1999	2006	1991	1986	2009	2009	1992	2000	2007	2004	2004	2011	2013
Soil threats																					
1. Erosion	X													X			X			X	X
2. Organic matter decline	X				X															X	X
3. Compaction																					
4. Biodiversity decline	X	X		X	X		X			X			X	X	X				X	X	X
5. Salinization																					
7. Contamination		X		X		X	X	X	X	X		X	X			X		X	X	X	X
6. Sealing																				X	X
Soil functions																					
1. Biomass production	X													X						X	X
2. Storing, filtering and transformation	X	X	X	X	X	X	X	X		X		X	X			X		X			X
3. Habitat and gene pool	X	X		X	X		X			X			X	X	X				X	X	X
4. Physical and cultural environment	X			X			X		X											X	X
5. Source of raw materials	X								X	X										X	X
6. Carbon pool	X												X	X						X	X
7. Geological and archeological archive	X							X									X				X

Table 3. The number of existing European legislative policies that cover soil threats and soil functions and the resultant gaps in legislation. Policies are divided according to whether the policy "prevents acceleration of threat" or "reduces threat" and "prevents reduction of function" or "improves function".

Soil threats	Gap	Prevents acceleration of threat	Reduces threat	Policies involved
1. Erosion		2		• Renewable energy, mining waste
			2	• Floods, CAP
2. Decline in organic matter		2		• CAP, GMO
3. Compaction	X			
4. Loss of biodiversity		8		• Plant protection product, biocidal products, GMO[a], environmental liability, carbon storage, pesticide use, mining waste[a], renewable energy[a]
5. Salinization	X		2[a]	• Habitat, CAP
6. Contamination		12		• Waste, landfill, mining waste, pesticide use, plant protection products, biocidal products, environmental liability, carbon storage, water framework, air quality framework, sewage sludge, industrial emissions
			2	• Renewable energy, environmental liability
7. Sealing	X			

Soil functions	Gap	Prevents reduction of function	Improves function	Policies involved
1. Biomass production		2		• CAP, Renewable energy
2. Storing, filtering and transformation		13		• Nitrates[b], pesticide use[b], waste, landfill, mining waste, plant protection products, biocidal products, GMO, industrial emissions, carbon storage, water framework, sewage sludge
			1	• CAP[c]

Table 3. *Cont.*

Soil threats	Gap	Prevents acceleration of threat	Reduces threat	Policies involved
3. Habitat and gene pool		8		• Plant protection product, biocidal products, GMO[a], environmental liability, carbon storage, pesticide use, mining waste[a], renewable energy[a]
			2[a]	• Habitat, CAP
4. Physical and cultural environment for mankind		3		• Landfill[d], pesticide use, CAP
5. Source of raw materials		2		• Mining waste, landfill
6. Carbon pool		2		• CAP, renewable energy
7. Geological and archaeological archive		2		• CAP, floods

[a] None of these policies is directly linked to soil, but they all focus on biodiversity in general. [b] The Nitrates Directive, Pesticide use Directive (and CAP) focus on off-site impacts that improve this function in some areas of farms, but these activities also contribute to reducing the function by, e.g., the application of pesticides. Measures to prevent the reduction of this function in these directives are related to, e.g., buffer strips and not to the soil as a whole. [c] CAP includes measures for maintaining soil organic matter and soil structure, which indirectly improve soil Function 2. [d] The Landfill Directive prevents the reduction of this soil function in one area by preventing the location of landfills near residential and recreational areas and reduces the function (establishing residential and recreational areas) in other areas where a landfill is located.

3. Results and Discussion

3.1. States: Soil Threats

The existing soil-related policies that were identified in the cross-policy analysis as addressing soil threats and soil functions are presented in Table 2. The relationships of these existing policies to the states of soil threats are outlined below, as well as a discussion of the relevance of each soil threat and the extent to which it is covered in existing legislation. A summary of the results of the cross-policy analysis is presented in Table 3.

Soil Threat 1: Erosion

Soil erosion causes adverse on-site effects that include damage to land-based production (reducing crop yields) and loss of topsoil that is rich in nutrients and organic matter. It also causes adverse off-site effects that include blocking infrastructure and drainage channels, property damage, pollution of water bodies and destruction of wildlife habitats. The main pressures affecting the state of soil erosion are conversion to arable land, inappropriate land management, deforestation, overgrazing, forest fires and construction activities. Sites are especially at risk when incomplete plant coverage coincides with strong winds (wind erosion) or heavy rainfall (water erosion) [35,36].

Erosion is covered by four binding laws and two EC communications (Table 2). The Common Agricultural Policy (CAP) supports agricultural production, which tends to accelerate soil erosion. However, the CAP includes incentives for landowners to implement land management practices that limit soil erosion. A framework was established that set Good Agricultural and Environmental Conditions for land (GAEC). Those standards are intended to prevent soil erosion. Assessment and management of flood risk are set as targets in the Floods Directive. Management of flood risk is not related directly to soil erosion, but flood risk management implies the management of erosion control. The Renewable Energy Directive encourages conservation of areas that provide watershed protection and erosion control. The Mining Waste Directive states that construction of a new waste facility or modification of an existing waste facility must include measures that ensure that soil erosion that is caused by water or wind is minimized to the degree that is technically possible and economically viable, but the directive does not mention specific measures.

With regard to the non-binding EC communications, the Resource Efficiency Roadmap includes a milestone of the area of land in the EU that is subject to soil erosion of more than 10 tons per hectare per year should be reduced by at least 25% by 2020 and encourages Member States to implement actions that are needed for reducing erosion. The 7 EAP states that more efforts to reduce soil erosion are encouraged.

Hence, only the Floods Directive and a few measures of the CAP are the legislations that aim to reduce soil erosion. The Renewable Energy and Mining Waste Directives include only measures that are intended to prevent the acceleration of erosion (Table 3). On top of this, the Renewable energy Directive encourages the cultivation of crops (corn, oil seed, sunflowers) that are more erosive compared to wheat and prevents residues returning to the soil, which are an accelerator of soil erosion. More legislative emphasis on reducing soil erosion seems warranted because of the serious consequences of soil erosion.

Soil Threat 2: Decline in organic matter

Soil organic matter plays a vital role and is often defined as the most important indicator of soil quality, because it affects such physical, chemical and biological processes as water retention, nutrient cycling, contaminant retention and decay and providing habitat for soil organisms [37]. However, decline in soil organic matter is mentioned only in the CAP and the GMO Directive, as well as in the non-binding EC communications. The GAEC standards in the CAP are intended to maintain soil organic matter levels by means of appropriate practices, which include a ban on burning arable stubble. However, soil cultivation for agriculture itself reduces soil organic matter stocks [38]. The GMO

Directive requires an environmental risk assessment before releasing genetically-modified organisms into the environment, and the assessment is to include potential changes in the soil decomposition of organic material. However, the GMO Directive relates only to changes in the decomposition of organic matter that might occur as a result of releasing genetically-modified organisms. Both the GMO Directive and the CAP place more emphasis on preventing the acceleration of the loss of soil organic matter than on reducing its decline (Table 3). Of the non-binding EC communications, the Resource Efficiency Roadmap includes a milestone that by 2020, soil organic matter levels should not be decreasing overall and should increase for soils with currently less than 3.5% organic matter. Furthermore, it encourages Member States to implement actions that are needed for increasing and restoring organic matter content in soils. The 7 EAP also mentions the need to enhance efforts to increase soil organic matter.

Soil Threat 3: Compaction

None of the existing EU laws and neither of the two EC communications address the threat of soil compaction (Table 2). The GAEC of the CAP touch upon appropriate machinery use for maintaining soil structure, but do not specifically target soil compaction. Compaction results from the mechanical stress that is caused by heavy agricultural machinery, especially during fertilizer application and harvesting [39,40]. Compaction can also be caused by repeated trampling by grazing animals. These activities expose the soil to high pressure that reduces its porosity, aeration and biological activity. Consequences include reduced water infiltration, increased water run-off, increased erosion and reduced crop root growth. Decreased root growth may substantially decrease water and nutrient uptake efficiency, which decreases food production [41,42]. Topsoil is loosened annually by tilling, but soil compaction increases over time [23]. Therefore, site-adequate management practices that address these pressures are required to reduce the threat of soil compaction.

Soil Threat 4: Loss of biodiversity

The diversity of above-ground plants and animals is addressed frequently in existing policies, but there is no specific focus on soil biodiversity. The relevance of these policies, which are intended primarily to prevent acceleration of biodiversity loss, is discussed in Section 3.2 under "Soil Function 3: Habitat and gene pool."

Soil Threat 5: Salinization

None of the existing EU laws and neither of the two EC communications address the threat of soil salinization (Table 2). Salinization occurs naturally in some European soils, but the accumulation of salts results mainly from inappropriate irrigation practices [23]. Salinization is therefore expected to increase with the increased need for irrigation in response to climate change and anticipated increases in drought conditions, especially in Southern Europe [43], where salinization problems are already widespread [23]. Legislation that specifically targets the pressures of soil salinization would be useful, because salinization is likely to increase.

Soil Threat 6: Contamination

Contamination results from the use and presence of dangerous substances in industrial processes [44]. Contamination of soil is addressed in 13 existing legislative policies (Table 2), which is far more than any other soil threat. Directives regarding the disposal of wastes, which are the Waste, Landfill, Mining waste and Sewage sludge Directives, and directives regarding the application of chemicals, which are the Biocidal products Regulation and Plant protection and Pesticide use Directives, require that disposal and application of contaminants should be conducted in a manner that does not cause risks to soil. The Industrial emissions Directive addresses the prevention of emissions from entering soil. The Carbon storage Directive addresses the technology of CO_2 capture from industrial installations, its transport to storage sites and its injection into a suitable underground geological formation for permanent storage and ensures that there are no unwanted risks to the soil, such as deposition of impurities that are related to the technology. The Water Framework Directive addresses the identification and estimation of significant point-source pollution that originates from soils. Finally, the Air quality Framework Directive targets the effects of arsenic, cadmium, mercury, nickel and polycyclic aromatic hydrocarbons on human health and includes deposition limits to avoid accumulation of these substances in soil and related food chain impacts. All these directives are intended to prevent further acceleration of contamination, but none is focused on reducing this threat. The Renewable energy Directive recommends an increase in the amount of land that is available for biofuel cultivation by restoring heavily-contaminated land that cannot be used in its present state for agricultural purposes. The Environmental liability Directive also introduces the "polluter-pays" principle to prevent further soil contamination and measures for remediating land damage.

The two EC communications also address contamination. The Resource Efficiency Roadmap provides an inventory of contaminated sites and a schedule for remedial work by 2015, and it establishes a goal of remedial work on contaminated sites being well underway by 2020. The 7 EAP targets contamination by encouraging remediation of contaminated sites to be well underway by 2020.

However, 11 of the 13 directives aim only at preventing acceleration of soil contamination, because these directives derive from sectors that are pressures to soil contamination by applying waste or chemicals to soils. Only two directives (Table 3) address remediation of contaminated soil. The Renewable Energy Directive addresses the restoration of contaminated soil, and the Environmental liability Directive introduces remediation of land damage. Remediation of existing soil contamination is a relevant topic for legislation.

Soil Threat 7: Sealing

Sealing is both a soil function and a soil threat. Sealing serves urbanization (see *"Soil Function 4: Physical and cultural environment"*), but all other soil functions are lost when soil is permanently sealed in the course of urban construction. Increasing land take due to urbanization threatens fertile soils throughout Europe [45].

None of the existing EU legislation addresses the threat of soil sealing (Table 2). However, the European Commission does acknowledge soil sealing in its two communications (Resource Efficiency Roadmap and 7 EAP), which both set targets of no net land take by 2050. An upcoming EC communication, which is titled "Land as a Resource", also includes this goal. The Resource Efficiency Roadmap further mentions the aim of reducing the annual land take from 1000 km^2 per year to 800 km^2 per year at the EU level by 2020. The Commission also launched "Guidelines on best practice to limit, mitigate or compensate soil sealing" (SWD(2012)101 final/2) in 2012 to confront the challenge of increasing land take by urban construction. Until now, there have been only strategies to address this soil threat, but there has been no binding legislation. Soil sealing would therefore be another important threat to address in legislative policies. Urban soils are, in general, a gap in existing legislation.

In summary, despite targets that are set in existing legislation for erosion control, organic matter decline, minimizing contamination and minimizing loss of biodiversity, these threats continue to cause soil degradation in all of the EU Member States. This raises questions about whether the existing legislation is sufficiently comprehensive and is effectively implemented in the Member States. Additionally, soil compaction, soil salinization and soil sealing are not addressed in binding legislation. Several laws do mention "sustainable management", which is a term that indirectly covers all threats to soil functions (including salinization, compaction and sealing). An earlier analysis of the indirect effects of existing legislative policies on the conservation of soils for agricultural production highlighted that existing policies have the potential to address all recognized soil threats across the EU [46]. However, we believe that it is crucial that policies address soil threats and soil functions directly to ensure that all soil threats and soil functions are managed. This includes targeting specific soil threats and functions to simplify and optimize the implementation of new soil management procedures that are intended to prevent soil degradation. There are many agricultural management options for preventing or reducing soil threats in Europe [38,47]. It goes beyond the scope of this paper to go into detail about existing conservation management practices. However, there is a lack of implementation, and the use of such general terms as "sustainable management" may hinder the establishment of specific goals for conserving soil resources.

3.2. Impacts: Soil Functions and Grand Societal Challenges

The seven soil threats are very closely linked to the seven soil functions, because each soil threat leads to decreased functions of soil (Figure 1). Hence, in maintaining soil functions, all soil threats must be addressed. Shifting the policy paradigm from soil threats to soil functions addresses the values of soil for society and, therefore, better justifies policy action for maintaining and supporting soil functions. This is because it also lays focus on the public good character of soil processes and services, rather than private goods, which are only of value to the land owners. However, the concept of soil functions is less specific than the concept of soil degradation processes. We therefore believe that the focus of soil protection should be based on soil functions and that the targets should be based on soil threats, because it is the threats that impact soil functions (Figure 1).

Soil Function 1: Food and biomass production

The most obvious function of soil is the production of food for people and feed for farm animals. This relates to the grand societal challenges of food and energy security and sustainable agriculture (Figure 1). The need to produce food is increasing, because globally, productive arable soils will have to satisfy the needs of nine billion people by 2050, with an estimated increase in the demand for food production by around 70% [48,49] or even 100% [50]. In addition to providing food and feed, soil is the resource for meeting the increased demand of the growing world population for non-food biomass that is dedicated to energy and fiber products. One of the key challenges for soil protection is to simultaneously conserve soil while increasing productivity to ensure food security and provide bio-energy.

The CAP is the most important document for addressing agricultural food and biomass production. The CAP promotes increased production of agricultural products, and this has been its main focus. However, the focus has changed to include more measures for sustainability. The "greening" of the new CAP (which is planned for the period 2014–2020) will include additional support for farmers that implement management practices or establish ecological focus areas that benefit the climate or the environment. This coincides with GAEC standards that are intended to contribute to the maintenance of the landscape, water protection, climate action and management practices that increase the levels of soil organic matter and prevent soil erosion (see above). The Renewable energy Directive provides direct support for farmers by requiring that the GAEC standards also apply to biofuel production. This Directive attempts to assess the impacts of the competing demands of food and biofuel production. The Directive encourages restoration of severely degraded or heavily contaminated land that cannot be used in its present state for food production purposes.

In addition to these laws, the two EC communications, (the Resource Efficiency Roadmap and 7 EAP) both mention that high biomass production must be maintained to meet the increasing demand for agricultural products. Both communications promote sustainable management of agricultural production to ensure protection of soil resources. The 7 EAP includes a specific target that all land in the EU is to be managed sustainably, and soil is to be protected adequately by 2020. The Resource Efficiency Roadmap recommends the development of "innovation partnerships" that meet resource efficiency goals that pertain to productive and sustainable agriculture.

Because agricultural productivity is so specific, it is addressed only in policies that address biomass production, although almost all directives mention arable land. The incentives for GAEC that are included in the CAP and in the Renewable energy Directive are intended to prevent a reduction of this function. However, because all soil threats are applicable to agricultural productivity (Figure 1), it is crucial to target reductions in all soil threats to maintain and improve this soil function in the future. In addition, contrary to the targets of the Renewable energy Directive, the analysis of its implementation has shown that it creates trade-offs that compromise soil quality and reduce the soil function of biomass production in the long run [51].

Soil Function 2: Storing, filtering, transformation

The storing, filtering and transformation function of soil refers to the role of soil as a storage reservoir for nutrients and wastes, as a filter for water and air contaminants and as a transformation medium for chemicals. Water quality is particularly dependent on this soil function, because chemical inputs to soil may cause severe water pollution if they are not captured by the soil [52]. This function serves all of the grand societal challenges that are related to soil, which include food and energy security, climate action and resource efficiency (Figure 1). Storing, filtering and transformation is the single soil function that is most commonly addressed by existing policies (Tables 2 and 3). These three processes are discussed separately here.

Many of the directives address the storage function, including waste deposition on land (Waste, Landfill, Mining waste, Sewage sludge and Urban waste Directives, although the latter directive does not directly address soil or land), industrial emissions (Industrial emissions Directive) and the technology of CO_2 capture and geological storage (Carbon storage Directive). These directives, which all utilize the storage function of soil, target the avoidance of soil contamination for preventing a reduction in the function, but they do not target the maintenance of the soil function.

The filtering function is addressed in the Water Framework, Nitrates, Air quality Framework and Pesticide use Directives and in the CAP. These laws address soil as a filter for water and air purification. The Water Framework Directive requires identification and estimation of significant point-source pollution from urban and industrial sites and regional pollution from agricultural land, but only the Nitrates Directive and the CAP specifically address water pollution from agricultural sources. Agricultural management practices (GAEC), such as introducing buffer strips to protect waterways from pollution from agricultural runoff, are mentioned in the CAP, Nitrates Directive and Pesticide use Directive. These directives directly target maintenance of the filtering function of soil, but their focus is on off-site impacts, *i.e.*, avoiding contamination of soil or water bodies. No policy targets on-site impacts.

The transformation function is addressed in the Plant protection products Directive and Biocidal products Regulation (in relation to products that have disinfectant, preservative and pest-controlling properties). These directives address toxic chemicals and ensure the authorization of only those products that have no unacceptable effects on soil. It requires that ecotoxicological studies be carried out with respect to degradation, adsorption/desorption, mobility and the possibility of destruction or decontamination following the release of the products into soil. The GMO Directive requires an environmental risk assessment of the effects of releasing genetically-modified organisms on biogeochemical cycles, specifically on carbon and nitrogen recycling. These directives, which recognize the regulating function of soil, are intended to minimize the effects on soil and, therefore, can then be described as preventing a reduction of the function. However, they do not target the maintenance or improvement of this soil function.

The 7 EAP includes recommendations to integrate water protection into planning and decisions that are related to land use by reducing nutrient release from inefficient fertilizer management and inadequate wastewater treatment. The 7 EAP encourages investment in research and improvements in the coherency and implementation of Union environmental legislation to achieve these protective measures. The communication also recommends phasing out the deposit of recyclable or recoverable waste in landfills and more sustainable and resource-efficient management of the nutrient (nitrogen and phosphorus) cycle by 2020.

The storing, filtering and transformation function is mentioned in most policies, because this soil function is relevant to many sectors of land use. The goal in all of these policies is to protect soil resources (soil quality or avoiding soil contamination) by preventing the reduction of the function of soil as a medium that stores, filters or transforms contaminants to avoid risks to soil. However, simply avoiding contamination is not sufficient for improving the storing, filtering and transformation function in the future. This soil function is largely dependent on physical, chemical and biological properties, such as carbon content, soil pH and ground water level. Soil conservation measures are therefore needed for maintaining and improving this soil function over the long run. The CAP is intended to contribute to a minimum level of maintenance of this function by preventing soil erosion and maintaining soil organic matter and soil structure, particularly in high-threat areas.

Soil Function 3: Habitat and gene pool

Loss of biodiversity is receiving increased global awareness, but the biodiversity of soil is rarely considered. There is increased scientific interest in soil biodiversity, because soil provides habitats for many organisms, and many of functional traits of soil are yet to be discovered [53,54]. Such scientific interest includes platforms, such as globalsoilbiodiversity.org, as well as European Atlas of Soil Biodiversity (http://eusoils.jrc.ec.europa.eu/library/Maps/Biodiversity_Atlas/). Soil organisms play an important role in the release and/or retention of nutrients during the decomposition of organic matter. These organisms affect soil fertility and food production and, therefore, serve the grand societal challenge of food security (Figure 1). Soil biodiversity indisputably provides soil resistance and resilience against disturbance and stress, but the extent and dynamics of these effects are not completely understood [55]. Increasing attention is being given to the role of functional soil biodiversity, as contrasted with species diversity, for the provision and stability of soil processes and functions [53,54,56]. The soil fauna additionally serves as a large gene pool that could be a source of new drugs to fight infectious human diseases.

Several of the policies that we examined are intended to preserve biodiversity in general, but they do not mention soil biodiversity. The Habitat Directive includes the establishment of a coherent ecological network of special areas for the conservation of natural habitats and for the protection of wild fauna and flora within the EU (Natura, 2000). Member States are required to establish conservation measures to prevent the deterioration of natural habitats in these areas and to prevent the disturbance of the species for which the areas have been designated. The CAP supports agricultural practices and mitigation strategies that protect, improve, restore, preserve and enhance biodiversity. These include conservation practices in special areas (Habitat Directive) and establishment of ecological focus areas to safeguard and improve biodiversity on farms. These ecological focus areas should consist of areas that directly affect biodiversity, such as fallow land, landscape features, terraces, buffer strips, afforested areas and agroforestry areas, or that indirectly affect biodiversity by means of reduced use of inputs on the farm, such as areas that are covered by catch crops (fast-growing interseasonal crops) and green winter cover. Payments are given to farmers that convert to or maintain organic farming and to forest holders that provide environmentally-friendly or climate-friendly forest conservation services that are intended to enhance biodiversity. The CAP supports the exchange of best practices, training and capacity building and demonstration projects that relate to biodiversity. This is emphasized for projects that relate biodiversity and agroecosystem resilience, as contrasted with monocultures that are susceptible to crop failure or damage from pests and extreme climatic events. Although the Habitat Directive and the CAP are intended to improve biodiversity, they do not directly address soil biodiversity and, therefore, do not address the soil function, habitat and gene pool.

Many of the policies mention risk to fauna and flora. The GMO Directive addresses the long-term effects that the release of genetically-modified organisms have on the environment and on biological diversity and nontarget organisms. The Environmental liability and Mining waste Directive assesses the risks that are posed by harmful substances to organisms in the environment. The Renewable energy Directive does not target soil biodiversity directly, but it does target conservation of biodiversity in general by means of incentives when it can be proven that biofuel production does not originate in biodiverse areas (habitats). The Carbon storage Directive mentions that proximity to habitat conservation areas (as specified in the Habitat Directive) should be considered when choosing a new storage site. This directive additionally requires sensitivity tests on particular species that would be affected by leakage events. The tests involve the effects of elevated CO_2 concentrations, reduced soil pH and the effects of other substances that may be present in leaking CO_2 streams. Laws that apply to chemical substances, such as the Plant protection products Directive and the Biocidal products Regulation, ensure that authorization of chemicals occurs only after toxicity tests on the active substance, degradation products and additives show that there are no unacceptable effects on earthworms and other nontarget soil macro- and micro-organisms. The Plant protection products Directive additionally requires toxicity tests on soil microflora. The Pesticide use Directive promotes integrated pest management by means of such agricultural practices as crop rotation and biological control to suppress harmful organisms by low-pesticide pest management. Additionally, pesticides that are applied are required to be as target-specific as possible and have the fewest side effects on nontarget organisms and the environment. Note that none of these directives is intended to increase biodiversity. They are instead intended to reduce the deterioration of its function.

The Resource Efficiency Roadmap supports innovative solutions for the preservation of biodiversity and sets as a goal that improved efficiency in the transport sector will deliver reduced impacts on biodiversity by 2020. The Resource Efficiency Roadmap also supports increased biodiversity by means of good farming practices. Finally, the 7 EAP recommends the integration of biodiversity conservation into land-use planning and decisions. These non-binding EC communications also relate only to biodiversity in general and not specifically to soil biodiversity.

The habitat and gene pool function is frequently addressed in existing legislation (Table 2), but the targets of these laws are related to improving biodiversity in general and not to soil organisms (Table 3). The CAP includes measures for enhancing and improving biodiversity on farms, but it must be remembered that conventional agricultural production itself accelerates biodiversity decline. The policies that do target soil organisms address only the prevention of harm to nontarget organisms when, for example, plant protection chemicals are used. They do not address the decline in soil biodiversity or the need to maintain populations of organisms that are beneficial to soil. However, it is difficult to capture this aspect in legislation, because there is little knowledge of the significance of soil organisms and the diversity of functional traits among soil microbial communities. Neglecting soil biodiversity may have severe impacts on most of the other functions of soil (Figure 1). Addressing biodiversity in general and not soil biodiversity in these laws and communications neglects the abundance of soil organisms and their importance for soil quality [54].

Soil Function 4: Physical and cultural environment for mankind

The function of the physical and cultural environment for mankind relates to urbanization, recreational areas and nature tourism. This function is therefore strongly linked to land take, which is increasing rapidly in all Member States of EU [45]. This function does not relate directly to any of the grand societal challenges.

The CAP addresses rural development, which includes the creation and development of new economic activities that are related to healthcare and tourism in rural areas. The CAP also supports the development of local infrastructure and basic services in rural areas, which include leisure and culture services and renewal of villages. The Pesticide use Directive addresses this function by minimizing or prohibiting the use of pesticides in areas from which drinking water is extracted, along transport routes, on sealed or very permeable surfaces, in public parks, recreation grounds, school grounds and children's playgrounds and in proximity to healthcare facilities. This directive targets protection against pollution of areas that act as physical and cultural environments for mankind and, therefore, prevents the reduction of this soil function. The Landfill Directive also considers the distance from residential and recreational areas when locating landfills.

The Resource Efficiency Roadmap and the 7 EAP both target the growing issue of land take due to urbanization, and both have set a target of no net land take by 2050, as mentioned above under "Soil Threat 7: sealing".

This function is not mentioned extensively in soil-related policies. However, with the current trend of increasing land take [45], this function is not under threat, but is rather threatening the other functions, because this function often results in losses of other functions. For example, there are measures in the Landfill Directive that prevent the reduction of this soil function in one area by not locating landfills close to residential and recreational areas, but this reduces the soil function (establishing residential and recreational areas) in other areas where a landfill is already located. A recent study addresses how the impact of land take affects other functions, specifically food production [57]. They estimated that 19 EU countries lost approximately 0.81% of their potential agricultural production capacity between 1990 and 2006, with large variability between regions. Regions around the largest cities experienced the greatest loss of fertile soils [57].

Soil Function 5: Source of raw materials

Soil also functions as a source of minerals, fertilizers, gravel and other elements that are extracted or excavated by different industries. The grand societal challenge of resource efficiency and raw materials (Figure 1) requires proper and efficient use of this soil function. Development in the past century has been based on the ever-increasing use of natural resources. However, reduction in the current patterns of consumption is necessary if irreversible depletion of soil resources is to be avoided [58].

The Landfill and the Mining waste Directives both address preservation of soil as a source of raw materials. The Landfill Directive is intended to make the wasteful use of land unnecessary by encouraging prevention, recycling and recovery of waste and use of recovered materials in a resource-efficient way. The Mining waste Directive prohibits abandonment, dumping or uncontrolled deposition of extractive waste by putting it back into the space that was created by excavation after minerals have been extracted, by putting topsoil back in place after a waste facility is closed or by reusing topsoil elsewhere.

Under the Resource Efficiency Roadmap, the Commission will develop "innovation partnerships" for meeting resource efficiency goals that pertain to raw materials and will focus on Union research funding (EU Horizon 2020) and on key resource efficiency objectives that support innovative solutions for the management of natural resources and environmentally-friendly material extraction. The roadmap additionally sets milestones of no net land take by 2050, assuring a sustainable supply of phosphorus and reversing soil loss. It also promotes efficiency in the transport sector with optimal use of resources, such as raw materials, by 2020. The 7 EAP sets a goal of more sustainable and resource-efficient management of the nutrient (nitrogen and phosphorus) cycle.

The source of raw materials is not covered extensively in these soil-related policies. The mining industry promotes reuse of waste, but the mining directive is targeted only at the prevention of reduction of this soil function (Table 3). The rationale for improving soil that serves as a source of raw materials is unclear unless further extraction is prevented. In fact, a recent study considered the extraction of raw materials as a soil threat rather than a soil function [12].

Soil Function 6: Carbon pool

Soil has been estimated to store globally 1500 Gt of carbon [59] with 73 Gt of carbon stored in European topsoils [25] and 17.63 Gt in agricultural topsoils in Europe [60]. Soil carbon sequestration is especially important for the mitigation of the grand societal challenge of climate change (Figure 1). Carbon storage by soil is also very important for soil fertility, which ensures food and energy security (other grand societal challenges, Figure 1). Peatlands store particularly large amounts of carbon, and conversion of peatlands to arable land releases vast amounts of CO_2 into the atmosphere [60,61].

A comprehensive study has been carried out on estimating the effect of different agricultural management practices on the carbon sequestration of topsoils in Europe [60].

The CAP supports carbon sequestration in soil and maintenance of high organic matter levels in soil. However, an analysis of the CAP by Henriksen *et al.* [51] reports that although GAEC is an important component to encourage soil management practices for mitigating carbon stocks, there are failures of implementation in Member States. The Renewable energy Directive addresses the soil carbon pool by allowing land conversion for biofuel production only if the loss of soil carbon stock that is caused by conversion can be remediated by savings in greenhouse gas emissions that accrue from biofuel production within a reasonable period. The Commission provides incentives for sustainable biofuel production that minimizes the impacts of land use change. The directive attempts to avoid a net increase in arable land and related carbon losses by encouraging increased productivity on land that is already used for crops and encouraging the use of degraded land for biofuel production. Further, biofuel production is not allowed on land that has high carbon stocks, such as wetlands and forests. It should be noted that both the CAP and the Renewable energy Directive are intended to prevent the reduction of the soil carbon pool, but do not improve this function (Table 3). Additionally, the Renewable energy Directive implies that such carbon sources as crop residues, animal manure and other types of organic waste are not returned to the soil, which reduces the carbon pool of the soil.

To combat climate change, the Carbon storage Directive establishes a legal framework for environmentally safe geological storage of CO_2. However, this directive relates only to CO_2 storage in deep geological formations and is therefore not directly related to what is usually understood to be soil (for example, agricultural topsoil) that is involved in the soil carbon storage function. This is similar to the manner in which this directive addresses Soil Threat 2, organic matter decline of soils (Table 2).

The 7 EAP includes targeted priority objectives to sequester CO_2.

At a global level, the Kyoto protocol aims to protect and enhance greenhouse gas reservoirs and CO_2 sequestration technologies. The protocol states that countries shall formulate, implement and publish measures to mitigate climate change and to facilitate adequate adaptation to climate change. These efforts are to be based on the assessment of net changes in carbon stocks by sources and removals by sinks resulting from direct human-induced land use change and afforestation, reforestation and deforestation.

Soil Function 7: Geological and archaeological archive

Soil provides a geological and archaeological archive of natural and human history.

The CAP mentions preservation of the archaeological archive in the Rural Development Policy, which promotes protection of natural and cultural heritage by means of sustainable and responsible tourism in rural areas. The Floods Directive recommends establishment of a framework to assess and manage flood risks to reduce the adverse effects of floods on human health, the environment, cultural heritage and economic activity in the EU. The Landfill Directive addresses the protection of natural and cultural patrimony when locating landfills.

The geological and archaeological archive affects multiple sectors, but is not frequently mentioned in the policies that we reviewed. This may result from less public awareness of this function, because it is not mentioned in the media as often as, for example, biodiversity loss and climate change or because this soil function is not directly linked to the grand societal challenges (Figure 1).

In summary, all soil functions are addressed by existing legislation, but the usual focus is on the way in which soil currently serves a particular function and how to prevent a reduction of that particular function. Very few policies include targets that would improve the functions of soils over the long run (Table 3). Only nature conservation policies address maintenance and improvement of soil functions in a long-term perspective. The CAP includes some improvement strategies and does recommend a European innovation partnership (EIP) that would facilitate the establishment of pilot projects that would be related to soil functionality. However, as mentioned above, the CAP uses the nonspecific term "sustainable management" and does not directly address the specific threats to soil functions or provide specific targets for improving or maintaining soil functions. Other studies [62] have also criticized the subjectivity of the term "sustainable soil management." Further, the CAP is based on incentives that are given to farmers, and its provisions are not legally mandated. This results in the implementation of only some of the recommendations. It also implies that farmers would lose incentives that are important for their livelihoods if sustainable land use were to be made mandatory. Farmers also receive incentives for practices that prevent the acceleration of one soil threat even though those practices may cause the acceleration of other soil threats. For example, reduced tillage reduces erosion, but may increase contamination by increasing pesticide input [63].

4. Conclusions: Responses in Light of New Policy Targets

This cross-policy analysis shows that three of the seven soil threats, compaction, salinization and sealing, are not covered by existing legislation. Compaction and salinization are also not addressed in the EC communications. The decline in soil organic matter is barely mentioned. Biodiversity in general is addressed, but soil biodiversity is also barely mentioned. Soil erosion and especially soil contamination are two threats that are frequently addressed in existing legislation (Table 2). However, the analysis showed that almost all of the policies are intended to 'prevent acceleration of threats', but only a few target a reduction of the threats (Table 3).

The failure to address all seven soil threats threatens soil functions. The analysis also showed that all soil functions are addressed in the existing legislation, but nearly all of the directives are intended to prevent the reduction of a particular soil function. Few directives are intended to improve soil functions in the future (Table 3). It is therefore unclear if existing soil-related legislation is actually protecting the soil from soil threats and improving a soil function. Soil degradation is ongoing in Europe [23], which suggests that existing policies are not sufficient for maintaining soil functions. There appears to be a need for a new common, soil conservation policy at the European level. Soil degradation exists throughout the EU, but only a few Member States have enacted comprehensive national soil legislation [64]. The existing national soil protection laws of those Member States will not be threatened by common EU legislation, because Member States may adopt laws that are more protective than EU legislation. There are transboundary aspects of soil degradation even though soil is generally immobile; these include erosion, chemical contamination and international markets. European added value would also include the fact that common EU legislation would benefit internal market issues in cases where some Member States have strong soil conservation policies and others do not. Common legislation would also facilitate the export of expertise and technologies from the EU. The costs of inaction may surpass the costs of action within only a few years [65]. Furthermore, the cost of soil degradation is challenging due to both direct, indirect and non-use values of soil [62]. A further limitation for addressing the cost of soil degradation is the limited soil function monitoring [62]. When directives address soil functions individually, they neglect the multifunctionality of soil. Sustainable land use is often based on the multifunctionality of land or soil and is intended to maintain all soil functions [66]. Indeed, the specific functions of soil are site-specific and depend on the natural potential of soil to provide these functions. Often, these functions can be mutually exclusive, leading to trade-off situations. The multifunctionality of soil may be lost when soil functions are addressed separately in different directives.

A directive that is focused exclusively on soil might also be useful in the case of new technologies that affect soil (e.g., fracking). Common legislation could protect soil before specific laws that are related to new technologies can be passed.

Policy legislation and planning that maintain the non-economical functions of soil over the long run are required to ensure comprehensive soil functions. The policy legislation could be in the form of a Soil Framework Directive. This paper emphasizes that a common European soil conservation policy would provide added value to the EU by addressing the grand societal challenges that have been set forth by the European Commission. An IA based on soil functions provides the direct link to the societal value of soils and may better justify soil legislation. We believe that policies must address soil threats and functions directly to ensure that the threats and functions are targeted by new sustainable soil management practices. Because existing legislation fails to address soil threats and functions directly, a common European soil policy is needed to ensure the conservation of soil functions.

Acknowledgments: This paper is part of the research project, LIAISE (Linking Impact Assessment to Sustainability Expertise, www.liaisenoe.eu), which was funded by Framework Programme 7 of the European Commission and co-funded by the Dutch Ministry of Economic Affairs within the strategic research program "Sustainable spatial development of ecosystems, landscapes, seas and regions".

Author Contributions: The three authors of this paper have worked collaboratively since 2013 on the LIAISE research project, which is funded by the European Commission. All three authors have reviewed and commented on this manuscript.

Conflicts of Interest: The authors declare no conflict of interest.

References

1. European Commission (EC). Communication from the Commission to the Council, the European Parliament, the European Economic and Social Committee and Committee of the Regions "Towards a Thematic Strategy for Soil Protection" (COM(2002)179). Avaliable online: http://eur-lex.europa.eu/legal-content/EN/TXT/?uri=CELEX:52002DC0179 (accessed on 8 December 2014).
2. European Commission (EC). Communication from the Commission to the Council, the European Parliament, the European Economic and Social Committee and Committee of the Regions "Thematic Strategy for Soil Protection" (COM((2006)231). Avaliable online: http://eur-lex.europa.eu/legal-content/EN/TXT/?uri=CELEX:52006DC0231 (accessed on 8 December 2014).
3. European Commission (EC). Proposal for a Directive of the European Parliament and of the Council establishing a framework for the protection of soil and amending Directive 2004/35/EC (COM(2006)232). Avaliable online: http://eur-lex.europa.eu/LexUriServ/LexUriServ.do?uri=COM:2006:0232:FIN:EN:PDF (accessed on 8 December 2014).
4. Council of the European Union. Environment. Press Release 16183/07 (Presse 286). Avaliable online: http://www.consilium.europa.eu/ueDocs/cms_Data/docs/pressData/en/envir/97858.pdf (accessed on 8 December 2014).
5. Council of the European Union. Progress Report 6124/1/10 REV 1. Avaliable online: http://register.consilium.europa.eu/doc/srv?l=EN&f=ST%207100%202010%20INIT (accessed on 8 December 2014).
6. European Commission. Withdrawal of obsolete Commission proposals 2014/C 153/03. Avaliable online: http://eur-lex.europa.eu/legal-content/EN/TXT/?uri=OJ:C:2014:153:FULL (accessed on 8 December 2014).
7. European Commission. ANNEX to the communication from the commission to the European parliament, the council, the European economic and social committee and the committee of the regions on "Regulatory Fitness and Performance (REFIT): Results and Next Steps (COM(2013) 685). Avaliable online: http://ec.europa.eu/smart-regulation/docs/20131002-refit-annex_en.pdf (accessed on 8 December 2014).
8. European Commission (EC). Commission staff working document "Impact Assessment of the Thematic Strategy on Soil Protection" (SEC(2006)620). Avaliable online: http://ec.europa.eu/smart-regulation/impact/ia_carried_out/docs/ia_2006/sec_2006_1165_en.pdf (accessed on 8 December 2014).
9. European Commission (EC). Report from the Commission to the European Parliament, the Councel, the European Economic and Social Committee and the Committee of the Regions "The implementation of the Soil Thematic Strategy and ongoing activities" (COM(2012)46). Avaliable online: http://eusoils.jrc.ec.europa.eu/library/jrc_soil/policy/DGENV/COM%282012%2946_EN.pdf (accessed on 8 December 2014).
10. Robinson, D.A.; Lebron, I.; Vereecken, H. On the definition of the natural capital of soils: A framework for description, evaluation, and monitoring. *Soil Sci. Soc. Am. J.* **2009**, *73*, 1904–1911. [CrossRef]
11. Bouma, J.; McBratney, A. Framing soils as an actor when dealing with wicked environmental problems. *Geoderma* **2013**, *200–201*, 130–139. [CrossRef]
12. McBratney, A.; Field, D.J.; Koch, A. The dimensions of soil security. *Geoderma* **2014**, *213*, 203–213. [CrossRef]
13. Keesstra, S.; Mol, G. Soil Science in a Changing World, Wageningen Conference on Applied Soil Science. Available online: http://www.wageningensoilmeeting.wur.nl/UK/ (accessed on 8 December 2014).
14. Protection of soils functions—Challenges for the future. Avaliable online: http://proficiency-fp7.eu/index.php?option=com_content&view=article&id=242&Itemid=94 (accessed on 8 December 2014).

15. Blum, W.E.H. Soil protection concept of the Council of Europe and integrated soil research. In *Soil and Environment*; Kluwer Academic Publisher: Dordrecht, The Netherlands, 1993; pp. 37–47.

16. Bouma, J. Implications of the knowledge paradox for soil science. In *Advances in Agronomy*; Academic Press: Burlington, MA, USA, 2010; pp. 143–171.

17. Millennium Ecosystem Assessment. In *Ecosystems and Human Well-Being: A Framework for Assessment*; Island Press: Washington, DC, USA, 2003.

18. Carpenter, S.R.; de Fries, R.; Dietz, T.; Mooney, H.A.; Polasky, S.; Reid, W.V.; Scholes, R.V. Millennium Ecosystem Assessment: Research needs. *Science* **2006**, *314*, 257–258. [CrossRef] [PubMed]

19. Helming, K.; Diehl, K.; Geneletti, D.; Wiggering, H. Mainstreaming ecosystem services in European policy impact assessment. *Environ. Impact Assess. Rev.* **2013**, *40*, 82–87. [CrossRef]

20. Dunbar, M.B.; Panagos, P.; Montanarella, L. European perspective of ecosystem services and related policies. *Integr. Environ. Assess. Manag.* **2013**, *9*, 231–236. [CrossRef] [PubMed]

21. De Groot, R.; Wilson, M.A.; Boumans, R.M.J. A typology for the classification, description and valuation of ecosystem functions, goods and services. *Ecol. Econ.* **2002**, *41*, 393–408. [CrossRef]

22. Panagos, P.; van Liedekerke, M.; Jones, A.; Montanarella, L. European Soil Data Centre: Response to European policy support and public data requirements. *Land Use Policy* **2012**, *29*, 329–338. [CrossRef]

23. Jones, A.; Panagos, P.; Barcelo, S.; Bouraoui, F.; Bosco, C.; Dewitte, O.; Gardi, C.; Erhard, M.; Hervás, J.; Hiederer, R.; et al. *The State of Soil in Europe: A Contribution from JRC to the European Environmental Agency's Environment State and Outlook Report—SOER 2010*; Publications Office: Luxembourg, 2012.

24. Borrelli, P.; Panagos, P.; Ballabio, C.; Lugato, E.; Weynants, M.; Montanarella, L. Towards a pan-European assessment of land susceptibility to wind erosion. *Land Degrad. Dev.* **2014**. [CrossRef]

25. Jones, R.J.A.; Hiederer, R.; Rusco, E.; Montanarella, L. Estimating organic carbon in the soils of Europe for policy support. *Eur. J. Soil Sci.* **2005**, *56*, 655–671. [CrossRef]

26. Toth, G.; Adhikiri, K.; Várallyay, G.; Tóth, T.; Bódis, K.; Stolbovoy, V. Updated map of salt affected soils in the European Union. In *Threats to Soil Quality in Europe EUR 23438 EN*; Toth, G., Montanarella, L., Rusco, E., Eds.; Office for Official Publications of the European Communities: Luxembourg, 2008; pp. 65–77.

27. Jones, R.J.A.; Spoor, G.; Thomasson, A.J. Vulnerability of subsoils in Europe to compaction: A preliminary analysis. *Soil Tillage Res.* **2003**, *73*, 131–143. [CrossRef]

28. Panagos, P.; van Liedekerke, M.; Yigini, Y.; Montanarella, L. Contaminated Sites in Europe: Review of the Current Situation Based on Data Collected through a European Network. *J. Environ. Public Health* **2003**. [CrossRef]

29. Gabrielsen, P.; Bosch, P. *Environmental Indicators: Typology and Use in Reporting*; European Environmental Agency: Copenhagen, Denmark, 2003.

30. Rodrigues, S.M.; Pereira, M.E.; Ferreira da Silva, E.; Hursthouse, A.S.; Duarte, A.C. A review of regulatory decisions for environmental protection: Part I—Challenges in the implementation of National soil policies. *Environ. Int.* **2009**, *35*, 202–213. [CrossRef] [PubMed]

31. Bone, J.; Head, M.; Jones, D.T.; Barraclough, D.; Archer, M.; Scheib, C.; Flight, D.; Eggleton, P.; Voulvoulis, N. From chemical risk assessment to environmental quality management: The challenge for soil protection. *Environ. Sci. Technol.* **2011**, *45*, 104–110. [CrossRef] [PubMed]

32. Christensen, F.M.; Eisenreich, S.J.; Rasmussen, K.; Sintes, J.R.; Sokull-Kluettgen, B.; van de Plassche, E.J. European Experience in Chemicals Management: Integrating Science into Policy. *Environ. Sci. Technol.* **2011**, *45*, 80–89.

33. Temorshuizen, J.W.; Opdam, P. Landscape Services as a bridge between landscape ecology and sustainable development. *Landsc. Ecol.* **2009**, *24*, 1037–1052. [CrossRef]

34. Schulte, R.P.O.; Creamer, R.E.; Donnellan, T.; Farrelly, N.; Fealy, R.; O'Donoghue, C.; O'hUallachain, D. Functional land management: A framework for managing soil-based ecosystem services for the sustainable intensification of agriculture. *Environ. Sci. Policy* **2014**, *38*, 45–58. [CrossRef]

35. Lal, R. Soil degradation by erosion. *Land Degrad. Dev.* **2001**, *12*, 519–539. [CrossRef]

36. Prager, K.; Schuler, J.; Helming, K.; Zander, P.; Ratinger, T.; Hagedorn, K. Soil degradation, farming practices, institutions and policy responses: An analytical framework. *Land Degrad. Dev.* **2011**, *22*, 32–46. [CrossRef]

37. Reeves, D.W. The role of soil organic matter in maintaining soil quality in continuous cropping systems. *Soil Tillage Res.* **1997**, *43*, 131–167. [CrossRef]

38. Creamer, R.E.; Brennan, F.; Fenton, O.; Healy, M.G.; Lalor, S.T.J.; Lanigan, G.J.; Regan, J.T.; Griffiths, B.S. Implications of the proposed Soil Framework Directive on agricultural systems in Atlantic Europe—A review. *Soil Use Manag.* **2010**, *26*, 198–211. [CrossRef]

39. Horn, R.; van den Akker, J.J.H.; Arvidson, J. Subsoil compaction-distribution, processes and consequences. In *Advances in GeoEcology*; Catena Verlag: Reiskirchen, Germany, 2000.

40. Krümmelbein, J.; Horn, R.; Pagliai, M. Soil degradation. In *Advances in GeoEcology*; Catena Verlag: Reiskirchen, Germany, 2013.

41. Arvidsson, J. Nutrient uptake and growth of barley as affected by soil compaction. *Plant Soil* **1997**, *208*, 9–19. [CrossRef]

42. Sadras, V.O.; O'Leary, G.J.; Roget, D.K. Crop responses to compacted soil: Capture and efficiency in the use of water and radiation. *Field Crop Res.* **2005**, *91*, 131–148. [CrossRef]

43. Falloon, P.; Betts, R. Climate impacts on European agriculture and water management in the context of adaptation and mitigation—The importance of an integrated approach. *Sci. Total Environ.* **2010**, *408*, 5667–5687. [CrossRef] [PubMed]

44. Vanheusden, B. Recent developments in European policy regarding brownfield remediation. *Environ. Pract.* **2009**, *11*, 256–262. [CrossRef]

45. Tóth, G. Impact of land-take on the land resource base for crop production in the European Union. *Sci. Total Environ.* **2012**, *435–436*, 202–214. [CrossRef] [PubMed]

46. Louwagie, G.; Gay, S.H.; Sammth, F.; Ratinger, T. The potential of European Union policies to address soil degradation in agriculture. *Land Degrad. Dev.* **2009**, *22*, 5–17. [CrossRef]

47. Kibblewhite, M.G.; Miko, L.; Montanarella, L. Legal frameworks for soil protection: Current development and technical information requirements. *Curr. Opin. Environ. Sustain.* **2012**, *4*, 573–577. [CrossRef]

48. Food and Agriculture Organization (FAO). How to feed the world in 2050. In Proceedings of the Expert Meeting on How to Feed the World in 2050, Rome, Italy, 24–26 June 2009.

49. Makowski, D.; Nesme, T.; Papy, F.; Doré, T. Global agronomy, a new field of research: A review. *Agron. Sustain. Dev.* **2013**, *34*, 293–307. [CrossRef]

50. Tilman, D.; Balzerb, C.; Hill, J.; Befort, B.L. Global food demand and the sustainable intensification of agriculture. *Proc. Nat. Acad. Sci.* **2011**, *108*, 20260–20264. [CrossRef] [PubMed]

51. Henriksen, C.B.; Hussey, K.; Holm, P.E. Exploiting soil-management strategies for climate mitigation in the European Union: Maximizing "win-win" solutions across policy regimes. *Ecol. Soc.* **2011**. [CrossRef]

52. Keesstra, S.D.; Geissen, V.; Mosse, K.; Piiranen, S.; Scudiero, E.; Leistra, M.; van Schaik, L. Soil as a filter for groundwater quality. *Curr. Opin. Environ. Sustain.* **2012**, *4*, 507–516. [CrossRef]

53. Lavelle, P.; Decaëns, T.; Auberts, M.; Barot, S.; Blouin, M.; Burau, F.; Margerie, P.; Mora, P.; Rossi, J.P. Soil invertebrates and ecosystem services. *Eur. J. Soil Biol.* **2006**, *42*, S3–S15. [CrossRef]

54. Cluzeau, D.; Guernion, M.; Chaussod, R.; Martin-Laurent, F.; Villenave, C.; Cortet, J.; Ruiz-Camacho, N.; Pernin, C.; Mateille, T.; Philippot, L.; *et al.* Integration of biodiversity in soil quality monitoring: Baselines for microbial and soil fauna parameters for different land-use types. *Eur. J. Soil Biol.* **2012**, *49*, 63–72. [CrossRef]

55. Brussaard, L.; Pulleman, M.M.; Ouédraogo, E.; Mando, A.; Six, J. Soil fauna and soil function in the fabric of the food web. *Pedobiologia* **2007**, *50*, 447–462. [CrossRef]

56. Barrios, E. Soil biota, ecosystem services and land productivity. *Ecol. Econ.* **2007**, *64*, 269–285. [CrossRef]

57. Gardi, C.; Panagos, P.; van Liedekerke, M.; Bosco, C.; de Brogniez, D. Land take and food security: Assessment of land take on the agricultural production in Europe. *J. Environ. Plan. Manag.* **2014**. [CrossRef]

58. Salvati, L.; Zitti, M. Natural resource depletion and the economic performance of local districts: Suggestions from a within-country analysis. *Int. J. Sustain. Dev. World Ecol.* **2008**, *15*, 518–523. [CrossRef]

59. Scharlemann, J.P.W.; Tanner, E.V.J.; Hiederer, R.; Kapos, V. Global soil carbon: Understanding and managing the largest terrestrial carbon pool. *Carbon Manag.* **2014**, *5*, 81–91. [CrossRef]

60. Lugato, E.; Bampa, F.; Panagos, P.; Montanarella, L.; Jones, A. Potential carbon sequestration of European arable soils estimated by modelling a comprehensive set of management practices. *Global Change Biol.* **2014**, *20*, 3557–3567. [CrossRef]

61. Marmo, L. EU strategies and policies on soil and waste management to offset greenhouse gas emissions. *Waste Manag.* **2008**, *28*, 685–689. [CrossRef] [PubMed]

62. Robinson, D.A.; Fraser, I.; Dominati, E.J.; Davíðsdótti, B.; Jónsson, J.O.G.; Jones, L.; Jones, S.B.; Tuller, M.; Lebron, I.; Bristow, K.L.; *et al.* On the value of soil resources in the context of natural capital and ecosystem service delivery. *Soil Sci. Soc. Am. J.* **2014**, *78*, 685–700. [CrossRef]

63. Lahmar, R. Adoption of conservation agriculture in Europe—Lessons of the KASSA project. *Land Use Policy* **2010**, *27*, 4–10. [CrossRef]

64. Kutter, T.; Louwagie, G.; Schuler, J.; Zander, P.; Helming, K.; Hecker, J.M. Policy measures for agricultural soil conservation in the European Union and its member states: Policy review and classification. *Land Degrad. Dev.* **2011**, *22*, 18–31. [CrossRef]

65. Nkonya, E.; von Braun, J.; Mirzabaev, A.; Le, Q.B.; Kwon, H.Y.; Kirui, O. Economics of land degradation initiative: Methods and approach for global and national assessments. Avaliable online: http://ssrn.com/abstract=2343636 (accessed on 8 December 2014).

66. Helming, K.; Pérez-Soba, M. Landscape scenarios and multifunctionality: Making land use impact assessment operational. Avaliable online: http://www.ecologyandsociety.org/vol16/iss1/art50/ (accessed on 8 December 2014).

Review

The State of Soil Degradation in Sub-Saharan Africa: Baselines, Trajectories, and Solutions

Katherine Tully [1,2,*], Clare Sullivan [2], Ray Weil [3] and Pedro Sanchez [2]

[1] Department of Plant Science and Landscape Architecture; University of Maryland, College Park, Maryland, MD 20742, USA
[2] Agriculture and Food Security Center, Earth Institute at Columbia University, Palisades, NY 10964, USA; csullivan@ei.columbia.edu (C.S.); psanchez@ei.columbia.edu (P.S.)
[3] Department of Environmental Science and Technology; University of Maryland, College Park, Maryland, MD 20742, USA; rweil@umd.edu
* Correspondence: kltully@umd.edu; Tel.: +1-301-405-1469; Fax: +1-301-314-9306

Academic Editor: Marc A. Rosen
Received: 4 February 2015; Accepted: 5 May 2015; Published: 26 May 2015

Abstract: The primary cause of soil degradation in sub-Saharan Africa (SSA) is expansion and intensification of agriculture in efforts to feed its growing population. Effective solutions will support resilient systems, and must cut across agricultural, environmental, and socioeconomic objectives. While many studies compare and contrast the effects of different management practices on soil properties, soil degradation can only be evaluated within a specific temporal and spatial context using multiple indicators. The extent and rate of soil degradation in SSA is still under debate as there are no reliable data, just gross estimates. Nevertheless, certain soils are losing their ability to provide food and essential ecosystem services, and we know that soil fertility depletion is the primary cause. We synthesize data from studies that examined degradation in SSA at broad spatial and temporal scales and quantified multiple soil degradation indicators, and we found clear indications of degradation across multiple indicators. However, different indicators have different trajectories—pH and cation exchange capacity tend to decline linearly, and soil organic carbon and yields non-linearly. Future research should focus on how soil degradation in SSA leads to changes in ecosystem services, and how to manage these soils now and in the future.

Keywords: soil degradation; sub-Saharan Africa; baselines; indicators; sustainability; resilience

1. New Perspectives for Examining Soil Degradation in Sub-Saharan Africa

Soil degradation is a major global problem, the effects of which may be felt most strongly in developing countries where large proportions of the population reap their livelihoods directly from the soil. In this review, we will focus on soil degradation in sub-Saharan Africa (SSA), where declines in crop productivity have been linked to hunger and poverty [1,2]. While the reality of hunger in SSA is undeniable, the nature and extent of soil degradation, and the role it plays in the vicious cycle of poverty, is still under debate [3]. Across SSA, 75 percent of the population depended on subsistence farming at the end of the last century [4,5]. Livelihoods are diversifying [6] and urbanization is on the rise [7], but in the near-term, soils in SSA must currently sustain a largely subsistence population. Using the Brundtland Commission's definition of "sustainability", sustainable soils meet the needs of present populations without preventing future generations from meeting their needs [8]; thus, soil degradation can be defined in contrast to this, as the processes by which soils can no longer maintain the provisioning, supporting and regulating ecosystem services required by current and future generations. In order to reverse soil degradation, it is critical to understand the factors that affect the stability and resilience of soils.

Unfortunately, there are few data on soil degradation across SSA, so rigorous assessment frameworks are lacking to guide research on the topic. In this review, we will highlight the handful of studies that have evaluated soil degradation in SSA in a comprehensive way by clearly defining the (1) temporal and (2) spatial scale of analysis and (3) examining multiple degradation indicators. We then provide a description of useful methods for measuring degradation in remote regions. Finally, we will provide a brief overview of practices that may reverse soil degradation in SSA.

1.1. Time Horizons

Long-term data are crucial for evaluating soil degradation, as a snapshot of soil properties can be misleading. Soil phosphorus (P) levels in tropical forests, for example, can fluctuate within a day [9], year [10], and across centuries [11,12]. Capturing one point in time could incorrectly suggest soil P depletion or P surplus. Humans can drive change in soils. Their activities, such as farmer management practices, play a large role in soil degradation and may vary greatly between seasons and across years [13,14]. Thus, longitudinal studies that follow specific sites for years provide the most reliable data on the changes in soil properties over long time scales. Unfortunately, longitudinal studies require continuity of access to study sites, funding, and infrastructure. While difficult to secure in any region, this is especially true in SSA, where land tenure, political systems, and local markets are frequently unstable, and there has been low and inconsistent investment in national universities and research institutions.

Chronosequences are often used in place of longitudinal studies and substitute space for time. A primary assumption of chronosequence studies, with respect to soil degradation, is that the soil properties at sites characterized by different times since conversion to agriculture were initially the same when under natural vegetation. This approach further assumes that differences among these sites represent the trajectory of change in soil properties during periods of cultivation. While this approach can be useful, it is limited by (1) the fact that farmers tend to clear the best land first; (2) ability to find sites that have similar soil textures and horizon structures; and (3) selection of an appropriate benchmark or baseline site. We will examine a number of chronosequences to evaluate and contextualize their findings.

In order to understand the extent of soil degradation in SSA, we need clear baselines from which to examine the differences in physical and chemical properties. Studying fossil plants (e.g., pollen grains and macrofossils) allows scientists to reconstruct the history of forest loss [15], and river sediments to provide insights into erosion rates over several centuries [16]. Still, there is a paucity of data on early forest cover and practically no data on historical soil fertility in SSA (even from this last century). Appropriate selection of a baseline or reference state is particularly crucial for any study on degradation. When a forest becomes a farm, a land use shift occurs, and suddenly, the controls on ecosystem structure and function change as the system settles into a different state (stability domain) [17,18]. For example, monitoring the system on any stable branch before or after the switch would lead one to conclude that little change occurred, but monitoring during the rapid state change might suggest "catastrophic" losses in SOC [17]. Thus resilience, like soil degradation, must be evaluated over a long time period in order to observe the ability or inability of the ecosystem to continue to perform its desired functions when confronted with stress or external shocks [19].

Sub-Saharan Africa itself underwent a major land use change about 3000 years ago when much of the Central African rainforest was rapidly replaced by savannas. Though often linked to climate change, recent evidence suggests that the transformation may have been related to a major population expansion of the Bantu people across Central Africa, which led to the clearance of vast swaths of land for shifting cultivation and charcoal production [20]. Such strong ecosystem shifts indicate that ecosystem resilience itself can be changed or degraded by both natural and human forcings. At the same time, the persistence of ecosystems and societies suggests that resilient systems must be adaptive systems [21,22]. The resilience conceptual framework is particularly useful for evaluating soil degradation in SSA as both degradation and resilience must be evaluated within its spatial, temporal, economic, environmental, and cultural context [23].

1.2. Spatial Scales

Sub-Saharan Africa is an enormous region of 24.6 million km^2, with a huge range of soil and land management types [24]. The predominant soils (Table 1) are Arenosols (21.5%), Cambisols (10.8%), and Ferralsols (10.4%), and Leptosols (17.5%). The type and degrees of soil constraints vary widely. Nearly 40% of soils in SSA are low in nutrient capital reserves (<10% weatherable minerals), 25% suffer from aluminum toxicity, and 18% have a high leaching potential (low buffering capacity; [25]; Table 3). A region's initial soil fertility will affect the extent of soil degradation—with regions of low soil fertility degrading more quickly than regions with higher natural soil fertility. If (plant-available) soil nutrient stocks are initially high, the process of nutrient depletion can take a long time, but the absolute amount of nutrients lost will be high. However, if nutrient stocks are low to begin with, this process can reach critical levels within a few years. Further, inherent soil properties will play a large role in resilience and sustainability of a particular land use (e.g., how long continuous agriculture remains productive). For example, anion exchange capacity in subsoils will affect the ability of soils to retain and efficiently recycle nutrients (in particular, anions like NO_3^-; [26,27]). These subsoil properties are highly spatially variable [28,29] and often ignored in soil degradation studies—only two out of 18 studies in Table 4 reported subsoil properties.

Table 1. Distribution of soil types in Africa based on the Harmonized World Soil Database. Modified from [24].

	Million ha in Africa	Percent of Land in Africa *
Acrisol	87.8	2.9
Alisols	20.3	0.7
Andosols	4.0	0.1
Arenosols	650.3	21.5
Chernozems	1.0	<0.1
Calcisols	161.0	5.3
Cambisols	325.4	10.8
Durisols	0.9	<0.1
Fluvisols	82.2	2.7
Ferralsols	312.4	10.3
Gleysols	52.5	1.7
Gypsisols	37.5	1.2
Histosols	4.4	0.1
Kastanozems	2.7	0.1
Leptosols	530.0	17.5
Luvisols	105.1	3.5
Lixisols	126.8	4.2
Nitisols	60.4	2
Phaeozems	12.1	0.4
Planosols	27.7	0.9
Plinthosols	146.1	4.8
Podzols	2.9	0.1
Regosols	93.5	3.1
Solonchaks	32.6	1.1
Solonets	36.0	1.2
Stagnosols	0.5	<0.1
Technosols	0.0	<0.1
Umbrisols	5.6	0.2
Vertisols	102.0	3.4

* Note that percentages do not add up to 100% as soil may be affected by multiple soil modifiers.

Soil degradation occurs at multiple scales: a farm field (individual), a farming community (social system), or landscape (biophysical system). There is no single scale at which it must be studied, but it is critical that the chosen spatial scale of analysis can encompass the type of soil degradation being described. For example, the presence of gullies in farms is usually indicative of a change in land use upstream (at the head of the watershed) such as heavy grazing or excessively mechanized agriculture, which leads to erosion or contamination downstream [30]. In SSA, this raises some interesting cultural concerns, because uplands and foothills will surely be managed by different households (landholdings are small in SSA). In some cases, neighboring areas are managed by different ethnic groups, with pastoralists of one ethnic group grazing cattle upslope from agriculturalists of a different ethnic group. Clearly, solving landscape-level erosion issues requires community cooperation across agroecological zones that may cross ethnic and cultural lines.

Most studies in the literature compare and contrast management practices [31–34] or examine one farming practice across different regions [5,35]. There are relatively few studies that attempt to examine soil degradation at a scale that can encompass the spatial and temporal heterogeneity of farmed landscapes in SSA. Although a great deal of soil data exists for Africa, there is little standardization in the sampling design or analytical tests conducted. The Africa Soils Information System is an example of how this situation may be remedied in the future by standardized protocols that examine change at large spatial scales through time [36].

1.3. Multiple Indicators

When evaluating soil degradation, it is important to define the particular ecosystem function, management practice, and/or livelihood outcome you are trying to sustain [19], which usually cannot be captured by one soil property or indicator. Certain soil properties may be considered "degraded" for a particular crop, but not for another [37,38]. For example, higher soil residue cover may prevent N losses during the non-growing season (good for the environment), but lead to reduced available N during the following growing season (bad for yields [39,40]). While some indicators of degradation are incontrovertible (e.g., gully formation), others are evaluated subjectively (e.g., livestock walk longer to reach water; [41]). It was this subjectivity that led to the heated debates of the 1990s surrounding soil degradation in SSA. Some studies suggested that SSA agriculture was inherently unsustainable, and indicated losses of productivity due to erosion and declines in soil fertility at continental [42,43] and global scales [44]. However, estimations of the extent and rate of degradation was limited by an overall lack of biophysical data on Africa, and thus relied heavily on estimations of one indicator (namely, erosion, which was modeled not measured) and interpolation when scaling-up to regions and countries [3]. Many refuted the claim that farmers were to blame for the increased rates of soil degradation and suggested that more attention should be paid to farmer knowledge and adaptability [45–48]. It is not the goal of this review to resolve this debate, rather, we offer a critical examination of the works that have followed in its wake. We find that even decades later, there are very few studies that have comprehensively measured soil degradation in SSA.

2. Soil Degradation in Sub-Saharan Africa

2.1. Drivers of Degradation

Sociopolitical and economic drivers determine (1) where; (2) which; and (3) how many people live in a given region. In many cases, the poorest people in SSA are pushed into unproductive lands, or areas with minimal infrastructure and accessibility [49]. One of the most extreme examples of this is Tanzania's Ujamaa "villagization" campaign of 1973–1976, where over five million rural residents were relocated from their dispersed family homesteads into concentrated settlements [50]. The social and ecological effects of this major resettlement campaign are evinced in the replacement of fallow cycles with intensified, continuous cropping systems.

The tenure system often determines how land is managed and used and thus is often implicated as a primary driver of degradation [51,52]. For example, in smallholder systems in East Africa, investments in soil fertility are more likely when there is security in tenure or ownership [53]. For those who have tenure, policies that raise the farm-gate prices of commodities are critical means for encouraging good land management strategies since they provide farmers with both resources and incentives [48]. Smallholder farmers in SSA often operate at the economic "margin" where agricultural investments are a lower household priority than school fees, medical treatment, or funeral costs [53]. Farmer wealth and ethnicity often determines whether soil degradation can be addressed on the farm. Wealthier farmers, who have more access to resources, are better equipped to cope with soil degradation [54].

Gender roles have direct input on household foods security and nutritional levels [55]. Men are often forced to seek jobs in urban areas leaving women to tend to the land, but without the primary decision-making power. Women and men also tend to invest differently in soil fertility management, with women more likely to adopt organic amendments like manure and men more likely to purchase mineral fertilizer [56]. Population density in farming communities will also have a large impact, either positive or negative, on degradation potential. High population density usually means little land available for rotation into natural vegetation fallow. However, low population density may result in labor shortages and long distance from homestead to fields. Labor shortage is a primary reason why labor-intensive conservation measures have low adoption rates in many regions of SSA [57].

2.2. Types of Degradation in Sub-Saharan Africa

Soils can be altered physically, chemically, or biologically as the result of natural processes (Table 2). For example, soil itself forms over millennia through physical and chemical weathering of rocks (morphogenesis/soil formation). Wind erosion shifts the dunes in sparsely vegetated deserts, and transports dust to other continents. Humans, however, are accelerating many of these natural processes, causing the degradation of soils.

Physical degradation can occur when excessive soil tillage breaks down soil aggregates; thus rapidly decomposing organic matter, loosening the soil in excess and making it vulnerable to wind and water erosion. Cultivation on steep slopes, clearing of vegetation (especially leaving land bare between cultivation cycles), and poorly managed grazing are the primary factors accelerating soil erosion in SSA [58]. High rates of topsoil loss contribute to downstream sedimentation and degradation of local and regional water bodies. For example, in Tigray, Ethiopia, reservoirs designed to improve water access with a 20-year lifespan, lost half of their storage capacity in only five years due to sedimentation [59]. Tillage itself—independent of wind and water—also moves a great deal of soil downslope. This is especially evident on steep, short slopes where hand or animal traction tillage moves the soil preferentially in the easier downslope direction [60]. Poorly managed grazing in pastureland can also contribute significant amounts of sediment downstream [61]. Poor management of both grazing and tillage can lead to compaction of surface or subsurface soil layers [62], and in turn to reduced infiltration [63].

Table 2. Major types of soil degradation and the conditions under which they are most commonly found. Although the table separates physical, chemical and biological degradation, in reality soils are complex systems in which these processes interact and influence one another. The first three processes listed, erosion by water, wind and tillage, together dominate soil degradation on the vast majority of land area degraded. (Modified from [64]).

Category	Specific degradation processes	State factors		Socioeconomic drivers
		Parent material and topography	Climate	
Physical	Soil erosion by water	Slope	Humid to semi-arid regions	Tillage agriculture, deforestation and improper grazing
	Soil erosion by wind	Less vegetation	Semi-arid to arid regions	Disturbance of soil, vegetation or bio-crust by agricultural tillage and poorly-managed grazing
	Soil erosion by tillage	Hilly landscapes		Continuous cultivation, especially with tillage
	Surface sealing	Low organic matter sandy or silty soils		Urbanization, compaction, tillage
	Soil compaction	Clayey soils	Humid regions	Heavy machinery, grazing
	Reduced capacity to store water	Low organic matter		Compaction, erosion, removal of mulch or residue
Chemical	Nutrient depletion	Low inherent fertility		Low input agriculture, grazing, excessive forest harvest
	Acidification	Old, weathered soils	Humid regions	Excessive N fertilization, leaching, sulfur and nitrogen oxidation
	Dispersion/alkalization	Excessive monovalent ions, exposure and incorporation of calcareous subsoil material into surface horizon		Poor quality irrigation water, loss of perennial vegetation, tillage
	Salinization	Shallow water table	Arid to semi-arid regions	Excessive irrigation
	Toxic Contamination			Urbanization, mining, industrial waste spillage or disposal
Biological	Depletion of soil organic matter	Sandy texture, steep slopes, deep water table	High temperatures, limited rainfall	Degradation of vegetation, excessive tillage, lack of sufficient organic amendments and plant residues; excessive biomass removal by harvest, grazing or fire; erosion of sloping surface soil by tillage, wind and water
	Loss of soil biological diversity	Sandy texture, steep slopes, root limiting subsoil layers (fragipans, cemented layers, aluminum toxicity, calcic horizons)	High temperatures	Mono-cropping, deforestation and poorly managed grazing
	Loss of plant, animal and microbial biomass	Side slopes, shallow bedrock, root limiting subsoil layers (fragipans, cemented layers, aluminum toxicity, calcic horizons)		Reduced plant growth and subsequent addition of litter, roots and exudates limits carbon fuel for food web; exposure to extremes of dryness and temperature by removal of plant litter; destruction of macropores, aggregates and other habitat by tillage, compaction and erosion.

Unlike physical degradation, chemical soil degradation it not easily observed by the naked eye. Nutrient depletion is the primary form of soil degradation in SSA. For decades, across SSA, nutrient outputs have exceeded inputs, exhausting soil nutrient pools. Partial nutrient balances (or budgets) are typically used to describe the stocks and fluxes (ins and outs) of a soil [65]. They have been calculated for many different regions and countries [66], and are often used in Africa to evaluate management practices that promote nutrient surpluses or deficits [42,67–69]. In many SSA farming systems, certain soils suffer from nutrient depletion even if the whole farm or farming community does not. This pattern of nutrient depletion has been documented in many studies that show how nutrients are transported from "out fields" to fields near the homestead in the form of crops harvested and animal manure deposited [68,70].

Soils in SSA also suffer from declining cation exchange capacity, cation imbalances, and declining soil pH (which can lead to Al toxicity; Table 3). Secondary soil acidification can occur due to long-term application of relatively high rates of N fertilizers (mostly in South Africa) or continuous cropping without organic inputs [71]. In certain coastal area (e.g., Senegal, Gambia), lowering of the water table for crop production has led to formation of active acid sulfate soils and extreme acidity (pH < 3.5) [72]. Alkalization can also occur when perennial vegetation is lost, or when calcareous subsoil material is incorporated into the topsoil as a result of erosion or tillage [73]. Other forms of chemical degradation such as salinization, while common in other tropical soils, is less common than alkalization in SSA [74] (Table 3).

Table 3. Prevalence of soil constraints in sub-Saharan Africa based on the fertility capability soil classification (FCC) system [25,75].

Soil Constraint	Modifier	Million ha in SSA	Percent of Land in SSA *
Low nutrient capital reserves	k	942.06	39.94
Al toxicity	a	588.27	24.94
High P fixation	i	200.35	8.49
Steep sloped (>30%)	s	55.62	2.36
Poor drainage	g	159.95	6.78
High leaching potential	e	425.05	18.02
Calcareous reaction	b	158.11	6.70
Salinity	s	19.09	0.81
Alkalinity	n	52.06	2.21
Allophane	x	2.83	0.12
Shrink-swell	v	132.65	5.62
Total area		2358.79	

* Note that percentages do not add up to 100% as soil may be affected by multiple soil modifiers.

Biological degradation is closely linked to chemical degradation. Both the balance of different nutrients and their chemical forms are also important to soil fertility [76,77]. Population pressures in some countries have reduced or eliminated natural fallow periods, reducing nutrient and organic matter inputs [3,78,79] and thus causing declines in soil biological activity and soil species diversity [80–82] Reductions in organic matter can reduce porosity [83,84] and infiltration capacity and therefore change water and nutrient cycles, plant productivity, and even the energy balance of a system [85,86]. The abundance and biodiversity of soil organisms is reduced as a result of intensive grazing, biomass burning (either of crop residue or for land clearing) [87], tillage and bed preparation [88], leaving soils bare, mono-cropping, especially in maize growing areas, and excess fertilizer application [82,89]. Such changes in the soil diversity (or functional diversity) of soil biota can affect the availability of nutrients [90,91] and alter pest and disease pressure [81] as well as the complexity of food-webs [81] with consequences for ecosystem resilience.

3. Synthesis of Knowledge

While the African subcontinent is often at the nexus of discussions on soil degradation, a relatively small number of studies rigorously assess it. We define rigorous assessments as studies having:

(1) A temporal dimension, as degradation is a dynamic process;

(2) A spatial scale of analysis that is meaningful both for assessing degradation and for providing soil management recommendation for smallholder farmers; and

(3) Multiple criteria of assessment that reflect the use of the soil because degradation results from a complex set of processes and cannot be captured in a single measure.

We identified 18 studies that meet these criteria (see Table 4). We classified these studies into three groups: longitudinal studies, chronosequences, and integrated assessments.

3.1. Methods for Data Synthesis

Information on the temporal and spatial scale, indicators measured, *etc.* from each study is reported in Table 4. We also extracted data from 15 of those studies that reported soils data. We extracted data from four studies in annual crops (e.g., maize) that reported cation exchange capacity (CEC) from soils collected from 0–10 or 0–15 cm depth. In all four studies, CEC was measured at pH 5.5–7.5, and calculated by summing the base cations. Study sites had similar clay contents (~20%) and bulk densities (66 g cm^{-3}) and did not report data from an uncultivated site, thus we report raw CEC data. Thirty-year trends in soil pH are reported for red soils near Holetta Research Center, Ethiopia. These data are previously unpublished (Appendix A). Soil organic carbon (SOC) data were extracted from three published studies plus unpublished data from the Holetta red soils (R. Weil; Appendix A), all of which used the Walkley-Black method for SOC determination. To normalize the data from different soil types and agroecological zones, we calculated the percent SOC remaining and plotted against time since conversion. Data on maize yields were reported in tons ha^{-1} from two regions: western Kenya and southwestern Nigeria. In some cases, the farm field age was not reported, thus we used reported sampling dates and the date of forest clearance to calculate the time since forest conversion. To avoid any site or sampling bias, we plotted maize yield data separately for the two regions. When data were reported in graphical form, they were extracted using GraphClick 3.0 (Arizona Software, 2008). Figures and statistics were performed in the R statistical package [92].

Table 4. Published studies examining soil degradation across large spatial and temporal scales using multiple indicators.

Reference	Study Type	Select Indicators of Degradation Quantitative	Qualitative	Temporal scale	Spatial scale	Baseline (Reference)	Depth	Region	Trajectory
[93]	Chrono	Particle size, Water holding capacity, SOM, Exch. Ca, Exch. K, Exch. Mg, total N, Ext. P, pH, and CBC	NA	15 years	Landscape		0–20 cm	Nigeria	Downward
[94]	Chrono	Soil spectra, total C, Exch. Mg, Exch. Ca, Exch. K, total N, pH, ECEC, Clay, Silt, and Sand	NA	100 years	Landscape	Humid tropical forest	0–20 cm	Kenya	Downward
[95]	Chrono	Total N, pH, SOM, Sand, Silt, Clay, Bulk density, Tree density, Tree species	NA	50 years	Landscape	Tropical dry Afro-montane forest (deforested/heavy harvesting)	0–100 cm	Ethiopia	Downward
[16]	Long	Soil erosion (water-induced), Sediment flux, River discharge, and Coral Ba/Ca	NA	300 years	River basin (66,800 km^2)	None	NA	Kenya	Downward
[78]	Long; Integ	Land use and land cover, Trees in fields, CEC, Exch. Ca, Exch. K, Exch. Mg, total N, Ext. P, pH, and SOC	Farmer mgmt, perception of change, veg cover	15 years (imagery); 8 years (soils)	Multi-scale (Landscape and farm field)	1981—imagery; 1988—soils	0–20 cm	Burkina Faso	Minimal change to upward (field scale), Possibly downward (landscape scale)
[96]	Long	Exch. Ca, Exch. Mg, ECEC, SOC, pH, bulk density, maize grain yield	NA	13 years	Landscape	Tropical forest	0–15 cm	Nigeria	Mixed dependent on management strategies: Decline without fallow or addition of organic input
[97]	Chrono	Total N, Ext. P, SOM, Maize biomass, Plant tissue (N, P, K, Ca, Mg, Mn, Cu and Zn), Socioeconomic survey	Crop yield, Indicator plants, Soil softness and Soil color	57 years	Landscape	Tropical dry Afromontane forest (deforested/heavy harvesting)	0–20 cm	Ethiopia	Downward (maize biomass)
[98]	Chrono	CEC (effective and potential), pH, SOC, Grain and stover yield, Plant tissue: N, P, K, Ca, and Mg	NA	100 years	Landscape	Humid tropical forest	0–10 cm	Kenya	Downward (non-linear)
[99]	Long	Land cover classes, Precipitation, Socioeconomic survey, Soil chemical properties	Incidence of soil erosion	40 years	Landscape	Baseline (1966)	NA	Tanzania	Spatially heterogeneous (Downward in some zones)
[100]	Long	CEC, Exch. Ca, Exch. K, Exch. Mg, pH, total N, Ext. P, SOC, Bulk density, Infiltration, Penetrometer resistance, Soil moisture retention, Water stable aggregates, and Yield	NA	8 years	Farm field (Field trial)		0–20 cm	Nigeria	Downward (dependent on management)

Table 4. *Cont.*

Reference	Study Type	Select Indicators of Degradation Quantitative	Select Indicators of Degradation Qualitative	Temporal scale	Spatial scale	Baseline (Reference)	Depth	Region	Trajectory
[79,101]	Chrono	Soil depth, Base Saturation, % of CEC, C:N, Exch. Ca, Exch. K, Exch. Na, Total N, Ext. P, pH, SOC, Bulk density, Particle size analysis, Pore space, 13C and 15N, carbon fractions	Qualitative land evaluation for maize	53 years	Landscape	Tropical dry Afro-montane forest (deforested/heavy harvesting)	0–20 cm; 60–70 cm, 90–100 cm	Ethiopia	Downward (C-exponential) in topsoil, C & N increase in subsoil
[102]	Chrono	Active C, CEC, Exch. Ca, EC, Exch. K, Exch. Mg, pH, Total N, Ext. P, S, SOM, Zn, Sand, Silt, Clay, Water stable aggregation (WSA), Available water capacity (AWC), Penetrometer resistance, Crop yield	NA	77 years	Landscape	Humid tropical forest	0–15 cm, 0–45 cm	Kenya	Downward in most properties, slope of trajectory less severe with better soil management
[103]	Chrono	Mineral N, P fractions, P sorption capacity, Fertilizer recovery, Maize yield, Maize nutrient concentration	NA	100 years	Landscape	Humid tropical forest	0–10 cm	Kenya	Downward trend in soil fertility; yield increased dependent on nutrient additions
[104]	Chrono	Soil C & N concentration, Isotopic signature of soil C, Infiltrability, Bulk density, Proportion of macro and micro-aggregates in soil	Crop yield estimates	120 years	Landscape	Humid tropical forest	0–15 cm	Kenya	Downward
[105]	Long	EC, Exch. K & Exch. Mg, Ext. P, pH, SOM, and Plant tissue analysis (N, P, K, Ca, Mg, S, Zn, B, Mn, Fe, Cu and Al)	NA	7 years	Sub-national	Baseline (1991)	0–15 cm	Gambia	Minimal change
[106]	Chrono	13C, Near-edge X-ray absorption fine structure, SOC,	NA	103 years (Kenya); 90 years (South Africa)	Landscape	Humid tropical forest (Kenya); Subtropical grassland (South Africa)	0–10 cm (Kenya); 0–20 cm (South Africa)	Kenya; South Africa	Downward (exponential)
[41]	Chrono; Integ	N, P, K, SOC, Woody and herbaceous species, Land cover change	Soil properties Livestock Yield, Pests, Trees	50 years (soil); 15 years (imagery)	Landscape	Grass strips adjacent to fields	NA	Botswana and Swaziland	Downward
[107]	Chrono	CEC, Exch. Ca, Exch. K, Exch. Mg, pH, total N, Ext. P, SOC, Clay, Silt, SFI, Surface reflectance, Soil spectra	Soil quality - poor, average, good	50 years	Landscape	Rainforest	0–20 cm	Madagascar	Downward

145

3.2. Longitudinal Studies

We identified six studies that go beyond the traditional long-term trials to examine soil degradation in SSA. In sum, these studies indicate that rates of soil degradation vary through time (are non-linear) and that not all indicators behave the same way. The longest study is the best example of this, which uses coral barium to calcium ratios from the Malindi reef to evaluate sediment transport (erosion) from the Sabaki river basin in Kenya [16]. Sediment flux was relatively low and consistent from 1700 to 1905, but rises after 1905, corresponding to the start of British settlement and land clearing, and periodic spikes that can be traced back to historical changes in land management. This study clearly shows that picking one point (or a small portion) along the timeline does not capture the dynamics of soil degradation. While a study in Nigeria showed steady declines in pH, soil organic carbon (SOC), and available P (over eight years; [100]), a similar study in Gambia (over 1159 fields) showed no changes in any of those soil properties (over six years) [105]. Seemingly conflicting results may be due to the fact that sites are at different points along a non-linear curve. For example, a 13-year study in Nigeria showed non-linear trends in many indicators, with SOC and maize yields declining in the first seven years of the study (similar to [100]), and reaching a steady state for the remainder of the study (similar to [105]; Figure 1d). On the other hand, soil pH, exchangeable calcium and magnesium, and effective CEC all declined linearly with each year of continuous cultivation [96]; Figure 1a,c). A final study showed different conclusions about degradation could be drawn from different indicators. The comparison of land-cover maps for the Monduli District in northeast Tanzania showed a 94% increase in agricultural, but only a 16% decline in vegetation between the 1960s and the 1990s. Using only one of these indicators would easily lead one to different conclusions regarding the extent of degradation. Between the 1991 and 1999, however, was the rapid increase (by almost 1700%) in the presence of gullies and bare land, (equivalent to 1400 ha per year across 400,000 ha [99]).

3.3. Chronosequences (Space-for-Time)

Chronosequences are the most common method for studying soil degradation. Typically, forests are used as the baseline, with only the upper few cm of soil considered. Thus, cultivated soils almost always appear degraded in comparison. Most of the studies were located in the same region using Kenya's Kakamega and Nandi forests as the baseline and measured soil properties in continuous maize farms cleared between 50 and 100 years ago [94,98,102,103,106,108]. Similar to the longitudinal studies, chronosequences tended to show non-linear declines in topsoil properties with time since forest conversion to agriculture. Soil infiltrability [93], SOM [93,102,106], Soil P [103], pH [102,107], and total C and N [107,108] all showed marked declines in cultivated compared to forested baselines.

Soil type varies widely across SSA ([74]; Table 3), and thus it is possible that some results may be confounded by differences in inherent soil properties. For example, soil texture in the soil profile is a property not likely to change considerably with either management or time, and thus similarity in the texture (and color) profile is a good indication that the soils are comparable across space and time. Further, soils in chronosequence sites should belong to the same Great Group in Soil Taxonomy [109]. If one is examining erosion, the criteria should also be adjusted for topsoil loss. For an excellent example of how soil profiles are used to validate a chronosequence (in Brazil), see [110]. Almost all the studies examined only the top 10 cm, comparing the rich A horizon of a forest soil to the Ap horizon of an agricultural soil (mixture of the A and B horizons). This is a serious limitation of many of the studies presented here, as only one study presented texture data to 100 cm [95] and another to 40 cm [79].

The studies that examined multiple depths also found non-linear declines in topsoil C and N with increasing farm age, eventually reaching steady state after several decades [79,95,101] (Figure 1c). However, they also showed that a good portion of this C (70%) may be transferred to the deeper soil layers [80], and total C stocks (0–1 m) remain stable for many decades [95]. Non-linear declines in (unfertilized) maize yields, served as an indicator of soil degradation in many studies. Yields declined rapidly immediately following forest conversion to agriculture (first 14 years; [96,100]), but reached a steady state after 35 years [103], 77 years [102] and after 100 years of cultivation ([106]; Figure 1d).

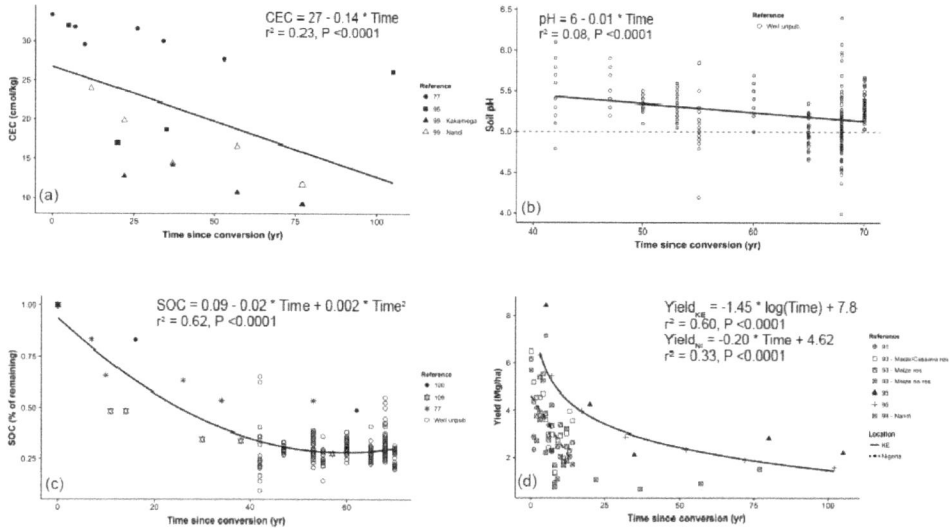

Figure 1. Selected indicators of soil degradation as a function of time since conversion. (**a**) Cation exchange capacity (CEC; 0–10 cm); (**b**) pH in water (1:1 slurry); (**c**) percent remaining soil organic carbon (SOC); and (**d**) maize yields with increasing time since forest conversion. Where data were reported in graphical form, points were extracted using GraphClick 3.0 (Arizona Software, 2008). In panel (**b**), dashed line represents the point below which aluminum toxicity can occur (pH = 5.0). In panel (**d**), two trend lines are reported for the two study regions: Yield$_{KE}$ refers to the best-fit equation for maize yields from Kenya and Yield$_{NI}$ the equation for maize yields from Nigeria. Number corresponds to the source study in References section.

3.4. Integrated Assessments

Studies that actively involve community members have the potential to improve their relevance and application, and are more likely to have broad impact on land management and system resilience. Farmers and scientists measure soil degradation differently with the former often relying on visual assessments of crop performance and yield and the latter on chemical analyses. Still, in some cases, there is good agreement between farmers knowledge and scientific indicators of soil degradation (SOM and maize yields; [97]). There was significant overlap between scientific and local understanding of soil degradation indicators (e.g., crop yield, plant stunting and presence of weeds) in Swaziland and Botswana [41] and Ethiopia [111], however no data on soil properties other than color and texture were collected.

Where scientists manage soils to maximize fertility and improve production, farmers optimize soil use for livelihood priorities. Thus, degradation may be difficult to discern from integrated assessments, which evaluate specific priorities. For example, the replacement of forest by cropland can be used as a landscape scale indicator of degradation [78], even if at the field-scale, farmers report no declines in yield. Similarly, farmers may report improving maize yields when soil properties (C, N, and pH) remain unchanged [48].

Clearly the goal is to reverse degradation, and therefore farmer perceptions must not be overlooked, as they are a primary actor on agricultural landscapes. Farmers provide invaluable information on the location and type of degradation they observe on their lands as well as describe solutions. Still, to rigorously assess the trajectory or extent of degradation, quantitative data on soil properties must be collected.

3.5. Synthesis Summary

Overall, the longitudinal and the chronosequence studies indicate that most indicators of soil degradation decline with time since conversion. However, the rate of change differs among them, emphasizing the importance of evaluating multiple indicators when assessing degradation. We found that soil chemical properties (CEC, exchangeable bases, pH) decline linearly with farm age (Figure 1a,b). On the other hand, soil biological properties (SOC, maize yields) tend to decline rapidly at first and then reach a steady state (Figure 1c,d). Differing responses have consequences for thresholds and system resilience. For example, chemical thresholds may be easier to define and their consequences for ecosystem functioning more predictable. For example, aluminum toxicity can occur in soils with a pH (in water) below 5.5, depending on the percentage of aluminum saturation, at which point crop yields may suffer substantially [112]. On the other hand, losses of SOC will have different consequences depending on other biophysical conditions. That is, a dramatic loss of SOC in a sandy soil may lead to a regime change as the primary mechanism for water retention is removed [113–115]. Soil moisture in a clayey soil, on the other hand, which has a higher water holding capacity, may not be as sensitive to SOC loss. As agriculture in SSA is primarily rain-fed, any changes in soil moisture regimes will have serious consequences for crop yields and food security outcomes. The integrated assessments indicate that some farmers are good and others are poor quantitative estimators of soil degradation, and that soils and yield should always be monitored in tandem with farmer perceptions in order to make accurate assessments of degradation. Farmers are the primary actors and stakeholders on the SSA landscape; their perspective must not be ignored, especially when it comes to developing strategies for reversing degradation and improving food security.

4. Methods for Monitoring Soil Degradation in Sub-Saharan Africa

Clearly, long-term monitoring is needed as reporting changes in degradation indicators (especially biological indicators like SOC) on a stable branch suggest little change, while monitoring only during the rapid decline suggest dramatic losses [17]. While there have been major logistical barriers to measuring soil physical and chemical properties in SSA due to a lack of resources, recent growth in investment and technical expertise in SSA is leading to better environmental monitoring. Sample preservation, transportation, and traditional chemical analysis are limited in the region. Here, we offer practical methods for evaluating soil degradation in spite of the logistical barriers encountered in remote regions.

4.1. Visual Indicators

Visual assessment can provide much detail on the state and potential drivers of soil degradation. Root exposure in trees and shrubs are other indicators of soil erosion that can be quickly assessed. Crop productivity often declines as you move uphill (even on very gentle slopes) as soil moves downslope (Figure 2). Erosion "pins" can be deployed easily at the beginning of a cropping season to measure the amount of sheet erosion occurring within a given time period [116].

4.2. Management Indicators

Biomass removal is a common practice in smallholder systems where weeds and crop residues are uprooted from the farm field and tossed to the field edges. Relocation of this biomass translates to relocation of valuable nutrients and organic matter to the field edges and nutrient mining in the middle of the farm fields. In contrast, rice threshing often occurs in the middle of the drained paddy, which concentrates nutrients (mainly K) in the center of the field (Figure 3).

(A) (B)

Figure 2. (**A**) Difference in size maize plants in (**B**) a field experiencing soil degradation due to erosion near Mwandama, Malawi. Reduced stature of maize (**B**) appears to be a matter of perspective however, when plants from each end of the field are compared side-by-side (**A**), it is clear that small slope can have dramatic effects on crop productivity due to the movement of water, soil, and nutrients. Photo credit, R. Weil.

Figure 3. Aerial photograph of rice paddies after harvest in Tanzania. Difference in soil color in the middle of the fields is indicative of variation in soil nutrient availability within rice paddies, which is caused by the movement of biomass to the middle of the field during threshing. Photo credit, R. Weil.

4.3. Physical Indicators

The soil aggregate stability is a key indicator as it integrates physical, chemical, and biological information into a single measurement. It is closely related to soil organic matter composition [117], biological activity [118], infiltration capacity [119], and erosion resistance [120]. The micro-sieve method developed by [121] is a simple, field-ready assessment of aggregate stability that can provide detailed information on management-induced changes to soil structure.

4.4. Chemical Indicators

Soil organic matter content is another integrative measure of soil degradation. Active carbon (C) can be determined in the field using a dilute permanganate extraction and can serve as a good proxy for soil organic matter [122]. If laboratory facilities are available, we suggest measuring total organic matter, pH and other important plant nutrients (total N, inorganic N, available and total P, total S, exchangeable Ca, Mg, K). Further, most soil tests are performed on the top 15 cm of soil, with subsoil properties largely ignored. We suggest that studies examine both the A horizon (typically 0–15 cm) and the upper subsoil (usually a B horizon at 20–50 cm). Sampling soil increments solely by a set depth may confound changes in horizon thickness and allow a single sample to cross boundaries between contrasting horizons. In fact, the thickness of the A horizon is a valuable measure of degradation where a clear color change marks the boundary of the horizon. Likewise, if a profile is characterized by a clay accumulation or an old erosional surface or stone line, the depth from the surface or from the bottom of the A horizon to the top of the subsoil layer may also be indicative of soil truncation and degradation (but could also indicate a shallow soil). Assessing nutrient depletion solely on topsoil soil properties may be especially misleading for some elements. For example, K may be low in the topsoil, but be in sufficient quantities of the subsoil [123,124]. Other important indicators will depend on the location. For example, in regions vulnerable to salinization, such as arid or semi-arid landscapes or irrigated agriculture, electrical conductivity and pH should be more systematically measured.

4.5. Biological Indicators

Net productivity can be indicative of overall ecosystem health. In an agricultural system, it is important to consider the biomass generated in both the intentional and unintentional species present (e.g., crop and weeds). Crop yields are sensitive to minor changes in management practices, and in poorly managed farms, yields may suffer to the benefit of weed populations. In such a case, low crop productivity may suggest soil degradation when, in fact, the high weed productivity would tell a different story. The species of weeds present can serve as a proxy for certain soil properties. For example, witchweed (*Striga spp.*) is a parasitic weed that plagues cereal crops across East Africa. This weed often occurs when soil N levels are low and is often used as a visual indicator of low soil available N [41]. Further, some fern species, native to tropical forests, are indicators of extreme acidity if found in farm fields [125].

5. Positive Trajectories and Conclusions

The conversion from forest to managed land substantially alters soil physical, chemical, and biological properties, however the extent of these changes is mediated by the new land use practice. In our review thus far, we have focused on continuous (typically unfertilized) agriculture in SSA, which offers little opportunity for the rehabilitation of soils. The majority of the available literature on degradation describes longitudinal or chronosequence studies along a degradation gradient from a forest or unmanaged baseline. However, a growing body of research in SSA uses the same study design to examine land management practices that may improve soil conditions (aggrade soils) from a degraded baseline. Such practices include (but are not limited to) communal grazing [126,127], tree plantations [93,128], and fallowing [96,129].

Many studies have compared soil properties among different management treatments in SSA, with indications that some are better suited to smallholder farming systems, can be practiced across a large range of climates and soil types, and are more readily adopted by farmers. Extensive research has been conducted into the broader frameworks of integrated soil fertility management [130–137], conservation agriculture [138–143], erosion control [144–148], and improved grazing management [149–151]. There is also a wealth of information on the benefits of specific practices such as short legume rotations (improved fallows) [152–158], agroforestry systems [159–165], and no-till systems [166–170]. Most of these studies, however, are short-term and geographically limited. We know that one management cannot fit all soil types, landscapes, or cultures. Still, these evidence-based practices hold great potential for supporting sustainable soil management, and broad improvement will require a coherent policy framework to support their wider adoption and long-term investment by farmers. Fortunately, a growing global demand for good quality, low-cost soils data has been moving forward [36,85]. Such integrated research efforts are necessary to inform national and international efforts that invest in agricultural intensification across SSA [171–173]. Land management strategies will only be successful if they can adapt to future demands for food and other ecosystem services. Future research efforts should focus on how soil degradation leads to changes in soil ecosystem services, and what land management strategies make systems resilient and, thus, more sustainable.

Acknowledgments: We thank Stephen Wood and Todd Rosenstock for their comments on this paper.

Author Contributions: Katherine Tully and Clare Sullivan wrote the manuscript. Ray Weil contributed data and concepts. Pedro Sanchez edited for content and provided guidance.

Appendix   Appendix: Methods Used by R. Weil for Collecting Thirty-Year Trends on Soil Properties in Red Soils near Holetta Research Center, Ethiopia

Soil archives at the Holetta Research Center, Ethiopia were searched for historical soil data from farmer fields near the station. Archived data were only present in hardcopy and were entered into a database, which excluded soil samples that were collected on the research station as they were likely from manipulated trials. Originally, soil samples that were collected between 0–30 cm were included and soils with a P_2O_5 concentration greater than 25 ppm were excluded as it was this was used as a marker of past fertilizer application. However, only 8 samples had high P concentrations, and their inclusion in statistical models did not change the patterns observed. The archived data contained 338 records that met these criteria collected between 1972 and 2000. We report data on soil organic carbon (Walkey-Black method) and pH (1:1 soil to water slurry) for this time period.

Conflicts of Interest: The authors declare no conflict of interest.

References

1. Sanchez, P.A.; Swaminathan, M.S. Hunger in Africa: The link between unhealthy people and unhealthy soils. *Lancet* **2005**, *365*, 442–444. [CrossRef] [PubMed]
2. Sanchez, P.A. Soil fertility and hunger in Africa. *Science* **2002**, *295*, 2019–2020. [CrossRef] [PubMed]
3. Koning, N.; Smaling, E. Environmental crisis or "lie of the land?" The debate on soil degradation in Africa. *Land Use Policy* **2005**, *22*, 3–11. [CrossRef]
4. Sanchez, P.; Palm, C.; Sachs, J.; Denning, G.; Flor, R.; Harawa, R.; Jama, B.; Kiflemariam, T.; Konecky, B.; Kozar, R.; *et al.* The African Millennium Villages. *Proc. Natl. Acad. Sci. USA* **2007**, *104*, 16775–16780. [CrossRef] [PubMed]
5. Nziguheba, G.; Palm, C.A.; Berhe, T.; Denning, G.; Dicko, A.; Diouf, O.; Diru, W.; Flor, R.; Frimpong, F.; Harawa, R.; *et al.* The African Green Revolution. In *Advances in Agronomy*; Elsevier: Amsterdam, The Netherlands, 2010; Volume 109, pp. 75–115.
6. Barrett, C.B.; Reardon, T.; Webb, P. Nonfarm income diversification and household livelihood strategies in rural Africa: Concepts, dynamics, and policy implications. *Food Policy* **2001**, *26*, 315–331. [CrossRef]
7. United Nations Human Settlements Programme (UN-Habitat). *The State of African Cities: Re-Imagining Sustainable Urban Transitions*; UN-Habitat: Nairobi, Kenya, 2014.

8. World Commission of Environment and Development. *Our Common Future: Report of the World Commission on Environment and Development*; Oxford University Press: Oxford, UK, 1987; pp. 1–300.

9. Vandecar, K.L.; Lawrence, D.; Wood, T.; Oberbauer, S.F.; Das, R.; Tully, K.; Schwendenmann, L. Biotic and abiotic controls on diurnal fluctuations in labile soil phosphorus of a wet tropical forest. *Ecology* **2009**, *90*, 2547–2555. [CrossRef] [PubMed]

10. McGrath, D.A.; Comerford, N.B.; Duryea, M.L. Litter dynamics and monthly fluctuations in soil phosphorus availability in an Amazonian agroforest. *Forest Ecol. Manag.* **2000**, *131*, 167–181. [CrossRef]

11. Crews, T.E.; Kitayama, K.; Fownes, J.H.; Riley, R.H.; Herbert, D.A.; Mueller-Dombois, D.; Vitousek, P.M. Changes in Soil Phosphorus Fractions and Ecosystem Dynamics across a Long Chronosequence in Hawaii. *Ecology* **1995**, *76*, 1407–1424. [CrossRef]

12. Walker, T.W.; Syers, J.K. The fate of phosphorus during pedogenesis. *Geoderma* **1976**, *15*, 1–19. [CrossRef]

13. Zingore, S.; Murwira, H.K.; Delve, R.J.; Giller, K.E. Influence of nutrient management strategies on variability of soil fertility, crop yields and nutrient balances on smallholder farms in Zimbabwe. *Agric. Ecosyst. Environ.* **2007**, *119*, 112–126. [CrossRef]

14. Tully, K.L.; Wood, S.A.; Almaraz, M.; Neill, C.; Palm, C.A. The effect of mineral and organic nutrient inputs on yields and nitrogen balances in western Kenya. *Agric. Syst.* **2015**. submitted for publication.

15. Hamilton, A.C.; Taylor, D. History of climate and forests in tropical Africa during the last 8 million years. *Clim. Chang.* **1991**, *19*, 65–78. [CrossRef]

16. Fleitmann, D.; Dunbar, R.B.; McCulloch, M.; Mudelsee, M.; Vuille, M.; McClanahan, T.R.; Cole, J.E.; Eggins, S. East African soil erosion recorded in a 300 year old coral colony from Kenya. *Geophys. Res. Lett.* **2007**, *34*, L04401. [CrossRef]

17. Scheffer, M.; Carpenter, S.; Foley, J.A.; Folke, C.; Walker, B. Catastrophic shifts in ecosystems. *Nature* **2001**, *413*, 591–596. [CrossRef] [PubMed]

18. Holling, C.S. Resilience and Stability of Ecological Systems. *Annu. Rev. Ecol. Syst.* **1973**, *4*, 1–23. [CrossRef]

19. Folke, C. Resilience: The emergence of a perspective for social—Ecological systems analyses. *Global Environ. Chang.* **2006**, *16*, 253–267. [CrossRef]

20. Bayon, G.; Dennielou, B.; Etoubleau, J.; Ponzevera, E.; Toucanne, S.; Bermell, S. Intensifying Weathering and Land Use in Iron Age Central Africa. *Science* **2012**, *335*, 1219–1222. [CrossRef] [PubMed]

21. Carpenter, S.; Walker, B.; Anderies, J.M.; Abel, N. From Metaphor to Measurement: Resilience of What to What? *Ecosystems* **2001**, *4*, 765–781. [CrossRef]

22. Levin, S.A. *Fragile Dominion: Complexity and the Commons*; Kluwer Academic Publishers: Dordrecht, The Netherlands, 1999.

23. Warren, A. Land degradation is contextual. *Land Degrad. Dev.* **2002**, *13*, 449–459. [CrossRef]

24. Dewitte, O.; Jones, A.; Spaargaren, O.; Breuning-Madsen, H.; Brossard, M.; Dampha, A.; Deckers, J.; Gallali, T.; Hallett, S.; Jones, R.; *et al.* Harmonisation of the soil map of Africa at the continental scale. *Geoderma* **2013**, *211–212*, 138–153. [CrossRef]

25. Sanchez, P.A.; Palm, C.A.; Buol, S.W. Fertility capability soil classification: A tool to help assess soil quality in the tropics. *Geoderma* **2003**, *114*, 157–185. [CrossRef]

26. Lohse, K.A.; Matson, P. Consequences of nitrogen additions for soil losses from wet tropical forests. *Ecol. Appl.* **2005**, *15*, 1629–1648. [CrossRef]

27. Kinjo, T.; Pratt, P.F. In some acid soils of Mexico and South America. II. In competition with chloride, sulfate and phosphate. III. Desorption, movement and distribution in Andepts. *Soil Sci. Soc. Am. J.* **1971**, *35*, 722–732. [CrossRef]

28. Tittonell, P.; Muriuki, A.; Klapwijk, C.J.; Shepherd, K.D.; Coe, R.; Vanlauwe, B. Soil Heterogeneity and Soil Fertility Gradients in Smallholder Farms of the East African Highlands. *Soil Sci. Soc. Am. J.* **2013**. [CrossRef]

29. Almaraz, M.; Tully, K.L.; Neill, C.; Palm, C.A.; Porder, S. Nitrogen dynamics in soil profiles from intensifying agricultural fields in sub-Saharan Africa: The role of soil type. **2015**, unpublished work.

30. Scoones, I. Wetlands in drylands: Key resources for agricultural and pastoral production in Africa. *Ambio* **1991**, *20*, 366–371.

31. Mekonnen, K.; Buresh, R.J.; Jama, B. Root and inorganic nitrogen distributions in sesbania fallow, natural fallow and maize fields. *Plant Soil* **1997**, *188*, 319–327. [CrossRef]

32. Chintu, R.; Mafongoya, P.L.; Chirwa, T.S.; Mwale, M.; Matibini, J. Subsoil nitrogen dynamics as affected by planted coppicing tree legume fallows in eastern Zambia. *Ex. Agric.* **2004**, *40*, 327–340. [CrossRef]

33. Vanlauwe, B.; Kihara, J.; Chivenge, P.; Pypers, P.; Coe, R.; Six, J. Agronomic use efficiency of N fertilizer in maize-based systems in sub-Saharan Africa within the context of integrated soil fertility management. *Plant Soil* **2010**, *339*, 35–50. [CrossRef]

34. Tittonell, P.; Vanlauwe, B.; Leffelaar, P.A.; Rowe, E.C.; Giller, K.E. Exploring diversity in soil fertility management of smallholder farms in western Kenya. *Agric. Ecosyst. Environ.* **2005**, *110*, 149–165. [CrossRef]

35. Palm, C.A.; Smukler, S.M.; Sullivan, C.C.; Mutuo, P.K.; Nyadzi, G.I.; Walsh, M.G. Identifying potential synergies and trade-offs for meeting food security and climate change objectives in sub-Saharan Africa. *Proc. Natl. Acad. Sci. USA* **2010**, *107*, 19661–19666. [CrossRef] [PubMed]

36. Shepherd, K.D.; Shepherd, G.; Walsh, M.G. Land health surveillance and response: A framework for evidence-informed land management. *Agric. Syst.* **2015**, *132*, 93–106. [CrossRef]

37. Letey, J.; Sojka, R.E.; Upchurch, D.R. Deficiencies in the soil quality concept and its application. *J. Soil Water Conserv.* **2003**, *58*, 180–187.

38. Karlen, D.L.; Mausbach, M.J.; Doran, J.W. Soil quality: A concept, definition, and framework for evaluation. *Soil Sci. Soc. Am. J.* **1997**, *61*, 4–10. [CrossRef]

39. Wyland, L.J.; Jackson, L.E.; Schulbach, K.F. Soil-plant nitrogen dynamics following incorporation of a mature rye cover crop in a lettuce production system. *J. Agric. Sci.* **1995**, *124*, 17–25. [CrossRef]

40. Dean, J.E.; Weil, R.R. Brassica Cover Crops for Nitrogen Retention in the Mid-Atlantic Coastal Plain. *J. Environ. Qual.* **2009**, *38*, 520–528. [CrossRef] [PubMed]

41. Stringer, L.C.; Reed, M.S. Land degradation assessment in Southern Africa: Integrating local and scientific knowledge bases. *Land Degrad. Dev.* **2007**, *18*, 99–116. [CrossRef]

42. Stoorvogel, J.J.; Smaling, E.M. A. *Assessment of Soil Nutrient Depletion in Sub-Saharan Africa,1983–2000*; Winand Staarting Center-DLO: Wageningen, The Netherlands, 1990.

43. Lal, R. Erosion-Crop Productivity Relationships for Soils of Africa. *Soil Sci. Soc. Am. J.* **1995**, *59*, 661–667. [CrossRef]

44. Oldeman, L.R.; Hakkeling, R.T.A.; Sombroek, W.G. *World Map of the Status of Human-Induced Soil Degradation*; ISRIC: Wageningen, The Netherlands; FAO: Nairobi, Kenya, 1991; pp. 1–35.

45. Mazzucato, V.; Niemeijer, D. Rethinking Soil and Water Conservation in a Changing Society: A Case Study from Burkina Faso. Ph.D. Thesis, Wageningen University, Wageningen, The Netherlands, 20 June 2000; pp. 1–421.

46. Scherr, S.J. Economic factors in farmer adoption of agroforestry: Patterns observed in Western Kenya. *World Dev.* **1995**, *23*, 787–804. [CrossRef]

47. Forsyth, T. Science, myth and knowledge: Testing himalayan environmental degradation in Thailand. *Geoforum* **1996**, *27*, 375–392. [CrossRef]

48. Tiffen, M.; Mortimer, M.; Gichuki, F. *More People, Less Erosion*; John Wiley & Sons, Ltd.: Chichester, UK, 1994; pp. 1–326.

49. Barbier, E.B. The economic linkages between rural poverty and land degradation: Some evidence from Africa. *Agr. Ecosyst. Environ.* **2000**, *82*, 355–370. [CrossRef]

50. McCall, M. Environmental and agricultural impacts of Tanzania's villagization programme. In *Population and Development Projects in Africa*; Clark, J.I., Khogali, M., Kosinski, L.A., Eds.; Cambridge University Press: Cambridge, UK, 1985; pp. 123–140.

51. Thomas, D.S.G.; Sporton, D.; Perkins, J. The environmental impact of livestock ranches in the Kalahari, Botswana: Natural resource use, ecological change and human response in a dynamic dryland system. *Land Degrad. Dev.* **2000**, *11*, 327–341. [CrossRef]

52. Rohde, R.F.; Moleele, N.M.; Mphale, M.; Allsopp, N.; Chanda, R.; Hoffman, M.T.; Magole, L.; Young, E. Dynamics of grazing policy and practice: Environmental and social impacts in three communal areas of southern Africa. *Environ. Sci. Policy* **2006**, *9*, 302–316. [CrossRef]

53. Mafongoya, P.L.; Bationo, A.; Kihara, J.; Waswa, B.S. Appropriate technologies to replenish soil fertility in southern Africa. *Nutr. Cycl. Agroecosyst.* **2006**, *76*, 137–151. [CrossRef]

54. Boserup, E. *The Conditions of Agricultural Growth*; Adine Publishing Company: Chicago, IL, USA, 1965.

55. Hoddinott, J.; Haddad, L. Does female income share influence household expenditures? Evidence from Côte d'Ivoire. *Oxford Bull. Econ. Stat.* **1995**, *57*, 77–96. [CrossRef]

56. Nkonya, E.; Moore, K. *Smallholder Adoption of Integrated Soil Fertility Management*; USAID&Agrilinks: Washington, DC, USA, 2015; p. 51.

57. Nkonya, E. *Soil Conservation Practices and Non-Agricultural Land Use in the South-Western Highlands of Uganda*; The International Food Policy Research Institute (IFPRI): Washington, DC, USA, 2002; pp. 1–31.

58. Tamene, L.; Vlek, P.L.G. Soil Erosion Studies in Northern Ethiopia. In *Land Use and Soil Resources*; Braimoh, A.K., Vlek, P.L.G., Eds.; Springer Netherlands: Dordrecht, The Netherlands, 2008; pp. 73–100.

59. Tamene, L.; Park, S.J.; Dikau, R.; Vlek, P.L.G. Reservoir siltation in the semi-arid highlands of northern Ethiopia: Sediment yield–catchment area relationship and a semi-quantitative approach for predicting sediment yield. *Earth Surf. Process. Landforms* **2006**, *31*, 1364–1383. [CrossRef]

60. Kimaro, D.N.; Deckers, J.A.; Poesen, J.; Kilasara, M.; Msanya, B.M. Short and medium term assessment of tillage erosion in the Uluguru Mountains, Tanzania. *Soil Tillage Res.* **2005**, *81*, 97–108. [CrossRef]

61. Collins, A.L.; Walling, D.E.; Sichingabula, H.M.; Leeks, G.J.L.G. Suspended sediment source fingerprinting in a small tropical catchment and some management implications. *Appl. Geogr.* **2001**, *21*, 387–412. [CrossRef]

62. Taddese, G.; Saleem, M.A.M.; Abyie, A.; Wagnew, A. Impact of grazing on plant species richness, plant biomass, plant attribute, and soil physical and hydrological properties of vertisol in East African highlands. *Environ. Manag.* **2002**, *29*, 279–289. [CrossRef]

63. Van N du Toit, G.; Snyman, H.A.; Malan, P.J. Physical impact of grazing by sheep on soil parameters in the Nama Karoo subshrub/grass rangeland of South Africa. *J. Arid Environ.* **2009**, *73*, 804–810. [CrossRef]

64. Weil, R.R.; Brady, N. *Nature and Properties of Soils*, 15 ed.; 2015; unpublished work.

65. Nye, P.H.; Greenland, D.J. *The Soil under Shifting Cultivation*, 51st ed.; Commonwealth Agricultural Bureau, Farnham Royal: Berks, Great Britain, 1960; p. 156.

66. Vitousek, P.M.; Naylor, R.; Crews, T.; David, M.B.; Drinkwater, L.E.; Holland, E.; Johnes, P.J.; Katzenberger, J.; Martinelli, L.A.; Matson, P.A.; *et al.* Nutrient Imbalances in Agricultural Development. *Science* **2009**, *324*, 1519–1520. [CrossRef] [PubMed]

67. Oenema, O.; de Vries, W. Approaches and uncertainties in nutrient budgets: Implications for nutrient management and environmental policies. *Eur. J. Agron.* **2003**, *20*, 3–16. [CrossRef]

68. Henao, J.; Baanante, C.A. *Estimating Rates of Nutrient Depletion in Soils of Agricultural Lands of Africa*; Intl Fertilizer Development Center: Muscle Shoals, AL, USA, 1999.

69. Cobo, J.G.; Dercon, G.; Cadisch, G. Nutrient balances in African land use systems across different spatial scales: A review of approaches, challenges and progress. *Agric. Ecosyst. Environ.* **2010**, *136*, 1–15. [CrossRef]

70. Amede, T.; Belachew, T.; Geta, E. *Reversing the Degradation of Arable Land in the Ethiopian Highlands*; IIED: London, UK, 2001; p. 23.

71. Juo, A.S.R.; Dabiri, A.; Franzluebbers, K. Acidification of a kaolinitic Alfisol under continuous cropping with nitrogen fertilization in West Africa. *Plant Soil* **1995**, *171*, 245–253. [CrossRef]

72. Barbiéro, L.; Mohamedou, A.O.; Roger, L.; Furian, S.; Aventurier, A.; Rémy, J.C.; Marlet, S. The origin of Vertisols and their relationship to Acid Sulfate Soils in the Senegal valley. *CATENA* **2005**, *59*, 93–116. [CrossRef]

73. Van Asten, P.J.A.; Barbiéro, L.; Wopereis, M.C.S.; Maeght, J.L.; van der Zee, S.E.A.T.M. Actual and potential salt-related soil degradation in an irrigated rice scheme in the Sahelian zone of Mauritania. *Agric. Water Manag.* **2003**, *60*, 13–32. [CrossRef]

74. *Soil Atlas of Africa*; Jones, A. (Ed.) Publications Office of the European Union: Luxembourg, 2013.

75. HarvestChoice Updating Soil Functional Capacity Classification System. Available online: http://harvestchoice.org/node/1435 (accessed on 24 January 2015).

76. Vlek, P.L.G. The role of fertilizers in sustaining agriculture in sub-Saharan Africa. *Fert. Res.* **1990**, *26*, 327–339. [CrossRef]

77. Bationo, A.; Lompo, F.; Koala, S. Research on nutrient flows and balances in West Africa: State-of-the-art. *Agr. Ecosyst. Environ.* **1998**, *71*, 19–35. [CrossRef]

78. Gray, L.C. Is land being degraded? A multi-scale investigation of landscape change in southwestern Burkina Faso. *Land Degrad. Dev.* **1999**, *10*, 329–343. [CrossRef]

79. Lemenih, M.; Karltun, E.; Olsson, M. Assessing soil chemical and physical property responses to deforestation and subsequent cultivation in smallholders farming system in Ethiopia. *Agric. Ecosyst. Environ.* **2005**, *105*, 373–386. [CrossRef]

80. Bossio, D.A.; Girvan, M.S.; Verchot, L.; Bullimore, J.; Borelli, T.; Albrecht, A.; Scow, K.M.; Ball, A.S.; Pretty, J.N.; Osborn, A.M. Soil Microbial Community Response to Land Use Change in an Agricultural Landscape of Western Kenya. *Microb. Ecol.* **2005**, *49*, 50–62. [CrossRef] [PubMed]

81. Brussaard, L.; de Ruiter, P.C.; Brown, G.G. Soil biodiversity for agricultural sustainability. *Agr. Ecosyst. Environ.* **2007**, *121*, 233–244. [CrossRef]

82. Wood, S.A.; Almaraz, M.; Bradford, M.A.; McGuire, K.L.; Naeem, S.; Palm, C.A.; Tully, K.L.; Zhou, J. Farm management, not soil microbial diversity, controls nutrient loss from smallholder tropical agriculture. *Front. Micr.* **2015**, *6*, 1–10.

83. Swift, M.J.; Anderson, J.M. Biodiversity and ecosystem function in agricultural systems. In *Biodiversity and Ecosystem Function*; Schulze, E.D., Mooney, H.A., Eds.; Springer: Berlin, Germany, 1994; pp. 15–41.

84. Beare, M.H.; Reddy, M.V.; Tian, G.; Srivastava, S.C. Agricultural intensification, soil biodiversity and agroecosystem function in the tropics: The role of decomposer biota. *Appl. Soil Ecol.* **1997**, *6*, 87–108. [CrossRef]

85. Palm, C.; Sanchez, P.; Ahamed, S.; Awiti, A. Soils: A Contemporary Perspective. *Annu. Rev. Environ. Resourc.* **2007**, *32*, 99–129. [CrossRef]

86. Manlay, R.J.; Feller, C.; Swift, M.J. Historical evolution of soil organic matter concepts and their relationships with the fertility and sustainability of cropping systems. *Agric. Ecosyst. Environ.* **2007**, *119*, 217–233. [CrossRef]

87. Mathieu, J.; Rossi, J.P.; Mora, P.; Lavelle, P.; Martins, P.F.D.S.; Rouland, C.; Grimaldi, M. Recovery of soil macrofauna communities after forest clearance in eastern Amazonia, Brazil. *Conserv. Biol.* **2005**, *19*, 1598–1605. [CrossRef]

88. Mutema, M.; Mafongoya, P.L.; Nyagumbo, I.; Chikukura, L. Effects of crop residues and reduced tillage on macrofauna abundance. *J. Org. Syst.* **2013**, *8*, 5–16.

89. Wood, S.A.; Bradford, M.A.; Gilbert, J.A.; McGuire, K.L.; Palm, C.A.; Tully, K.; Zhou, J.; Naeem, S. Agricultural intensification and the functional capacity of soil microbes on smallholder African farms. *J. Appl. Ecol.* **2015**. [CrossRef]

90. Bradford, M.A.; Wood, S.A.; Bardgett, R.D.; Black, H.I.J.; Bonkowski, M.; Eggers, T.; Grayston, S.J.; Kandeler, E.; Manning, P.; Setala, H.; *et al.* Discontinuity in the responses of ecosystem processes and multifunctionality to altered soil community composition. *Proc. Natl. Acad. Sci. USA* **2014**, *111*, 14478–14483. [CrossRef] [PubMed]

91. De Vries, F.T.; Bardgett, R.D. Plant–microbial linkages and ecosystem nitrogen retention: lessons for sustainable agriculture. *Front. Ecol. Environ.* **2012**, *10*, 425–432. [CrossRef]

92. R Development Core Team 2008. *R: A Language and Environment for Statistical Computing*; R Foundation for Statistical Computing: Vienna, Austria, 2008.

93. Adejuwon, J.O.; Ekanade, O. A comparison of soil properties under different landuse types in a part of the Nigerian Cocoa Belt. *CATENA* **1988**, *15*, 319–331. [CrossRef]

94. Awiti, A.O.; Walsh, M.G.; Shepherd, K.D.; Kinyamario, J. Soil condition classification using infrared spectroscopy: A proposition for assessment of soil condition along a tropical forest-cropland chronosequence. *Geoderma* **2008**, *143*, 73–84. [CrossRef]

95. Demessie, A.; Singh, B.R.; Lal, R. Soil carbon and nitrogen stocks under chronosequence of farm and traditional agroforestry land uses in Gambo District, Southern Ethiopia. *Nutr. Cycl. Agroecosyst.* **2013**, *95*, 365–375. [CrossRef]

96. Juo, A.S.R.; Franzluebbers, K.; Dabiri, A.; Ikhile, B. Changes in soil properties during long-term fallow and continuous cultivation after forest clearing in Nigeria. *Agr. Ecosyst. Environ.* **1995**, *56*, 9–18. [CrossRef]

97. Karltun, E.; Lemenih, M.; Tolera, M. Comparing farmers' perception of soil fertility change with soil properties and crop performance in Beseku, Ethiopia. *Land Degrad. Dev.* **2011**, *24*, 228–235. [CrossRef]

98. Kimetu, J.M.; Lehmann, J.; Ngoze, S.O.; Mugendi, D.N.; Kinyangi, J.M.; Riha, S.; Verchot, L.; Recha, J.W.; Pell, A.N. Reversibility of Soil Productivity Decline with Organic Matter of Differing Quality Along a Degradation Gradient. *Ecosystems* **2008**, *11*, 726–739. [CrossRef]

99. Kiunsi, R.B.; Meadows, M.E. Assessing land degradation in the Monduli District, northern Tanzania. *Land Degrad. Dev.* **2006**, *17*, 509–525. [CrossRef]

100. Lal, R. Deforestation and land-use effects on soil degradation and rehabilitation in western Nigeria. II. Soil chemical properties. *Land Degrad. Dev.* **1998**, *7*, 87–98. [CrossRef]

101. Lemenih, M.; Karltun, E.; Olsson, M. Soil organic matter dynamics after deforestation along a farm field chronosequence in southern highlands of Ethiopia. *Agric. Ecosyst. Environ.* **2005**, *109*, 9–19. [CrossRef]

102. Moebius-Clune, B.N.; van Es, H.M.; Idowu, O.J.; Schindelbeck, R.R.; Kimetu, J.M.; Ngoze, S.; Lehmann, J.; Kinyangi, J.M. Long-term soil quality degradation along a cultivation chronosequence in western Kenya. *Agr. Ecosyst. Environ.* **2011**, *141*, 86–99. [CrossRef]

103. Ngoze, S.; Riha, S.; Lehmann, J.; Verchot, L.; Kinyagi, J.; Mbuga, D.; Pell, A. Nutrient constraints to tropical agroecosystem productivity in long-term degrading soils. *Global Change Biol.* **2008**, *14*, 2810–2822. [CrossRef]

104. Nyberg, G.; Bargués Tobella, A.; Kinyangi, J.; Ilstedt, U. Soil property changes over a 120-yr chronosequence from forest to agriculture in western Kenya. *Hydrol. Earth Syst. Sci.* **2012**, *16*, 2085–2094. [CrossRef]

105. Peters, J.B. Gambian soil fertility trends, 1991–1998. *Commun. Soil Sci. Plan.* **2000**, *31*, 2201–2210. [CrossRef]

106. Solomon, D.; Lehmann, J.; Kinyagi, J.; Amelung, W.; Lobe, I.; Pell, A.; Riha, S.; Ngoze, S.; Verchot, L.; Mbuga, D.; *et al.* Long-term impacts of anthropogenic perturbations on dynamics and speciation of organic carbon in tropical forest and subtropical grassland ecosystems. *Global Change Biol.* **2007**, *13*, 511–530. [CrossRef]

107. Vågen, T.-G.; Shepherd, K.D.; Walsh, M.G. Sensing landscape level change in soil fertility following deforestation and conversion in the highlands of Madagascar using Vis-NIR spectroscopy. *Geoderma* **2006**, *133*, 281–294. [CrossRef]

108. Awiti, A.O.; Walsh, M.G.; Kinyamario, J. Dynamics of topsoil carbon and nitrogen along a tropical forest–cropland chronosequence: Evidence from stable isotope analysis and spectroscopy. *Agr. Ecosyst. Environ.* **2008**, *127*, 265–272. [CrossRef]

109. Sanchez, P.A.; Palm, C.A.; Davey, C.B.; Szott, L.T.; Russell, C.E. Tree Crops as Soil Improvers in the Humid Tropics? In *Attributes of Trees as Crop Plants*; Cannell, M.G.R., Jackson, J.E., Eds.; Institute of Terrestrial Ecology: Huntingdon, UK, 1985; pp. 327–358.

110. Russell, C.E. *Nutrient Cycling and Productivity of Native and Plantation Forests at Jari Florestal, Pará, Brazil*; Institute of Ecology, University of Georgia: Athens, GA, USA, 1983.

111. Tesfahunegn, G.B.; Tamene, L.; Vlek, P.L.G. A participatory soil quality assessment in Northern Ethiopia's Mai-Negus catchment. *CATENA* **2011**, *86*, 1–13. [CrossRef]

112. Sanchez, P.A. *Properties and Management of Soils in the Tropics*; John Wiley & Sons Inc.: Hoboken, NJ, USA, 1976.

113. Hudson, B. Soil organic matter and available water capacity. *J. Soli Water Conserv.* **1994**, *49*, 189–194.

114. Weil, R.R.; Magdoff, F. Significance of soil organic matter to soil quality and health. In *Soil Organic Matter in Sustainable Agriculture*; Magdoff, F., Weil, R.R., Eds.; CRC Press: Boca Raton, FL, USA, 2004; pp. 1–42.

115. Lotter, D.W.; Seidel, R.; Liebhardt, W. The performance of organic and conventional cropping systems in an extreme climate year. *Am. J. Alt. Agric.* **2003**, *18*, 146–154. [CrossRef]

116. Vigiak, O.; Okoba, B.O.; Sterk, G.; Groenenberg, S. Modelling catchment-scale erosion patterns in the East African Highlands. *Earth Surf. Process. Landforms* **2005**, *30*, 183–196. [CrossRef]

117. Tisdall, J.M. Formation of soil aggregates and accumulation of soil organic matter. In *Structure and Organic Matter Storage in Agricultural Soils*; Carter, M.R., Stewart, B.A., Eds.; CRC Press: Boca Raton, FL, USA, 1996; pp. 57–96.

118. Wander, M.M.; Traina, S.J.; Stinner, B.R.; Peters, S.E. Organic and Conventional Management Effects on Biologically Active Soil Organic Matter Pools. *Soil Sci. Soc. Am. J.* **1994**, *58*, 1130–1139. [CrossRef]

119. Pierson, F.B.; Blackburn, W.H.; Vactor, S.S.; Wood, J.C. Partitioning small scale spatial variability of runoff and erosion on sagebrush rangeland. *J. Am. Water Resour. As.* **1994**, *30*, 1081–1089. [CrossRef]

120. Blackburn, W.H.; Pierson, F.B. Sources of Variation in Interrill Erosion on Rangelands. In *Variability in Rangeland Water Erosion Processes*; Blackburn, W.H., Pierson, F.B., Schulman, G.E., Zartman, R., Eds.; SSSA Special Publication; Soil Science Society of America: Madison, WI, USA, 1994; pp. 1–9.

121. Herrick, J.E.; Whitford, W.G.; de Soyza, A.G.; van Zee, J.W. Field soil aggregate stability kit for soil quality and rangeland health evaluations. *CATENA* **2001**, *44*, 27–35. [CrossRef]

122. Lucas, S.T.; Weil, R.R. Can a Labile Carbon Test be Used to Predict Crop Responses to Improve Soil Organic Matter Management? *Agron. J.* **2012**, *104*, 1160–1170. [CrossRef]

123. Khan, S.A.; Mulvaney, R.L.; Ellsworth, T.R. The potassium paradox: Implications for soil fertility, crop production and human health. *Renew. Agric. Food Syst.* **2013**, *29*, 3–27. [CrossRef]

124. Marín-Spiotta, E.; Silver, W.L.; Ostertag, R. Long-term patterns in tropical reforestation: Plant community composition and aboveground biomass accumulation. *Ecol. Appl.* **2007**, *17*, 828–839. [CrossRef] [PubMed]

125. Murage, E.W.; Karanja, N.K.; Smithson, P.C.; Woomer, P.L. Diagnostic indicators of soil quality in productive and non-productive smallholders' fields of Kenya's Central Highlands. *Agr Ecosyst. Environ.* **2000**, *79*, 1–8. [CrossRef]

126. Verdoodt, A.; Mureithi, S.M.; Ye, L.; van Ranst, E. Chronosequence analysis of two enclosure management strategies in degraded rangeland of semi-arid Kenya. *Agric. Ecosyst. Environ.* **2009**, *129*, 332–339. [CrossRef]

127. Fterich, A.; Mahdhi, M.; Mars, M. Impact of grazing on soil microbial communities along a chronosequence of Acacia tortilis subsp. raddiana in arid soils in Tunisia. *Eur. J. Soil Biol.* **2012**, *50*, 56–63. [CrossRef]

128. Dawoe, E.K.; Quashie-Sam, J.S.; Oppong, S.K. Effect of land-use conversion from forest to cocoa agroforest on soil characteristics and quality of a Ferric Lixisol in lowland humid Ghana. *Agroforest. Syst.* **2013**, *88*, 87–99. [CrossRef]

129. Mando, A.; Ouattara, B.; Somado, A.E.; Wopereis, M.C.S.; Stroosnijder, L.; Breman, H. Long-term effects of fallow, tillage and manure application on soil organic matter and nitrogen fractions and on sorghum yield under Sudano-Sahelian conditions. *Soil Use Manag.* **2005**, *21*, 25–31. [CrossRef]

130. Vanlauwe, B.; Diels, J.; Sanginga, N.; Merckx, R. Long-term integrated soil fertility management in South-western Nigeria: Crop performance and impact on the soil fertility status. *Plant Soil* **2005**, *273*, 337–354. [CrossRef]

131. Zingore, S.; Murwira, H.K.; Delve, R.J.; Giller, K.E. Soil type, management history and current resource allocation: Three dimensions regulating variability in crop productivity on African smallholder farms. *Field Crops Res.* **2007**, *101*, 296–305. [CrossRef]

132. Vanlauwe, B.; Bationo, A.; Chianu, J.; Giller, K.E.; Merckx, R.; Mokwunye, U.; Ohiokpehai, O.; Pypers, P.; Tabo, R.; Shepherd, K.D.; *et al.* Integrated soil fertility management. *Outlook Agric.* **2010**, *39*, 17–24. [CrossRef]

133. Vanlauwe, B.; Giller, K. Popular myths around soil fertility management in sub-Saharan Africa. *Agric. Ecosyst. Environ.* **2006**, *116*, 34–46. [CrossRef]

134. Defoer, T. Learning about methodology development for integrated soil fertility management. *Agric. Syst.* **2002**, *73*, 57–81. [CrossRef]

135. Zingore, S.; Tittonell, P.; Corbeels, M.; van Wijk, M.T.; Giller, K.E. Managing soil fertility diversity to enhance resource use efficiencies in smallholder farming systems: a case from Murewa District, Zimbabwe. *Nutr. Cycl. Agroecosyst.* **2011**, *90*, 87–103. [CrossRef]

136. Chivenge, P.; Vanlauwe, B.; Six, J. Does the combined application of organic and mineral nutrient sources influence maize productivity? A meta-analysis. *Plant Soil* **2011**, *342*, 1–30. [CrossRef]

137. Bationo, A.; Waswa, B.; Kihara, J.; Six, J. Advances in integrated soil fertility management in sub Saharan Africa: Challenges and opportunities. *Nutr. Cycl. Agroecosyst.* **2006**. [CrossRef]

138. Bot, A.; Benites, J. *Conservation agriculture: Case studies in Latin American and Africa*; Food & Agriculture Organization: Rome, Italy, 2001.

139. Erenstein, O.; Sayre, K.; Wall, P.; Dixon, J.; Hellin, J. Adapting No-Tillage Agriculture to the Conditions of Smallholder Maize and Wheat Farmers in the Tropics and Sub-Tropics. Goddard, T., Zoebisch, M., Gan, Y., Ellis, W., Watson, A., Sombatpanit, S., Eds.; In *No-till Farming Systems*; Special Publication 3 World Association of Soil and Water Conservation (WASWC): Bangkok, Thailand, 2007; pp. 253–274.

140. Chivenge, P.; Murwira, H.; Giller, K.; Mapfumo, P.; Six, J. Long-term impact of reduced tillage and residue management on soil carbon stabilization: Implications for conservation agriculture on contrasting soils. *Soil Till. Res.* **2007**, *94*, 328–337. [CrossRef]

141. Palm, C.; Blanco-Canqui, H.; DeClerck, F.; Gatere, L.; Grace, P. Conservation agriculture andecosystem services: An overview. *Agric. Ecosyst. Environ.* **2014**, *187*, 87–105. [CrossRef]

142. Verhulst, N.; Govaerts, B.; Verachtert, E.; Castellanos-Navarrete, A.; Mezzalama, M.; Wall, P.; Deckers, J.; Sayre, K. Conservation agriculture, improving soil quality for sustainable production systems? In *Food Security and Soil Quality*; Lal, R., Stewart, B.A., Eds.; CRC Press: Boca Raton, FL, USA, 2010; pp. 137–208.

143. Thierfelder, C.; Cheesman, S.; Rusinamhodzi, L. Benefits and challenges of crop rotations in maize-based conservation agriculture (CA) cropping systems of southern Africa. *Int. J. Agric. Sustain.* **2013**, *11*, 108–124. [CrossRef]

144. Nyssen, J.; Poesen, J.; Haile, M.; Moeyersons, J.; Deckers, J. Tillage erosion on slopes with soil conservation structures in the Ethiopian highlands. *Soil Till. Res.* **2000**, *57*, 115–127. [CrossRef]

145. Valentin, C.; Poesen, J.; Li, Y. Gully erosion: Impacts, factors and control. *CATENA* **2005**, *63*, 132–153. [CrossRef]

146. Gebrernichael, D.; Nyssen, J.; Poesen, J.; Deckers, J.; Haile, M.; Govers, G.; Moeyersons, J. Effectiveness of stone bunds in controlling soil erosion on cropland in the Tigray Highlands, northern Ethiopia. *Soil Use Manag.* **2005**, *21*, 287–297. [CrossRef]

147. Astatke, A.; Jabbar, M.; Tanner, D. Participatory conservation tillage research: an experience with minimum tillage on an Ethiopian highland Vertisol. *Agric. Ecosyst. Environ.* **2003**, *95*, 401–415. [CrossRef]

148. Bryan, R.B. *Soil Erosion, Land Degradation and Social Transition: Geoecological Analysis of A Semi-Arid Tropical Region, Kenya*; Catena-Verlag: Cremlingen Destedt, Germay, 1994.

149. Beukes, P.C.; Cowling, R.M. Non-Selective Grazing Impacts on Soil-Properties of the Nama Karoo. *J. Range Manag.* **2003**, *56*, 547–552. [CrossRef]

150. Gebremeskel, K.; Pieterse, P.J. Impact of grazing around a watering point on soil status of a semi-arid rangeland in Ethiopia. *Afr. J. Ecol.* **2007**, *45*, 72–79. [CrossRef]

151. Mekuria, W.; Veldkamp, E.; Tilahun, M.; Olschewski, R. Economic valuation of land restoration: The case of exclosures established on communal grazing lands in Tigray, Ethiopia. *Land Degrad. Dev.* **2011**, *22*, 334–344. [CrossRef]

152. Bünemann, E.K.; Smithson, P.C.; Jama, B.; Frossard, E.; Oberson, A.; Oberson, A. Maize productivity and nutrient dynamics in maize-fallow rotations in western Kenya. *Plant Soil* **2004**, *264*, 195–208. [CrossRef]

153. Ojiem, J.O.; Vanlauwe, B.; de Ridder, N.; Giller, K.E. Niche-based assessment of contributions of legumes to the nitrogen economy of Western Kenya smallholder farms. *Plant Soil* **2007**, *292*, 119–135. [CrossRef]

154. Oikeh, S.O.; Carsky, R.J.; Kling, J.G.; Chude, V.O.; Horst, W.J. Differential N uptake by maize cultivars and soil nitrate dynamics under N fertilization in West Africa. *Agric. Ecosyst. Environ.* **2003**, *100*, 181–191. [CrossRef]

155. Adiku, S.G.K.; Jones, J.W.; Kumaga, F.K.; Tonyigah, A. Effects of crop rotation and fallow residue management on maize growth, yield and soil carbon in a savannah-forest transition zone of Ghana. *J. Agric. Sci.* **2009**, *147*, 313–322. [CrossRef]

156. Chikowo, R.; Mapfumo, P.; Nyamugafata, P.; Giller, K.E. Mineral N dynamics, leaching and nitrous oxide losses under maize following two-year improved fallows on a sandy loam soil in Zimbabwe. *Plant Soil* **2004**, *259*, 315–330. [CrossRef]

157. Ndufa, J.K.; Gathumbi, S.M.; Kamiri, H.W.; Giller, K.E.; Cadisch, G. Do mixed-species Legume fallows provide long-term maize yield benefit compared with monoculture legume fallows? *Agron. J.* **2009**, *101*, 1352–1362. [CrossRef]

158. Bado, B.V.; Bationo, A.; Cescas, M.P. Assessment of cowpea and groundnut contributions tosoil fertility and succeeding sorghum yields in the Guinean savannah zone of Burkina Faso (West Africa). *Biol. Fert. Soils* **2006**, *43*, 171–176. [CrossRef]

159. Shepherd, K.D.; Ohlsson, E.; Okalebo, J.R.; Ndufa, J.K. Potential impact of agroforestry on soil nutrient balances at the farm scale in the East African Highlands. *Fert. Res.* **1996**, *44*, 87–99. [CrossRef]

160. Buresh, R.J. Soil improvement by trees in sub-Saharan Africa. *Agroforest. Syst.* **1998**, *38*, 51–76. [CrossRef]

161. Odhiambo, H.O.; Ong, C.K.; Deans, J.D.; Wilson, J.; Khan, A.; Sprent, J.I. Roots, soil water and crop yield: Tree crop interactions in a semi-arid agroforestry system in Kenya. *Plant Soil* **2001**, *235*, 221–233. [CrossRef]

162. Akinnifesi, F.K.; Makumba, W.; Sileshi, G.; Ajayi, O.C.; Mweta, D. Synergistic effect ofinorganic N and P fertilizers and organic inputs from Gliricidia sepium on productivity of intercropped maize in Southern Malawi. *Plant Soil* **2007**, *294*, 203–217. [CrossRef]

163. Van Noordwijk, M.; Lusiana, B. WaNuLCAS, a model of water, nutrient and light capture in agroforestry systems. In *Agroforestry for Sustainable Land-Use Fundamental Research and Modelling with Emphasis on Temperate and Mediterranean Applications*; Springer Netherlands: Dordrecht, The Netherlands, 1999; Volume 60, pp. 217–242.

164. Sileshi, G.; Akinnifesi, F.K.; Ajayi, O.C.; Place, F. Meta-analysis of maize yield response to woody and herbaceous legumes in sub-Saharan Africa. *Plant Soil* **2008**, *307*, 1–19. [CrossRef]

165. Nyamadzawo, G.; Nyamugafata, P.; Wuta, M.; Nyamangara, J.; Chikowo, R. Infiltration and runoff losses under fallowing and conservation agriculture practices on contrasting soils, Zimbabwe. *Water South Africa* **2012**, *38*, 233–240.

166. Munodawafa, A. Assessing nutrient losses with soil erosion under different tillage systems and their implications on water quality. *Phys. Chem. Earth Pt. A/B/C* **2007**, *32*, 1135–1140. [CrossRef]

167. Agbede, T. Nutrient availability and cocoyam yield under different tillage practices. *Soil Till. Res.* **2008**, *99*, 49–57. [CrossRef]

168. Enfors, E.; Barron, J.; Makurira, H.; Rockström, J.; Tumbo, S. Yield and soil system changes from conservation tillage in dryland farming: A case study from North Eastern Tanzania. *Agric. Water Manag.* **2011**, *98*, 1687–1695. [CrossRef]

169. Mchunu, C.N.; Lorentz, S.; Jewitt, G.; Manson, A.; Chaplot, V. No-Till Impact on Soil and SoilOrganic Carbon Erosion under Crop Residue Scarcity in Africa. *Soil Sci. Soc. Am. J.* **2011**, *75*, 1503–1512. [CrossRef]

170. Ouédraogo, E.; Mando, A.; Stroosnijder, L. Effects of tillage, organic resources and nitrogen fertiliser on soil carbon dynamics and crop nitrogen uptake in semi-arid West Africa. *Soil Till. Res.* **2006**, *91*, 57–67. [CrossRef]

171. Sanchez, P.A.; Denning, G.L.; Nziguheba, G. The African Green Revolution moves forward. *Food Sec.* **2009**, *1*, 37–44. [CrossRef]

172. Sanchez, P.A. En route to plentiful food production in Africa. *Nature Plants* **2015**, *1*, 1–2. [CrossRef]

173. Alliance for a Green Revolution in Africa (AGRA). *Building on the New Momentum in African Agriculture: AGRA in 2008*; Alliance for a Green Revolution in Africa: Nairobi, Kenya, 2009.

sustainability

MDPI

Review

Monitor Soil Degradation or Triage for Soil Security? An Australian Challenge

Andrea Koch [1,*], Adrian Chappell [2,†], Michael Eyres [3,†] and Edward Scott [3]

1　United States Studies Centre, University of Sydney, Institute Building, H03, Sydney NSW 2006, Australia
2　CSIRO Land and Water National Research Flagship, G.P.O. Box 1666, Canberra ACT 2601, Australia; adrian.chappell@csiro.au
3　Injekta Field Systems, Level 1, 69 Fullarton Road, Kent Town SA 5067, Australia; michael@injekta.com.au (M.E.); ed@injekta.com.au (E.S.)
*　Correspondence: andrea.koch@sydney.edu.au; Tel.: +61-408-030-081, Fax: +61-2-9351-6877
†　These authors contributed equally to this work.

Academic Editor: Marc A. Rosen
Received: 11 March 2015; Accepted: 9 April 2015; Published: 24 April 2015

Abstract: The Australian National Soil Research, Development and Extension Strategy identifies soil security as a foundation for the current and future productivity and profitability of Australian agriculture. Current agricultural production is attenuated by soil degradation. Future production is highly dependent on the condition of Australian soils. Soil degradation in Australia is dominated in its areal extent by soil erosion. We reiterate the use of soil erosion as a reliable indicator of soil condition/quality and a practical measure of soil degradation. We describe three key phases of soil degradation since European settlement, and show a clear link between inappropriate agricultural practices and the resultant soil degradation. We demonstrate that modern agricultural practices have had a marked effect on reducing erosion. Current advances in agricultural soil management could lead to further stabilization and slowing of soil degradation in addition to improving productivity. However, policy complacency towards soil degradation, combined with future climate projections of increased rainfall intensity but decreased volumes, warmer temperatures and increased time in drought may once again accelerate soil degradation and susceptibility to erosion and thus limit the ability of agriculture to advance without further improving soil management practices. Monitoring soil degradation may indicate land degradation, but we contend that monitoring will not lead to soil security. We propose the adoption of a triaging approach to soil degradation using the soil security framework, to prioritise treatment plans that engage science and agriculture to develop practices that simultaneously increase productivity and improve soil condition. This will provide a public policy platform for efficient allocation of public and private resources to secure Australia's soil resource.

Keywords: soil; soil security; agriculture; erosion; no-till; conservation agriculture; Australia

---

1. Soil Security, Soil Degradation and Agriculture in Australia

Soil security is achieved when the condition of soil enables it to support the ongoing production of food and fiber and continue its role in cycling of fresh water, climate regulation, and overall ecosystem resilience [1]. The concept of soil security emanated from deep scientific concern about global soil degradation and its impact on sustainable development. It emphasizes the critical role of soil in achieving food, water and energy security, biodiversity and climate change mitigation, and the ongoing provision of ecosystem services—all significant global challenges [2]. Soil security is necessary to achieve sustainable development and long-term agricultural productivity.

Soil is secured through agricultural land management practices that are matched to the functional capability of the soil and which improve and maintain soil condition ("Soil condition" can be used interchangeably with "soil quality" and is the official term used by the Australian Government to describe *"the capacity of a soil to function, within land use and ecosystem boundaries, to sustain biological productivity, maintain environmental health, and promote plant, animal, and human health"* [3,4]). This requires consideration and understanding of the five dimensions of soil security that encompass the biophysical, social and economic aspects of soil, and policy and legal frameworks that support them, *i.e.*:

- Capability—the potential functionality of any given soil, how it can be expected to perform and what it may produce
- Condition—the contemporary state of the soil referenced to its capability, and an outcome of how it is managed
- Capital—the economic value of the soil resource and the services that flow from it
- Connectivity—the social dimension, concerned with the connection of the land manager with their soil, and the resources and knowledge they have to manage the soil according to its capability; as well as broader societal recognition of the soil resource
- Codification—the public policy and legal frameworks required to support the securing of soil [1].

Land management practices that lead to soil degradation put soil into a state of insecurity leading to short and long term implications for sustainable development. Ultimately this may lead to the soil becoming degraded to a point that it may never return to its original state and function. In order to achieve soil security, land managers must understand the inherent economic value and biophysical capability and condition of their soil, have the connectedness of knowledge and resources to manage it for productivity whilst improving or maintaining its condition, and be supported by public policy that enables them to do this—encapsulating the five dimensions laid out above [1].

The Australian National Soil Research, Development and Extension Strategy (hereafter, the "Strategy") was developed through a cross-jurisdictional collaborative effort by Federal and State agencies, Commonwealth Scientific and Industrial Research Organization (CSIRO), the academic research community and industry representation through the relevant research and development corporations and farmer groups [5]. The Strategy calls for securing soil as a contribution to the current and future competitiveness of Australian agriculture and recognizes the critical need for soil knowledge, information and data to increase agricultural productivity, profitability and sustainability in Australia.

Australian farmers and policy makers have battled with soil degradation since the country was first settled. The transition from natural to agricultural land has historically been associated with the removal of natural vegetation for cultivation and grazing. However, the native vegetation had generally protected the soil from erosion by wind and water. Consequently, soil erosion accelerated where historical agriculture expanded and particularly where ploughing was used. Other soil degradation processes (e.g., acidification, salinization, loss of soil organic matter, compaction) have also increased, but globally the area affected by soil degradation is dominated by soil erosion [6].

European cropping and grazing methods were first applied across Australia at the time of settlement in the late 1700s. Agriculture and pastoral grazing then spread across the continent over the next 150 years, [7] reaching a critical mass by the early 20th century as shown in Figure 1.

Due to a fundamental lack of understanding and knowledge of Australia's old, weathered and largely infertile soil, inappropriate agricultural methods mined the soil resource with disastrous consequences. Pioneering European farmers had little context and scarce resources in an environment that was completely foreign to their experience. The innovation of "frontier farming" fed the growing colony and its export economy [8] but caused widespread unintended consequences. During the 1930s and 1940s, large tracts of land in Australia became "dust bowls" [9–11]. The resulting soil erosion led to significant Australian policy responses in the early 20th century and an agriculture sector that has continually sought to implement soil conservation practices [9–12]. This had led to success in reducing soil degradation over time, but as we will show, there is still much work to be done.

Sustainability **2015**, *7*, 4870–4892

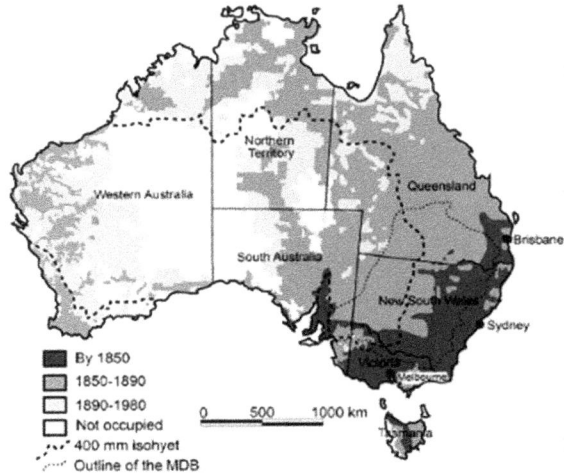

Figure 1. Pastoral development of Australia through time [8] (reproduced with permission).

Soil security is a recently developed concept. Progress towards soil security can only be judged by evaluating historical efforts and outcomes of soil conservation in an agricultural context. In this paper we outline key issues of soil degradation at the continental scale in Australia within the context of agricultural land use and land management changes. We provide evidence of the relationship between soil erosion and agriculture over time, and show how adoption of conservation agriculture, specifically no-till cropping, has helped to address soil erosion in cropping systems and why conservation agriculture is a practical example of soil security. We continue by showing that despite the gains, this has not solved erosion in current cropping systems and outline a range of impacts on soil condition as a result of no-till. We contend that the next step-change in cropping should be based on analysis of soil at the paddock level. This will lead to agricultural practices that provide the win-win of increased productivity and improved soil condition—both key requirements for soil security. Finally, we return to the continental scale to suggest that complacency towards soil erosion will greatly impact Australia's broader ability to secure soil and increase food production. We contend that monitoring the continued degradation of Australia's soil will not lead to soil security. Instead, we propose the adoption of a triaging approach to soil degradation that is based on current science and improved agricultural practices supported by appropriate public policy to ensure that public and private sector resources are applied appropriately and rapidly to safeguard the nation's soil resources.

2. Australian Land Use and Soil Degradation

It is important to understand some key aspects of Australian agriculture to contextualize historical and contemporary soil degradation patterns. Just over half (52 percent) of Australia is used for agriculture. Of that, 86 percent (340 million hectares) is used predominantly for grazing and about eight percent (32 million hectares) is used mainly for cropping [13] (Figure 2).

Agriculture is mainly located in the south-eastern and south-western regions of Australia, with the remaining 81 percent of land consisting of rangeland *i.e.*, native vegetation with erratic and/or small rainfall that precludes agriculture [14].

Pastoralism is the main agricultural pursuit in the rangelands. Australia's economic growth *"rode on the sheep's back"* due to heavy grazing by sheep for wool and meat in the rangelands and agricultural areas during early settlement and through to the 20th century [8]. Today, rangeland livestock is predominantly beef cattle. This contemporary land use pattern may be set to change, with a Federal Government plan for significant development of northern Australia including large new irrigation/dam projects for agriculture [15]. New trade agreements with Asian nations will increase opportunities for significant increases in livestock production. This has potential implications for soil security in the north, which has been and continues to be impacted by soil degradation [16].

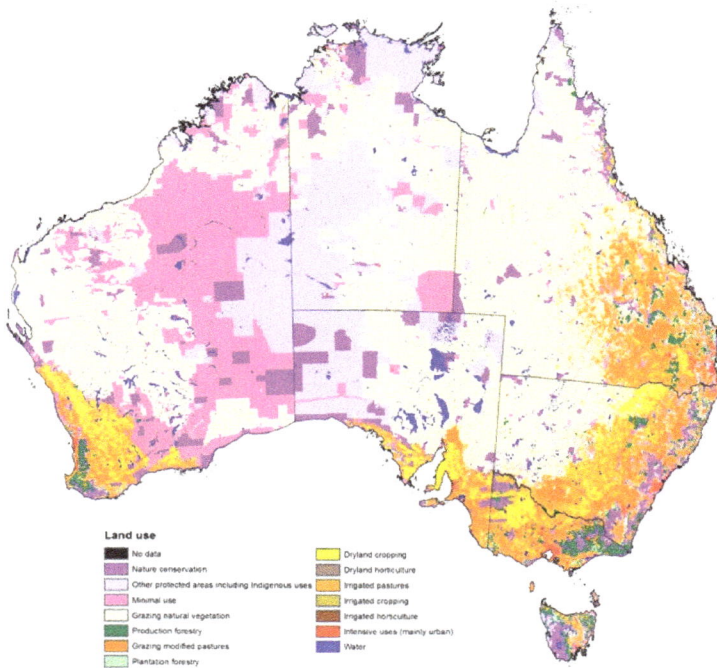

Figure 2. Land use of Australia for 2005–2006, national scale land use map developed by ABARES in 2010. Source: Australian Bureau of Agricultural and Resource Economics and Sciences [17].

At the continental scale, soil degradation in Australia is broadly characterized by four processes; soil acidification, soil carbon loss, water and wind erosion [16,18]. These processes are used to indicate the status of the soil or its condition. Policy focus on these processes is driven from economic and environmental considerations.

Soil acidification is estimated to affect over 50 percent of cropping and/or intensively grazed regions, with soil acidity in many agricultural regions continuing to deteriorate [16]. The Western Australian Government estimates that soil acidity is already impacting state farm gate returns to the tune of $400 million per annum through lost production. This region produces half of Australia's wheat (*Triticum aestivum* L.) crop and supplies 80 percent of the wheat exports [19].

Australian agricultural soil has lost between 40 and 60 percent of its soil carbon content since settlement [20]. Soil carbon dynamics have received much attention over recent years as governments have considered the potential of soil carbon sequestration as carbon sink in climate change policy. Prior to this, increasing soil organic carbon (SOC) was a key program focus for State based Catchment Management Authorities and Natural Resource Management agencies across the country, in efforts to improve soil condition. Community based programs administered through Landcare and the Federal Caring for Country program have also provided focus for improving SOC [21].

It is estimated that water erosion is now outstripping soil formation rates across Australia by a factor of several hundred and in some areas, several thousand [16]. There is a dearth of measured water erosion data across continental Australia, but a modeling study in 2003 predicted sheetwash and rill erosion rates for the entire Australian continent using the revised universal soil loss equation (RUSLE) and spatial data layers that provided a range of environmental factors [22]. The study concluded that northern Australia is at higher risk for water erosion than the south and that there is significant seasonal variation between summer and winter. It estimated the average erosion rate to be 4.1 ton/ha/year across the continent, and that about 2.9×10^9 tonnes of soil is moved annually, representing 3.9 per cent of global soil erosion from 5 per cent of world land area [22]. One of the limitations of using the RUSLE model within the context of commenting on soil erosion by all processes is the inability to account for deposition and wind erosion.

As we show later, the prevalence and severity of wind erosion has subsided since the "dust bowl" years of the late 19th and early 20th centuries, however the strongest determinant of wind erosion may be climate [16]. Climate change in Australia is expected to result in more extreme wind and flooding events. Both have been experienced across the continent during the past decade and anecdotally caused significant dust storms and soil loss through water erosion [16].

The issue of soil degradation by erosion is compounded by the selective nature of the removal processes and therefore the type of material that is removed. This is particularly important for soil organic carbon, which is critical for soil aggregation, moisture holding and feedback processes for soil fertility. Consequently, preferential removal of SOC by erosion increases susceptibility of the soil to further erosion and depletion of soil nutrients.

Overall, these continental scale indicators have considerably different spatial and temporal frequencies; they may be discrete in some areas and overlap in other areas and likely have interactions that stem from a common cause (*i.e.*, soil erosion). In any case, soil erosion dominates in areal extent. The complement of water and wind erosion covers the majority of the Australian continent.

Soil erosion has and continues to impact agriculture, but it has lost focus as an issue. Constitutionally, soil conservation is the responsibility of State Governments in Australia, and while many States have long term legislation aimed at reducing soil erosion, over recent decades there has been a gradual decline in the resources invested to address soil degradation including erosion, compared with competing natural resource management issues [21]. Should this trend continue, Australian soil in many areas will become increasingly insecure.

Erosion is the highest existential threat to soil, and the biggest risk to achieving soil security. In the next section, we describe soil erosion over time to show how it continues to be the primary soil degradation issue at the continental scale. By mitigating soil erosion, many other forms of soil degradation can also be alleviated. For example, soil erosion by wind and water is partly responsible for the loss of soil organic carbon. Similarly, the loss of topsoil and/or the preferential loss of nutrient- and carbon-rich fines may have aggravated the acidification process. We also show that strategies to address soil erosion in cropping systems have produced unintended changes in soil conditions (such as stratification of nutrients, herbicide accumulation, compaction, localized acidity, and aluminum toxicity) that are causing paddock scale issues for soil management.

3. Soil Erosion in Australia—A Historical Snap Shot

3.1. Soil Erosion Rates by Land Use—1950s to 1990s

For a single point within the landscape, the ^{137}Cs technique provides a reliable retrospective estimate of net (time-integrated) soil redistribution (erosion and deposition) due to the combined effect of wind, water and tillage. The ^{137}Cs technique has been used to estimate soil erosion in studies around Australia since the 1980's. In the early 1990's a national survey of ^{137}Cs-derived net (1950–1990) soil redistribution from transects across all states in Australia showed that soil losses were significantly greater under conditions of intensive agriculture and on rangelands compared with uncultivated pasture and forest [23]. These differences were attributed to soil disturbance, lack of ground cover and susceptibility to water and wind erosion (erosivity and erodibility). Sixty per cent of sites had net soil losses greater than 1 tonne per hectare per annum; well above the tolerable soil loss threshold of

Sustainability 2015, 7, 4870–4892

0.5 tonnes per hectare per annum followed by Loughran *et al.* [23]. This analysis showed that despite significant landholder awareness of erosion events, serious unsustainable soil losses had occurred throughout much of Australia between the mid-1950s and 1990. Further analysis of these data by Chappell *et al.* [24,25] showed that nearly five times more soil was lost from cultivated land than from uncultivated land in Australia over that period (Figure 3).

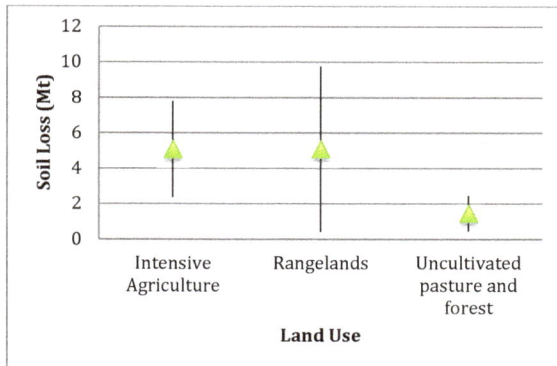

Figure 3. Comparative soil loss (1950s–1990) by land use.

3.2. 1990s to Present—Changing Cropping Practices and the Impact on Soil Erosion on Agricultural Land

Many Australian grain farmers are now well versed in the nature and benefits of conservation tillage, *i.e.*, seeding with no prior cultivation [12]. Stubble management, and more recently, "retained stubble" management has had marked influence on soil erosion [26] by increasing surface roughness, reducing near surface wind speed by up to 80 percent [27] and controlling mechanical dispersion and structural degradation due to rainfall impact and surface water run-off [28].

From the 1980s onwards, there was rapid and widespread adoption of conservation tillage practices. The adoption rates for no-till practices across the grain growing areas of south-eastern and south-western Australia are summarized in Figure 4 [29].

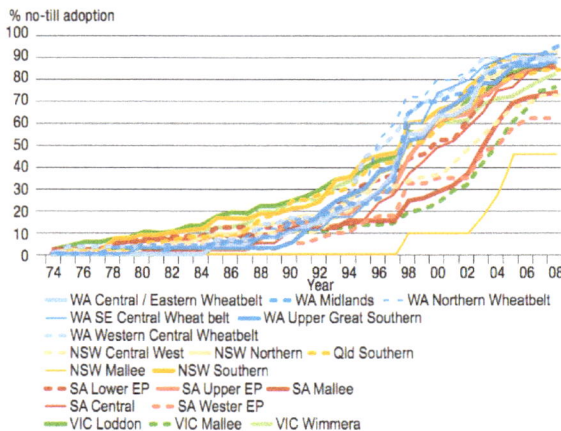

Figure 4. Cumulative adoption of no-till (decision to first use no-till) across Australian cropping areas [29].

There is evidence to suggest that the wide-scale adoption of no-till and other conservation agriculture practices has had a marked effect in reducing soil erosion in the cropping zones. For example, net (1990–2010) soil erosion in south-eastern agricultural Australia erosion had declined on average (Figure 5) from −9.7 ton/ha/year to +3.9 ton/ha/year [30].

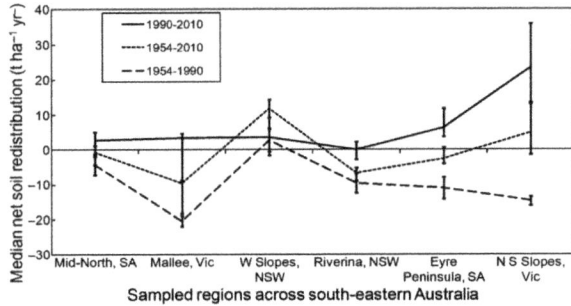

Figure 5. Change in soil redistribution median and interquartile range for south-eastern Australia (1954–2010) [30].

The regional decline in soil erosion was attributed to the widespread adoption of soil conservation measures and in particular conservation agriculture over the last 30 years. Notably this average decline included considerable spatial variation with an interquartile range of −1.6 to +10.7 ton/ha/year, likely due to variability in the adoption of conservation agriculture.

A study by Marx *et al.* [31] also indicates the success and efficacy of conservation agriculture in reducing wind erosion since the 1990s. Using dust deposited in a Snowy Mountains mire, they reconstructed the wind erosion history and the expansion of dust sources associated with the progression of European farming practices across south-eastern Australia (Figure 6). They identified a rapid increase in dust deposition after 1879 (B) and a rapid decrease in dust deposition after 1989 (H). Three phases of dust deposition were defined, which they described as: (1) pre-European 1700–1879; (2) agricultural expansion 1880–1989; and (3) agricultural stabilization 1990 to 2006.

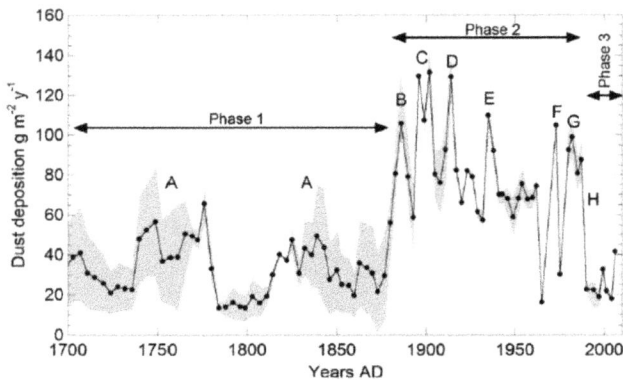

Figure 6. Dust deposition rates in the Snowy Mountains based on core data plotted from 1700 to 2006 [31]. Reproduced with permission.

The widespread adoption of no-till and conservation agriculture and the resulting reduction in soil erosion provides a case study in the practical application of soil security. Research into the uptake of no-till as shown in Figure 4, also found that farmers were motivated mostly by the reduction of fuel and labour costs, and soil conservation including soil moisture management and improved soil structure [29]. This indicates an awareness of the biophysical capability and condition issues of soil by farmers, and a desire to improve the situation. Further to this, there was a perceived increase in capital value of the soil. Consequently half of the surveyed farmers indicated they would be willing to pay more for neighbouring land that been cropped under no-till and conservation agriculture [29]. Further analysis of this case study framed by the five dimensions of soil security may inform future planning to develop approaches that secure soil.

4. Soil Security Issues Stemming from Success

The successful reduction of soil erosion in some Australian cropping systems is well acknowledged. There are two arising issues that we now address. The first is concerned with the codification (policy) dimension of soil security, the second with the connectivity dimension.

4.1. False Belief that Erosion has been Solved

First, the success in reducing soil erosion through such a widespread practice change has led to a belief in some quarters that soil erosion is no longer a major problem for farmers. As a result, State Government programs to address soil erosion have been scaled back over recent decades [21]. This has led to a level of confidence that soil erosion has been addressed, but erosion levels are still above tolerable and regenerative limits and will continue to impact agriculture, particularly under a changing climate, unless preventing further soil erosion remains a focus.

At a tolerable loss rate of 0.5 tonnes per hectare per annum [23], 55% of Australia could be considered net stable but many areas, particularly in agricultural Australia, are at a considerable risk (e.g., 70% chance) of exceeding that threshold [25]. With the anticipated development of agriculture in northern Australia it is important to quantify soil erosion and respond quickly to any deterioration in the soil resource. Figure 7 shows the probability of soil erosion exceeding the threshold at different locations across Australia. Large areas of Australia were at risk in 1990 and the most susceptible areas to soil erosion were the main agricultural regions of Australia.

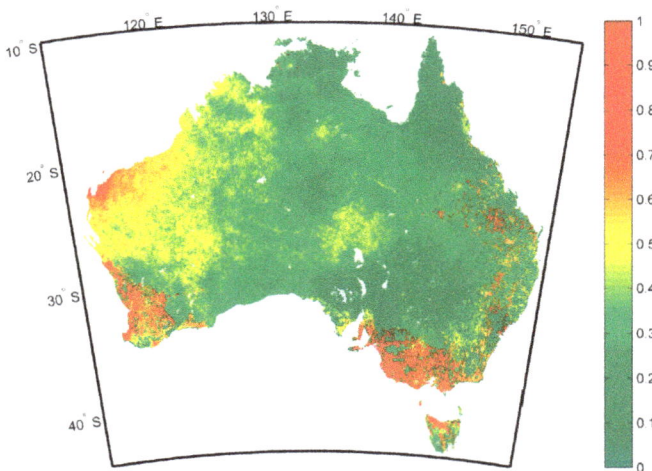

Figure 7. Probability (0–1) of exceeding the threshold −0.5 ton ha^{-1} year^{-1} of tolerable soil erosion [25]. Reproduced with permission.

Conservation tillage has likely gone part way to reducing intolerable levels of soil erosion in cropping areas, but soil erosion has not been eradicated. Estimates that show a regional decline in soil erosion between 1990 and 2010 also show considerable spatial variability, indicating that many sampled locations have not reduced soil erosion, most notably in the Mallee region (see case study below). Continued policy complacency towards this trend may lead to reduced agricultural productivity and sustainability.

4.2. Unintended Consequences of Conservation Tillage

A second issue associated with the perceived success in controlling soil erosion is that evidence has emerged that conservation tillage is producing unintended consequences for soil resources at the paddock/field level. This includes the physical and chemical restructuring of the soil profile, leading to new constraints for plant growth and crop yield. But perhaps a greater threat is the declining connectivity of farmers with the soil they manage, resulting in a lack of focus on soil and complacency towards soil management, leading to a "blind spot" in the land management toolkit.

With conservation tillage, soil is often cultivated (and fractured) to a depth greater than with conventional tillage systems; many tillage passes are used to work the soil more vigorously but to a shallower depth. This is certainly the case with knife-edged tillage which uses tractors with far more horsepower (and weight) per tyne than ever before. This deeper tillage can lead to soil disturbance and compaction [28], and may produce either positive or negative consequences depending on the soil type and condition. Conservation tillage machinery provides a well-engineered set of instruments that successfully achieve the outcome they were designed to achieve—causing minimal surface soil disturbance whilst maintaining a suitable seedbed for seed planting and growth. However limitations arise when the soil capability is not well understood and soil condition is not further managed to accommodate the new tillage strategy. Due to the reduced physical intervention with the surface soil compared to conventional tillage systems, more emphasis is placed on the inherent soil chemical, physical and biological condition and how it can support the given management system. For example soil compaction is often generally attributed to vehicular or animal traffic, however certain soil types are more susceptible to this type of decline in condition and resilience. This places those soil resources at a higher risk for this form of degradation.

There is also evidence of the stratification of soil layers due to conservation tillage, with an unintended effect of more defined nutrient distributions within the various soil layers [26]. This stratification is leading to layers of altered soil chemical condition down through the profile in duplex, texture contrast soil types or even homogeneous soil textures. Other problems include acidification, alkalization, salinity and sodicity. This altered soil chemistry can directly impact the potential for above and below ground crop performance and lead to greater risk of land degradation in the post crop phase. For an example, see the Victoria Mallee Case Study below.

Case Study 1. Victorian Mallee

The Mallee region in south-western Victoria is at high risk of wind erosion due to the removal of native vegetation and its replacement with agricultural ground cover, which is often below 50% [32]. The region also has a sandy soil type (relict aeolian dune-swale landforms) and small mean annual rainfall (270–370 mm). The dune-swales are often duplex soils with sandy textured topsoil over a dispersive medium clay. At a typical study site in this region, wind erosion was compounded by water erosion during rare high intensity rainfall in the summer of December 2011. Despite the use of stubble retention water erosion removed the sandy topsoil and revealed the sub surface effect of conservation tillage equipment tines (Figure 8). The site also had a sealing, dispersive subsoil with low soil porosity (*i.e.*, smaller hydraulic conductivity than the topsoil), degraded soil structure and poor root penetration despite the shallow subsoil depth. This case study demonstrates the impact of conservation tillage on soil structure and the lack of understanding of the complexity of agricultural soil profiles.

Figure 8. Rill Erosion in no-till cropping system after extensive and intensive rainfall event. Evidence of striations in clay subsoil from no-till machinery sealing the dispersive clay subsoil. Manangatang Vic.

These problems described above can be managed effectively if the farmer is aware of them. The adoption of conservation tillage has led to Australian agricultural management systems that are managed from the "top-down" due to a focus on plant breeding and growing plants that are more suited to the conservation tillage. Consequently, there is less focus on the "bottom-up" specific soil constraints that prevent plants from successfully performing in the soil environment. Prior to no-till and chemical farming, land managers had to understand their soils to greater depth and detail to implement effective weed management, improve plant germination and early growth, and achieve effective soil moisture management without chemical intervention [33].

Although gaining improved outcomes in reducing soil erosion and improving productivity, the new approach has reduced the previously close connection between farmers and the soil they manage.

Today, land managers have the most influence on soil performance at one critical point in the season (at planting). From then on the crop management relies heavily on the extensive use of chemicals [34]. This system leads to greater reliance on plant information as a guide for future performance with little focus on soil condition and performance for more effective farm management strategies. Achieving the win-win of increased productivity and reduced soil degradation in cropping lands is possible, but only with a greater agronomic understanding of the soil throughout the profile in order to optimize conservation tillage practices. Soil is more likely to be investigated in the top 10–15cm but there is much to be gained by understanding the subsoil horizons and their often marked impact on plant performance.

5. Soil Management for Improved Productivity and Soil Security in Cropping Systems

Some farmers are now reconnecting with the soil and embracing soil management as the basis for agricultural production and as a key indicator to land management performance [35]. Soil Use Efficiency (SUE) is a diagnostic and management tool that is being used as part of this approach.

Assessments of SUE use individual and inter-related factors (inherent and dynamic) affecting soil condition (*i.e.*, chemical, physical and biological soil properties and processes as well as soil nutrient availability and nutrient uptake potential) as effective reference points for improving crop productivity on individual and varying soil types. SUE is effectively an interpretation of soil condition and land suitability to generate field information powerful enough to affect net farm productivity. SUE is the practical application of soil capability and condition, the two biophysical dimensions of soil security.

Water Use Efficiency (WUE) and Nutrient Use Efficiency (NUE) guidelines are being adopted, but these strategies provide measurements as an output and include too few management guidelines to further improve land management systems. Optimization of WUE and NUE requires an understanding of the boundary conditions of SUE. This will enable Australian agriculture to progress and help farmers utilize conservation tillage practices to their full capacity by renewing the focus on the fundamental underlying agricultural performance factor—the soil, and the individual components of site-specific soil management.

The starting point for understanding and implementing SUE is the excavation of soils to depth in paddocks to gain an understanding of how plants are interacting with soils throughout the soil profile, and to identify any constraints that the soil profile is placing on plant growth. This is augmented with detailed and extensive soil physical and chemical assessments to provide details of chemical constraints such as acidity, sodicity, alkalinity, or elemental toxicity. Collectively, these assessments provide the farmer with a clear understanding of what is happening in relation to soil performance.

Once this soil analysis is complete further site-specific actions can be implemented within the furrow. The best intervention point for plant production is with the usage of tillage implements when sowing. Sowing equipment can be used to tailor the application of inputs suitable to both soil type and condition for optimum plant performance based on soil capability and condition. The approach replaces the current "plants down" approach to soil management (*i.e.*, soil adaptability) with a "soil-up" approach that is key to increasing agricultural productivity. This can be referred to as soil management based agriculture.

Many cropping systems are utilizing variable rate nutrient applications (nitrogen and phosphorus) applied across the landscape according to soil types, which doesn't necessarily account or allow for the condition of the soil throughout the profile. When conservation tillage practices are combined with the SUE approach, a Vertical Rate guideline can be established. This effectively incorporates the impact of the individual soil horizons on plant accessible water and nutrients (including oxygen and carbon dioxide).

SUE and Vertical Rate guidelines are already being utilized in management strategies. For example in addressing the key issues of soil acidification (which can often be caused by aluminium hydrolysis), utilization of strategic lime applications and the integration of furrow applied high pH calcium based liquid soil conditioners offset the impact of aluminium toxicity and localize the increase of pH in the root zone to support the growth of juvenile plants. Deep banding of manures, composts, lime and gypsum has also been integrated in to some cropping systems, leading to improvement in the subsoil condition by reducing the constraints of poor soil structure, sodicity, salinity, or other issues.

Crop rotation is also a widely utilized mechanism of soil management to depth. By rotating cereals, brassicas, legumes and pasture, the soil is given capacity to respond and recover from intensive cropping systems. Due to varying root morphology and architecture the utilization of the soil profile can vary. Plants with tap root systems provide biopores that are capitalized on by future crops and water infiltration processes. These approaches are being proven on a case-by-case basis to reduce soil erosion and to increase agricultural productivity—a soil security "win-win". An example is provided in the South Australian Case Study below.

Case Study 2. South Australian Mallee—Soil Use Efficiency at Work

In the South Australian Mallee the evidence of soil management and its impact on the landscape is severe. Figure 9 shows the impact of wind erosion after a prior "conservation tillage" crop despite stubble retention. Plant density was poor and wind erosion caused soil loss over the summer periods. The sandy soil is hydrophobic—water repellent—and the infiltration of water into the soil profile is slow and non-uniform, impacting seed germination and plant density. In the following planting season, the land manager has strategically ploughed the surrounding dune area to reinstate the crest of the dune. Hydrophilic sand was redistributed across the surface and integrated with the hydrophobic sand. At the following planting, wetting agents and biodegradable surfactants were used to reduce the hydrophobic properties of the soil. A comprehensive fertility programme was delivered to the seed in furrow with a liquid delivery system at planting. A top-dressing program was also implemented throughout the season. This resulted in a greater germination and plant density on the area, leading to greater soil stabilisation and retention of topsoil in the following summer period (Figure 10).

Figure 9. Evidence of wind erosion at trig point near Bow Hill, South Australia, 6 February 2013, in a continuous No-Till cropping system in a dune swale system.

Figure 10. Impact of stubble retention and improved management practices on soil stabilization at the same site from Eastern end. 26 February 2014.

6. Soil Erosion Still Remains a Major Threat to Soil Security

Despite the advances in reducing soil erosion in cropping systems, soil erosion remains a major problem in Australia and in particular in the rangelands and is a major threat to future soil security. A well-established dust emission model was used to show that between 2000 and 2011 mean dust (< 22 μm) emission was 1.34 TgC/year and 0.11 TgC/year for rangeland and agricultural Australia, respectively [36]. Despite smaller SOC dust emission and SOC contents than agricultural Australia, the largest loss of SOC dust emission is from rangeland Australia because of the large area affected by wind erosion and dust emission. Using net soil erosion for rangeland Australia (−0.22 ton/ha/year) and the area of rangeland (666 Mha), the net total amount of soil removed was 147 Mt/year. Using net soil erosion for agricultural Australia (−1.48 ton/ha/year) and the area of agricultural land (53.6 Mha) affected by soil redistribution, the net total amount of soil removed was 79 Mt/year (Figure 11).

This indicates that despite smaller erosion rates in rangelands, their large area contributes more to the net loss of soil across the continent than the larger erosion rates from agricultural regions. The sum of 226 Mt/year (226 Tg/year or 0.23 Pg/year) is broadly consistent with the global regional estimates of gross water and tillage erosion from croplands and pasture in Australia [37]. Estimates of ^{137}Cs-derived net (1950's–1990) soil redistribution by wind, water and tillage from rangeland and agricultural Australia amount to about 3% and <1%, respectively of the global total [25] removing approximately 2% of the total SOC stock (0–10 cm) from the land surface over this ca 40-year period [38].

Policy for cropping land introduced in the 1980s makes a clear linkage between changed agricultural practice and soil conservation, and has been successful due to widespread implementation by farmers and as outlined previously, provides a useful case study of the practical application of the soil security framework. Unfortunately there has been no such effective counterpart in rangelands. Soil erosion in rangelands appears to be set to continue and potentially accelerate under a changing climate. There is also evidence that abnormal climate events can still have devastating effects on soil. The following two case studies illustrate the impact of unusual but extreme wind and flooding events. This suggests that for small magnitude, large frequency climate events conservation agriculture may be coping. However, large magnitude, small frequency climate events demonstrate how marginal the protection is against wind and water erosion, particularly in rangelands, where wide-scale adoption of management strategies that secure soil is yet to happen.

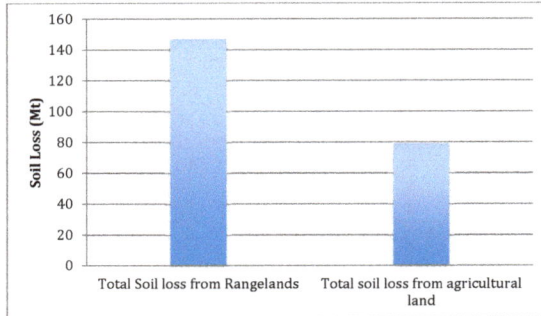

Figure 11. Soil loss by land use 2000–2011.

Case Study 3. Red Dawn

The dust storm on 22–23 September 2009, dubbed by the Press as "Red Dawn", was the largest recorded in 50 years and caused the largest reduction of visibility ever recorded in Sydney (Figure 12). The cause was attributed to extensive drought conditions and extreme winds [39]. The sources of the dust were the lower Lake Eyre Basin in South Australia, the grazing lands of north-western NSW and the mining areas around Cobar and Broken Hill in addition to the Channel Country of Queensland [39]—all rangeland areas.

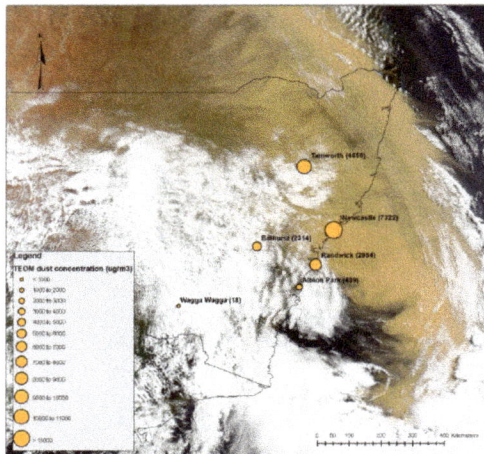

Figure 12. MODIS image showing "Red Dawn" dust storm extending from south of Sydney to NSW border with Queensland [39]. Reproduced with permission.

The economic impact of Red Dawn to the state of New South Wales was estimated to be A\$299 million, compared with a conservative estimate of the annual average cost to Australia of dust storms of A\$9 million [40].

Case Study 4. The Gascoyne Flood, 2010 and 2011

The Gascoyne River catchment covers an area of about 80,400 km in the southern rangelands of Western Australia. The town of Carnarvon is located at the mouth of the river, along with about 1000 hectares of irrigated horticulture. The catchment supports an extensive pastoral industry mainly grazing beef cattle (although historically, predominantly sheep) [41].

Widespread and extreme floods occurred during the period December 2010 and during the summer of 2010–2011, causing an estimated A\$90 million in damage to the horticulture industry. Soil loss through water erosion was estimated to have been at least 5,625,000 tonnes—the plume of red soil into the ocean was visible from space [41] (Figure 13).

Figure 13. Gascoyne River mouth sediment plume—10.00 am, 22 December 2010. Image processed and enhanced by Landgate, Satellite Remote Sensing Services; Erosion cell [41].

Contributing factors responsible for this massive soil loss from the Gascoyne catchment were the poor condition with reduced groundcover (perennial vegetation), a series of poor growing seasons and continuous stocking of livestock [41]. This case study provides a snapshot of perhaps the most insecure soil in Australia, a situation in the making since the days of settlement.

There is a potential pathway to soil security in this region. Innovative pastoralist-led initiatives to regenerate the landscape have had proven success in-so-far as they have been able to be applied with limited private resources [42]. However, without public policy support and significant investment, wide-scale regeneration and subsequent return to agricultural productivity will be difficult.

Climate change projections are for warmer, drier weather with more time in drought with little change in wind speeds, which will increase the susceptibility of Australian soil to wind erosion [43], and reduced, rather than enhanced soil security. In addition, extremes in rainfall (intensity and scarcity) [43] are likely which means that water erosion may also increase in the future.

7. Triage for Soil Security

The Australian National Soil Research, Development and Extension Strategy emphasizes the need for mapping, modeling and monitoring of soil conditions, but as important as those activities are, none of the diagnostic activities will in themselves lead to soil security. As stated in the introduction, soil security can only be achieved when all five dimensions of soil security are addressed. This requires agricultural land management practices that are matched to the functional capability of the soil, and which improve and maintain soil condition, and implies that farmers are well connected with their soil and have the knowledge and resources and public policy support to apply them.

We have demonstrated in the case studies above that when an integrated approach is used, improvements in soil security are possible e.g., the impact of conservation tillage in cropping zones, and the marked effect this has had in reducing soil erosion. However, soil degradation, and specifically soil erosion, still poses an existential threat to soil and to soil security. Soil condition monitoring provides indicators of land degradation, however it does not provide a solution. Soil security provides a positive framework for developing solutions. We contend therefore that monitoring the continued degradation of Australia's soil will not lead to soil security. We propose therefore, an approach to determining soil that is at risk, and prioritizing the application of treatment plans to secure that soil—*i.e.*, "triage" for soil security.

The word "triage" is used almost exclusively in medical contexts to describe a decision system that prioritizes the allocation of scarce resources to patients. According to Iserson and Moskop [44] medical triaging is based on the satisfaction of three key conditions: a scarcity of medical resources, the assessment of needs for each patient by a triage officer based on a medical examination, and establishment of a system or plan, usually based on an algorithm or set of criteria to determine treatment and/or the priority of treatment for the patient.

The French root word, "trier", means "to sort", and was originally used to describe the sorting of agricultural products [44]. It is not unfitting therefore to think of soil in any form of degraded state and which impacts the production of agricultural products, as a patient requiring medical attention. Soil in this parlous state is insecure—the question is whether it needs minor treatment or intensive care in order to make it secure. The analogy is extended further (Table 1).

Table 1. Application of the principles of medical triage to triage for degraded soil.

Medical Triage	Triage for Soil Security
1. Scarcity of medical resources.	1. Scarcity of public and private resources to halt and/or reverse soil degradation and secure soil.
2. Assessment of needs of each patient by a triage officer based on a medical examination.	2. Assessment of the severity of soil degradation at a determined unit level, based on expert assessment and condition monitoring.
3. Establishment of a system or plan, usually based on an algorithm or set of criteria to determine treatment and/or the priority of treatment for the patient.	3. Establishment of a system or plan based on the severity of the degradation and its risk to soil security, to determine a treatment plan and to prioritise the treatment of soil that is at high risk of loss. The plan must meet the criteria laid out by the five dimensions of soil security.

In order for a triage officer to make use of an established plan to treat the patient, the plan must first exist and there must be a decision to use the plan. In this context, triage planning for soil security would involve development and adoption of a national approach, utilized and implemented at State level to prioritize and determine the treatment of degraded soils in particular agricultural contexts (paraphrasing Iserson and Moskop [44]). By linking "prioritized treatment plans" with approaches to agriculture the requirements for soil security can be met by simultaneously increasing agricultural productivity and improving soil condition, supported by public policy that enables the required extension and education programs for implementation, the requirements for soil security can be met.

This triage framework would address the issues outlined above and make the best use of limited resources for supporting soil that needs "intensive care". This would include soil that is at greatest risk of wind and water erosion. In the future, how will soil erosion respond to climate changes, will land management change cope with soil erosion responses or cause a reversal and increase in soil degradation? Unless careful consideration is given to how ground cover protects the soil from wind and water erosion, the next phase of agricultural expansion in northern Australia could cause an increase in soil erosion that could be a large unmitigated risk for soil security. For example, cover viewed from above (fractional cover) of the soil surface is a poor representation of the protection against wind and water erosion since both processes operate laterally requiring lateral cover. A highly prioritised treatment plan for "at risk soil" would focus on agricultural and grazing strategies that increase lateral vegetation cover. However, as demonstrated above, going down into the soil itself, and applying SUE to management practices will take agriculture and soil security to the next level that is required for long term sustainability.

The development of such a plan would require a significant coordinated effort, similar to that made to develop the National Strategy. The Strategy lays the groundwork for research, development and extension that is required. The charter for its implementation could be easily widened to develop a triage system for soil security. Involving innovative agronomy and farmer groups in this process will ensure that an agricultural approach to solutions is maintained.

8. Recommendations and Conclusions

We conclude with two recommendations that will contribute to the future security of Australian agricultural soils. At the farm scale, we recommend the continued development and adoption of soil management-based agriculture, including diagnostic and management tools such as SUE and Vertical Rate guidelines, as a practical approach for securing soil and alleviating or reversing soil degradation, while at the same time increasing productivity. Public policy support for this approach should include a renewed focus on soil security by State Governments and full implementation of the National Soil Research Development and Extension Strategy.

At the continental scale we recommend the development of a triage approach to mitigate soil degradation, framed by the five dimensions of soil security. This will provide an early warning system for soil loss, identify soils with the highest risk, and provide focus on areas where soil is insecure. Research, development and extension activities will be needed to find ways in which agricultural management practices can reduce soil loss, benefit from soil condition improvement and deal with the associated "creeping" issues of acidification and soil carbon loss, while at the same time increasing productivity.

The National Strategy now forms part of the backdrop for a changing national agriculture policy that includes development of agriculture in northern Australia and the requirement for a competitive and sustainable agriculture sector in the future. By integrating a soil security focus into this context, Australia's soil resource will continue to provide food, fiber, water, and environmental services and the economic returns required by the Australian people.

Author Contributions: Each author contributed to preparation of various components that were combined in this paper. Andrea Koch coordinated the effort and combined the components.

Conflicts of Interest: Andrea Koch declares no conflict of interest. Adrian Chappell declares no conflict of interest. Edward Scott and Michael Eyres are both employed by Injekta Field Systems, a soil advisory business in conservation tillage based agricultural systems and both authors do not have a conflict of interest in relation to the information provided in this paper at date of submission.

References

1. McBratney, A.B.; Field, D.J.; Koch, A. The dimensions of soil security. *Geoderma* **2013**, *213*, 203–213. [CrossRef]
2. Koch, A.; McBratney, A.; Adams, M.; Field, D.; Hill, R.; Crawford, J.; Minasny, B.; Lal, R.; Abbott, L.; O'Donnell, A.; *et al.* Soil Security: Solving the Global Soil Crisis. *Glob. Policy* **2013**, *4*, 434–441. [CrossRef]

3. Cork, S.; Eadie, L.; Mele, P.; Price, R.; Yule, D. *The Relationships between Land Management Practices and Soil Condition and the Quality of Ecosystem Services Delivered from Agricultural Land in Australia*; Kiri-ganai Research Pty Ltd.: Canberra, Australia, 2012.

4. Doran, J.W.; Zeiss, M.R. Soil health and sustainability: Managing the biotic component of soil quality. *Appl. Soil Ecol.* **2000**, *15*, 3–11. [CrossRef]

5. Department of Agriculture. The National Soil Research, Development and Extension Strategy, Securing Australia's Soil, for Profitable Industries and Healthy Landscapes. Available online: daff.gov.au/natural-resources/soils (accessed on 23 February 2015).

6. Lal, R. Soil degradation by erosion. *Land Degrad. Dev.* **2001**, *12*, 519–539. [CrossRef]

7. Australian Bureau of Statistics. 1301.0—Year Book Australia, Changing Patterns of Land Use in Australia, 1988. Available online: http://www.abs.gov.au/ausstats/abs@.nsf/featurearticlesbytitle/92916F88103CF060CA2569DE001F1085?OpenDocument (accessed on 25 February 2015).

8. Henzell, T. *Australian Agriculture: Its History and Challenges*; CSIRO Publishing: Collingwood, Australia, 2007.

9. Carey, B. Saving our soil: Soil Conservation in Queensland since thethe 1930s. Available online: http://www.slq.qld.gov.au/audio-video/webcasts/queensland-memory (accessed on 23 February 2015).

10. Breckwoldt, R. *The Dirt Doctors a Jubilee History of the Soil Conservation Service of NSW*; Soil Conservation Service of NSW: Sydney, Australia, 1988.

11. Tideman, A.F. The Struggle for Landcare in South Australia. Available online: http://www.pir.sa.gov.au/__data/assets/file/0017/151082/NRMHist_Struggle_V16.pdf (accessed on 23 February 2015).

12. D'Emden, F.H.; Llewellyn, R.S.; Burton, M.P. Adoption of conservation tillage in Australian cropping regions: An application of duration analysis. *Technol. Forecast. Soc. Chang.* **2006**, *73*, 630–647. [CrossRef]

13. ABS. 4627.0—Land Management and Farming in Australia, 2012–2013. Available online: http://www.abs.gov.au/AUSSTATS/abs@.nsf/Latestproducts/4627.0Main%20Features12012-13?opendocument&tabname=Summary&prodno=4627.0&issue=2012--13&num=&view= (accessed on 23 February 2015).

14. Safstrom, R.D.; Waddell, P.-J. Using economic, social and ecological spatial patterns to guide policy development in the Pilbara and Southern Rangelands of Western Australia. *Rangel. J.* **2013**, *35*, 231–239. [CrossRef]

15. Australian Government White Paper on Developing Northern Australia, 2014. Available online: https://northernaustralia.dpmc.gov.au/ (accessed on 10 March 2015).

16. State of the Environment Committee. *Australia State of the Environment 2011*; Independent Report to the Australian Government Minister for Sustainability, Environment, Water, Population and Communities; DSEWPaC: Canberra, Australia, 2011.

17. Australian Bureau of Agricultural and Resource Economics and Sciences. *Land Use and Land Management Information for Australia: Workplan of the Australian Collaborative Land Use and Management Program (ACLUMP)*; ABARES: Canberra, Australia, 2010.

18. National Land & Water Resources Audit. *The National Land and Water Resources Audit 2002–2008. Achievements and Challenges*; National Land & Water Resources Audit: Canberra, Australia, 2008.

19. Herbert, A. *Opportunity Costs of Land Degradation Hazards in the South-West Agriculture Region*; Resource Management Technical report 349; Western Australian Agriculture Authority: Perth, Australia, 2009.

20. Sanderman, J.; Farquharson, R.; Baldock, J. *Soil Carbon Sequestration Potential: A Review for Australian Agriculture*; Commonwealth Scientific and Industrial Research Organisation: Canberra, Australia, 2010.

21. Campbell, A. Managing Australia's Soils: A Policy Discussion Paper. http://www.clw.csiro.au/aclep/documents/Soil-Discussion-Paper.pdf (accessed on 16 April 2015).

22. Lu, H.; Prosser, I.P.; Moran, C.J.; Gallant, J.C.; Priestley, G.; Stevenson, J.G. Predicting sheetwash and rill erosion over the Australian continent. *Aust. J. Soil Res.* **2003**, *41*, 1037–1062. [CrossRef]

23. Loughran, R.J.; Elliott, G.L.; McFarlane, D.J.; Campbell, B.L. A survey of soil erosion in Australia using caesium-137. *Aust. Geogr. Stud.* **2004**, *42*, 221–233. [CrossRef]

24. Chappell, A.; Hancock, G.; Viscarra Rossel, R.; Loughran, R. Spatial uncertainty of the ^{137}Cs reference inventory for Australian soil. *J. Geophys. Res. Earth Surf.* **2011**, *116*. [CrossRef]

25. Chappell, A.; Viscarra Rossel, R.; Loughran, R. Spatial uncertainty of ^{137}Cs-derived net (1950s–1990) soil redistribution for Australia. *J. Geophys. Res. Earth Surf.* **2011**. [CrossRef]

26. Scott, B.J.; Eberbach, P.L.; Evans, J.; Wade, L.J.; Graham, E.H. *Centre Monograph No. 1: Stubble Retention in Cropping Systems in Southern Australia: Benefits and Challenges*; Clayton, E.H., Burns, H.M., Eds.; Industry & Investment NSW: Orange, Australia, 2010.

27. Wang, X.B.; Enema, O.; Hoogmed, W.B.; Perdok, U.D.; Cai, D. Dust storm erosion and its impact on soil carbon and nitrogen losses in northern China. *Catena* **2006**, *66*, 221–227. [CrossRef]

28. Zhang, G.S.; Chan, K.Y.; Oates, A.; Heenan, D.P.; Huang, G.B. Relationship between soil structure and runoff/soil loss after 24 years of conservation tillage. *Soil Tillage Res.* **2007**, *92*, 122–128. [CrossRef]

29. Llewellyn, R.S.; D'Emden, F.H. *Adoption of No-tillage Cropping Practices in Australian Grain Growing Regions*; Grains Research and Development Corporation: Kingston, Australia, 2010; pp. 1–31.

30. Chappell, A.; Sanderman, J.; Thomas, M.; Read, A.; Leslie, C. The dynamics of soil redistribution and the implications for soil organic carbon accounting in agricultural south-eastern Australia. *Glob. Chang. Biol.* **2012**, *18*, 2081–2088. [CrossRef]

31. Marx, S.K.; McGowan, H.A.; Kamber, B.S.; Knight, J.M.; Denholm, J.; Zawadzki, A. Unprecedented wind erosion and perturbation of surface geochemistry marks the Anthropocene in Australia. *J. Geophys. Res.* **2014**, *119*, 45–61.

32. Hopley, J.; Perry, E.; Clark, R.; MacEwan, R. *Wind Erosion Threat and Agricultural Land Cover in the Mallee*; Report for the Mallee CMA; Department of Environment and Primary Industries: Melbourne, Australia, 2013.

33. Reicosky, D.C.; Allmaras, R.R. Advances in tillage research in North American cropping systems. *J. Crop. Prod.* **2003**, *8*, 75–125. [CrossRef]

34. Anderson, G. *The Impact of Tillage Practices and Crop Residue (Stubble) Retention in the Cropping System of Western Australia*; Department of Agriculture and Food: South Perth, Australia, 2009.

35. Valzano, F.; Murphy, B.; Koen, T. *The Impact of Tillage on Changes in Soil Carbon Density with Special Emphasis on Australian Conditions*; National Carbon Accounting System Technical Report; Australian Greenhouse Office: Canberra, Australia, 2005.

36. Chappell, A.; Webb, N.P.; Butler, H.; Strong, C.; McTainsh, G.H.; Leys, J.F.; Viscarra Rossel, R. Soil organic carbon dust emission: An omitted global source of atmospheric CO_2. *Glob. Chang. Biol.* **2013**, *19*, 3238–3244. [CrossRef] [PubMed]

37. Doetterl, S.; van Oost, K.; Six, J. Towards constraining the magnitude of global agricultural sediment and soil organic carbon fluxes. *Earth Surf. Proc. Land.* **2012**, *37*, 642–655. [CrossRef]

38. Chappell, A.; Webb, N.P.; Viscarra Rossel, R.A.; Bui, E. Australian net (1950s–1990) soil organic carbon erosion: Implications for CO_2 emission and land-atmosphere modelling. *Biogeosciences* **2014**, *11*, 5235–5244. [CrossRef]

39. Leys, J.F.; Heidenrich, S.K.; Strong, C.L.; McTainsh, G.H.; Quigley, S. PM_{10} concentrations and mass transport during "Red Dawn"—Sydney 23 September 2009. *Aeolian Res.* **2011**, *3*, 327–342. [CrossRef]

40. Tozer, P.; Leys, J. Dust storms—What do they really cost? *Rangel. J.* **2013**, *35*, 131–142. [CrossRef]

41. Waddell, P.A.; Thomas, P.W.E.; Findlater, P.A. *A Report on the Gascoyne River Catchment Following the 2010/11 Flood Events*; Department of Agriculture and Food: South Perth, Australia, 2012.

42. Soils for Life. Restoring the Gascoyne Rangeland—Commitment, Cooperation and Hard Work. Accessible online: http://www.soilsforlife.org.au/case-studies.html (accessed on 10 March 2015).

43. CSIRO and Bureau of Meteorology. Climate Change in Australia Information for Australia's Natural Resource Management Regions: Technical Report. Accessible online: http://www.climatechangeinaustralia.gov.au/media/ccia/2.1.5/cms_page_media/168/CCIA_2015_NRM_TechnicalReport_WEB.pdf (accessed on 16 April 2015).

44. Iserson, K.V.; Moskop, J.C. Triage in Medicine, Part I: Concept, History, and Types. *Ann. Emerg. Med.* **2007**, *49*, 275–281. [CrossRef] [PubMed]

sustainability

MDPI

Review

Soil Degradation in India: Challenges and Potential Solutions

Ranjan Bhattacharyya [1,*], Birendra Nath Ghosh [2], Prasanta Kumar Mishra [2], Biswapati Mandal [3], Cherukumalli Srinivasa Rao [4], Dibyendu Sarkar [5], Krishnendu Das [6], Kokkuvayil Sankaranarayanan Anil [7], Manickam Lalitha [7], Kuntal Mouli Hati [8] and Alan Joseph Franzluebbers [9]

[1] Centre for Environment Science & Climate Resilient Agriculture, NRL Building, Indian Agricultural Research Institute, New Delhi 110 012, India

[2] Central Soil & water Conservation Research & Training Institute, Dehradun 248 195, India; bnghosh62@rediffmail.com (B.N.G.); pkmbellary@gmail.com (P.K.M.)

[3] Bidhan Chandra Krishi Viswa-Vidayala, Kalyani, West Bengal 741 235, India; mandalbiswapati@gmail.com

[4] Central Research Institute on Dryland Agriculture, Hyderabad, Telangana 500 059, India; cherukumalli2011@gmail.com

[5] ICAR Research Complex for North Eastern Hill Region, Imphal, Manipur 795 004, India; dsarkar04@rediffmail.com

[6] National Bureau of Soil Survey & Land Use Planning, Kolkata Regional Center, Kolkata 700 091, India; das_krishnendu@hotmail.com

[7] National Bureau of Soil Survey & Land Use Planning, Bangalore Regional Center, Bangalore 560 024, India; anilsoils@yahoo.co.in (K.S.A.); mslalit@yahoo.co.in (M.L.)

[8] Division of Soil Physics, Indian Institute of Soil Science, Bhopal 462 038, India; kuntalmouli@gmail.com

[9] USDA-ARS, Plant Science Research Unit, Raleigh, NC 27695, USA; alan.franzluebbers@ars.usda.gov

* Correspondence: ranjan_vpkas@yahoo.com or ranjan@iari.res.in; Tel.: +91-7838781447

Academic Editor: Marc A. Rosen

Received: 16 November 2014; Accepted: 27 February 2015; Published: 25 March 2015

Abstract: Soil degradation in India is estimated to be occurring on 147 million hectares (Mha) of land, including 94 Mha from water erosion, 16 Mha from acidification, 14 Mha from flooding, 9 Mha from wind erosion, 6 Mha from salinity, and 7 Mha from a combination of factors. This is extremely serious because India supports 18% of the world's human population and 15% of the world's livestock population, but has only 2.4% of the world's land area. Despite its low proportional land area, India ranks second worldwide in farm output. Agriculture, forestry, and fisheries account for 17% of the gross domestic product and employs about 50% of the total workforce of the country. Causes of soil degradation are both natural and human-induced. Natural causes include earthquakes, tsunamis, droughts, avalanches, landslides, volcanic eruptions, floods, tornadoes, and wildfires. Human-induced soil degradation results from land clearing and deforestation, inappropriate agricultural practices, improper management of industrial effluents and wastes, over-grazing, careless management of forests, surface mining, urban sprawl, and commercial/industrial development. Inappropriate agricultural practices include excessive tillage and use of heavy machinery, excessive and unbalanced use of inorganic fertilizers, poor irrigation and water management techniques, pesticide overuse, inadequate crop residue and/or organic carbon inputs, and poor crop cycle planning. Some underlying social causes of soil degradation in India are land shortage, decline in per capita land availability, economic pressure on land, land tenancy, poverty, and population increase. In this review of land degradation in India, we summarize (1) the main causes of soil degradation in different agro-climatic regions; (2) research results documenting both soil degradation and soil health improvement in various agricultural systems; and (3) potential solutions to improve soil health in different regions using a variety of conservation agricultural approaches.

Keywords: land degradation; soil erosion; conservation agriculture; agroforestry; nutrient management; sustainable crop intensification

1. Sources of Land Degradation

Land degradation is not being adequately addressed, but is of vital importance to raise awareness so that future land management decisions can lead to more sustainable and resilient agricultural systems. Of India's total geographical area (328.7 Mha), 304.9 Mha comprise the reporting area with 264.5 Mha being used for agriculture, forestry, pasture and other biomass production. The severity and extent of soil degradation in the country has been previously assessed by many agencies (Table 1). According to the National Bureau of Soil Survey and Land Use Planning [1] ~146.8 Mha is degraded. Water erosion is the most serious degradation problem in India, resulting in loss of topsoil and terrain deformation. Based on first approximation analysis of existing soil loss data, the average soil erosion rate was ~16.4 ton ha^{-1}year^{-1}, resulting in an annual total soil loss of 5.3 billion tons throughout the country [2]. Nearly 29% of total eroded soil is permanently lost to the sea, while 61% is simply transferred from one place to another and the remaining 10% is deposited in reservoirs.

Table 1. Extent of land degradation in India, as assessed by different organizations.

Organizations	Assessment Year	Reference	Degraded Area (Mha)
National Commission on Agriculture	1976	[3]	148.1
Ministry of Agriculture-Soil and Water Conservation Division	1978	[4]	175.0
Department of Environment	1980	[5]	95.0
National Wasteland Development Board	1985	[6]	123.0
Society for Promotion of Wastelands Development	1984	[7]	129.6
National Remote Sensing Agency	1985	[8]	53.3
Ministry of Agriculture	1985	[9]	173.6
Ministry of Agriculture	1994	[10]	107.4
NBSS&LUP	1994	[11]	187.7
NBSS&LUP (revised)	2004	[12]	146.8

Soil degradation has become a serious problem in both rainfed and irrigated areas of India. India is losing a huge amount of money from degraded lands (Table 2). This cost is documented by declining crop productivity, land use intensity, changing cropping patterns, high input use and declining profit [13–16]. Reddy [17] valued the loss of production in India at Rupees (Rs) 68 billion in 1988–1989 using the National Remote Sensing Agency (NRSA) dataset. Additional losses resulting from salinization, alkalinization and waterlogging were estimated as Rs 8 billion. Of late, in a comprehensive study made on the impact of water erosion on crop productivity, it was revealed that soil erosion due to water resulted in an annual crop production loss of 13.4 Mt in cereal, oil seeds and pulse crops equivalent to ~US$162 billion [18].

Table 2. Estimates on the annual direct cost of land degradation in India.

Parameters	NRSA [19]	ARPU [20]	Sehgal and Abrol [21]
Area affected by soil erosion (Mha)	31.5	58.0	166.1
Area affected by salinization, alkalinization and waterlogging (Mha)	3.2	-	21.7
Total area affected by land degradation (Mha)	34.7	58.0	187.7
Cost of soil erosion in lost nutrients (Rs billion)	18.0	33.3	98.3
Cost of soil erosion in lost production (Rs billion)	67.6	124.0	361.0
Cost of salinization, alkalinization and waterloggingin lost production (Rs billion)	7.6	-	87.6
Total direct cost of land degradation (Rs billion)	75.2	-	448.6

Apart from faulty agricultural activities that led to soil degradation (discussed in the next Section), other human-induced land degradation activities include: land clearing and careless management of forests, deforestation, over-grazing, improper management of industrial effluents and wastes, surface mining, and industrial development. Each of these factors are discussed briefly, but offering greater detail is beyond the scope of this review.

1.1. Overgrazing, Deforestation and Careless Forest Management

Overgrazing and deforestation have caused degradation in eight Indian states which now have >20% wasteland (Source: Wasteland atlas of India by national remote sensing agency; NRSA). Loss of vegetation occurs due to cutting beyond the silviculturally permissible limit, unsustainable fuelwood and fodder extraction, encroachment by agriculture into forest lands, forest fires and overgrazing, all of which subject the land to degradation forces. A cattle population of 467 million grazes on 11 Mha of pastures, implying an average of 42 head per hectare of land compared to a sustainable threshold level of 5 animals per hectare [22]. High livestock density in arid regions causes overgrazing, resulting in decreased infiltration and accelerated runoff and soil erosion. Due to overgrazing, soil loss is 5 to 41 times greater than normal at the mesoscale and 3 to 18 times greater at the macroscale [23]. Tendency of cultivation on slopes in the 1990s led to deforestation and land degradation [24]. Impoverishment of the natural woody cover of trees and shrubs is a major factor responsible for wind and water erosion. This occurs because the per capita forest land in the country is only 0.08 ha compared to a requirement of 0.47 ha to meet basic needs, thus creating excessive pressure on forest lands.

1.2. Urban Growth, Industrialization and Mining

An increase in industrialization, urbanization and infrastructure development is progressively taking away considerable areas of land from agriculture, forestry, grassland and pasture, and unused lands with wild vegetation. Opencast mining is of particular focus because it disturbs the physical, chemical, and biological features of the soil and alters the socioeconomic features of a region. Negative effects of mining are water scarcity due to lowering of water table, soil contamination, part or total loss of flora and fauna, air and water pollution and acid mine drainage. Overburden removal from mine area results in significant loss of vegetation and rich topsoil [25]. Overburden removal is normally done by blasting or using excavators, resulting in generation of large volume of waste (soil, debris and other material). Open-pit mines produce 8 to 10 times as much waste as underground mines [26]. The magnitude and significance of impact on the environment due to mining varies from mineral to mineral and also on the potential of the surrounding environment to absorb the negative effects associated with geographical disposition of mineral deposits and size of the mining operations. Mineral production generates enormous quantities of waste/overburden and tailings/slimes and a huge land area is degraded (Table 3).

Table 3. Mineral Production, waste generation and land affected in 2005-06 (Data source: Sahu [25]).

Mineral	Production (Mt)	Overburden/Waste (Mt)	Estimated Land Affected (ha)
Coal	407	1493	10,175
Limestone	170	178	1704
Bauxite	12	8	123
Iron ore	154	144	1544
Others	9	19	-

1.3. Natural and Social Sources of land Degradation

Natural causes of land degradation include earthquakes, tsunamis, droughts, avalanches, landslides, volcanic eruptions, floods, tornadoes, and wildfires (discussed in more detail in Section 3). Some underlying social causes of soil degradation are land shortage, decline in per capita land availability, economic pressure on land, land tenancy, poverty, and population increase.

1.4. Land Shortage, Land Fragmentation and Poor Economy

In India, small land holdings are a prominent feature, particularly in rainfed regions. Some 80% of farmers' holdings are ≤2 ha, accounting for >50% of agricultural output. Average size of land holding declined from 2.3 ha to 1.3 ha during 1970–2000 with per capita land of 0.32 hectare in 2001 [27]. Small land holdings lead to severe economic pressures on farmers. Because of such pressure, labor, land and capital resources limit the use of green manuring or soil conservation structures. Therefore, land shortage and poverty, taken together, lead to non-sustainable land management practices as a direct source of degradation. This is also the underlying reason for two other direct causes of land degradation, improper crop rotations and unbalanced fertilizer use [28].

Despite several interventions by the Indian Government, land degradation is still a serious problem. Some programs have included Integrated Watershed Management in the catchment of flood prone areas-1980–1981; National Land Use and Development Council-1985; National Wasteland Development Board–1985; National Watershed Development Projects for Rainfed Areas-1985–1986; Reclamation and Development of Alkali and Acid soil-1985–1986; National Land Use Policy-1988; Integrated Wasteland Development Project-1989–1990; Constitution (74th amendment) Act-1992 (Regulation of Land Use) and National Rainfed Area Authority-2006. The United Nations Environmental Programme (UNEP) indicated that over the preceding 20 years the problem of land degradation had continued to worsen due to human activities and climate change causing prolonged or frequent droughts that aggravated land degradation. Other underlying causes included increasing population:land ratio (Agriculture share in GDP fell from 35% in 1981 to 13% in 2012); market and institutional failures; externality and tenurial system–insecure property rights.

1.5. Population Increase

India's land area is about 2.5% of the global land area, where as it supports more than 16% of the global human population and ~20% of the world's livestock population. Steady increases in human population, as well as livestock population, and the widespread incidence of poverty, are exerting heavy pressures on India's limited land resources. Urban sprawl is a consequence of increasing urban population. As urban population increases, infrastructure requirements including transportation, water and sewage facilities, housing, schools, commerce, health, and recreation all contribute to urban sprawl [29].

2. Agricultural Activities Leading to Land Degradation in India

"Most of the area under cultivation in India has been under cultivation for hundreds of years, and had reached its state of maximum impoverishment many years ago … In this connection it must be remembered that deficiency of combined nitrogen is the limiting factor throughout the greater part of India" (The Royal Commission on Agriculture in India Report, [30] (p. 76)). The Green Revolution brought about a technological breakthrough, leading to the use of short duration high yielding varieties that helped intensify land use within a year by increasing the area under irrigation and greatly increasing the use of chemicals such as fertilizers and pesticides. Agricultural production of India increased from 50 Mt to over 250 Mt, over the last five decades. This, however, had further consequences, including loss of plant biodiversity and environmental pollution. Widespread land degradation caused by inappropriate agricultural practices has a direct and adverse impact on the food and livelihood security of farmers. Basically, degradation is caused by erosion, which results in the loss of topsoil through the action of water and wind, or waterlogging, which results in soil salinization. Maheswarappa *et al.* [31] observed that (i) the C-sustainability index was high in 1960, and was indicative of the minimum usage of inputs prior to the onset of the Green Revolution and (ii) thereafter, the C-sustainability index decreased because of greater C-based inputs, in which a linear relationship exists between C inputs and C outputs.

Agricultural activities and practices can cause land degradation in a number of ways depending on land use, crops grown and management practices adopted. Some of the common causes of land degradation by agriculture include cultivation in fragile deserts and marginal sloping lands without any conservation measures, land clearing through clear cutting and deforestation, agricultural depletion of soil nutrients through poor farming practices, overgrazing, excessive irrigation, overdrafting (the process of extracting groundwater beyond the safe yield of the aquifer), urban sprawl and commercial development, and land pollution including industrial waste disposal to arable lands.

2.1. Low and Imbalanced Fertilization

Intensive farming practices, particularly with wheat (*Triticum aestivum* L.) and rice (*Oryza sativa* L.) in India, have virtually mined nutrients from the soil. The already imbalanced consumption ratio of 6.2:4:1 (N:P:K) in 1990–1991 has widened to 7:2.7:1 in 2000–2001 and 5:2:1 in 2009–2010 compared with a target ratio of 4:2:1. As food grain production increased with time, the number of elements deficient in Indian soils increased from one (N) in 1950 to nine (N, P, K, S, B, Cu, Fe, Mn, and Zn) in 2005–2006. Although the use of fertilizers has increased several fold, the overall consumption continues to be low in most parts of the country. Wide spread Zn deficiency, followed by S, Fe, Cu, Mn and B in are common throughout the country. Every year, ~20 Mt of the three major nutrients are removed by growing crops [32], but the corresponding addition through inorganic fertilizers and organic manures falls short of this harvest. Another estimate suggests that for the past 50 years, the gap between removals and additions of nutrients has been 8 to 10 Mt N + P_2O_5 + K_2O per year [33]. In addition nutrient loss through soil erosion is another reason for soil fertility depletion, accounting for an annual loss of 8 Mt of plant nutrients through 5.3 billion tons of soil loss [34].

2.2. Excessive Tillage and Use of Heavy Machinery

Excessive tillage coupled with use of heavy machinery for harvesting and lack of adequate soil conservation measures causes a multitude of soil and environmental problems. Decline in soil organic matter (SOM) leads to limited soil life and the poor soil structure. Puddling of soil for paddy rice degrades soil physical properties and has negative impacts on soil biology [35]. Poor physical condition of soil leads to poor crop establishment and waterlogging after irrigation. Intensive agriculture has also led to doubling of irrigated cropland over the past four decades, from 19% to 38% of the cropped area. Much of this water has been extracted from limited ground water resources. Improper use and maintenance of canal irrigation has contributed significantly to soil degradation problems like waterlogging and salinization. Excess nitrate has leached into groundwater due to heavy N fertilizer use. Unnecessary tillage for land preparation and planting, indiscriminate irrigation, and excessive fertilizer applications are the main sources of greenhouse gas (GHG) emission from agricultural systems.

2.3. Crop Residue Burning and Inadequate Organic Matter Inputs

The NBSS&LUP data [21] show that nearly 3.7 Mha suffer from nutrient loss and/or depletion of SOM. Burning of crop residues for cooking, heating or simply disposal is a pervasive problem in India and contributes to SOM loss. According to the Ministry of New and Renewable Energy [36], ~500 Mt of crop residues are generated every year and ~125 Mt are burned. Crop residue generation is greatest in Uttar Pradesh (60 Mt) followed by Punjab (51 Mt) and Maharashtra (46 Mt). Among different crops, cereals generate 352 Mt of residues followed by fibre crops (66 Mt), oilseeds (29 Mt), pulses (13 Mt) and sugarcane (*Saccharum officinarum*) (12 Mt). Rice (34%) and wheat (22%) are the dominant cereals contributing to crop residue generation [37].

2.4. Poor Irrigation and Water Management

Improper planning and management of irrigation systems and extraction of ground water in excess of the recharge capacity have resulted in a rise of the water table in most canal command areas. Specific issues of concern are inefficient use of irrigation water, poor land development, seepage from unlined water courses, non-conjunctive use of surface and ground water resources and poor drainage. Expansion of canal irrigation (like the Indira Gandhi Nahar Project, for instance) has been associated with widespread waterlogging and salinity problems in areas, such as in the Indo-Gangetic Plains (IGP). In arid, semi-arid and sub-humid regions, large areas have been rendered barren due to the development of saline-sodic soils because of poor irrigation and drainage management. Cracking of soil from poor irrigation management leads to bypass flow of water and subsequent nitrate leaching [29]. Cracks not closing properly leave a U-shaped trace, and upon drying these cracks can expand and cause soil shrinkage.

2.5. Poor Crop Rotations

Improper crop rotation coupled with lack of proper soil and water conservation measures are important reasons contributing to soil erosion in lands under cultivation. In addition, cultivation of marginal lands on steep slopes, in shallow or sandy soils, with laterite crusts, and in arid or semi-arid regions bordering deserts has resulted in land degradation. Agricultural production in marginal areas with low SOM due to unsuitable cropping patterns has been the major cause of accelerated wind and water erosion. Wind erosion is a serious problem in arid and semi-arid regions, in coastal areas with sandy soils, and in the cold desert regions of Leh in the extreme north of India.

2.6. Pesticide Overuse and Soil Pollution

Indiscriminate use of pesticides together with sewage sludge and composted municipal wastes leads to contamination of soil and water with toxic substances and heavy metals. Heavy metal pollution is due to improper disposal of industrial effluents and use of domestic and municipal wastes and pesticides. Some commercial fertilizers also contain appreciable quantities of heavy metals, which have undesirable effects on the environment. Indiscriminate use of agro-chemicals, such as fertilizers and pesticides, is often responsible for land degradation.

3. Extent and Causes of Soil Degradation by Region

The extent of land degradation in India, as estimated by NBSS&LUP and Indian Council of Agricultural Research (ICAR) is given in Table 4.

The Planning Commission of India has delineated 15 agro-climatic regions to form the basis for agricultural planning in the Eighth Plan. These are: 1. Western Himalayan Region, 2. Eastern Himalayan Region, 3. Lower Gangetic Plains Region, 4. Middle Gangetic Plains Region, 5. Upper Gangetic Plains Region, 6. Trans-Gangetic Plains Region, 7. Eastern Plateau & Hills Region, 8. Central Plateau & Hills Region, 9. Western Plateau & Hills Region, 10. Southern Plateau & Hills Region, 11. East Coast Plains & Hills Region, 12. West Coast Plains & Ghats Region, 13. Gujrat Plains and

Hills Region, 14. Western Dry Region, 15. The Island Region. Similar agro-climatic regions have been combined to form six major regions. Region-specific causes and extent of degradation are described in the online Supplementary Information.

Table 4. State-wise extent of various kinds of land degradation in India (Mha). Data source: NBSS&LUP-ICAR [12] on 1:250,000 scale. TGA is total ground area.

State	Water Erosion	Wind Erosion	Water Logging	Salinity/ Alkalinity	Soil Acidity	Complex Problem	Total Degraded Area	% of Degraded Area to TGA
Andhra Pradesh + Telengana	11.5	0	1.9	0.5	0.9	0.2	15.0	54.5
Goa	0.1	0	0.1	0	0	0	0.2	43.9
Karnataka	5.8	0	0.9	0.1	0.1	0.7	7.6	39.8
Kerala	0.1	0	2.1	0	0.1	0.3	2.6	67.1
Tamil Nadu	4.9	0	0.1	0.1	0.1	0.1	5.3	41.0
Manipur	0.1	0	0	0	1.1	0.7	1.9	42.6
Mizorum	0.1	0	0	0	1.1	0.7	1.9	89.2
Meghalaya	0.1	0	0	0	1.0	0	1.2	53.9
Assam	0.7	0	0	0	0.6	0.9	2.2	28.2
Arunachal Pradesh	2.4	0	0.2	0	2.0	0	0	53.8
Nagaland	0.4	0	0	0	0.1	0.5	1,0	60.0
Sikkim	0.2	0	0	0	0.1	0	0.2	33.0
Tripura	0.1	0	0.2	0	0.2	0.1	0.6	59.9
Himachal Pradesh	2.8	0	1.3	0	0.2	0	4.2	75.0
Jammu and Kashmir	5.5	0.1	0.2	0	0	0	7.0	31.6
Uttar Pradesh + Uttarakhand	11.4	0.2	2.4	1.4	0	0	15.3	52.0
Delhi	0.1	0	0	0	0	0.0	0.1	55.4
Haryana	0.3	0.5	0.1	0.3	0	0.2	1.5	33.2
Punjab	0.4	0.3	0.3	0.3	0	0	1.3	25.4
Bihar + Jharkhand	3.0	0	2.0	0.2	1.0	0	6.3	36.1
West Bengal	1.2	0	0.7	0.2	0.6	0.1	2.8	31.0
Union Territories	0.2	0	0	0	0	0.0	0.2	24.8
Gujarat	5.2	0.4	0.5	0.3	0	1.7	8.1	41.5
Rajasthan	3.2	6.7	0	1.4	0	0.1	11.4	33.2
Madhya Pradesh + Chhattisgarh	17.9	0	0.4	0	7.0	1.1	26.2	59.1
Maharashtra	11.2	0	0	1.1	0.6	0.3	13.1	42.4
Orissa	5.0	0	0.7	0	0.3	0.1	6.1	39.3
Grand Total (Mha)	93.7	9.5	14.3	5.9	16.0	7.4	146.8	-

4. Strategies to Mitigate Land Degradation

The salient mitigation techniques for reversing land degradation in India and their applicability in major agro-climates are given in Table 5.

Table 5. Major land degradation mitigation techniques in the agro-climatic zones of India.

Mitigation Technologies	Hilly Areas	Indo-Gangetic Plains	Dryland and Desert Areas	Southern Peninsular India	Central India	Coastal Areas
			Applicability			
Soil Erosion Control	√	√	√	√	√	√
Water Harvesting, Terracing and Other Engineering Structures	√	√	√	√	√	√
Landslide and Minespoil Rehabilitation and River Bank Erosion Control	√	√	√	√	√	√
Intercropping and Contour Farming	√	√	√	√	√	√
Subsoiling					√	
Watershed Approach	√	√	√	√	√	√
Participatory Resource Conservation and Management	√	√	√	√	√	√
Integrated Nutrient Management and Organic Manuring	√	√	√	√	√	√
Reclamation of Acid and Salt Affected Soils and Drainage (Desalinization)	√	√	√	√	√	√
Remediation of As contamination		√				√
Water Management and Pollution Control	√	√	√	√	√	√
Irrigation Management for Improving Input Use Efficiency	√	√	√	√	√	√
Judicious Use of Distillery Effluent	√	√	√	√	√	√
Reforestation, Grassland and Horticulture Development	√	√	√	√	√	√
Vegetative Barriers and Using Natural Geotextiles, Mulching and Diversified Cropping	√	√	√	√	√	√
Agroforestry	√	√	√	√	√	√
Conservation Agriculture (CA)	√	√	√	√	√	√
Intensive Cropping and Integrated Farming Systems (IFS)	√	√	√	√	√	√
Disaster (Tsunami) Management						√

4.1. Soil Erosion Control

Tolerance to soil loss (T) is defined as the upper threshold limit of soil erosion that can be allowed without degrading long term productivity of a particular soil. If soil erosion rates are greater than T, mitigation measures are needed to achieve sustainable productivity. T-values of the hilly regions of

India, as estimated by Mandal *et al.* [38], are given in Tables 6 and 7. It is projected that ~59% of land within the hilly region requires some form of erosion management to achieve T [38].

Table 6. Area under different erosion rates and soil loss tolerance limits in the northwestern Hills.

Erosion Categories Based on Soil Erosion (ton ha^{-1}year^{-1})	Very Low (<5)	Low (5 to 10)	Moderate (10 to 20)	Severe (20–40)	Very Severe (>40)	Others
Area (Mha) under each category	1.7 (5.2) *	2.5 (7.5)	3.3 (9.8)	1.9 (5.8)	4.5 (13.7)	19.2 (58.0)
T-value (ton ha^{-1}year^{-1})	2.5	5.0	7.5	10.0	12.5	Rocks/ unreported
Area (Mha) under each T value	0.4 (1.2)	0.3 (0.8)	3.5 (10.6)	9.0 (27.2)	1.3 (3.9)	18.7 (56.3)

* Values in the parentheses are percentages of total area. Data source: Mandal *et al.* [38].

Table 7. Area under different potential erosion rates and soil loss tolerance limits in the northeastern Hills (Source: Mandal *et al.* 38). * Values in the parentheses are percentages of area.

Erosion Categories Based on Soil Erosion (ton ha^{-1}year^{-1})	Very Low (<5)	Low (5 to 10)	Moderate (10 to 20)	Severe (20-40)	Very Severe (>40)	Others
Area (Mha) under each category	1.2 (4.5) *	5.8 (21.2)	4.6 (16.8)	3.6 (13.0)	8.2 (29.8)	4.1 (14.8)
T-value (ton ha^{-1}year^{-1})	2.5	5.0	7.5	10.0	12.5	Rocks /unreported
Area (Mha) under each T value	-	0.1 (0.3)	4.7 (17.1)	13.1 (47.7)	5.8 (21.0)	3.8 (13.9)

* Values in the parentheses are percentages of total area.

Soil conservation measures, such as contour ploughing, bunding, use of strips and terraces, can decrease erosion and slow runoff water. Mechanical measures, e.g., physical barriers such as embankments and wind breaks, or vegetation cover (and use of vegetative buffer strips and geotextiles) and soil husbandry are important measures to control soil erosion [39]. In addition, conservation agriculture (CA), agro-forestry, integrated nutrient management (INM) and diversified cropping also conserve soil and water. These are discussed sequentially as physical, chemical and biological means of soil conservation and land degradation mitigation in the following sections.

4.2. Water Harvesting, Terracing and Other Engineering Structures

Mechanical soil and water conservation measures are required for controlling soil erosion, retaining maximum rainfall within the slope and safe disposal of excess runoff from the top to the foot hills of India. These structures are often used in case of extreme soil degradation. The measures are: **Bunding-**small earthen barriers built on agricultural lands with slopes ranging from 1%–6% slope. Bunds are used in agriculture to collect surface *run-off*, increase water infiltration and prevent soil erosion. **Graded bunds-**constructed in medium to high rainfall areas of ~600 mm year^{-1}. **Contour bunds-** either mechanical or vegetative barrier created across the slope. A study conducted at Doon valleys in the northwestern hills region indicted that contour bunds decreased runoff 25%–30% compared to field bunds [40]. **Bench terrace and half moon terrace-**adopted where soil depth is >1.0 m. Half-moon terraces are level circular beds having 1 to 1.5 m diameter cut into half-moon shape on the hill slopes. Beds are used for planting and maintaining saplings of fruit and fodder trees in horticulture/agro-forestry land uses. **Grassed waterways**-channels laid out preferably on natural drainage lines in the watershed. **Water harvesting ponds**-dug-out embankment type of water harvesting structure used for creating seasonal and perennial ponds at the foot of a micro-watershed for irrigation and fish farming purposes.

In vertisols (of central India), graded broad bed and furrow system of land configuration improves surface drainage and allows better water infiltration. It also facilitates drainage of excess water through grassed waterways. However, the broad bed and furrow system is not as effective for shallower Vertic soils, as it encourages runoff. Runoff and soil loss were lower from broad bed and furrow land surface management practices than from a flat on grade system (Table 8). The broad bed and furrow system decreased soil loss to a greater extent (31% to 55%) than its effect on runoff volume (24% to 32%) compared with that of flat on grade system.

Table 8. Seasonal rainfall, runoff and soil loss from different land configuration, broad-bed and furrow (BBF) and flat on grade (FOG) (Data source: Mandal *et al.* [41]).

Year	Rainfall (mm)	Runoff (mm)		Soil Loss (ton ha^{-1})	
		BBF	FOG	BBF	FOG
2003	1058.0	163.0 (15.4%)	214.9 (20.3%)	2.0	2.9
2004	798.2	124.0 (15.5%)	183.3 (23.0%)	0.7	1.5
2005	946.0	177 (18.7%)	246 (26.1%)	1.4	3.1
2006	1513.0	502 (33.2%)	873 (57.7%)	3.5	6.4

Values within parentheses indicate the percent of total rainfall.

4.3. Landslide and Minespoil Rehabilitation and River Bank Erosion Control

High soil erosion rates were checked and brought within permissible limits (Table 9) by using bioengineering treatments on landslide affected (Nalotanala watershed; area ~60 ha) and minespoil affected (Sahastradhara watershed; area ~64 ha) areas. Restoration of limestone minespoil areas resulted in improved water quality through a reduction in Ca content (Table 10). For river bank erosion control, bio-engineering technologies such as spurs, retaining walls and earthen embankments may be used in conjunction with suitable vegetation such as giant cane (*Arundo donax*), five-leaf chaste trees (*Vitex negundo*), morning glory (*Ipomoea* sp.), Bamboo (*Bambusa vulgaris*), napiergrass (*Pennisetum purpureum*) or munja (*Saccharum munja*) [40].

Table 9. Effect of bioengineering measures on landslide (1964–1994) and minespoil rehabilitation (1984–1996) project [40].

Particulars	Landslide Project		Minespoil Project	
	Before Treatment	After Treatment	Before Treatment	After Treatment
Sediment load (ton ha^{-1} year^{-1})	320	6	550	8
Vegetative cover (%)	<5	>95	10	80

Table 10. Water quality parameters (mg L^{-1}) for treated and untreated minespoils (Data source: CSWCR&TI Vision [40]).

	Ca	Mg	SO$_4$
Treated mine	74	34	138
Untreated mine	188	39	240

4.4. Intercropping and Contour Farming

Agronomical practices like use of cover crops, mixed/inter/strip cropping, crop rotation, green manuring and mulch farming are vital practices associated with integrated nutrient management. Growing soybean (*Glycine max*)/groundnut (*Arachis hypogoea*)/cowpea (*Vigna radiata*) with maize (*Zea mays*)/jowar (*Sorghum bicolor*)/bajra (*Pennisetum glaucum*) is a common example of intercropping in the drylands [39]. Strip cropping is a combination of contouring and crop rotation in which alternate strips of row crops and soil conserving crops are grown on the same slope, perpendicular to the wind or water flow in drylands and hilly regions, respectively. Intercropping cowpea with maize (2 rows of cowpea with 1 row of maize) decreased runoff by 10% and soil loss by 28% compared to pure maize. Minimum runoff (36% of rainfall) was recorded under barnyard millet (*Echinochloa frumentacea* L.) followed by black soybean (*Glycine max* L.) and maize which was 37% and 42%, respectively. Black soybean and maize alone had maximum soil loss of 7.1 and 6.7 ton ha^{-1}, respectively, followed by barnyard millet (4.8 ton ha^{-1}). The practice of line sowing of wheat and mustard (*Brassica juncea* L.) crops and maintaining row ratio of 8:1 ensured optimum use of space and soil moisture, increased wheat equivalent yield by 14% and net returns by 30% compared to mixed sowing (Table 11) [42,43].

Table 11. Water use efficiency, yield and net return as affected by different technologies and crop rotation in farmers' fields of Uttarakhnad, Jammu and Kashmir and Himachal Pradesh.

Intercropping	Crops	Water Use Efficiency (kg ha^{-1}mm^{-1})			Yield (t ha^{-1})			Net Return (INR ha^{-1})		
		C	T	% Increase	C	T	% Increase	C	T	% Increase
Maize + cowpea (1:2) − wheat	Maize	3.19	5.60	76	2.21 *	3.67 *	66	4448	11,690	163
	Wheat	5.30	8.31	57	1.13	1.64	46	3176	6149	88
Maize − wheat + mustard (9:1)	Maize	3.00	4.34	45	1.94	2.75	42	3248	8658	163
	Wheat	6.33	9.66	50	1.31 *	1.93**	47	4455	9041	105
Maize – potato – onion (irrigated)	Maize	3.09	4.52	46	1.95	2.86	46	3361	9135	172
	Potato	53.70	76.50	42	17.10	23.50	33	9775	19,250	97
	Onion	18.87	25.45	35	12.05	15.10	25	38,700	51,050	32

Source: Ghosh [44], C-Conventional, T-Intercropping/crop rotation *—Maize equivalent yield; ** wheat equivalent yield. 60 INR (Indian Rupees) ~1USD (2014).

When crops like maize, sorghum and castor (*Ricinus communis* L.) are cultivated along with legumes such as groundnut, green gram (*Vigna radiata* L.), black gram (*Vigna mungo* L.), soybean and cowpea in inter-row spaces, sufficient cover on the ground is ensured and erosion hazards decreased [45]. Pathak *et al.* [46] reported several soil conservation measures based on rainfall in a particular area (Table 12).

Table 12. Soil and water conservation measures to be taken up based on seasonal rainfall in the Peninsular India (Source: Pathak *et al.* [46]).

Seasonal Rainfall (mm)	Soil and Water Conservation Measures	
<500	Contour cultivation with conservation furrows, ridging, Sowing across slope, Mulching, Scoops, Off season tillage, Inter row water harvesting system, Small basins, Field bunds, Khadin	Tied ridges, contour bunds
500–750		Zingg terrace, modified contour bunds and broad bed furrow
750–1000	Broad bed furrow (vertisols), field bunds, vegetative bunds and graded bunds	Conservation furrows, sowing across slope, conservation tillage, Lock and spill drains, small basins, nadizingg terrace
>1000		Level terrace and zingg terrace (conservation bench terrace)

4.5. Subsoiling

Low infiltration rate is one of the major problems of black soils (Vertisols) in central India. In Vertisols, improved tillage practices, particularly deep tillage (subsoiling with chisel plough), can improve soil water storage by greater infiltration and minimizing water stress. A study with three tillage treatments consisting of conventional tillage (CT), CT + subsoiling in alternate years, and CT + subsoiling in every year showed that the basic infiltration rate and soil water storage in the 90 cm profile were greater in CT + subsoiling every year than in CT [47].

4.6. Watershed Approach

Integrated watershed management, which involves soil and water conservation coupled with suitable crop management, is another excellent strategy for mitigating soil erosion. Development and management of watershed resources to achieve optimum production without causing deterioration in the resources base is integrated watershed management. It involves construction of check dams along gullies, bench terracing, contour bunding, land leveling and planting of grasses. These strategies will increase percolation of water, decrease runoff and improve water availability. Several reviews are available on the performance of watershed development projects [48,49], as well as their limitations. An operational research project on watershed management at Fakot by the CSWCR&TI during 1975–1986 is a successful example of participatory integrated watershed management approach [50]. Conservation agriculture along with above-said practices has great potential to reverse soil loss.

4.7. Participatory Resource Conservation and Management

A case study in Netranahalli Watershed (Karnataka) in the Southern Peninsular India stressed the importance of involvement of communities for conservation of natural resources (mainly soil and water) and their management. Improvement in ground water levels, soil and moisture conservation, development of irrigation facilities, water regeneration capacity, forestry and horticulture development, change in land use pattern and cropping pattern, improvement in animal health, employment and income generation were noticed by Adhikari *et al.* [51]. In a joint programme by Bangalore Regional Centre of NBSS&LUP, Nagpur and Tamil Nadu State Department of Agriculture at Shivagangai district, Tamil Nadu, water harvesting (in better maintained existing tanks), recharging (in aquifers) and providing drainage facilities (in lowlands) prevented water erosion and decreased incidence of salinity and sodicity [52].

4.8. Integrated Nutrient Management and Organic Manuring

Integrated nutrient management, *i.e.*, the application of NPK mineral fertilizers along with organic manure, increases crop productivity, improves SOC content, and decreases soil loss. In the northwestern hill region, integrated nutrient management improved soil health and SOC storage in all cropping systems. Kundu *et al.* [53] and Bhattacharyya *et al.* [54] observed that about 19% and 25% of gross C input contributed to greater SOC content after 30 years of rainfed or after nine years of irrigated soybean-wheat production, respectively. Annual farmyard manure addition improved labile (movable; short-lived) and long-lived C pools [55]. Nearly 16% (mean of all treatments) of the estimated added C was stabilized into SOC both in the labile and recalcitrant pools, preferentially in the 0–30 cm soil layer (Figure 1). However, the labile:recalcitrant SOC ratios of applied C stabilized was largest in the 15–30 cm soil layer (Figure 2). The labile pool constituted about 62% of the total SOC in the 0–45 cm soil layer and about 50% of the applied C stabilized in the labile pool (Figure 3). The integrated nutrient management approach of 5 ton ha^{-1} of farmyard manure +50% recommended fertilizer led to an additional grain yield of 2.65 ton ha^{-1} in paddy-maize cropping system (reference). Under rainfed conditions, C retention rate varied from 0.61 to 1.8 ton ha^{-1} year^{-1} in different crop rotations, which also enhanced crop yield (Table 13). However, with green manuring, wheat had greater water use (289 mm) than wheat in a wheat-fallow system (273 mm) or wheat (270 mm) rotated with maize [56].

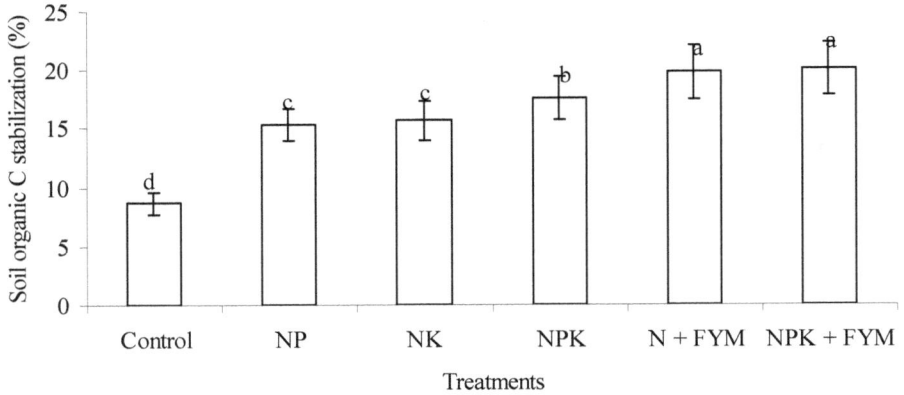

Figure 1. Soil organic C (SOC) stabilization in the 0 to 45 cm soil layer as affected by 32 years of continuous annual fertilization under soybean-wheat cropping in a sandy clay loam soil of the Indian Himalayas. Bars with the same lowercase latter indicate that the values are not significantly different (at $P < 0.05$ according to Tukey's HSD tests). Error bars indicate standard errors. (Source: Bhattacharyya *et al.* [55]).

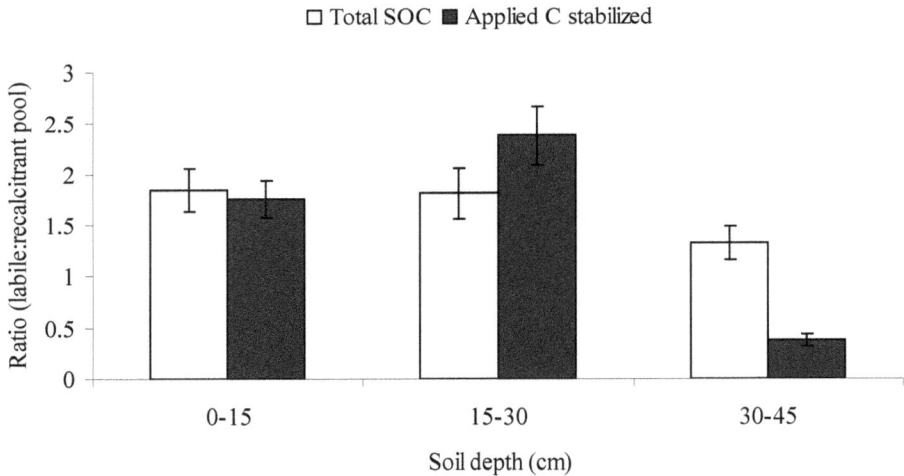

Figure 2. Ratios of labile and recalcitrant pools of total SOC and applied C stabilized in soils by depth after 32 years of cropping with different fertilization (error bars indicate standard error of mean).

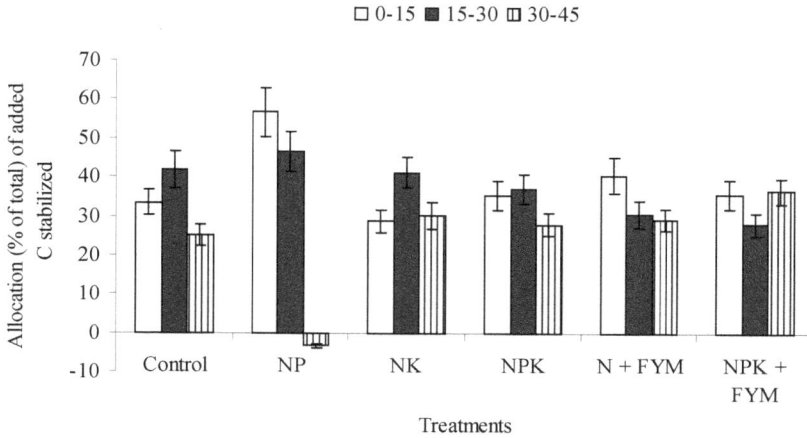

□ 0-15 ■ 15-30 ▥ 30-45

Figure 3. Depth (cm) distribution of total estimated added C stabilized after 32 years of fertilization (error bars indicate the standard error of mean; Source: Bhattacharyya *et al.* [55]).

Table 13. Fertilization impacts on carbon retention in the 0–15 cm layer and crop yield change in the Indian Himalayas (Data Source: Bhattacharyya *et al.* [54,57–59]).

Rainfed Management Practices	Duration of Adoption (year)	Carbon Retention Over Control (Mg ha^{-1}year^{-1})	Yield Change Over	Yield Change over Unfertilized Control/Two Irrigations (Mg ha^{-1}year^{-1})
NPK + FYM application-rainfed	32	0.87	Unfertilized control	2.3 (S) & 1.17 (W)
NPK + FYM application-irrigated	9	1.28	Unfertilized control	0.80 (S) & 1.74 (W)
FYM at 15 Mg ha^{-1}	3	1.63	Unfertilized control	6.2 (GP) & 7.1 (FB) & 0.55 (BC)
FYM at 10 Mg ha^{-1}	3	1.80	Unfertilized control	3.5 (GP) & 1.3 (R)
Four irrigations in wheat	4	0.35	Two irrigations	0.17 * (R) & 0.44 (W)

S—soybean, W—wheat, SEY—soybean equivalent yield, FB—French bean, GP—Garden pea, BC—baby corn. * indicates not significant.

In the lower Indo-Gangetic Plains, Mandal *et al.* [60] observed 25%–38% greater accumulation of total SOC with NPK + FYM/compost than the control (no fertilizer). The order of such accumulation under different cropping systems was rice-mustard-sesame > rice-fallow-rice > rice-wheat-fallow > rice-wheat-jute (*Corchorus* sp.) > rice-fallow-barseem (*Trifolium alexandrinum*), over the control (Table 14). Mandal *et al.* [60] further observed that the amount of residue C inputs in the rice-wheat-fallow system (3.33 Mg ha^{-1} year^{-1}) was similar to that in the rice-fallow-berseem (3.17 Mg ha^{-1} year^{-1}), but the rate of annual C accumulation in the former was more than double (0.27 Mg C ha^{-1} year^{-1}) than that in the latter (0.13 Mg C ha^{-1} year^{-1}). Higher N content in crop residues of berseem (2.6%) and jute (1.8%) but lower lipids and lignin in rice-fallow-berseem and rice-wheat-jute systems may have accelerated decomposition and thus hastened loss of C. Crop residues from rice and wheat, which have low N content, are likely to be more efficient in C sequestration than the residue of crops like berseem and jute, which have higher N content. Likewise, in the drylands, Srinivasarao *et al.* [61] found that integrated nutrient management could improve C accumulation rates up to 0.45 ton ha^{-1}year^{-1} in a groundnut based cropping system (Table 15).

Table 14. Effects of balanced fertilization (NPK and NPK + FYM or compost) on C build up in soils under different cropping systems (Data source: Mandal *et al.* [60]).

Cropping System	C Build-Up (%) in Treatments over the Control Plots		C Build-Up Rate (Mg C ha^{-1}year^{-1}) over the Control Plots	
	NPK	NPK + FYM	NPK	NPK+FYM
R-M-S	51.8 a	55.7 a	1.91 a	2.05 a
R-W-F	16.8 c	23.4 c	0.27 b	0.37 c
R-F-B	9.3 d	24.7 c	0.13 c	0.36 c
R-W-J	14.9 c	32.3 b	0.11 c	0.25 d
R-F-R	33.5 b	54.8 a	0.28 b	0.45 b

Build-up = [(NPK//NPK + FYM – Control)/Control] × 100; Build-up rate = [(NPK//NPK + FYM – Control)/year]; R-M-S, rice-mustard-sesame; R-W-F, rice-wheat-fallow; R-F-B, rice-fallow-berseem; R-W-J, rice-wheat-jute; R-F-R, rice-fallow-rice, FYM, farmyard manure.

Table 15. Carbon accumulation rate in soil (0–20 cm) and potential carbon emission reduction (CER) under different INM practices (Data source: Srinivasarao *et al.* [61]).

Production Systems	Suggested INM Practice	C Accumulation (ton ha^{-1}year^{-1})		Potential CER from the Suggested Practice	
		Farmers' Practice	Suggested Practice	ton ha^{-1}	Value (US $)
Groundnut-based (in Alfisols)	50% RDF + 4 ton groundnut shell ha^{-1}	0.08	0.45	0.370	1.85
Groundnut–finger millet (in Alfisols)	FYM 10 ton + 100% RDF (NPK)	−0.138	0.241	0.379	1.90
Finger millet–finger millet (in Alfisols)	FYM 10 ton + 100% RDF (NPK)	0.046	0.378	0.332	1.66
Sorghum-based (in Vertisols)	25 kg N ha^{-1} (through FYM) + 25 kg N ha^{-1} (through urea)	0.101	0.288	0.187	0.94
Soybean-based (in Vertisols)	6 ton FYM ha^{-1} + 20 kg N + 13 kg P	−0.219	0.338	0.557	2.79
Rice-based (in Inceptisols)	100% organic (FYM)	−0.014	0.128	0.142	0.71
Pearl millet-based (in Aridisols)	50% N (inorganic fertilizer) + 50% N (FYM)	−0.252	−0.110	0.142	0.71

CER at US$ 5 ton^{-1} C (prevailing market price of CER for agroforestry and other related practices); RDF, Recommended dose of fertilizer; FYM, Farmyard manure.

Integrated nutrient management also decreases soil loss. Runoff and soil loss increased with increase in slope from 0.5% to 2.0% at Bellary (Table 16). However, in the treatments with application of recommended rate of fertilizer along with farm-yard manure, it was comparatively low. Plots under coir pith compost and integrated nutrient management improved maize yield and rainwater use efficiency from 4.95 to 5.79 kg ha^{-1} mm at Ayalur watershed, Tamil Nadu (Table 17).

Table 16. Runoff and soil loss under different crops on varying slopes at research farm, Bellary (Source: CSWCR&TI Annual Report [62]).

Treatments	Runoff (mm)						Soil Loss (ton ha^{-1})					
	Sorghum			Chickpea			Sorghum			Chickpea		
	0.5	1.0	2.0	0.5	1.0	2.0	0.5	1.0	2.0	0.5	1.0	2.0
	Slope (%)											
With fertilizer	52.3	66.78	94.8	48.71	64.45	84.56	2.45	4.04	5.67	2.01	2.72	4.79
Without fertilizer	63.16	66.85	101.79	49.06	65.64	92.99	2.72	4.79	6.08	2.19	3.31	5.35

Table 17. Effect of coir pith compost and integrated nutrient management (INM) on maize —a Case study in Ayalur watershed, Tamil Nadu (Source: Kannan *et al.* [63]).

Particulars	Farmers' Practice	INM	Control	Coir Pith
Yield (ton ha^{-1})	4.5	5.5	4.2	4.9
Additional yield (ton ha^{-1})	-	1.0	-	0.7
Additional cost (Rs)	-	2747	-	-
Additional benefits(Rs)	-	8000	-	-
Rain WUE (kg ha^{-1}mm^{-1})	10	12.2	4.95	5.79

4.9. Reclamation of Acid and Salt Affected Soils and Drainage (Desalinization)

Liming is the most desirable practice for amelioration of acid soils. Lime raises soil pH, thereby increasing the availability of plant nutrients and reducing toxicity of Fe and Al [64–66]. Sharma and Sarkar [64] and Bhat *et al.* [66] recommended low dose of lime (*i.e.*, one-tenth to one-fifth of lime requirement) applied along with fertilizers in furrows at the time of sowing. Bhat *et al.* [67] also tested low-cost locally available basic slag, a by-product of a steel factory as an ameliorant for acidic red and lateritic soils of West Bengal under mustard-rice.

Management of saline soils involves tillage, irrigation and leaching. Inversion tillage can decrease potential soluble salt accumulation in the root zone compared to zero tillage [68]. However, deep tillage may bring more salts to the soil surface and root zone. The most efficient method is through application of high quality irrigation water (low electrical conductivity) and growing of salinity tolerant crops. Tolerant crops also support formation of stable soil aggregates, which help to improve soil tilth. Rice is the potential crop for reclamation of sodic soils. Salt affected soils are reclaimed by leaching followed by application of green manures. Gypsum is the major chemical used for reclamation of alkali soils. Other amendments used are: phosphogypsum or acid formers like pyrites, sulphuric acid, aluminium sulphate and sulphur. The treated field should be kept submerged with good quality water to facilitate reaction and subsequent leaching. In addition, proper drainage through deep and open drains can be adopted wherever problems persist. Restoration of salt-affected soils can also lead to a significant increase in SOC pool. Garg [69] observed a dramatic increase in SOC of a sodic soil planted with perennials (e.g., mesquite) after 8 years. Bhojvaid and Timmer [70] also reported an increase in SOC pool by restoration of salt-affected soils and a similar potential exists [71].

To reclaim non-saline sodic soil, incorporation of relatively soluble Ca salt like gypsum, phosphogypsum, iron salt like pyrite, $CaCl_2$, sulfuric acid (H_2SO_4), or other acid formers like sulphur (S), lime-sulphur (9% Ca + 25% S), ferric sulfate and aluminium sulfate to replace exchangeable Na from the clay complex, along with recommended water and crop management practices, have been researched by many [72–74]. Reclaiming acid sulfate soils may follow approaches like: (i) pyrite and soil acidity can be removed by leaching after drying and aeration; (ii) pyrite oxidation can be limited or stopped and existing acidity inactivated by maintaining a high water table, with or without (iii) additional liming and fertilization with phosphorus, though liming may often be uneconomical in practical use. For coastal acid sulfate soils of Sundarbans, application of lime, superphosphate and rock phosphate have been found useful [75].

Since 1970 in India, there has been significant commercial development using various desalination technologies, including distillation, reverse osmosis and electrolysis [76]. Desalination mostly uses fossil fuels. Many facilities in coastal region are using reverse osmosis for desalinization. For example, at Kalpakkam reactor, Tamil Nadu, 1.8 million liters of water is being produced per day. Installation of one-way sluice gates on the river banks or any other suitable location to drain out excess water from the land during low tides in river, use of subsurface tile drains combined with moling perpendicular to the former [77], and open surface drains along with moling perpendicular to it [78] are some of the practices.

4.10. Remediation of As Contamination

Mitigation of As contamination could be achieved by replacing *boro* rice requiring more ground water with summer legumes and pulses, decreased irrigation coupled with addition of zinc sulfate, greater use of organic/green manures that moderate As toxicity in soils and plants [79,80], and phytoremediation employing hyper-accumulating plants like brake fern (*Pteris vittata*) and water hyacinth (*Eichornia crassipes*). Blue-green algae also have ability to decontaminate as of paddy soils through accumulation in its biomass and subsequent removal.

4.11. Water Management and Pollution Control

Promoting water conservation and efficient water management along with expansion of irrigation facilities, drip irrigation and sub-surface irrigation in some areas holds promise. Domestic and municipal wastes, sludges, pesticides, industrial wastes, *etc.* need to be used with utmost caution to avoid the possibility of pollution of soil. Mined land can be better reclaimed by proper back filling and spreading topsoil over the surface [81]. Reclaimed land after mining can be used for planting trees. The use of geo-textiles, permeable fabrics which separate, filter, reinforce, protect or drain the soil, will help the re-vegetation process [57,82].

Sen and Oosterbaan [83] presented a practical working method on integrated water management for Sundarbans through surface gravity induced drainage during the rainy season (through land shaping)-cum-excess rainwater storage for irrigation during dry season. Ambast and Sen [84] developed a user-friendly software 'RAINSIM' primarily for small holdings in the Sundarbans region based on hydrological processes, as well as in different agro-climatic regions. The software may be used for (i) computation of soil water balance; (ii) optimal design of water storage in the "on-farm reservoir" concept for converting up to 20% of the watershed; (iii) design of surface drainage in deep waterlogged areas to decrease water congestion in 75% of the area; and (iv) design of a simple linear program to propose optimal land allocation.

4.12. Irrigation Management for Improving Input Use Efficiency

Scheduling of irrigation based on critical stages of crops, or atmospheric demand stimulates optimum plant growth and increases the transpiration component of evapo-transpiration loss of water, thereby improves crop productivity and decreases soil degradation. Compared to flooding, sound surface irrigation methods like sprinkler or drip give better input efficiency. It not only improves the yield levels and input use efficiencies but also saves considerable volume of water. Besides this, in Vertisols, a considerable volume of irrigation water is lost beyond the root zones through bypass flow when irrigation water is applied through flooding. Loss of water through bypass flow in Vertisols could also be decreased by adopting irrigation application based on the atmospheric demand of water, *i.e.*, by adopting (irrigation water/cumulative pan evaporation) based scheduling. Bandyopadhyay *et al.* [85] reported that irrigation scheduling at 0.8 irrigation water/cumulative pan evaporation significantly improved the soil water extraction, root length density and grain yield of wheat over irrigation at 0.6 IW/CPE. Besides, integrated use of 75% of the recommended doses of NPK to wheat with farm-yard manure at 5 Mg ha^{-1} or poultry manure at 1.5 Mg ha^{-1} or phosphocompost at 5 Mg ha^{-1} to the rainy season crops (like soybean or sorghum) significantly improved the root length density, yield and water use efficiency of wheat over application of 100% NPK to both crops, leading to a saving of 25% fertilizer NPK in both the seasons and improvement of the use efficiency of the fertilizer nutrients.

4.13. Judicious Use of Distillery Effluent

In the vicinity of distillery industries, irrigation water as distillery effluent can be applied judiciously as a waste by-product and this technique has a considerable impact on mitigating land degradation. Both spent wash and post-methanated effluent were tested in a field experiment on soybean-wheat system for five years in a Vertisol. The SOC of the surface (0–15 cm) layer and aggregate stability were greater with application of both techniques. Proportion of macroaggregates was greater with spent wash than with post-methanated effluent, as well as compared with no distillery effluents and NPK+farm-yard manure treatments. Macroaggregate-associated C was also greater in spent water treated plots. Plots receiving waste by-produts had greater SOC, mean-weight diameter of aggregates, and percent macro- and macroaggregate-associated C than farmers' typical practice [86].

4.14. Reforestation, Grassland and Horticulture Development

In the hills, the majority of the upper slope is covered with horticultural crops using half-moon terraces and contour bunds and the remaining one-third of the lower section is used for cultivation of cereals, or oil crops with bench terraces. The following crops may be grown: (1) Fruit trees in half-moon terraces (triangular system of planting) on contour; (2) Pineapple (*Ananas comosus* L.) in two rows planted closer together in contour bunds; (3) Vegetables like bean (*Phaseolus vulgaris* L.), cowpea (*Vigna sinensis* L.), guar or clusterbean (*Cyamopsis tetragonoloba* L.), pea (*Pisum sativum* L.) and good cover crops like sweet potato in the interspaces of the contour and (4) Ginger (*Zinziber officinale* L.) and turmeric (*Curcuma longa* L.) grown in the inter-space area of contours. Grewal [87] found that under Eucalyptus-Bhabar grass (*Eulaliopsis binata*) system, soil loss was negligible (0.07 ton ha^{-1}) (Table 18). Likewise, reforestation and grassland development in wastelands have great potential to decrease land degradation.

Table 18. Soil loss under different land use systems in Shivaliks (Source: Grewal [87]).

Land Use Systems	No. of Years of Observations	Soil Loss (ton ha^{-1})	Runoff (% of Total Rainfall)	N Loss (kg ha^{-1})	K Loss (kg ha^{-1})
Eucalyptus-Bhabar grass	6	0.1	0.1	0.5	0.9
Acacia catechu–forage grass	3	0.2	2.0	7.0	0.5
Leucaena-Napier grass (*Pennisetum purpureum* L.)	3	0.3	4.4	6.6	1.2
Teak (*Tectona grandis* L.)-*Leucaena*-Bhabar	3	0.4	3.3	2.1	0.6
Eucalyptus-*Leucaena*-Turmeric	5	0.6	2.6	2.5	0.7
Poplar (*Liriodendron tulipifera*)-*Leucaena*-Bhabar	5	1.5	4.8	5.9	1.1
Sesamum (*Sesamum indicum*)-Rapeseed (*Brassica napus*)	3	2.7	20.5	42.5	3.0
Cultivated fallow	3	5.6	23.0	51.3	5.0

4.15. Vegetative Barriers and Using Natural Geotextiles, Mulching and Diversified Cropping

In general, results from the Himalaya region indicate that vegetative barriers can decrease runoff by 18%–21% and soil loss by 23%–68% on slopes varying from 2%–8% (Table 19). Vegetative barriers of *Guinea* grass (Scheme 1), *Khuskhus* and *Bhabar* were effective (after 3–4 years) in reducing soil loss by 6–8 ton ha^{-1}year^{-1} and runoff by 33%–38% [40]. Maize and wheat yield increased ~32 and 10%, respectively, due to conserved moisture in the hilly region [88]. Pigeonpea (*Cajanas cajan*), because of its very good canopy cover (95%–98%) as a vegetative barrier, was effective in reducing runoff (28%–29%) and soil loss (2.1 to 2.6 ton ha^{-1}) in a finger millet (*Eleusine coracana* L.)/kodo millet (*Paspalum scrobiculatum* L.)-lentil (*Lens esculentus* L.) cropping sequence. Pigeonpea improved SOC along with addition of 22 to 41 kg of N ha^{-1} in the soil. The practice increased maize yield 5%–10% and wheat yield 10%–15% in the hills.

Table 19. Effect of grass barriers on yield, runoff and soil loss in different slopes of the northwestern hill region (Data Source: CSWCR&TI Vision [40]).

Particulars	Slope (%)					
	2	4			8	
	Guinea Grass	Guinea Grass	Khus khus	*Bhabar*	Guinea Grass	Khus khus
Runoff (% of total rainfall)	25.8	33.3	35.1	37.9	38.90	40.52
Soil loss (ton ha^{-1}year^{-1})	3.27	6.12	6.72	8.34	9.45	9.87
Maize yield (kg ha^{-1})	2530	2460	2444	2296	2285	2180
Wheat yield after maize (kg ha^{-1})	2852	2693	2555	2362	2415	2385
Dry grass yield (kg ha^{-1}year^{-1})	1675	1540	542	1090	1375	485

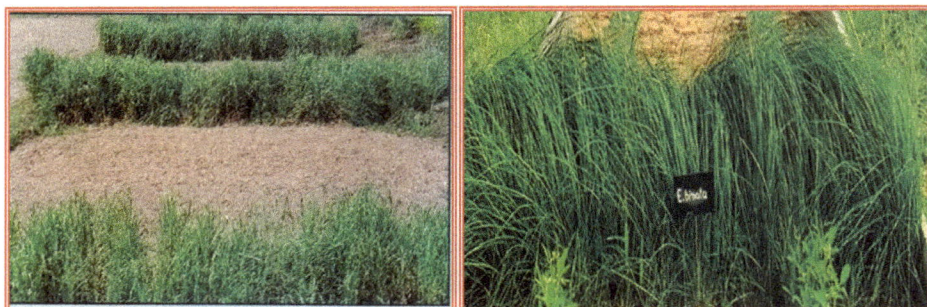

Scheme 1. Vegetative barrier of Guinea grass and Bhabar grass (Source: CSWCR&TI Vision [40]).

Research results in neighbouring countries, like China, Thailand and Vietnam indicated that even short-term use of biological geotextiles (maize stalk mats, bamboo mats, borassus and buriti mats and wheat straw mats) in highlands significantly improved biomass and decreased runoff and soil loss [89–92]. Borassus mats (geotextile mats manufactured from *Borassus aethiopum* L.) and buriti mats (geotextiles manufactured from *Mauritia flexuosa* L.) were also tested for soil conservation in a loamy sand soils of the UK. The results are very novel and worth mentioning here. Results reveal that: (i) borassus mats significantly decreased soil splash erosion; (ii) complete cover by borassus mats is unnecessary and only 10% mat cover as buffer strips had similar erosion control to completely covered plots; and (iii) borassus mat-covered plots maintained SOM and other selected soil properties [93–96]. Vegetative barriers are also used to mitigate soil degradation in non-arable areas. Vegetative barriers of tree species are effective in controlling runoff and soil loss on 4% slope [97]. Total sediment deposited along hedgerows (3-year period) and tree rows (9-year period) ranged from 184 to 256 ton ha^{-1}, equivalent to 15 to 20 mm soil depth (Table 20).

Table 20. Sediment deposition along vegetative barriers at Dehradun (Source: Narain *et al.* [97]).

Vegetative Barrier	No. of Years	Sediment Deposited (ton ha^{-1}year^{-1})	Average Deposition (ton ha^{-1}year^{-1})	Soil Loss s(ton ha^{-1}year^{-1})
Leucaena hedge in turmeric field	3	47.3	15.8	7.6
Leucaena hedge in maize field	3	184.0	61.3	12.1
Leucaena trees in maize field	9	256.5	28.5	8.8
Eucalyptus trees in maize field	9	185.6	20.6	5.8
Leucaena trees in turmeric field	9	90.1	10.1	6.8
Eucalyptus trees in turmeric field	9	103.7	11.5	7.1

In a novel attempt in Bangladesh, implementation of jute geotextiles aided by native vegetation cover was investigated in 2009. Combined presence of jute geotextiles and vegetation cover decreased erosion rates by ~95% and runoff by ~70% with respect to bare plots (that had ~18 ton ha^{-1} year^{-1} soil loss) on a 20% land slope [98].

4.16. Agroforestry

Agroforestry systems are an appropriate management tool for both acid and salt-affected soils, because perennial woody vegetation recycles nutrients, maintains soil organic matter, and protects soil from surface erosion and runoff [99]. Four multipurpose tree species were compared with a control plot (without tree plantation) for soil fertility status in an acid soil of India. The presence of trees improved the physico-chemical and microbial biomass parameters by storing greater SOC [100]. Tree vegetation in an agroforestry system serves two major purposes: (i) the fine root system holds soil in place, reducing susceptibility to erosion; and (ii) plant stems decrease the flow velocity of runoff, enhancing sedimentation.

Nair [101] stated three environmental benefits of agroforestry systems: water-quality enhancement, C sequestration, and soil improvement. These benefits are based on the perceived ability of (i) vegetative buffer strips to decrease surface transport of agrochemical pollutants; (ii) large volumes of aboveground and belowground biomass of trees to store C deeper in the soil profile; and (iii) trees enhance soil productivity through biological N$_2$ fixation, efficient nutrient cycling, and deep capture of nutrients. Legume-based agroforestry has the capacity to support biological N fixation to enhance subsequent soil N availability and therefore improve soil fertility and crop yields [102].

Biosaline (agro) forestry is the cultivation of trees and/or crops on salt-affected soils. Some tree species are less susceptible to soil salinity and sodicity than agricultural crops and hence the cultivation of these trees can help regenerate these soils. In alkaline waste lands, mechanical impedance is a major cause of poor root proliferation. This problem could be overcome by planting *Prosopis juliflora*, which has roots to vertically penetrate a hard pan [103]. Mishra *et al.* [104] opined that soil erosion can be decreased in alkaline soils with *Prosopis juliflora* and *Casuarina equisetifolia* due to the formation of stable soil aggregates in the surface layers. Kaur *et al.* [105] analyzed the role of agroforestry systems (*Acacia, Eucalyptus* and *Populus* along with rice–berseem (*Trifolium alexandrinum* L.)) to improve soil organic matter, microbial activity and N availability and observed that: (i) microbial biomass C and N were greater by 42% and 13%, respectively, in tree-based systems than mono-cropping; (ii) soil organic C increased by 11%–52% due to integration of trees along with crops after 6–7 years.

In India, many tolerant species for saline soils have been tried since long (Table 21), like: *Prosopis juliflora, Salvadorapersica, S. oleoides, Tamarixericoides, T. troupii, Salsolabaryosma etc.*, successful on sites with ECe > 35 dS m^{-1}, *Tamarixarticulata, Acacia farnesiana, Parkinsonia aculeate* on sites with moderate salinity (ECe 25–35 dS m^{-1}), *Casuarina (glauca, obesa, equiselifolia), Acacia tortilis, A. nilotica, Callistemon lanceolata, Pongamia pinnata, Eucalyptus camaldulensis, Albizia lebbeck* on sites with moderate salinity (ECe 15–25 dS m^{-1}), trees like *Casuarina cunninghamiana, Eucalyptus tereticornis, Acacia catcechu, A. ampliceps, A. eburnea, A. leucocephala, Dalbergia sissoo, etc.* on sites with lower salinity (ECe 10–15 dS m^{-1}).

Table 21. Ameliorative effects of tree plantation on salt affected soils of India.

Region	Tree Species	Soil Depth (cm)	Original		After		References
			pH	EC (dS m^{-1})	pH	EC (dS m^{-1})	
Karnataka	*Acacia nilotica* (age 10 years)	0–15	9.2	3.73	7.9	2.05	Basavaraja *et al.* [106]
Karnal	*Eucalyptus tereticornis* (age 9 years)	0–10	10.06	1.90	8.02	0.63	Mishra *et al.* [107]
Lucknow and Bahraich in north India.	*Terminaliaarjuna Prosopisjuliflora Tectonagrandis*	0–15	9.60 ± 0.42	1.47 ± 0.45	8.40 ± 0.27 8.70 ± 0.33 6.15 ± 0.23	0.31 ± 0.07 0.42 ± 0.06 0.06 ± 0.006	Singh and Kaur [108]

4.17. Conservation Agriculture (CA)

Conservation agriculture refers to a set of principles, grounded in sound science that is gradually being adopted globally. The concept includes: (1) causing minimum disturbance to the soil surface by using no- or minimum-tillage; (2) keeping the soil surface covered all the time through practices such as retention of crop residue, mulching, or growing cover crops; (3) adopting crop sequences or rotations that include agroforestry in spatial and temporal scales; and (4) controlled traffic [109]. Collectively these practices lead to an increase in water stable aggregates, greater SOC concentrations, and protection from wind and water erosion. Conservation agriculture-based crop management technologies include zero tillage (ZT) with residue recycling; laser assisted precision land levelling, direct drilling into the residues and direct seeding.

In the Himalayan region, year-round ZT under irrigated rice-wheat system with two irrigations at critical growth stages [110], year-round ZT with integrated nutrient management under an irrigated rice-wheat system [58], and 10-cm stubble retention (under CA) of rice and wheat crops for maximum yield and fodder production [111] are novel technologies (Table 22). Zero tillage enhanced macroaggregate-associated SOC and intra-aggregate particulate organic C under a rainfed finger millet-lentil system (Figure 4), but only in the topsoil [57–59]. Plots with minimum tillage (MT; a 50% tillage reduction) improved SOC stock in the 0–15 cm layer, as well as soybean yield. Under direct-seeded rice-wheat systems, adoption of ZT with two irrigations in each crop improved topsoil physical properties and SOC content after four years with similar mean crop yields as with CT using four irrigations [112]. Conservation tillage improved soil aggregate stability and labile C pools in the surface layer, across different cropping systems both under rainfed and irrigated conditions in the Himalayas [57]. Introduction of a legume crop improved C retention in surface soils under conservation tillage even with only short-term adoption.

Table 22. Impacts of conservation tillage practices on carbon retention in the 0–15 cm layer and crop yield change in the Indian Himalayas.

Rainfed Management Practices	Duration of Adoption (year)	Carbon Retention over Control/CT(Mg ha^{-1}year^{-1})	Yield Change over	Yield Change (Mg ha^{-1}year^{-1})
Zero tillage-irrigated	4	0.20	CT	−0.09 * (R) & −0.23 * (W)
Zero tillage-rainfed	4	0.61	CT	−0.44 (SEY)

SEY = Soybean equivalent yield, R = Rice and W = Wheat. * indicates not significant.

As mentioned earlier, CA has emerged as a new paradigm to achieve goals of sustainable agricultural production in South Asia [113]. Another technology is controlled traffic farming using permanent tram lines. For this system, all equipment on the farm needs a standardized track width. Soil between tram lines has better structure and is free of compaction, while the heavily compacted tramlines provide better trafficability and traction [114]. As a result, tillage cost is decreased and yield increase in the cropping area exceeds the loss of land due to tramlines. In the Indo-Gangetic Plain region, bed planting under CT and ZT generally saves irrigation water [115] and labor requirements without sacrificing crop productivity [116,117].

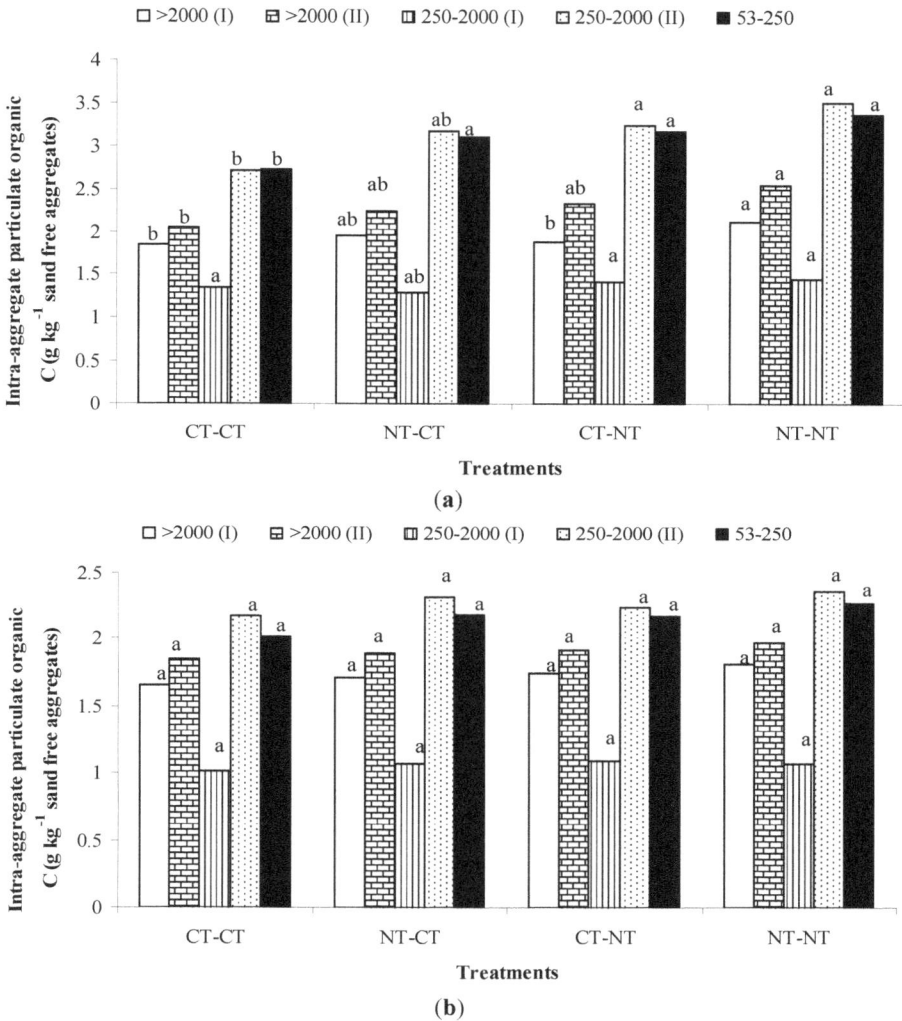

(a)

(b)

Figure 4. Intra-aggregate particulate organic matter-carbon (iPOM-C; g kg^{-1} of sand-free aggregates) in aggregate-size fractions at the (**a**) 0- to 5-cm and (**b**) 5- to 15-cm soil layers as affected by tillage practices after six years of rainfed cropping. Bars followed by a similar letter between treatments within an aggregate size class are not significantly different at $P < 0.05$ level of significance according to Tukey's HSD mean separation test. "(I)" and "(II)" in legend refer to coarse (250–2000 μm) and fine (53–250 μm) iPOM in the respective aggregate sizes (Source: Bhattacharyya *et al.* [57]).

In another study in the region, Das *et al.* [118] observed that plots under permanent broad bed with 20% cotton residue and 40% wheat residue retention had significantly higher economic profitability and crop productivity than farmers' practice under a CT cotton-wheat cropping system. In this study, Das *et al.* [118] concluded that 2-year mean seed cotton yield under ZT permanent broad-bed sowing with residue retention was about 24% and 51% greater compared with ZT narrow-bed sowing without residue retention (2.91 Mg ha^{-1}) and CT (2.59 Mg ha^{-1}), respectively (Table 23). Mean water productivity of the system in the permanent broad bed with residue retention (12.58 kg wheat grain ha^{-1}mm^{-1}) was 12%–48% greater compared with CT, narrow bed with and without residues, broad bed, and ZT plots (Table 24). Net return of the permanent broad bed plots with residue retention was 36% and 13% greater compared with CT and narrow bed plots, but was similar to other treatments. Some of the challenges that follow from continuous ZT practice are management of perennial weeds and strategies to combat yield reduction. Yields of ZT crops are often decreased by 5% to 10% on sandy loam soils of India compared with under CT in the initial years [119].

Table 23. Productivity (Mg ha^{-1}) of cotton, wheat and system productivity (Mg ha^{-1}) in terms of wheat equivalent yield (WEY) as affected by tillage, bed planting and residue management practices in the western Indo-Gangetic Plains (Data source: Das *et al.* [118]).

Treatments *	2010–2011			2011–2012			2012–2013		
	Seed Cotton Yield	Wheat Grain Yield	System Productivity (WEY)	Seed Cotton Yield	Wheat Grain Yield	System Productivity (WEY)	Seed Cotton Yield	Wheat Grain Yield	System Productivity (WEY)
CT	2.44 c	4.85 a	10.30 b	2.73 c	4.29 b	11.16 c	2.70 c	4.46 b	12.25 b
PNB	2.71 bc	4.55 a	10.60 b	3.10 bc	4.37 b	12.17 bc	3.08 ab	4.83 ab	13.72 ab
PNB + R	2.96 b	4.61 a	11.23 ab	3.33 b	4.60 ab	12.97 b	3.38 a	4.98 a	14.74 a
PBB	3.13 ab	4.82 a	11.81 ab	3.42 ab	4.19 bc	12.80 b	3.11 ab	4.75 ab	13.72 ab
PBB + R	3.28 a	4.85 a	12.16 a	3.93 a	4.77 a	14.67 a	3.46 a	4.89 a	14.88 a
ZT + R	-	-	-	4.00 a	4.44 ab	14.50 a	3.21 ab	4.73 ab	13.99 ab
ZT	-	-	-	3.95 a	4.00 c	13.93 ab	3.02 bc	4.63 ab	13.35 ab

* Means followed by a similar lowercase letter within a column are not significantly different (at $P < 0.05$) according to Tukey's HSD Test. ZT = Zero tillage; CT = Conventional tillage; ZT + R = Zero tillage + residue retention; PNB = Permanent narrow bed; PBB = Permanent broad bed; PNB+R = Permanent narrow bed + residue retention; PBB + R = Permanent broad bed + residue retention.

Table 24. Impacts of tillage, bed planting and residue management practices on water productivity (kg wheat grain/ha.mm) under the cotton-wheat system (Source: Das *et al.* [118]).

Treatments *	2011–2012		2012–2013	
	Total Water Applied in the System (mm)	System Water Productivity	Total Water Applied in the System (mm)	System Water Productivity
CT	1417	8.65 d	1331	8.38 d
PNB	1297	10.58 b	1208	10.07 c
PNB + R	1282	10.50 bc	1181	10.98 bc
PBB	1260	10.89 b	1160	11.03 bc
PBB + R	1222	12.18 a	1130	12.98 a
ZT + R	1312	10.66 b	1247	11.62 b
ZT	1387	9.62 c	1310	10.63 bc

Means followed by a similar lowercase letter within a column are not significantly different (at $P < 0.05$) according to Tukey's HSD Test. ZT = Zero tillage; CT = Conventional tillage; ZT + R = Zero tillage + residue retention; PNB = Permanent narrow bed; PBB = Permanent broad bed; PNB+R = Permanent narrow bed + residue retention; PBB + R = Permanent broad bed + residue retention.

Unlike conventional farming methods, CA minimizes soil disturbance and recycles crop residues. Soil bulk density may be decreased, soil aggregation may be improved, and SOC may increase to reverse land degradation with CA. Specific results from four years of wheat-based cropping system in the western Indo-Gangetic Plains indicate that ZT had higher C retention potential than CT in the 0–30 cm soil layer with 8.6% and 10.2% of the gross C input retained under CT and ZT, respectively (Figure 5; Das *et al.* [120]).

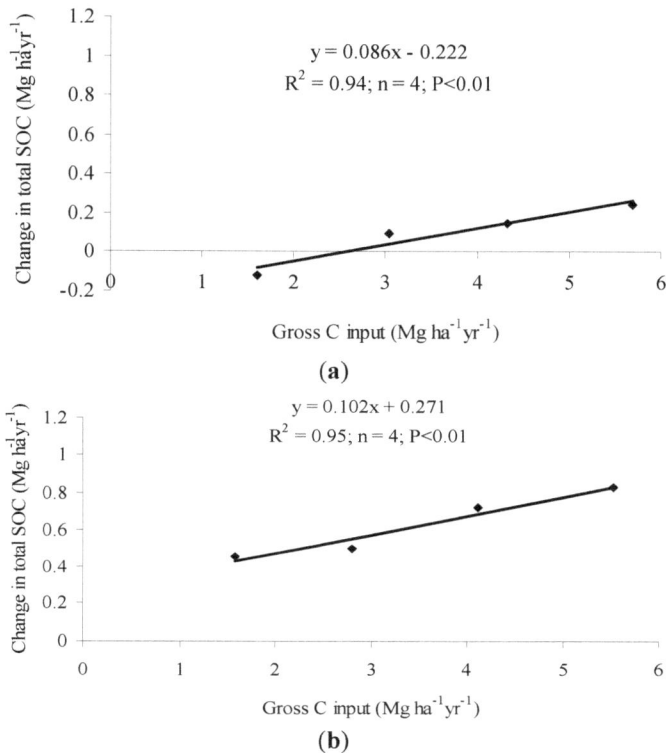

Figure 5. Total soil organic C (SOC) retention potential of residue management practices under (**a**) conventional tillage and (**b**) zero tillage under a wheat based cropping system in the Indo-Gangetic Plains (Source: Das *et al.* [120]).

In another study in the Indo-Gangetic Plains, topsoil under ZT with bed planting had greater concentration of macroaggregates (0.25–8 mm) and mean weight diameter with a concomitant lower silt + clay sized particles than under CT with bed planting and CT with flat planting after 4 years. Soil with both cotton/maize and wheat residue retention had greater macroaggregate concentration and mean weight diameter and similar bulk density than with residue removal (Figure 6). Soil aggregation is improved with larger aggregates and greater mean weight diameter [110].

Conservation agriculture has the potential to decrease sub-surface compaction and improve least limiting water range [121]. During the third year of a study at New Delhi, soil penetration resistance exceeded 2 MPa in the 15–30 cm soil layer beneath puddled and transplanted rice in rotation with wheat under CT, but under direct seeded rice with brown manuring and ZT penetration resistance values were <1.5 MPa throughout the 0–60 cm profile. Soil bulk density was lower under the ZT system than under the CT system in the 0–30 cm soil layer (Table 25). Retaining crop residues with permanent broad beds had significantly lower penetration resistance than with permanent narrow-beds with residue and other tillage and residuce management in a cotton-wheat system [121]. Retaining crop residue resulted in lower BD values in the 15–30 cm soil layer under the cotton-wheat system than removing them (Table 26). Retaining crop resicue also had ~12% higher least limiting water range than with CT (10.1%) in the 15–30 cm layer. In the 0–15 cm soil layer, retaining residues under ZT, permanent broad beds, and permanent narrow beds had 13%, 24% and 11% higher mean least limiting water range, respectively, than the same tillage systems without residue retention. This indicates that

ZT without residue addition had deleterious impact on soil water availability and structural property and should be avoided. Overall, among the treatments, PBB + R and DSR + BM-ZTW were the best management practices for improved soil physical environment under cotton-wheat and rice-wheat systems, respectively, and therefore should be adopted.

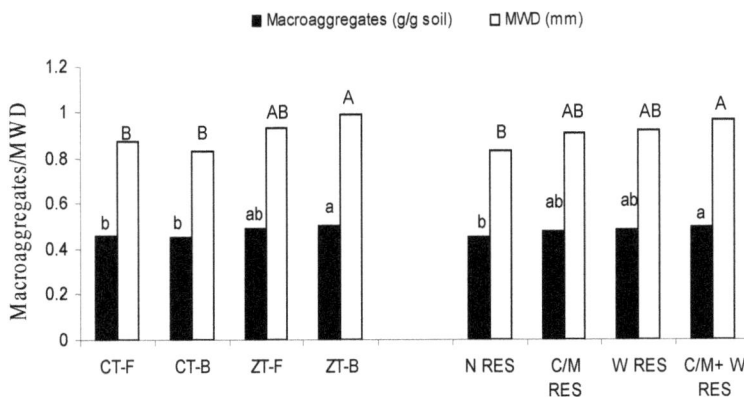

Figure 6. Impacts of conservation agriculture (CA) on soil aggregation in the 0–5 cm layer in the upper IGP (Source: Bhattacharyya *et al.* [110]).

Table 25. Soil bulk density of plots with different conservation agricultural practices in the rice–wheat system (Source: Mishra *et al.* [121]).

Conservation Agricultural Practices	\multicolumn{3}{c}{Bulk Density (Mean of Two Sampling Events) during Rice (Mg m^{-3})}		
	0–15 cm	15–30 cm	30–45 cm
PTR − CTW	1.45 b	1.70 a	1.72 a
DSR + BM − ZTW	1.47 ab	1.63 b	1.70 a
DSR − ZTW	1.50 a	1.64 ab	1.72 a

PTR − CTW = Puddled transplanted rice-conventionally tilled wheat; DSR − ZTW = Direct seeded rice-zero tilled wheat and DSR + BM − ZTW = Direct seeded rice + Brown manuring-zero tilled wheat; Means with similar lowercase letters within a soil depth and crop growth period are not significant at $P < 0.05$ according to Tukey's HSD test.

Adoption of CA, as a complete package, is one of the major strategies for increasing SOC stock. Although crop residue incorporation initially leads to immobilization of inorganic N, addition of 15–20 kg N ha^{-1} with straw incorporation eventually increases yield of rice and wheat. Incorporation/retention of rice residue in the soil returns essential organic C and N back to the field to favorably impact soil structural status. Surface residue placement had greater C retention than residue incorporation in a maize-wheat-greengram cropping system [120]. Zero tillage in particular can complicate manure application and may also contribute to nutrient stratification within the soil profile from repeated surface applications without mechanical incorporation. Conservation agriculture has tremendous potential to improve water use efficiency of crops and decrease water loss. If CA could be adopted on half of Haryana's rice-wheat area, the practice would decrease diesel use by 17.4 million liters per year. Using a conversion factor of 2.6 kg CO_2 per liter of diesel burned, this would represent a reduction of more than 25,000 tons each year in CO_2 emissions [116]. Because ZT takes immediate advantage of residual moisture from the previous rice crop, as well as cutting down on subsequent irrigation, water use is decreased by about 10 cm-hectares, or approximately 1 million liters ha^{-1}year^{-1}.

Table 26. Soil bulk density of plots with different conservation agricultural practices in the cotton-wheat system during crop growth period (Source: Mishra *et al.* [121]).

Bulk Density (Mean of Four Sampling Events in Two Years) during Cotton (Mg m^{-3})			
Conservation Agricultural Practices	**0–15 cm**	**15–30 cm**	**30–45 cm**
CT	1.52 bc	1.65 ab	1.70 a
PNB	1.48 c	1.62 b	1.71 a
PBB	1.50 c	1.63 b	1.70 a
ZT	1.63 a	1.68 a	1.70 a
PNB + R	1.43 d	1.56 c	1.70 a
PBB + R	1.44 d	1.57 c	1.69 a
ZT + R	1.57 b	1.60 bc	1.70 a

Means with similar lowercase letters within a crop growth period are not significantly different at *P* < 0.05. CT = conventional tillage; PNB = Permanent narrow bed; PNB + R = Permanent narrow bed + residue retention; PBB = Permanent broad bed; PBB + R = Permanent broad bed + residue retention; ZT = Zero tillage; ZT + R = Zero tillage + residue retention

With greater aggregation due to fewer disturbances by tillage operations and addition of surface residues, total pore space in soil under CA increases. In a study from central India, CA generally improved water retention properties of soil through its effect on pore size distribution and soil structure. Volumetric soil water retention of the surface 0–15 cm soil was greater in ZT and decreased tillage systems than in CT (Table 27). Similarly at permanent wilting point, CA treatments retained more water than with CT. Difference in water storage between tillage treatments was less at permanent wilting point (2.5%) than at field capacity (4.2%). Conservation agriculture increased soil-water retention more at lower suctions due to increase in micro-pores and inter-aggregate pores caused by enhanced SOC content and higher activity of soil fauna e.g., earthworms and termites. At higher tensions close to permanent wilting point (1.5 MPa), nearly all pores were filled with air and surface area and thickness of water films on soil particle surfaces determined moisture retention. Following addition of organic matter, specific surface area of soils increased resulting in increased water holding capacity at higher tensions [122].

Table 27. Effect of different tillage systems on soil water retention of Vertisols (Source: Hati *et al.* [122]).

Tillage	Soil Water Retention (%) (*v/v*)		
	Field Capacity (0.033 MPa)	**Permanent Wilting Point (1.5 MPa)**	**Available Water Capacity**
Conventional Tillage	33.5	22.6	10.9
Mouldboard tillage	35.4	24.7	10.7
Reduced Tillage	36.5	24.6	11.9
No Tillage	37.7 a	25.1	12.6
LSD (*P* = 0.05)	2.4	2.1	NS

Despite many benefits of CA practices as mentioned above, the adoption rate in India is very low. Farmers prefer to follow a partial adoption of CA practices, *i.e.*, transplanted rice in puddled soil in the *Kharif* season and CA (ZT with residue retention) for wheat in the *Rabi* season due to several factors, including (i) poor germination and low crop productivity under direct seeded rice, because puddling and waterlogged condition helps to reduce soil pH in alkaline soils and thus improve soil chemical health during the rice growing season; (ii) availability of rain and irrigation water for raising a good rice crop under puddled condition; and (iii) less care is needed for transplanted rice in puddled soil compared with direct seeding. Some farmers even grow CT maize/jowar in the *Kharif* season (for better weed control, aeration and reduction in surface compactness/crusting) and raise wheat under CA in the *Rabi* season. However, repeated puddling aggravates other problems like soil compaction, development of salt affected soils and decline in water table in the area (due to high evaporative demand in this climate). Looking at these facts and due to constant efforts by several institutions, some farmers of the district have started adopting full or complete CA (*i.e.*, direct-seeded rice followed by ZT wheat), but the duration of adoption is less than 3–4 years.

In drylands, Jat *et al.* [123] opined that the major constraints to the use of CA include insufficient amounts of residues due to water shortage and degraded nature of soil resource, competing uses of crop residues, resource poor smallholder farmers, and lack of in-depth research. Even then, CA holds considerable promise in the arid region, because it can control soil erosion by wind and water, reduce compaction and crusting. Due to limited production of biomass, competing uses of crop residues and shortage of firewood, farmers often find it hard to use crop residues to cover soil surface in dryland eco-systems, where only a single crop is grown in a year. With CA (soil cover with crop residues), it is sometimes possible to grow a second crop with residual soil moisture in the profile. It is, however, better to use the chopped biomass of semi-hard woody perennial plants instead of crop residues to cover the soil surface [37].

4.18. Intensive Cropping, Diversified Cropping and Integrated Farming Systems

There is already a greater emphasis on crop diversification due to growing concerns about the unsustainability of the rice–wheat system throughout the Indo-Gangetic Plains. The water requirement for rice is about 80% greater than for other crops. Growing non-rice crops in some areas and summer cropping with legumes such as green gram, cowpea (*Vigna unguiculata* L.) or dhaincha (*Sesbania* sp.) are essential for conserving resources and improving productivity. In Punjab, Haryana and Rajasthan, >95% of the area of rice, as well as a large portion of wheat, is under irrigation. Water use efficiency could be greatly increased with cover crops or growing of non-rice based cropping systems. Productivity of waterlogged soils in the eastern Indo-Gangetic Plains could be increased by practicing a raised-sunken bed system. Soybean can be grown on raised beds, while paddy rice can be grown in sunken beds during the rainy season. Productivity of rice fallows in eastern India could be increased by growing *rabi* legumes on raised beds and wheat in sunken beds.

In a study to develop sustainable agricultural intensification with CA in an alkaline soil in Karnal, Gathala *et al.* [124] compared four novel scenarios. Maize under ZT was as productive and almost as profitable as rice during the rainy (*kharif*) season, while using 90%–95% and 88%–91% less irrigation water than puddle transplanted rice and ZT rice, respectively (Table 28). Maize can therefore be an alternative to rice in areas with extreme labor and water scarcity. Avoiding puddling and use of ZT (for rice or maize) with full residue retention increased profitability and yield of the succeeding wheat crop each year (by 0.5 to 1.2 ton ha^{-1}) than farmers' practice. Inclusion of green gram in the cropping system resulted in greater system productivity and profitability than without green gram.

Table 28. Drivers of agricultural change, crop rotation, tillage, crop establishment method and residue management of the four scenarios as studied by Gathala *et al.* [124].

Scenario	Crop Rotation	Tillage	Crop Establishment	Residue Management
Farmers' practice; S1	Rice–Wheat	CT-CT *	Transplanted–Broadcasted	Removal
To deal with increasing food demand; S2	Rice–Wheat–Green gram	CT-ZT-ZT	Transplanted–ZT-ZT	Anchored–Removal–Incorporation
To deal with rising scarcity of labor, water and energy and degrading soil health (CA based); S3	Rice–Wheat–Green gram	ZT-ZT-ZT	Direct Drilling	Retention–Anchored–Retention
Futuristic intensified and diversified cropping system (CA based); S4	Maize–Wheat–Green gram	ZT-ZT-ZT	Direct Drilling /Planting	Retention–Anchored–Retention

* CA—Conservation agriculture, CT—Conventional tillage, ZT—Zero tillage.

In India, 65% of farming households are considered marginal in sustainability (<1 ha). These farms comprise nearly 400 million people and nearly 40% of them are vulnerable, marginalized and food insecure. Hence, integrated farming systems have emerged as a well-accepted, single window, and sound strategy for harmonizing simultaneously joint management of land, water, vegetation, livestock, and human resources. The goals of integrated farming systems are to meet soils' productive potential and reduce risks of environmental degradation. By including tree crops with a high quality of leaf litter and root binding ability, erodibility from rainfall/runoff can be reduced and physico-chemical conditions improve. Soil health can be managed and sustained through organic inputs.

4.19. Disaster (Tsunami) Management

The following management aspects are important in case of a tsunami: (i) traditional disaster detection systems should be integrated with current scientific techniques; (ii) early warning systems need to be installed in coastal regions; (iii) protection against tsunamis can be achieved through construction of sea walls, beach defenses, tree plantations, and making buffer zones like raised land masses and forests; (iv) awareness about tsunamis and their impact in coastal areas has to be created not only among the public but also among officials; and (v) enforcement of by-laws and 'Coastal Regulation Zone Norms' should be strictly implemented to minimize tsunami damage. A tsunami early warning system for the Indian Ocean was installed. The Indian Ocean Tsunami Warning System was agreed to in a United Nations conference.

5. Conclusions

Appropriate mitigation strategies of the nearly 147 Mha of existing degraded land in the sub-continent of India are of the utmost importance. With changing climate, land degradation is expected to only increase due to high intensity storms, extensive dry spells, and denudation of forest cover. Combating further land degradation and investing in soil conservation is a major task involving promotion of sustainable development and nature conservation. An integrated watershed approach should be given maximum attention to combat land degradation and environmental problems particularly in fragile areas. Sustainable agricultural intensification using innovative farming practices have tremendous potential of increasing productivity and conserving natural resources, particularly by sequestering SOC (both labile and recalcitrant) and improving soil quality. Conservation agriculture (CA) coupled with other technologies like micro-irrigation, fertigation, and management of problem soils using specific and necessary technologies hold great promise to increase productivity of crops and fruits and reverse soil degradation. Novel CA practices include: permanent broad bed with residue retention under maize/cotton/pigeon pea-wheat cropping systems and seasonal tillage alterations under rainfed and rice-based agro-ecosystems. These practices need to be evaluated in micro-environments of different agro-climatic regions with different farming practices for wider adaptability on a watershed basis. For sure, the non-edible (to animals) agricultural residues must not be burnt and should be used for mulching along with growing of cover crops, preferably legumes. Improved grazing practices, irrigation management, control on urban sprawl and control and management on mining are a few other solutions for preventing land degradation. Domestic and municipal wastes, sludges, pesticides, industrial wastes, *etc.* need to be used if possible to close nutrient cycles, but with caution to avoid the possibility of soil pollution. Future research should focus on enhancing nutrient and water use efficiencies and reduction in the pesticide use under CA.

For promotion of CA practices across diverse agro-ecologies, appropriate policy and institutional and technology support would be a prerequisite. Suitable economic incentives should be given to internalize land degradation wherever feasible. Many CA practices like rainwater harvesting through farm pond renovation or construction and its recycling are both capital and labor-intensive, which resource-poor farmers in rainfed areas may not be able to afford and hence need to be supported. Initial incentives, to procure appropriate machinery and to offset any economic loss due to residue retention or production loss, are also important to motivate irrigated farmers to follow CA. Involvement is needed of local communities at every stage in the implementation of resource conserving technologies, judicious irrigation water management, wasteland reclamation, watershed development, and afforestation. A well-defined integrated land use policy to include rural fuelwood and fodder grazing is urgently needed at the implementation level to guide sustainable management of land and forest with a scientific backing. Finally, another critical challenge is controlling fragmentation of land holdings. This could be achieved by providing security of land rights and land tenure and encouraging the efficient use of marginal lands.

Acknowledgments: The authors are grateful to the anonymous reviewer and the Guest Editor for their valuable and editing comments.

Author Contributions: Ranjan Bhattacharyya and Alan Joseph Franzluebbers conceived and designed the whole review work based on invitation from the Guest Editor. Ranjan Bhattacharyya, Birendra Nath Ghosh, Prasanta Kumar Mishra, Biswapati Mandal, ChSrinivasa Rao, Dibyendu Sarkar, Krishnendu Das, Kokkuvayil Sankaranarayanan Anil, Manickam Lalitha, Kuntal Mouli Hati wrote the paper. Alan Joseph Franzluebbers and Ranjan Bhattacharyya edited the manuscript.

Conflicts of Interest: The authors declare no conflict of interest.

Supplementary Materials: Supplementary materials can be accessed at: http://www.mdpi.com/2071-1050/7/4/3528/s1.

References

1. National Bureau of Soil Survey & Land Use Planning (NBSS&LUP). *Soil Map (1:1 Million Scale)*; NBSS&LUP: Nagpur, India, 2004.
2. Dhruvanarayan, V.V.N.; Ram, B. Estimation of soil erosion in India. *J. Irrig. Drain. Eng.* **1983**, *109*, 419–434. [CrossRef]
3. NCA. *Report of the National Commission on Agriculture. National Commission of Agriculture*; Government of India: New Delhi, India, 1976; pp. 427–472.
4. MoA. *Indian Agriculture in Brief*, 17th ed.Directorate of Economics and Statistics, Ministry of Agriculture, Department of Agriculture and Cooperation, Ministry of Agriculture and irrigation, Government of India: New Delhi, India, 1978.
5. Vohra, B.B. *A Policy for Land and Water*; Department of Environment, Government of India: New Delhi, India, 1980; Volume 18, pp. 64–70.
6. NWDB. *Ministry of Environment and Forests, National Wasteland Development Board Guidelines for Action*; Government of India: New Delhi, India, 1985.
7. Bhumbla, D.R.; Khare, A. *Estimate of Wastelands in India. Society for Promotion of Wastelands Development*; Allied: New Delhi, India, 1984; p. 18.
8. NRSA. *Waste Land Atlas of India*; Government of India Balanagar: Hyderabad, India, 2000.
9. MoA. *Indian Agriculture in Brief*, 20th ed.Directorate of Economics and Statistics, Ministry of Agriculture, Department of Agriculture and Cooperation, Ministry of Agriculture, Goverment of India: New Delhi, India, 1985.
10. MoA. *Indian Agriculture in Brief*, 25th ed.Directorate of Economics and Statistics, Ministry of Agriculture, Department of Agriculture and Cooperation, Goverment of India: New Delhi, India, 1994.
11. National Bureau of Soil Survey and Land Use Planning (NBSS&LUP). *Global Assessment of Soil Degradation (GLASOD) Guidelines*; NBSS&LUP: Nagpur, India, 1994.
12. National Bureau of Soil Survey and Land Use Planning (NBSS&LUP). *Annual Report 2005, Nagpur*; NBSS&LUP: Nagpur, India, 2005.
13. Joshi, P.K.; Agnihotri, A.K. An Assessment of the Adverse Effects of Canal Irrigation in India. *Ind. J. Agric. Ecol.* **1984**, *39*, 528–536.
14. Parikh, K.; Ghosh, U. *Natural Resource Accounting for Soil: Towards and Empirical Estimate of Costs of Soil Degradation for India'*; Discussion Paper No. 1995, 48; Indira Gandhi Institute of Development Research, IDEAS: Mumbai, India, 1995.
15. Joshi, P.K.; Wani, S.P.; Chopde, V.K.; Foster, J. Farmers Perception of Land Degradation: A Case Study. *Econ. Polit. Wkly.* **1996**, *31*, A89–A92.

16. Srinivasarao, C.H.; Venkateswarlu, B.; Lal, R.; Singh, A.K.; Sumanta, K. Sustainable management of soils of dry land ecosystems for enhancing agronomic productivity and sequestering carbon. *Adv. Agron.* **2013**, *121*, 253–329.

17. Reddy, V.R. Land degradation in India: Extent, costs and determinants. *Econ. Polit. Wkly.* **2003**, *38*, 4700–4713.

18. Sharda, V.N.; Dogra, P.; Prakash, C. Assessment of production losses due to water erosion in rainfed areas of India. *Indian J. Soil Water Conserv.* **2010**, *65*, 79–91. [CrossRef]

19. NRSA. *National Remote Sensing Agency, IRS-Utilisation Programme: Soil Erosion Mapping*; Project Report National Remote Sensing Agency: Hyderabad, India, 1990.

20. ARPU. *Agro-Climatic Regional Planning: An Overview*; Agro-Climatic Regional Planning Unit: New Delhi, India, 1990.

21. Sehgal, J.; Abrol, I.P. *Soil Degradation in India: Status and Impact*; Oxford and IBH: New Delhi, India, 1994; p. 80.

22. Sahay, K.B. Problems of livestock population. Available online: http://www.tribuneindia.com/2000/20000411/edit.htm (accessed on 20 March 2015).

23. Sharma, K.D. Assessing the impact of overgrazing on soil erosion in arid regions at a range of spatial scales. Human impact on erosion and sedimentation. In Proceedings of the International Symposium of the Fifth Scientific Assembly of the International Association of Hydrological Sciences (IAHS), Rabat, Morocco, 23 April–3 May 1997; pp. 119–123.

24. MoEF. *National Forestry Action Programme of the Ministry of Environment and Forests*; Goverment of India: New Delhi, India, 1999; Volume 1, p. 79.

25. Sahu, H.B.; Dash, S. Land degradation due to Mining in India and its mitigation measures. In Proceedings of the Second International Conference on Environmental Science and Technology, Singapore, 26–28 February 2011.

26. Anon. *Dirty Metal, Mining Communities and Environment, Earthworks*; Oxfam America: Washington, WA, USA, 2006; p. 4.

27. Mythili, M. Intensive Agriculture and Its Impact on Land Degradation. 2013. Available online: http://coe.mse.ac.in/pdfs/coebreifs/Mythili.pdf (accessed on 13 October 2014).

28. FAO. *Land Degradation in South India; Its Severity, Causes and Effects on People*; World Soil Resources Report; Food and Agriculture Organization: Rome, Italy, 1994; Volume 78, p. 100.

29. Barman, D.; Sangar, C.; Mandal, P.; Bhattacharjee, R.; Nandita, R. Land degradation: Its Control, Management and Environmental Benefits of Management in Reference to Agriculture and Aquaculture. *Environ. Ecol.* **2013**, *31*, 1095–1103.

30. *Royal Commission on Agriculture in India Report*; Agricole Publishing Academy: New Delhi, India, 1928; pp. 75–76.

31. Maheswarappa, H.P.; Nanjappa, H.V.; Hegde, M.R.; Biddappa, C.C. Nutrient content and uptake by galangal (*Kaempferia galanga* L.) as influenced by agronomic practices as intercrop in coconut (*Cocos nucifera* L.) garden. *J. Spices Arom. Crops* **2011**, *9*, 65–68.

32. Tandon, H.L.S. Assessment of Soil Nutrient Depletion. In Proceedings of the FADINAP Regional Seminar on Fertilization and the Environment, Chiangmai, Thailand, 7–11 September 1992.

33. Tandon, H.L.S. *Fertilizers in Indian Agriculture—From 20th to 21st Century*; FDCO: New Delhi, India, 2004; p. 240.

34. Prasad, R.N.; Biswas, P.P. Soil Resources of India. In *50 Years of Natural Resource Management*; Singh, G.B., Sharma, B.R., Eds.; Indian Council of Agricultural Research: New Delhi, India, 2000.

35. Hobbs, P.; Sayre, K.; Gupta, R. The role of conservation agriculture in sustainable agriculture. *Phil. Trans. R. Soc. B* **2008**, *363*, 543–555. [CrossRef] [PubMed]

36. MNRE. *Annual Report of the Ministry of New and Renewable Energy*; Government of India: New Delhi, India, 2009.

37. NAAS. *Management of Crop Residues in the Context of Conservation Agriculture*; Policy Paper No. 58; National Academy of Agricultural Sciences: New Delhi, India, 2012; p. 12.

38. Mandal, D.; Sharda, V.N.; Tripathi, K.P. Relative efficacy of two biophysical approaches to assess soil loss tolérance for Doon Valley soils of India. *J. Soil Water Conserv.* **2010**, *65*, 42–49. [CrossRef]

39. Srinivasarao, C.H.; Venkateswarlu, B.; Lal, R. Long-term manuring and fertilizer effects on depletion of soil organic stocks under Pearl millet-cluster vean-castor rotation in Western India. *Land Degrad. Dev.* **2014**, *25*, 173–183. [CrossRef]

40. *CSWCR&TI Vision, 2030*; Vision 2030 of the Central Soil and water Conservation Research and Training Institute, Allied publisher: Dehradun, India, 2011; pp. 1–46.

41. Mandal, K.G.; Hati, K.M.; Misra, A.K.; Bandyopadhyay, K.K.; Tripathy, A.K. Land surface modification and crop diversification for enhancing productivity of a Vertisol. *Int. J. Plant Prod.* **2013**, *7*, 455–472.

42. Singh, G.; Ram, B.; Narain, P.; Bhushan, L.S.; Abrol, I.P. Soil erosion rates in India. *Ind. J. Soil Conserv.* **1992**, *47*, 97–99.

43. Sharma, N.K.; Ghosh, B.N.; Khola, O.P.S.; Dubey, R.K. Residue and tillage management for soil moisture conservation in post maize harvesting period under rainfed conditions of north-west Himalayas. *Ind. J. Soil Conserv.* **2013**, *42*, 120–125.

44. Ghosh, B.N.; Sharma, N.K.; Dadhwal, K.S. Integrated nutrient management and intercropping/cropping system impact on yield, water productivity and net return in valley soils of north-west Himalayas. *Ind. J. Soil Conserv.* **2011**, *39*, 236–242.

45. Rao, J.V.; Khan, I.A. *Research Gaps in Intercropping Systems under Rainfed Conditions in India, an On Farm Survey*; CRIDA: Hyderabad, India, 2003.

46. Pathak, P.; Mishra, P.K.; Rao, K.V.; Wani, S.P.; Sudi, R. Best opt-options on soil and water conservation. In Best Bet Options for Integrated Watershed Management, Proceedings of the Comprehensive Assessment of Watershed Programs in India (ICRISAT. Pantancheru 502 324), Andhra Pradesh, India, 23–27 July 2009; Wani, S.P., Venkateswarlu, B., Sahrawat, K.L., Rao, K.V., Ramakrishna, Y.S., Eds.; pp. 75–94.

47. Rao, K.P.C.; Steenhuis, T.S.; Cogle, A.L.; Srinivasan, S.T.; Yule, D.F.; Smith, G.D. Rainfall infiltration and runoff from an Alfisol in semi-arid tropical India. II. Tilled systems. *Soil Tillage Res.* **1998**, *48*, 61–69. [CrossRef]

48. Palanisami, K.; Suresh Kumar, D.; Chandrasekharan, B. (Eds.) *Watershed Management: Issues and Policies for 21st Century*; Associated Publishing Company: New Delhi, India, 2002; p. 341.

49. Joy, K.J.; Parnjpe, S.; Shah, A.; Badigar, S.; Lele, S. Scaling up of watershed development projects in India: learning from the first generation projects. In Proceedings of the Fourth IWMI-Tata Annual Partners Meet, International Water Management Institute, Anand, India, 24–26 February 2005; pp. 133–134.

50. Sharda, V.N. Land degradation and watershed management issues in Himalayan region: Status and Strategies. In *Workshop on "Mountain Agriculture in Himalayan Region: Status, Constraints and Potential"*; ICAR: Dehradun, India, 2011; pp. 1–22.

51. Adhikari, R.N.; Patil, A.; Raizada, D.; Ramajayam, M.; Prabhavathi, N.; Mondal, B.; Mishra, P.K. *Participatory Resource Conservation and Management in Semi-arid India—A Case Study from Netranahalli Watershed (Karnataka)*; Technical Bulletin; Central Soil & Water Conservation Research & Training Institute, Research Centre: Bellary, Karnataka, India, 2013; p. 65.

52. Natarajan, A.; Janakiraman, M.; Manoharan, S.; Kumar, A.K.S.; Vadivelu, S.; Sarkar, D. Assessment of land degradation and its impact on land resources of Sivagangai block, Tamil Nadu, India, 2010. In *Land Degradation, and Desertification: Assessment, Mitigation and Remediation*; Zdruli, P., Pagliai, M., Kapur, S., Faz Cano, A., Eds.; Springer Science: Berlin, Germany; pp. 235–252.

53. Kundu, S.; Bhattacharyya, R.; Ved-Prakash; Ghosh, B.N.; Gupta, H.S. Carbon sequestration and relationship between carbon addition and storage under rainfed soybean–wheat rotation in a sandy loam soil of the Indian Himalayas. *Soil Tillage Res.* **2007**, *92*, 87–95. [CrossRef]

54. Bhattacharyya, R.; Prakash, V.; Pandey, S.C.; Kundu, S.; Srivastva, A.K.; Gupta, H.S. Effect of fertilization on carbon sequestration in soybean-wheat rotation under two contrasting soils and management practices in the Indian Himalayas. *Aust. J. Soil Res.* **2009**, *47*, 592–601. [CrossRef]

55. Bhattacharyya, R.; Kundu, S.; Srivastva, A.K.; Gupta, H.S.; Prakash, V.; Bhatt, J.C. Long term fertilization effects on soil organic carbon pools in a sandy loam soil of the Indian Himalayas. *Plant Soil* **2011**, *341*, 109–124. [CrossRef]

56. Sharda, V.N.; Sharma, N.K.; Mohan, S.C.; Khybry, M.L. Green manuring for conservation and production in western Himalayas: 2. Effect on moisture conservation, weed control and crop yields. *Indian J. Soil Conserv.* **1999**, *27*, 31–35.

57. Bhattacharyya, R.; Tuti, M.D.; Bisht, J.K.; Bhatt, J.C.; Gupta, H.S. Conservation tillage and fertilization impacts on soil aggregation and carbon pools in the Indian Himalayas under an irrigated rice-wheat rotation. *Soil Sci.* **2012**, *177*, 218–228. [CrossRef]

58. Bhattacharyya, R.; Tuti, M.D.; Kundu, S.; Bisht, J.K.; Bhatt, J.C. Conservation tillage impacts on soil aggregation and carbon pools in a sandy clay loam soil of the Indian Himalayas. *Soil Sci. Soc. Am. J.* **2012**, *76*, 617–627. [CrossRef]

59. Bhattacharyya, R.; Pandey, S.C.; Bisht, J.K.; Bhatt, J.C.; Gupta, H.S.; Titi, M.D.; Mahanta, D.; Mina, B.L.; Singh, R.D.; Chandra, S.; *et al.* Tillage and irrigation effects on soil aggregation and carbon pools in the Indian sub-Himalayas. *Agron. J.* **2013**, *105*, 101–112. [CrossRef]

60. Mandal, B.; Majumder, B.; Bandopadhyay, P.K.; Hazra, G.C.; Gangopadhyay, A.; Samantaroy, R.N.; Misra, A.K.; Chowdhuri, J.; Saha, M.N.; Kundu, S. The potential of cropping systems and soil amendments for carbon sequestration in soils under long-term experiments in subtropical India. *Global Change Biol.* **2007**, *13*, 357–369. [CrossRef]

61. Srinivasarao, C.H.; Ravindra Chary, G.; Venkateswarlu, B.; Vittal, K.P.R.; Prasad, J.V.N.S.; Singh, S.R.S.K.; Gajanan, G.N.; Sharma, R.A.; Deshpande, A.N.; Patel, J.J.; *et al. Carbon Sequestration Strategies under Rainfed Production Systems of India*; Central Research Institute for Dryland Agriculture, Hyderabad (ICAR): Hyderabad, India, 2009; p. 102.

62. CSWCR&TI. *Annual Report of the Central Soil Water Conservation Research and Training Institute*; CSWCR&TI: Dehradun, India, 2012.

63. Kannan, K.; Khola, O.P.S.; Selvi, V.; Singh, D.V.; Moharnraj, R. *Agronomical Management Practices for Higher Productivity, Resource Use Efficiency and Farm Income in Semi-Arid Region-A Case Study in Ayalur Watershed*; Technical Bulletin; Central Soil & Water Conservation Research & Training Institute, Research Centre, Udhagamandalam: Tamil Nadu, India, 2013; p. 20.

64. Sharma, P.D.; Sarkar, A.K. *Managing Acid Soils for Enchancing Productivity*; Technical Bulletin; NRM Division, KAB-II, Pusa Campus: New Delhi, India, 2005; p. 23.

65. Fageria, N.K.; Baligar, V.C. Ameliorating soil acidity of tropical Oxisols by liming for sustainable crop production. *Adv. Agron.* **2008**, *99*, 345–399.

66. Bhat, J.A.; Mandal, B.; Hazra, G.C. Basic slag as a liming material to ameliorate soil acidity in Alfisols of sub-tropical India. *Am.-Euras. J. Agric. Environ. Sci.* **2007**, *2*, 321–327.

67. Bhat, J.A.; Kundu, M.C.; Hazra, G.C.; Santra, G.H.; Mandal, B. Rehabilitating acid soils for increasing crop productivity through low-cost liming material. *Sci. Total Environ.* **2010**, *408*, 4346–4353. [CrossRef] [PubMed]

68. Wilson, C.E.; Keisling, T.C., Jr.; Miller, D.M.; Dillon, C.R.; Pearce, A.D.; Frizzell, D.L.; Counce, P.A. Tillage influence on soluble salt movement in silt loam soils cropped to paddy rice. *Soil Sci. Soc. Am. J.* **2000**, *64*, 1771–1776. [CrossRef]

69. Garg, V.K. Interaction of tree crops with a sodic soil environment: Potential for rehabilitation of degraded environments. *Land Degrad. Dev.* **1998**, *9*, 81–93. [CrossRef]

70. Bhojvaid, P.P.; Timmer, V.R. Soil dynamics in an age sequence of *Prosopis juliflora* planted for sodic soil restoration in India. *Forest Ecol. Manage.* **1998**, *106*, 181–193. [CrossRef]

71. Gupta, R.K.; Rao, D.L.N. Potential of wasteland for sequestering carbon by reforestation. *Curr. Sci.* **1994**, *66*, 378–380.

72. Gupta, R.K.; Abrol, I.P. *Salt-Affected Soils. Their Reclamation and Management for Crop Production*; Lal, R., Stewart, B.A., Eds.; Springer: New York, NY, USA, 1990; pp. 223–288.

73. Dhanushkodi, V.; Subrahmaniyan, K. Soil management to increase rice yield in salt affected coastal soil-a review. *Int. J. Res. Chem. Environ.* **2012**, *2*, 1–5.

74. Prapagar, K.; Indraratne, S.P.; Premanandharajah, P. Effect of soil amendments on reclamation of saline-sodic soil. *Trop. Agric. Res.* **2012**, *23*, 168–176. [CrossRef]

75. Bandyopadhyay, A.K. Effect of Lime, Superphosphate, Powdered Oystershell, Rock phosphate and Submergence on Soil Properties and Crop Growth in Coastal Saline Acid Sulphate Soils of Sundarbans. In Proceedings of the International Symposium on Rice Production on Acid Soil of the Tropics, 26–30 June 1989.

76. Kumar, R.; Singh, R.D.; Sharma, K.D. Water resources of India. *Curr. Sci.* **2005**, *89*, 794–811.

77. Moukhtar, M.M.; El-Hakim, M.H.; Abdel-Mawgoud, A.S.A.; Abdel-Aal, A.I.N.; El-Shewikh, M.B.; Abdel-Khalik, M.I.I. Drainage and Role of Mole Drains for Heavy Clay Soils under Saline Water Table, Egypt. In Proceedings of the 9th ICID International Drainage Workshop, Utrecht, The Netherlands, 10–13 September 2003.

78. Abdel-Mawgoud, S.A.; El-Shewikh, M.B.; Abdel-Aal, A.I.N.; Abdel-Khalik, M.I.I. Open drainage and moling for desalinization of salty clay soils of Northeastern Egypt. In Proceedings of the 9th International Drainage Workshop, Utrecht, The Netherlands, 10–13 September 2003.

79. Mukhopadhyay, D.; Mani, P.K.; Sanyal, S.K. Effect of phosphorus, arsenic and farmyard manure on arsenic availability in some soils of West Bengal. *J. Indian Soc. Soil Sci.* **2002**, *50*, 56–61.

80. Ghosh, K.; Das, I.; Saha, S.; Banik, G.C.; Ghosh, S.; Maji, N.C.; Sanyal, S.K. Arsenic chemistry in groundwater in the Bengal Delta Plain: Implications in agricultural system. *J. Indian Chem. Soc.* **2004**, *81*, 1–10.

81. Elliot, P.; Grandner, J.; Allen, D.; Butcher, G. Completion Criteria for Alcoa of Australia Limited Bauxite Mine Rehabilitation. In Proceedings of 3rd International and 21st Annual Minerals Council of Australia Environmental Workshop, Newcastle, Australia, 14–18 October 1996.

82. Sharma, R.K.; Babu, K.S.; Chhokar, R.S.; Sharma, A.K. Effect of tillage on termites, weed incidence and productivity of spring wheat in rice-wheat system of North Western Indian plains. *Crop Prot.* **2004**, *23*, 1049–1054. [CrossRef]

83. Sen, H.S.; Oosterbaan, R.J. *Research on Water Management and Control in the Sunderbans, India*; Annual Report; ILRI: Wageningen, The Netherlands, 1992.

84. Ambast, S.K.; Sen, H.S. Integrated water management strategies for coastal ecosystem. *J. Indian Soc. Coastal Agric. Res.* **2006**, *24*, 23–29.

85. Bandyopadhyay, K.K.; Ghosh, P.K.; Hati, K.M.; Misra, A.K. Efficient utilization of limited available water in wheat through proper irrigation scheduling and integrated nutrient management under different cropping systems in a Vertisol. *J. Indian Soc. Soil Sci.* **2009**, *57*, 121–128.

86. Biswas, A.K.; Mohanty, M.; Hati, K.M.; Misra, A.K. Soil organic carbon and aggregate stability effects of applied distillery effluents on a vertisol in India. *Soil Tillage Res.* **2009**, *104*, 241–246. [CrossRef]

87. Grewal, S.S. Agroforestry systems for soil and water conservation in Shivaliks. In *Agroforestry in 2000 AD for the Semi-Arid and Arid Tropics*; NRC for Agroforestry: Janshi, India, 1993; pp. 82–85.

88. Ghosh, B.N. *Vegetative Barriers for Erosion Control in Western Himalayan Region*; CSWCR&TI: Dehradun, India, 2009; pp. 1–8.

89. Bhattacharyya, R.; Fullen, M.A.; Booth, C.A.; Kertesz, A.; Toth, A.; Kozma, K.Z.; Jakab, G.; Jankauskas, B.; Jankauskiene, G. Effeciveness of biological geotextiles on soil and water conservation in different agro-environments. *Land Degrad. Dev. Special Issue* **2011**, *22*, 495–504. [CrossRef]

90. Bhattacharyya, R.; Zheng, Y.; Li, Y.; Tang, L.; Panomtarachichigul, M.; Peukrai, S.; Dao, C.T.; Tran, H.C.; Truong, T.T.; Jankauskas, B.; *et al.* Effects of biological geotextiles on aboveground biomass production in selected agro-ecosystems. *Field Crops Res.* **2012**, *126*, 23–36. [CrossRef]

91. Fullen, M.A.; Subedi, M.; Booth, C.A.; Sarsby, R.W.; Davies, K.; Bhattacharyya, R.; Kugan, R.; Luckhurst, D.A.; Han, K.; Black, A.W.; *et al.* Utilizing biological geotextiles: Introduction to the BORASSUS Project and global perspectives. *Land Degrad. Dev. Spec. Issue* **2011**, *22*, 453–462. [CrossRef]

92. Smets, T.; Poesen, J.; Bhattacharyya, R.; Fullen, M.A.; Subedi, M.; Booth, C.A.; Kertesz, A.; Toth, A.; Szalai, Z.; Jakab, G.; *et al.* Evaluation of biological geotextiles for reducing runoff and soil loss under various environmental conditions using laboratory and field plot data. *Land Degrad. Dev. Special Issue* **2011**, *22*, 480–494. [CrossRef]

93. Bhattacharyya, R.; Fullen, M.A.; Booth, C.A. Using palm-mat geotextiles on an arable soil for water erosion control in the UK. *Earth Surf. Process. Landforms* **2011**, *36*, 933–945. [CrossRef]

94. Bhattacharyya, R.; Fullen, M.A.; Booth, C.A.; Smets, T.; Poesen, J.; Black, A. Using palm mat geotextiles for soil conservation. I. Effects on soil properties. *Catena* **2011**, *84*, 99–107. [CrossRef]

95. Bhattacharyya, R.; Fullen, M.A.; Davies, K.; Booth, C.A. Use of palm-mat geotextiles for rainsplash erosion control. *Geomorph.* **2010**, *119*, 52–61. [CrossRef]

96. Bhattacharyya, R.; Fullen, M.A.; Davies, K.; Booth, C.A. Utilizing palm leaf geotextile mats to conserve loamy sand soil in the United Kingdom. *Agric. Ecosyst. Environ.* **2009**, *130*, 50–58. [CrossRef]

97. Narain, P.; Singh, R.K.; Sindhwal, N.S.; Joshi, P. Agroforestry for soil and water conservation in the western Himalayan valley region of India. *Agrofor. Syst.* **1998**, *39*, 191–203. [CrossRef]

98. Mahmud, M.K.; Chowdhury, N.H.; Md-Manjur. Mitigation of soil erosion with Jute geotextile aided by vegetation cover: Optimization of an integrated tactic for sustainable soil conservation system (SSCS). *Glob. J. Res. Eng. Civil Struct. Eng.* **2012**, *12*, 8–14.

99. Nair, P.K.R. *An Introduction to Agro Forestry*; Kluwer Academic Publishers: Dordrecht, The Netherland, 1993.

100. Ramesh, P.; Panwar, N.R.; Singh, A.B.; Ramana, S.; Yadav, S.K.; Shrivastava, R. Status of organic farming in India. *Curr. Sci.* **2010**, *98*, 1190–1194.

101. Nair, T. India to launch a brave new initiative to save the Critically Endangered Gharial. *SPECIES–Mag. Spec. Surv. Comm.* **2011**, *21*, 53.

102. Rosenstock, T.S.; Tully, K.L.; Arias Navarro, C.; Neufeldt, H.; Butterbach Bahl, K.; Verchot, L.V. Agroforestry with N2-fixing trees: Sustainable development's friend or foe? *Curr. Opin. Environ. Sustain.* **2014**, *6*, 15–21. [CrossRef]

103. Singh, G. The Role of *Prosopis* in reclaiming high-pH soils and in meeting firewood and forage needs of small farmers. In *Prosopis: Semi-Arid Fuelwood and Forage: Tree Building Consensus for the Disenfranchised*; US National Academy of Science: Washington, DC, USA, 1996.

104. Mishra, A.; Sharmaa, S.D.; Pandeyb, R. Amelioration of degraded sodic soil by afforestation. *Arid Land Res. Manag.* **2004**, *18*, 13–23. [CrossRef]

105. Kaur, B.; Gupta, S.R.; Singh, G. Soil carbon, microbial activity and nitrogen availability in agroforestry systems on moderately alkaline soils in Northern India. *Appl. Soil Ecol.* **2000**, *15*, 283–294. [CrossRef]

106. Basavaraja, P.K.; Sharma, S.D.; Dhananjaya, B.N.; Badrinath, M.S. *Acacia nilotica*: A tree species for amelioration of sodic soils in Central dry zone of Karnataka, India. In Proceedings of the 19th World Congress of Soil Science, Soil Solutions for a Changing World, Brisbane, Australia, 1–6 August 2010; pp. 73–76.

107. Mishra, A.; Sharma, S.D.; Khan, G.H. Improvement in physical and chemical properties of sodic soil by 3, 6, and 9 year-old plantations of *Eucalyptus tereticornis* bio-rejuvenation of soil. *For. Ecol. Manag.* **2003**, *184*, 115–124. [CrossRef]

108. Singh, A.; Kaur, J. Impact of conservation tillage on soil properties in rice wheat cropping system. *Agric. Sci. Res. J.* **2012**, *2*, 30–41.

109. Food and Agriculture Organization (FAO). *"Climate-Smart" Agriculture. Policies, Practices and Financing for Food Security, Adaptation and Mitigation*; FAO: Rome, Italy, 2010; p. 41.

110. Bhattacharyya, R.; Das, T.K.; Pramanik, P.; Ganeshan, V.; Saad, A.A.; Sharma, A.R. Impacts of conservation agriculture on soil aggregation and aggregate-associated N under an irrigated agroecosystem of the Indo-Gangetic Plains. *Nutr. Cycl. Agro-Ecosyst.* **2013**, *96*, 185–202. [CrossRef]

111. VPKAS. *Annual Report of the Vivekananda Parvatiya Krishi Anusandhan Sansthan*; Vikas Publisher: Almora, India, 2011.

112. Bhattacharyya, R.; Kundu, S.; Pandey, S.; Singh, K.P.; Gupta, H.S. Tillage and irrigation effects on crop yields and soil properties under rice-wheat system of the Indian Himalayas. *Agric. Water Manag.* **2008**, *95*, 993–1002. [CrossRef]

113. Jat, M.L.; Saharawat, Y.S.; Gupta, R. Conservation agriculture in cereal systems of South Asia: Nutrient management perspectives. *Karnataka J. Agric. Sci.* **2011**, *24*, 100–105.

114. RWC-CIMMYT. *Addressing Resource Conservation Issues in Rice-Wheat Systems of South Asia: A Resource Book*; Rice-wheat consortium for the Indo-Gangetic Plains–International Maize and Wheat Improvement Centre: New Delhi, India, 2003; p. 305.

115. Gathala, M.K.; Ladha, J.K.; Saharawat, H.S.; Kumar, V.; Sharma, P.K. Effect of tillage and crop establishment methods on physical properties of a medium-textured soil under a seven-year rice–wheat rotation. *Soil Sci. Soc. Am. J.* **2011**, *75*, 1851–1862. [CrossRef]

116. Hobbs, P.R.; Gupta, R.K. Sustainable resource management in intensively cultivated irrigated rice-wheat cropping systems of Indo-Gangetic plains of south Asia: Strategies and options. In Proceedings of the International Conference on Managing Natural Resources for Sustainable Production in 21st Century, New Delhi, India, 14–18 February 2000; Singh, A.K., Ed.; Indian Society of Soil Science: New Delhi, India; pp. 584–592.

117. Ladha, J.K.; Kumar, V.; Alam, M.M.; Sharma, S.; Gathala, M.K.; Chandna, P.; Saharawat, Y.S.; Balasubramanian, V. Integrating crop and resource management technologies for enhanced productivity, profitability and sustainability of the rice–wheat system in South Asia. In *Integrated Crop and Resource Management in the Rice–Wheat System of South Asia*; Ladha, J.K., Singh, Y., Erenstein, O., Hardy, B., Eds.; IRRI: Los Banos, The Philippines, 2009; pp. 69–108.

118. Das, T.K.; Bhattacharyya, R.; Sudhishri, S.; Sharma, A.R.; Saharawat, Y.S.; Bandyopadhyay, K.K.; Sepat, S.; Bana, R.S.; Aggarwal, P.; Sharma, R.K.; *et al.* Conservation agriculture in an irrigated cotton–wheat system of the western Indo-Gangetic Plains: Crop and water productivity and economic profitability. *Field Crops Res.* **2014**, *158*, 24–33. [CrossRef]

119. Jat, M.L.; Gathala, M.K.; Saharawat, Y.S.; Terawal, J.P.; Gupta, R.; Singh, Y. Double no-till and permanent raised beds in maize-wheat rotation of north-western Indo-Gangetic plains of India: Effects on crop yields, water productivity, profitability and soil physical properties. *Field Crops Res.* **2013**, *149*, 291–299. [CrossRef]

120. Das, T.K.; Bhattacharyya, R.; Sharma, A.R.; Das, S.; Pathak, H. Impacts of conservation agriculture on total soil organic carbon retention potential under an irrigated agro-ecosystem of the western Indo-Gangetic Plains. *Eur. J. Agron.* **2013**, *51*, 34–42. [CrossRef]

121. Mishra, A.K.; Aggarwal, P.; Bhattacharyya, R.; Das, T.K.; Sharna, A.R.; Singh, R. Least limiting water range for two conservation agriculture cropping systems in India. *Soil Tillage Res.* **2015**, *150*, 43–56. [CrossRef]

122. Hati, K.M.; Chaudhary, R.S.; Mohanty, M.; Singh, R.K. Impact of conservation tillage on soil organic carbon content, its distribution in aggregate size fractions and physical attributes of Vertisols. In *IISS Contribution in Frontier Areas of Soil Research*; Kundu, S., Manna, M.C., Biswas, A.K., Chaudhary, R.S., Lakaria, B.L., Subba Rao, A., Eds.; Indian Institute of Soil Science: Bhopal, India, 2013; pp. 187–200.

123. Jat, R.A.; Wani, S.P.; Sahrawat, K.L. Conservation Agriculture in the Semi-Arid Tropics: Prospects and Problems. *Adv. Agron.* **2012**, *117*, 191–273.

124. Gathala, M.K.; Kumar, V.; Sharma, P.C.; Yashpal, S.; Saharawat, H.S.; Jat, M.S.; Kumar, A.; Jat, M.L.; Humphreys, E.; Sarma, D.K.; *et al.* Optimizing intensive cereal-based cropping systems addressing current and future drivers of agricultural change in the northwestern Indo-Gangetic plains of India. *Agric. Ecosyst. Environ.* **2013**, *177*, 85–97. [CrossRef]

sustainability

MDPI

Review

North American Soil Degradation: Processes, Practices, and Mitigating Strategies

R. L. Baumhardt [1,†,*], **B. A. Stewart** [2,†] and **U. M. Sainju** [3,†]

1   USDA-Agricultural Research Service, Conservation and Production Research Lab., P.O. Drawer 10, Bushland, TX 79012, USA

2   Dryland Agriculture Institute, West Texas A&M University, P.O. Box 60278, Canyon, TX 79016, USA; bstewart@wtamu.edu

3   USDA-Agricultural Research Service, Northern Plains Agricultural Research Lab., 1500 North Central Avenue, Sidney, MT 59270, USA; Upendra.Sainju@ars.usda.gov

*   Author to whom correspondence should be addressed; R.Louis.Baumhardt@ars.usda.gov; Tel.: +1-806-356-5766; Fax: +1-806-356-5750.

†   These authors contributed equally to this work.

Academic Editor: Marc A. Rosen

Received: 14 November 2014; Accepted: 27 February 2015; Published: 11 March 2015

Abstract: Soil can be degraded by several natural or human-mediated processes, including wind, water, or tillage erosion, and formation of undesirable physical, chemical, or biological properties due to industrialization or use of inappropriate farming practices. Soil degradation occurs whenever these processes supersede natural soil regeneration and, generally, reflects unsustainable resource management that is global in scope and compromises world food security. In North America, soil degradation preceded the catastrophic wind erosion associated with the dust bowl during the 1930s, but that event provided the impetus to improve management of soils degraded by both wind and water erosion. Chemical degradation due to site specific industrial processing and mine spoil contamination began to be addressed during the latter half of the 20th century primarily through point-source water quality concerns, but soil chemical degradation and contamination of surface and subsurface water due to on-farm non-point pesticide and nutrient management practices generally remains unresolved. Remediation or prevention of soil degradation requires integrated management solutions that, for agricultural soils, include using cover crops or crop residue management to reduce raindrop impact, maintain higher infiltration rates, increase soil water storage, and ultimately increase crop production. By increasing plant biomass, and potentially soil organic carbon (SOC) concentrations, soil degradation can be mitigated by stabilizing soil aggregates, improving soil structure, enhancing air and water exchange, increasing nutrient cycling, and promoting greater soil biological activity.

Keywords: soil erosion; compaction; salinization

---

1. Introduction

Soil degradation describes ongoing processes that generally limit agronomic productivity, result in undesirable or deteriorating physical, chemical or biological properties, enhance soil displacement due to wind or water driven erosion [1], and require reassignment of land resources. Soil degradation often interacts with terrain and climatic factors defining an ecosystem to reduce sustainable land productivity, which, eventually, threatens food security. Common examples of chemical and physical *in-situ* soil degradation include compaction (due to heavy machinery or repeated tillage operations), systematic loss of aggregate stabilizing soil organic matter (SOM), and soil salinization or acidification as a result of problematic drainage, nitrification, or chemical contamination. The greatest soil degradation threat, however, is wind- or water-induced erosion that displaces soil and depresses land productivity, and results in deteriorated physical properties, nutrient losses, and reshaped, potentially unworkable, field surface conditions. Both *in situ* deterioration and soil erosion are frequently a consequence of using unsuitable management practices because soil resource and climatic constraints are not well understood. A classic example in the semiarid Great Plains was the 1930s Dustbowl.

Two other human-induced causes of *in-situ* soil degradation and its resultant reduction in land productivity are industrial dislocation through mining operations and urban sprawl. The latter usually imposes no chemical or physical deterioration, but typically results in the irreversible reassignment of land resources for construction of housing and infrastructure as necessitated by population growth and related commerce. In the U.S., urban land use has increased by 400% from 6 to 24 million ha since 1945; however, this only accounts for ~3% of total land resources [2]. A larger critical issue associated with urban sprawl is that the continued expansion of infrastructure, such as interstate highway development currently exceeding 75,000 km, can promote suburban growth and results in agricultural land losses at a rate of ~120 ha for each added kilometer of interstate [3]. Soil degradation by land reassignment for urban growth is beyond the scope of this article, but redevelopment of existing urban land can conserve soil resources and have multiple additional benefits of rectifying traffic congestion and crime. For example, in 1982, decades after its 1890 establishment, Lubbock, Texas, redeveloped dilapidated and abandoned housing of one original ~130 ha residential area to partially meet housing needs of the growing ~300,000 population. That necrotic urban area, once occupied by 2% of the municipal population with 28% of the crime [4], now benefits from decreased crime, e.g., ~90% fewer burglaries than in 1983. Urban sprawl was delayed and Lubbock taxable property values increased from pre-redevelopment $27 million to $750 million upon eventual completion [5].

Mining to extract minerals, coal, or oil and gas is common throughout North America with methods that vary from open pits, as used for oil sands in Canada, to mountain top removal for coal from some Appalachian states [6]. In 2007, mined areas of the contiguous U.S. were included in ~27 million ha of miscellaneous land or 3% of the total land [2], but an earlier listing of mining activities by Lal *et al.* [7] estimated the disturbed area to be 4.4 million ha. Surface mining regulations in most of North America require topsoil and spoil reclamation to reverse soil degradation and approximate pre-mining conditions; however, the U.S. "Comprehensive Environmental Response, Compensation and Liability Act" of 1980 or Superfund targeted cleanup of related hazardous waste sites [8]. Superfund sites are replete with abandoned mineral mining and smelting locations [9] that introduce acidic water contaminated with various heavy metals into streams and the surrounding soil. Herron *et al.* [10] described successful site remediation that integrated multiple steps ending with revegetated soil caps protected by runoff diversion ditches for rainfall management (Figure 1). Mine related soil degradation also affects remote locations after land application of contaminated sediments that render treated land difficult to revegetate without amendments to correct reduced soil conditions and contaminant solubility [11].

Although soil resources can be degraded in many ways, our goal is to examine the problem from an agronomic perspective. The history of soil degradation in North America includes the catastrophic wind erosion during the 1930s U.S. Dust Bowl [12] and followed devastating water erosion in the southeastern U.S. referred to in 1910, for one example, as the Badlands of Mississippi [13]. Nevertheless, human recognition of soil degradation is very slow as evidenced by the 1909 U.S. Bureau of Soils Bulletin 55 described the soil resource as an "indestructible, immutable asset" [14]. That perception of the soil resource coupled with unsuitable production methods implementing repeated tillage to promote greater rain infiltration and, for semiarid production, to develop an evaporation limiting dust mulch [15] led to massive soil erosion losses during the Dust Bowl. Soil salinization as a result of irrigation together with compaction and the reduction of organic carbon due to tillage management practices represent consequences of still other agents that have degraded the soil productivity.

Figure 1. Mine spoil mitigation after installing soil cap that is protected from further contamination by stormwater runoff using collection and diversion channels.

Achieving the goal of sustainable management practices that remediate or prevent soil degradation requires a better understanding of interacting environmental conditions, production methods, and land resources. Soil degradation has been the topic of comprehensive reviews for over a quarter century and many correlated these interacting factors. Admittedly the nature of soil management and degradation is site-specific, but we submit that almost globally universal soil degradation agents will lead to common best management practices. Therefore, our objective is to highlight problematic process agents and successful integrated management solutions for mitigating and restoring the soil resources such that a management perspective meeting mutual soil stewardship goals may emerge.

2. Processes and Practices Associated with Soil Degradation

2.1. Tillage

The primary purpose of agriculture is to secure food resources, which has long relied on efforts to advance agricultural technology, ranging from preparing a seed bed with tillage sticks for improved soil contact to applying water by irrigation for stabilized crop production. In North America, advances in tillage technology can be traced to Thomas Jefferson's 1784 soil inverting "moldboard plow" design that John Deere produced and marketed during the 1830s [16]. Tillage was historically considered a beneficial practice that was generally considered necessary for weed control, preparing an ideal seedbed, and for increasing water infiltration. Good tillage was associated with good farming and became a revered part of the culture of agriculture [17] as exemplified by the seal of the U.S. Department of Agriculture contains a picture of a moldboard plow.

Although not mentioned as a benefit, tillage generally increased soil fertility by hastening the decomposition of soil organic matter (SOM) so N, P, K, S, and other nutrients required for plant growth are mineralized to forms that are readily available for use by plants. In contrast to the positive benefits that tillage has for soil fertility, the accelerated SOM loss ultimately contributes to increased soil erosion, loss of soil structure, decreased biological activity, and other factors that lower soil quality.

Tillage technology advanced rapidly in response to farm mechanization, including internal combustion powered tractors in the 1920s [18]. That is, when farm mechanization eliminated the need for animal traction it concurrently eliminated demand for pasture and forage production supporting the displaced draft animals. In lieu of the draft animal limits imposed on cultivated area that an individual farmer could manage, mechanization greatly expanded the amount of land exposed to soil degradation through tillage.

2.2. Degradation of Soil Organic Matter

The inherent amount of SOM in soils varies greatly depending on soil texture and environmental conditions. Agricultural soils in North America at the time they were developed from grass prairies or forestland had SOM concentrations ranging from ~1% to 10% (*w/w*). Although variable, the slowly decomposable portion of SOM, often called humus, contains 58% C and has a C/N ratio of 12/1, a C/P ratio of 50/1, and a C/S ratio of 70/1 [19]. Himes [19] estimated only 35% of the C in crop residues returned to soil was sequestered in the soil as humus. This is considerably higher than the 17% to 18% determined by Rasmussen and Albrecht [20] for wheat (*Triticum aestivum* L.) residues in dryland soils in Oregon but similar to the 35% they found for manure. Rasmussen and Albrecht [20] also showed in a 30-year study that soil organic matter declined under annual cropping of wheat even when N fertilizer was added, but the decline was not as great with N additions.

These studies [19,20] show why it is important to understand that while SOM is largely comprised of organic C, SOM also contains other elements such that the sequestration of C as SOM requires the simultaneously sequesters other elements. While the decomposition of SOM results in a loss of C from the system as CO_2, the other elements (*i.e.*, N, P, and S) are not immediately lost from the system. These elements not only become available for plant uptake and removal with harvested products, but also for potential loss through erosion and leaching depending on the particular element, and by denitrification in the case of N. Thus, sufficient nutrients may not be available for restoring soil organic matter when increasing C sources. Using the estimates of Himes [19], the decomposition of 1% soil organic carbon (SOC) from the top 15 cm of a soil would result in a loss of 12,992 kg ha^{-1} C from the soil as CO_2, but 1082 kg of N, 260 kg of P, and 186 kg of S would have been converted from organic compounds to inorganic compounds.

The SOM content of many soils in North America is only about 50% of the level present at the time they were converted from forests or prairies to farm lands. Many forest lands in the U.S. contained 6% (*w/w*) or greater SOM and the grasslands of the U.S. and Canada contained from 1% to more than 6% SOM depending on the texture and environment. Cultivation increases soil aeration that accelerates biological activity resulting in rapid losses of C as CO_2 and mineralization of N, P, K and other plant nutrients. Therefore, the degradation of SOM was responsible for supplying the more than ample amounts of N, P, S and other nutrients needed for crop production when North America was settled.

One major management practice that limits SOM in semiarid dryland cropping systems on the Great Plains is the use of fallowing to conserve soil water, control weeds, release plant nutrients, and increase succeeding crop yields [21]. Fallowing limits the amount of crop residue produced and returned to the soil. Additionally, SOM mineralization may be enhanced by greater microbial activity as a result of increased soil water and temperature [22]. In the northern Great Plains, conventional tillage (CT) of the wheat-fallow (WF) system has resulted in a decline of SOM by 30% to 50% of their original levels in the last 50 to 100 year [23]. Because fallowing also reduces annualized crop yields, Aase and Schaefer [24] concluded that the system had become inefficient, unsustainable, and uneconomical. A 30-year dryland cropping system study conducted to quantify the effects of tillage, cropping, and fallow on SOC to a depth of 120 cm in Culbertson, MT showed that most of the response was observed in the surface 0–7.5 cm layer and that tillage did not influence SOC (Figure 2). Conventional tillage with a spring wheat-fallow system (CT-WF), however, reduced SOC at 0–7.5 cm by 25% to 30% compared to continuous spring wheat (CW) under CT or no-till (NT), management or CT-CW and NT-CW, respectively (Figure 2A). The yearly rate of SOC decline within the 0–7.5 cm layer was almost double in CT-WF compared with in NT-CW and CT-CW (Figure 2B). Alternate-year fallow, therefore, can reduce SOC in the surface layer more rapidly than tillage in dryland cropping systems in the northern Great Plains. A similar study by Rasmussen and Albrecht [20] showed that SOM levels in dryland fallow systems could be increased by adding manure except when fallow was included in the rotation.

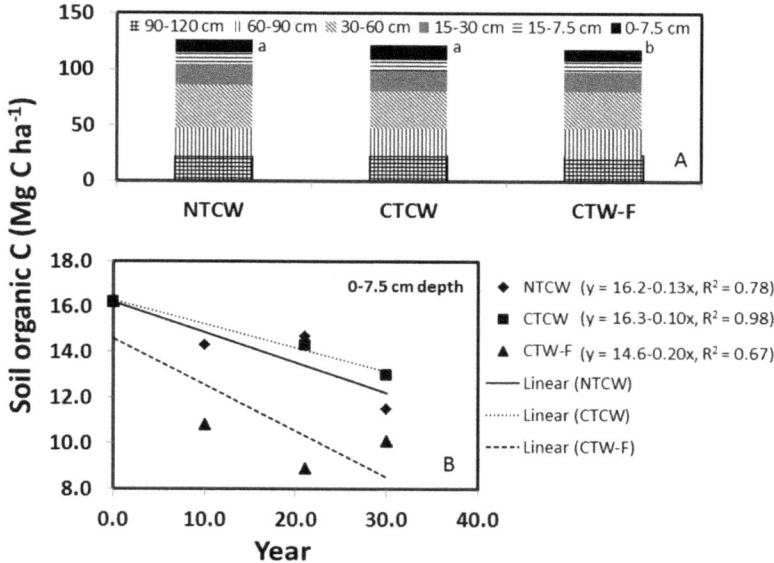

Figure 2. Effect of thirty years of tillage and cropping sequence on (**A**) soil organic C (SOC) contents at the 0–120 cm and (**B**) decline of SOC at 0–7.5 cm with year in dryland cropping systems in a field site, 10 km north of Culbertson, MT. NT-CW denotes no-till continuous spring wheat; CT-CW, conventional till continuous spring wheat; and CT-WF, conventional till spring wheat-fallow. Numbers followed by different letters at a depth in the bar are significantly different at $P = 0.05$ by the least square means test.

2.3. Degradation of Soil Physical Properties

Degradation of soil physical properties is closely linked to the loss of SOM because it serves as the glue to hold soil particles together to form aggregates. Aggregates provide structure that makes soils more resistant to erosion and compaction and increases the amount of plant available water they can hold. Hudson [25] showed that the volume of water held at field capacity decreased 3.6% (*v/v*) for each 1% decline in SOM. In all texture groups, decreasing the SOM content from 3.0% to 0.5% subsequently decreased the plant available water capacity by more than 50%. The loss of SOM also decreased the infiltration rate so that runoff increased, particularly during high intensity precipitation events. This resulted in water erosion as well as storing less water in the soil profile for plant use. The loss of SOM also makes the soils more vulnerable to wind erosion because individual soil particles are smaller and much more subject to erosion than aggregates.

Tillage incorporates plant residues but also disrupts aggregates, exposes new soil to wet-dry and freeze-thaw cycles, and affects microbial communities [26]. Tillage is more disruptive of larger aggregates, making SOC and soil N from larger aggregates more susceptible to mineralization [27,28]. Because particulate organic matter (POM) found in large aggregates is the main substrate for microorganisms, reduction in POM due to tillage can severely reduce soil aggregation [27,28]. Sainju *et al.* [29] concluded that fallowing reduces soil aggregation compared to continuous cropping by decreasing the amount of crop residue returned to the soil and by increasing soil organic matter mineralization due to enhanced microbial activity.

Soil organic matter is crucial to soil productivity because it affects soil physical, chemical and biological properties [23,27,30] including bulk density, aggregation, water holding and infiltration capacities, carbon sequestration, nutrient cycling, and microbial biomass and activity. Conventional tillage reduces SOM by disturbing soil and increasing aeration, which subsequently increases mineralization of SOC and soil organic nitrogen (SON) formed after incorporating crop residue [27]. Because of residue incorporation, Clapp *et al.* [31] observed that SOM level in Minnesota can occasionally be higher in subsurface than surface layers under conventional tillage compared with no-tillage.

Crop selection also influences soil aggregation. Soils from fields with legumes or bare soil will generally have smaller aggregates than from fields with non-legume vegetative cover. This occurs because lower amounts of crop residue are generally being returned to the soil and due to variation in the C/N ratio of residue. Crop residues with low C/N ratio decompose more rapidly than those with higher C/N ratio [32]. Sainju *et al.* [32] also noted that soil aggregation can also be lower in the surface than subsurface soils. Monoculture cropping systems can also reduce soil aggregate stability compared with diversified crop rotations [33]. Residue removal can also reduce soil aggregation, aggregate stability, macroporosity, aeration, and water infiltration compared with nonremoval [34].

Continuous monocropping can reduce crop yields due to greater disease and pest inoculum [35] that may, consequently, reduce the amount of residue returned to the soil for SOM [30]. The overall effect of N fertilization on SOM varies from increased levels due to greater biomass production and residue returned to the soil [36] to similar or decreased SOM due to increased mineralization as a result of reduced C/N ratio [37]. Removing residue for bioenergy production and by burning can seriously reduce SOM, since 5.2 to 12.5 Mg ha^{-1} of residue is needed to maintain SOC, depending on soil and climatic conditions [34,38].

2.4. Soil Degradation through Wind and Water Erosion

Wind and water erosion in North America increased rapidly with the expansion of cropland. In response, one of the most effective conservationists who sought to build public concern regarding soil erosion was Hugh Hammond Bennett. Often referred to as the "father of soil conservation," Bennett co-authored the highly influential publication entitled "Soil Erosion: A National Menace" [39] that influenced Congress to create the first federal soil erosion experiment stations in 1929 [40]. With the election of Franklin D. Roosevelt as President in 1932, conservation of soil and water became a national priority in the New Deal administration. The Soil Erosion Service was established in the Department of Interior in September 1933 with Bennett as Chief. The Soil Erosion Service established water erosion demonstration projects in critically eroded areas across the country to show landowners the benefits of conservation. Bennett's ability to influence public opinion is often illustrated by his effectiveness in getting support from the U.S. Congress. Beginning in 1932, persistent drought conditions throughout the Great Plains caused widespread crop failures resulting in serious wind erosion. A large dust storm on 11 May 1934 swept fine soil particles over Washington, D.C. and three hundred miles out into the Atlantic Ocean. More intense and frequent storms swept the Plains in 1935. On 6 March and again on 21 March, dust clouds passed over Washington and darkened the sky just as Congress commenced hearings on a proposed soil conservation law. Bennett seized the opportunity to explain the cause of the storms and to offer a solution. He penned editorials and testified to Congress urging the creation of a permanent soil conservation agency. The result was the Soil Conservation Act (PL 74-46), which President Roosevelt signed on 27 April 1935, creating the Soil Conservation Service (SCS) in the USDA [40]. In 1994, Congress changed SCS's name to the Natural Resources Conservation Service (NRCS) to better reflect the broadened scope of the agency's concerns.

Water erosion is dominant in the eastern portion of North America because of higher precipitation, but in the central and western areas where precipitation is lower and wind speeds are higher, wind erosion dominates. Figure 3 illustrates the areas in the U.S. where wind and water erosion rates are high enough that control practices are needed to minimize or prevent soil degradation.

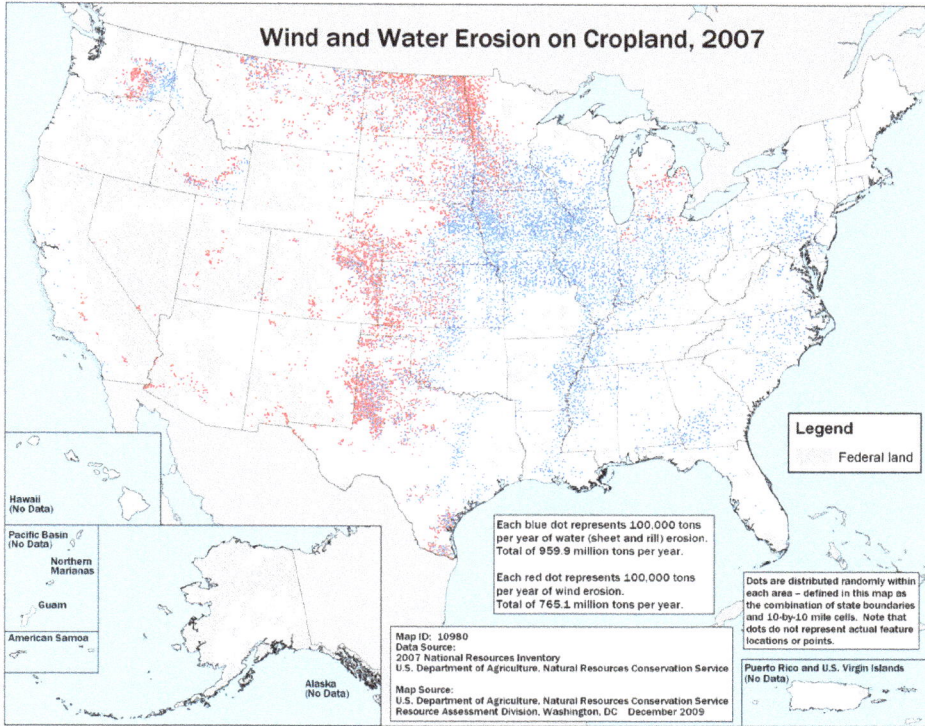

Figure 3. Distribution and amounts of water and wind erosion in the U.S. [41].

Soil degradation due to either wind or water is inextricably linked to loss of SOM. As a result, SOM in many North American soils have decreased by 30% to 40% [42], but with few exceptions, most croplands have been in crop production long enough that new equilibriums have been reached and SOM levels are no longer decreasing. In fact, several soils are showing some SOM increase, particularly where large amounts of crop residues are produced and limited or no tillage is used.

Soil organic matter consists of different pools of carbon compounds and some are considerably more active meaning that they are more amenable to decomposition processes. The major pools of SOM in virgin soils are often separated into plant residues, active SOM, and passive SOM as described by Brady and Weil [42]. Increases in SOM thus reflect greater amounts of C being added to the soil through larger root systems and more plant residue; two critical inputs causing SOC to reach a new equilibrium or even increase when tillage intensity is reduced. Finally, the large losses of SOM in North American soils, since they were converted to croplands, have reduced their inherent quality and productivity; crop yields on these soils have markedly increased because other factors (*i.e.*, improved cultivars, weed control, insect control, commercial fertilizers, and seeding methods) have more than offset SOM losses due to soil degradation.

2.5. Chemical Degradation

Chemical soil degradation can occur in response to various processes. Three principal consequences include: nutrient depletion, acidification, and salinization that are often associated with agricultural production systems. A fourth, contamination by heavy metals, industrial wastes, or radioactive material can be important, but is outside the scope of this contribution.

2.5.1. Nutrient Depletion

Declining SOM that depresses N mineralization will concomitantly decrease availability of P, K, and other nutrients and, for intensive crop production, increase dependence on fertility management or rotation alternatives depending on tillage. For example, Sainju *et al.* [29] reported that SOC, SON, and potential N mineralization were lower in CT-WF than NT-CW and CT-CW after 21 year in dryland cropping systems in eastern Montana although NO_3-N content was higher. Long-term nonlegume monocropping reduced N mineralization compared to crop rotation containing legumes and nonlegumes [43]. In other, long term rotation studies with sorghum (*Sorghum bicolor* (L.) Moench) and corn (*Zea mays* L.) summer crops, CT reduced soil Bray-P and cation exchange capacity at 0–5 cm compared with NT after 27 year under dryland spring wheat-sorghum/corn-fallow in Nebraska [44]. After 30 year of continuous wheat in Montana, tillage did not influence soil chemical properties (NT-CW *vs.* CT-CW) at 0–7.5 and 7.5–15 cm depth (Table 1) compared with the less intensively cropped CT-WF that, except for Ca and Mg, had generally lower values at 0–7.5 cm. The amount of nutrients removed through grain harvest can be higher in NT-CW and CT-CW than CT-WF due to increased cropping intensity and annualized yield [29,45]. Studies from the U.S. Corn Belt have shown that removing the residue for hay or bioenergy can have a similar adverse effect on soil fertility because residues contain plant nutrients that if not replaced have been shown to decrease crop yields [46] by as much as 1.8 to 3.3 Mg ha^{-1} after 50% to 100% straw removal [34].

Table 1. Effect of 30 years of tillage and cropping sequence combination on soil chemical properties under dryland spring wheat system in a field site, 10 km north of Culbertson, MT.

Chemical property	Soil Depth (cm)	Treatment †		
		NT-CW	CT-CW	CT-WF
Olsen-P (mg kg^{-1})	0–7.5	36.8 a ‡	40.0 a	25.0 b
	7.5–15	2.8 a	5.5 a	4.9 a
K (mg kg^{-1})	0–7.5	331 a	331 a	272 b
	7.5–15	279 a	282 a	186 b
Ca (mg kg^{-1})	0–7.5	989 b	894 b	1294 a
	7.5–15	1597 a	1606 a	2359 a
Mg (mg kg^{-1})	0–7.5	212 b	193 b	253 a
	7.5–15	340 b	350 b	433 a
Na (mg kg^{-1})	0–7.5	14.5 a	14.8 a	12.4 b
	7.5–15	14.4 a	14.3 a	15.5 a
SO$_4$-S (mg kg^{-1})	0–7.5	6.8 a	6.3 a	8.0 a
	7.5–15	3.5 a	3.5 a	8.1 a
Cation exchange capacity (cmol$_c$ kg^{-1})	0–7.5	14.3 a	14.5 a	11.9 b
	7.5–15	11.6 a	12.9 a	15.9 a

† Treatments are NT-CW, no-till continuous spring wheat; CT-CW, conventional till continuous spring wheat; and CT-WF, conventional till spring wheat-fallow; ‡ Values within a row followed by the same letter are not significantly different at $P = 0.05$ according to the least square means test.

2.5.2. Acidification

Replacing essential plant nutrients, which are no longer available because of SOM depletion, by applying NH_4-based fertilizers, can degrade soil by increasing acidity during hydrolysis that releases H ions [47]. Chen *et al.* [48] showed that N sources have different effects on soil acidity and ranked common fertilizer materials in the order $(NH_4)_2SO_4 > NH_4Cl > NH_4NO_3 >$ anhydrous $NH_3 >$ urea. Soil degradation through increasing acidity depresses the efficacy of subsequent fertilizer applications for sustaining crop yields [49], thereby resulting in inefficient use of fertilizers [43]. The long-term, 30 years, application of N fertilizer progressively reduced the 0–7.5 cm soil pH as cropping sequence intensified from WF to CW from an initial pH of 6.5 to 5.5 in CT-WF and 5.0 in NT-CW and CT-CW for dryland production in the northern Great Plains (Table 2). Likewise, tillage indirectly affects soil acidity as a result of enhanced soil water conservation using NT compared with CT that increases crop yields, the amount of required N fertilizer, and the removal of basic cations in harvested grain and biomass [43,50]. Soil acidification as a consequence of increased fertilization to intensify cropping systems productivity, as noted for the northern Great Plains, may exemplify an acceptable self-perpetuating production risk that requires additional neutralizing amendments as precipitation increases to the east. In contrast, this acidification provides a benefit for calcareous soils common to western North America.

Table 2. Effect of 30 years of no tillage (NT) and conventional tillage (CT) residue management with either continuous spring wheat (CW) or wheat-fallow (WF) cropping sequences on soil pH and bulk density at various soil depths for a field site 10 km north of Culbertson, MT.

Tillage and Cropping sequence	pH at the soil depth					
	0–7.5 cm	7.5–15 cm	15–30 cm	30–60 cm	60–90 cm	90–120 cm
NT-CW	5.33 ab †	6.50 ab	7.60 a	8.35 a	8.58 a	8.75 a
CT-CW	5.05 b	6.15 b	7.58 a	8.25 b	8.63 a	8.70 a
CT-WF	5.73 a	7.03 a	7.65 a	8.25 a	8.50 a	8.66 a
	Bulk density (Mg m^{-3}) at the soil depth					
	0–7.5 cm	7.5–15 cm	15–30 cm	30–60 cm	60–90 cm	90–120 cm
NT-CW	1.15 b †	1.48 a	1.49 a	1.67 a	1.52 a	1.64 a
CT-CW	1.26 b	1.38 a	1.43 a	1.55 a	1.51 a	1.68 a
CT-WF	1.45 a	1.48 a	1.53a	1.62 a	1.60 a	1.70 a

† Common parameter values within columns followed by the same letters are not significantly different at $P = 0.05$ according to the least square means test.

2.5.3. Salinization

Accumulating salts, including sodium, represents another problematic type of chemical soil degradation in North America that affects agronomic production, albeit limited to ~1% of the total land area [51]. A combination of geological, climatic, and cultural practices including cropping systems affect the development of saline seeps in some 800,000 ha of non-irrigated land in the northern Great Plains [52]. Seeps form when precipitation not used by plants moves below the root zone through the salt-laden substrata to impermeable layers and eventually flows from the recharge area to depressions where water evaporates leaving salt deposits enriched in Na, Ca, Mg, SO_4-S, and NO_3-N that retard crop growth [53]. In Canada, diversion of surface drainage from recharge areas and intensifying cropping systems to consume precipitation are recommended for mitigating management practice dependent "secondary salinity" problems [54]. Secondary salinity resulting from irrigation to supply part of the crop water use permitted intensification of cropping systems on arid and semi-arid land. This intensified production on ~7.5% of US farm land produced 55% of domestic crop value [55], but Postel [56] noted that salinity affected ~23% of that irrigated land. Where sufficient salt is applied to reduce crop yield, irrigation may be a "Faustian Bargain" degrading soil and requiring corrective management intervention, such as leaching or alternate crop selection.

3. Mitigation Strategies for Reversing Soil Degradation

The number of site-specific management strategies to mitigate degraded or degrading soil is diverse, but when considered from a broad perspective on potential solutions converge to a limited paradigm. Physical and chemical soil degradation through erosion, compaction, and acidification are commonly connected by absent biomass cover and declining soil organic matter as a result of tillage or moderated crop production. That is, residue preservation with reduced or no tillage is an avenue to increase soil organic matter while protecting soil from the erosion processes and mitigating soil compaction. Intensified cropping systems, likewise, increase biomass for greater soil organic matter to stabilize aggregates and render the soil less susceptible to erosion. The common management perspective for improving soil health, quality, and productivity is to reverse soil degradation by using residue retaining tillage practices and, where possible, intensifying cropping systems within rotations or by added cover crops.

Tillage and soil compaction also express a wide variety of site-specific interactions where more intensive cropping sequences offset tillage related compaction. For example, even though soil compaction generally increases as the frequency of conventional tillage increases, data from the Central Great Plains has shown that even in the absence of soil disturbing tillage (*i.e.*, no-tillage) compaction can increase as a result of soil consolidation during routine farm operations [53]. Another study in eastern Montana, comparing CT and NT after 30 year under dryland continuous wheat (CW) or wheat fallow (WF) cropping sequences showed that soil bulk density within the 0 to 7.5 cm depth increment was not different between NT-CW and CT-CW (Table 2). However, the bulk density was 13% to 21% greater for the same depth increment in CT-WF (1.45 Mg m^{-3}) than NT-CW and CT-CW (1.15 to 1.26 Mg m^{-3}). One reason suggested for this response was reduced root growth and lower soil organic C input (Figure 2).

3.1. Tillage Management

Moldboard plowing, which was the dominant tillage system for many years, buries essentially all plant residues beneath the soil surface. Conservation tillage was defined in 1984 by the USDA Soil Conservation Service (currently Natural Resources Conservation Service) as "any tillage system that maintains at least 30% of the soil surface cover by residue after planting primarily where the objective is to reduce water erosion". When wind erosion is a concern, the term refers to tillage systems that maintain at least 1000 pounds per acre (1120 kg ha^{-1}) of "flat small-grain residue equivalents" on the soil surface during critical erosion periods [57]. The significance of focusing on 30% cover originated from studies showing that this amount would reduce erosion by at least 50% compared to bare, fallow soil [58].

Compared with conventional tillage, both no tillage and conservation tillage limit soil disturbance and retain crop residue. The decreased tillage intensity subsequently increases SOM as aeration and mineralization are reduced [27]. This residue-retaining conservation tillage in the southern Great Plains practices also form mulches that reduce evaporation, increase soil water that, consequently, engendered greater crop yields [59] and related biomass to enhance SOC. The resulting greater soil organic matter promotes soil aggregation by enhancing the growth of fungi and hyphae that binds the particles together [27,28,60]. The larger stabilized surface aggregates limit soil susceptibility to wind erosion and improve rain infiltration for reduced runoff and, consequently, soil entrainment in eroding water [61].

To mitigate wind and water erosion, the types of tillage that can generally meet the goal of leaving enough crop residue on the soil surface after planting are no-till, ridge-till, and mulch-till [62]. The Conservation Technology Information Center (CTIC) classifies tillage methods that leave from 15% to 30% cover after planting as reduced tillage, and systems that leave from 0% to 15% as conventional tillage [62]. Based on that CTIC National Crop Management Survey of the USA data presented in Figure 4, tillage intensity has reduced steadily and significantly since 1990. That is, no-till has increased from less than 6% in 1990 to almost 24% in 2008, and when mulch-till and ridge-till amounts are included, tillage systems that meet the definition of conservation tillage have increased from about 26% to 42% during that 18-year period. Data are not readily available prior to 1990, but there was little or no widespread emphasis on reducing tillage intensity before the 1990s.

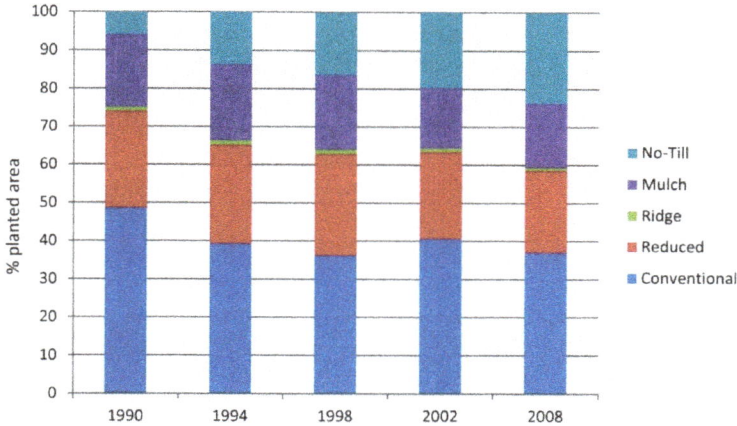

Figure 4. Types of tillage for planted area in the U.S., 1990 to 2008; no-till, mulch, and ridge tillage increased from 26% to 42% while providing >30% residue cover; reduced till 15% to 30% remained near a static 20%–25% of planted area, and conventional till, providing 0% to 15% residue cover on soil surface following planting, declined from ~48% to 37% of the planted area [62].

Beginning in 1982, there was significant effort to reduce soil erosion on all cropland [63]. From 1982 to 1997, sheet and rill erosion were reduced by 41% while wind erosion decreased by 43% (Figure 5) [64]. Since then, there continues to be a clear connection between tillage intensity and soil erosion even though both the reduction in tillage intensity and the reduction in soil erosion have declined. Ideally, no-tillage systems are best for mitigating soil degradation because they maximize the amount of crop residue remaining on the soil surface. Furthermore, in addition to reducing erosion, the residues reduce evaporation of water from the soil surface, which is particularly important in dry areas and during periods of drought. However, there are some disadvantages with no-till systems, such as increased dependence on herbicides and slow soil warming on poorly drained soils that have prevented adoption by many producers. Also, despite the numerically lower erodible fraction for soil managed with conservation tillage compared with CT, Van Pelt *et al.* [65] concluded that the protective mantle of crop residue is crucial to preventing erosion in the North American Central Great Plains. Conservation tillage also reduces soil compaction by increasing root growth and SOC [66], soil erosion by increasing surface residue cover [67], fuel costs for tillage, and potential global warming by increasing soil C sequestration [68] by conserving more soil water and increasing crop yields [29,44,47]. Although successful conservation tillage may require higher N fertilization because of enhanced N immobilization due to increased surface residue accumulation [69], benefits for mitigating soil degradation by increasing SOC and reducing soil compaction and erosion outweigh limitations. As a result, the conservation or no-tillage paradigm is recommended to improve soil and environmental quality and sustain crop yields.

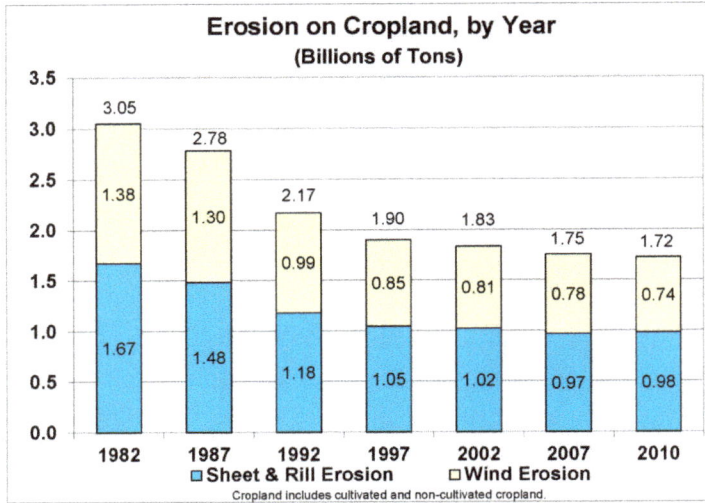

Figure 5. Combined mean water and wind erosion on U.S. cropland from 1982 until 2010 [64].

A similar, but somewhat more encouraging picture has emerged for Canada. Data presented in Figure 6 show that no-till increased from 7% in 1991 to 56% in 2011 while conventional tillage decreased from 68% in 1991 to 19% in 2011 [70]. Conservation tillage, defined in their analysis as having tillage intensity between no-till and conventional till, remained between about 25% and 30%. Similar to the U.S. where adoption of no-till increased from 6% in 1990 to 24% in 2008, adoption in Canada increased from 7% in 1991 to 56% in 2011. Overlapping within the 1991 to 2011 period of increasing no-till management, there was also a significant decrease in soil erosion risk in Canada from 1981 to 2006 associated with the decrease in tillage intensity (Figure 7). In contrast to the rate of soil erosion in the U.S. that has declined since 1997, the greater rate of reduction in soil erosion for Canada appears to be associated with both the increase in no-till area that reduces tillage intensity and the conversion of erodible land from annual crops to perennial forages and pastures [71].

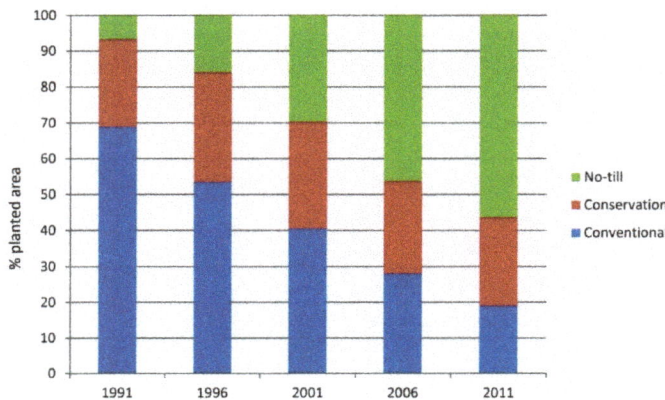

Figure 6. Types of tillage for planted area in Canada, 1991 to 2011; conservation tillage had tillage intensity between no-till and conventional tillage [70].

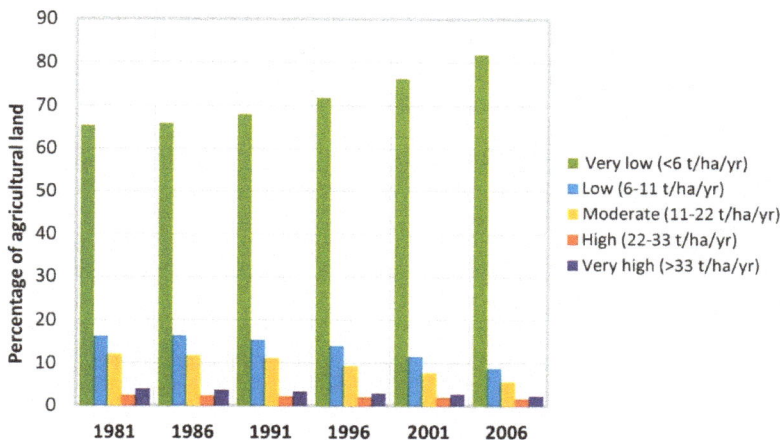

Figure 7. Soil erosion risk for cropland in Canada, 1981 to 2006 [71].

3.2. Cover Crops and Carbon Sequestration

Cover crops are defined as plant biomass grown for the purpose of providing a protective cover to prevent soil erosion and to limit nutrient loss by leaching or in runoff [72]. To this definition, Delgado *et al.* [73] added other management goals, including water conservation, nutrient scavenging and cycling management, and short duration livestock grazing. Grasses, legumes, and forbs grown for seasonal cover and conservation are not considered a production "crop". Cover crops in humid and subhumid regions of North America, and especially in areas with moderate winter conditions, such as the southeastern USA [74], are usually planted in the fall after summer cash crops are harvested. In semiarid regions with limited precipitation or regions with a short growing season, such as in the northern Great Plains, there are fewer opportunities for these crops. For cotton (*Gossypium hirsutum* L.) monocultures on the Texas High Plains that generally produce limited residue, Keeling *et al.* [75] introduced a chemically terminated wheat cover crop to control wind erosion. This practice increased mean irrigated cotton lint yield compared with conventional tillage, but establishment of the dryland cotton cash crop or wheat cover crop was problematic despite improved rain infiltration and greater crop water use [76]. Water is the most limiting factor in the central and southern Great Plains, so although growth of fall-planted cover crops may suffer due to low soil water availability, the real problem is that any use of soil water and N by cover crops may reduce cash crop yields compared with leaving the soil in a fallow condition [77,78]. Nevertheless, cover crops increase soil aggregation, water infiltration and water holding capacity [79], reduce soil erosion [80], and increase root growth of summer crops [32] over no cover crop.

Cover crops help mitigate soil degradation by improving nutrient management either by providing a nutrient source or by scavenging nutrients for eventual release from decomposing plant residues and recycling them to subsequent crops. The use of legume cover crops can supply N through fixation to increase crop yields compared with nonlegumes or no cover crop [30]. In contrast, nonlegume cover crops scavenge the soil for residual N following harvest of the primary crop, thereby reducing soil profile NO_3-N content and the potential for N leaching [81]. For example, a rye (*Secale cereale* L.) cover crop was projected to reduce NO_3 losses in drainage water within the Corn Belt states from a measured 11% [82] to a modeled 42.5% [83]. This could retain the N on site for use by subsequent crops and may have collateral benefits of reducing nutrient contamination that is one cause of hypoxia in the Gulf of Mexico. Growing a mixture of legume and nonlegume cover crops can maintain or increase SOC and SON concentrations by providing additional crop residue, which increases C and N inputs to the soil [22,37]. It can also help reduce N fertilizer requirements for subsequent summer crops [22,80].

In addition to providing protection against soil erosion and improving nutrient cycling, the use of cover crops and better management of crop residues have also been suggested as practices for enhancing carbon sequestration. The current focus on sequestering C in soils is to reduce CO_2 concentrations in the atmosphere and improve soil quality. Lal *et al.* [84] estimated that from 35 to 107 million Mg C could potentially be sequestered annually by conservation tillage and residue management on U.S. cropland. Although this might be possible, it is likely not feasible because sequestering 100 million Mg C would also sequester approximately 8 million Mg N and 2 million Mg P, which is about 75% and 100% of the amounts of these elements added each year in the U.S. through chemical fertilizers. Therefore, while efforts should continue to sequester C in soils, it is clear that the technology, practices, and policies needed to realize the estimated potential will be difficult to implement and the first priority should be to prevent further loss of SOC.

3.3. Intensified Cropping Systems

Traditionally, intensified farm production relied on established practices such as conventional tillage with monocropping and high rates of N fertilization to increase crop biomass yields. However, cropping systems in semi-arid regions of North America can be intensified by reducing the fallow frequency within crop sequences, such as by converting wheat-fallow (WF) to annually cropped wheat (Figure 8) or by introducing more productive summer crops into the rotation [85]. One example of the latter approach is the wheat-sorghum-fallow (WSF) rotation shown in Figure 9 [85]. Similar data from Saskatchewan showed that using fertilizer and crop sequences with progressively less frequent fallow periods increased annualized wheat grain and biomass yields that subsequently increased SOC [86]. Hansen *et al.* [87] also noted that cropping system intensification produced progressively greater biomass and SOC and, consequently, improved physical properties. Within the described W-F and WSF rotations a possibility exists for spring-planted cover crops to grow during early summer and partially replace fallow provided that normal cash crop production is unaffected by the redirected precipitation, especially in the semi-arid Great Plains. Where summer cover crops are grown in water conserving NT systems, aboveground biomass may be used for hay to improve cover crop economics [29,44,47]. The added biomass of intensified cropping systems that increases SOC and provides a protective cover can also decrease soil degradation by erosion. The benefits of residue to reduce soil entrainment by slowing wind or intercepting rain drop impact that leads to greater runoff combines with SOC stabilized aggregation to reduce soil erodibility [88,89].

In addition to increasing biomass for SOC and residue for soil protection, intensified cropping systems provide drainage and nutrient management alternatives in more humid North American climates. The cropping system intensification paradigm exemplifies a means to improved soil and environmental quality that also sustains crop yields.

Animal grazing can also be used to intensify farm production. Doing so can provide weed control, reduce feed cost, increase soil organic matter, and redistribute nutrients without depressing crop yields [49,90,91]. In the southeastern USA, moderate animal grazing can improve soil quality and productivity by enhancing soil organic matter and nutrient cycling, but excessive grazing can degrade soil properties by reducing SOM [92]. In the southern Great Plains, Baumhardt *et al.* [93] reported that surface compaction due to grazing cattle on vegetative dual purpose wheat without remediating tillage increased the soil profile penetration resistance. They also observed reduced water conservation resulting in depressed crop yields after three years compared with ungrazed no-tillage cropping systems production.

ANNUAL WHEAT

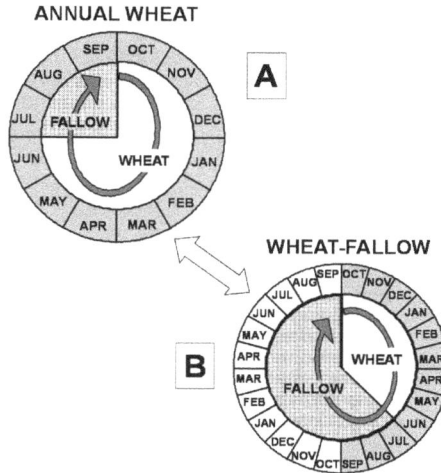

WHEAT-FALLOW

Figure 8. The annual wheat (**A**) and wheat-fallow (**B**) cropping sequences diagramed as a one or two year cycle, beginning with wheat establishment in October for the southern Great Plains [85]. In both sequences, wheat is harvested about nine months after planting and either fallowed briefly during July–September or after an additional 12 months if precipitation was insufficient for wheat establishment and growth.

WHEAT-SORGHUM-FALLOW

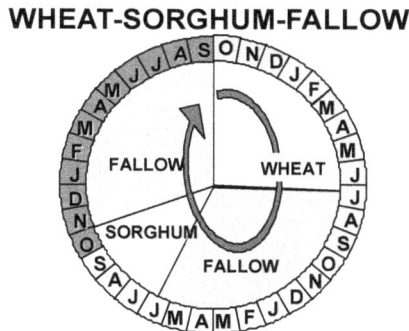

Figure 9. The wheat-sorghum-fallow (WSF) rotation diagramed as a three-year cycle beginning with wheat establishment in October and subsequent harvest 10-months later in July [85]. After delaying until June of the second year, grain sorghum is grown using stored soil water to augment summer rainfall. The soil is fallowed after sorghum harvest in November of the uncropped third year when the cycle repeats with wheat planting.

Excessive tillage and grazing used to intensify agricultural production can degrade soil by deforming or destroying soil structure [94]. This ultimately leads to compaction and a decrease in void space that, by definition, increases bulk density [95]. For example, in northwest Ohio, soil compaction was shown to reduce water movement, infiltration capacity, and root growth that, in turn, limited crop yield [96].

3.4. Engineering Strategies

For water erosion control, much of the early effort focused on using contour terraces. Although this worked well in many cases, there were major disadvantages because the terraces would frequently break during high intensity precipitation events. Also, as machinery became larger, contour terraces did not work well because space between terraces was highly variable. Parallel terraces were sometimes used to eliminate this problem, but this required more soil movement, made them more expensive to build, and often created soil fertility problems. Based on these experiences we maintain that long-term efforts to restore soil carbon by decreasing tillage intensities and retaining an appropriate amount of crop residue will be the most efficient approach for restoring degraded soils in North America.

4. Concluding Remarks

The U.S. Census Bureau [97] reported that the global population doubled from three billion in 1959 to six billion by 1999 and projected continued population growth to reach nine billion by 2044, which will require a corresponding increase in agricultural production to insure food security. Some 200 years after Thomas Malthus postulated failing global food security, Postel [98] observed in 1998 that water required for expanding overall crop production may be unavailable for degraded soils; thus, further threatening food security. In contrast, a 2013 report by Ausubel *et al.* [99] shows that the arable land required for sustaining crop production decreased by 65% during the period from 1961 to 2009, or the same period when the corresponding global population practically doubled. These very contradictory interpretations of resource productivity highlight unsettled future food security concerns, in part, because developing technologies have historically amplified agricultural production from fixed land resources to secure food demand. Soil degradation as a consequence of unsustainable management, however, may gradually decrease land productivity through *in-situ* soil salinization, compaction, declining SOM, and deteriorating aggregate stability.

In the concluding chapter of a soil degradation review, Lal and Stewart [100] advanced the case for separating "emotional rhetoric" of soil degradation from "precise scientific" results assessing soil resource condition and management. Principal agents that degrade soil, such as erosion or compaction frequently, follow the application of unsuitable agricultural management practices, including soil-inverting tillage. Eventually, scientific investigation advances improved management practices that reverse or mitigate soil degradation by negating the effects of causal processes or agents. Not surprisingly, soil erosion is mitigated through the use of cover crops and residue retaining tillage practices to promote aggregate stabilizing organic matter that, in turn, reduces soil susceptibility to erosion while providing crop mulches to intercept raindrop impact and prevent soil entrainment by wind or water. Improved soil and crop management practices must integrate unique differences in climate and soil specific properties, which deny the application of a common solution or priority for mitigating soil degradation. Preventing soil degradation, however, must control the universal processes or agents governing erosion, contamination, destabilization, and nutrient or SOM losses by crop production paradigms that emphasize resource stewardship.

Acknowledgments: The U.S. Department of Agriculture (USDA) prohibits discrimination in all its programs and activities on the basis of race, color, national origin, age, disability, and where applicable, sex, marital status, familial status, parental status, religion, sexual orientation, genetic information, political beliefs, reprisal, or because all or part of an individual's income is derived from any public assistance program. (Not all prohibited bases apply to all programs.) Persons with disabilities who require alternative means for communication of program information (Braille, large print, audiotape, *etc.*) should contact USDA's TARGET Center at (202) 720-2600 (voice and TDD). To file a complaint of discrimination, write to USDA, Director, Office of Civil Rights, 1400 Independence Avenue, S.W., Washington, DC 20250-9410, or call (800) 795-3272 (voice) or (202) 720-6382 (TDD). USDA is an equal opportunity provider and employer).

Author Contributions: All authors contributed equally to preparation of various components that were combined in this paper.

Conflicts of Interest: The authors declare no conflict of interest.

References

1. Oldeman, L.R. Global Extent of Soil Degradation. In *ISRIC Bi-Annual Report 1991–1992*; International Soil Reference and Information Centre: Wageningen, The Netherlands, 1992; pp. 19–36.
2. Nickerson, C.; Ebel, R.; Borchers, A.; Carriazo, F. *Major Uses of Land in the United States 2007, EIB-89*; U.S. Department of Agriculture Economic Research Service: Washington, DC, USA, 2011.
3. Mothorpe, C.; Hanson, A.; Schnier, K. The impact of interstate highways on land use conversion. *Ann. Reg. Sci.* **2013**, *51*, 833–870. [CrossRef]
4. Hunt, H.D. Overton's Overhaul. Available online: http://recenter.tamu.edu/tgrande/?m=123 (accessed on 4 March 2015).
5. Douglas, R.S. Overton nearing goal in revitalization efforts. Available online: http://lubbockonline.com/business/2013-08-21/overton-nearing-goal-revitalization-efforts (accessed on 18 July 2014).
6. EPA. Mid-Atlantic Mountaintop Mining. U.S. Environmental Protection Agency: 2013. Available online: http://www.epa.gov/region3/mtntop/ (accessed on 18 July 2014).
7. Lal, R.; Iivari, T.; Kimble, J.M. *Soil Degradation in the United States: Extent, Severity, and Trends*; CRC Press LLC.: Boca Raton, FL, USA, 2004; p. 204.
8. EPA. Superfund. U.S. Environmental Protection Agency: 2014. Available online: http://www.epa.gov/superfund/index.htm (accessed on 18 July 2014).
9. Scheppers, D.; Jamison, D.; Nabors, B. *Special Report to the Colorado General Assembly: The History, Status and Long-Term Funding Needs of the Colorado CERCLA Program*; Colorado Department of Public Health and Environment: Denver, CO, USA, 2012; p. 32.
10. Herron, J.; Stover, B.; Krabacher, P. *Reclamation Feasibility Report: Animas River below Eureka*; Colorado Division of Minerals and Geology: Denver, CO, USA, 2000; p. 148.
11. Neuman, D. Revegetation of Sediments from Montana's Milltown Dam. In Proceedings of the Mine Design, Operations & Closure Conference, Fairmont, MT, USA, 27 April 2014.
12. Baumhardt, R.L. The Dust Bowl Era. In *Encyclopedia of Water Science*; Stewart, B.A., Howell, T.A., Eds.; Marcel-Dekker: New York, NY, USA, 2003; pp. 187–191.
13. Dabney, S.M.; Shields, F.D.; Binger, R.L.; Kuhnle, R.A.; Rigby, J.R. Watershed management for erosion and sedimentation control case study Goodwin Creek, Panola County, MS. In *Advances in Soil Science, Soil Water and Agronomic Productivity*; Lal, R., Stewart, B.A., Eds.; Taylor & Francis: Boca Raton, FL, USA, 2012; pp. 539–568.
14. Whitney, M. U.S. Department of Agriculture Bureau of Soils Bulletin No. 55. Soils of the United States: Part I. Results of recent soil investigations & Part II. Classification and distribution of the soils of the United States. Available online: http://books.google.com/books?id=th8ZAQAAIAAJ (accessed on 21 July 2014).
15. Campbell, H.H. *A Complete Guide to Scientific Agriculture as Adapted to the Semi-Arid Regions*; The Campbell Soil Culture Co.: Lincoln, NE, USA, 1907.
16. Lal, R.; Reicosky, D.L.; Hanson, J.D. Evolution of the plow over 10,000 years and the rationale for no-till farming. *Soil Tillage Res.* **2007**, *93*, 1–12. [CrossRef]
17. Coughenour, C.M.; Chamala, S. *Conservation Tillage and Cropping Innovation: Constructing the New Culture of Agriculture*; Iowa State University Press: Ames, IA, USA, 2000; p. 360.
18. Binswanger, H. Agricultural mechanization: A comparative historical perspective. *World Bank Res. Obs.* **1986**, *1*, 27–56. [CrossRef]
19. Himes, F.L. Nitrogen, sulfur, and phosphorus and the sequestering of carbon. In *Soil Processes and the Carbon Cycle*; Advances in Soil Science; Lal, R., Kimble, J.M., Follett, R.F., Stewart, B.A., Eds.; CRC Press: Boca Raton, FL, USA, 1998; pp. 315–319.
20. Rasmussen, P.E.; Albrecht, S.L. Crop management effects on organic carbon in semi-arid Pacific Northwest soils. In *Management of Carbon Sequestration in Soil*; Advances in Soil Science; Lal, R., Kimble, J.M., Follett, R.F., Stewart, B.A., Eds.; CRC Press: Boca Raton, FL, USA, 1997; pp. 209–219.
21. Aase, J.K.; Pikul, J.L. Crop and soil response to long-term tillage practices in the northern Great Plains. *Agron. J.* **1995**, *87*, 652–656. [CrossRef]
22. Halvorson, A.D.; Wienhold, B.J.; Black, A.L. Tillage, nitrogen, and cropping system effects on soil carbon sequestration. *Soil Sci. Soc. Am. J.* **2002**, *66*, 906–912. [CrossRef]

23. Peterson, G.A.; Halvorson, A.D.; Havlin, J.L.; Jones, O.R.; Lyon, D.G.; Tanaka, D.L. Reduced tillage and increasing cropping intensity in the Great Plains conserve soil carbon. *Soil Tillage Res.* **1998**, *47*, 207–218. [CrossRef]

24. Aase, J.K.; Schaefer, G.M. Economics of tillage practices and spring wheat and barley crop sequence in northern Great Plains. *J. Soil Water Conserv.* **1996**, *51*, 167–170.

25. Hudson, B.D. Soil organic matter and available water capacity. *J. Soil Water Conserv.* **1994**, *49*, 189–194.

26. Beare, M.H.; Hendrix, P.F.; Coleman, D.C. Water-stable aggregates and organic matter fractions in conventional- and no-tillage soils. *Soil Sci. Soc. Am. J.* **1994**, *58*, 777–786. [CrossRef]

27. Cambardella, C.A.; Elliott, E.T. Carbon and nitrogen distribution in aggregates from cultivated and native grassland soils. *Soil Sci. Soc. Am. J.* **1993**, *57*, 1071–1076. [CrossRef]

28. Six, J.; Paustian, K.; Elliott, E.T.; Combrink, C. Soil structure and organic matter: I. Distribution of aggregate size classes and aggregate-associated carbon. *Soil Sci. Soc. Am. J.* **2000**, *64*, 681–689. [CrossRef]

29. Sainju, U.M.; Caesar-TonThat, T.; Lenssen, A.W.; Evans, R.G.; Kohlberg, R. Tillage and cropping sequence impact on nitrogen cycling in dryland farming in eastern Montana, USA. *Soil Tillage Res.* **2009**, *103*, 332–341. [CrossRef]

30. Kuo, S.; Sainju, U.M.; Jellum, E.J. Winter cover crop effects on soil organic carbon and carbohydrate. *Soil Sci. Soc. Am. J.* **1997**, *61*, 145–152. [CrossRef]

31. Clapp, C.E.; Allmaras, R.R.; Layese, M.F.; Linden, D.R.; Dowdy, R.H. Soil organic carbon and ^{13}C abundance as related to tillage, crop residue, and nitrogen fertilizer under continuous corn management in Minnesota. *Soil Tillage Res.* **2000**, *55*, 127–142. [CrossRef]

32. Sainju, U.M.; Whitehead, W.F.; Singh, B.P. Cover crops and nitrogen fertilization effects on soil aggregation and carbon and nitrogen pools. *Can. J. Soil Sci.* **2003**, *83*, 155–165. [CrossRef]

33. Raimbault, B.A.; Vyn, T.J. Crop rotation and tillage effects on corn growth and soil structural stability. *Agron. J.* **1991**, *83*, 973–985. [CrossRef]

34. Blanco-Canqui, H. Energy crops and their implications on soil and environment. *Agron. J.* **2010**, *102*, 403–419. [CrossRef]

35. Miller, P.R.; McConkey, B.; Clayton, G.W.; Brandt, S.A.; Staricka, J.A.; Johnston, A.M.; Lafond, G.P.; Schatz, B.G.; Baltensperger, D.D.; Neill, K.E. Pulse crop adaptation in the northern Great Plains. *Agron. J.* **2002**, *94*, 261–272. [CrossRef]

36. Omay, A.B.; Rice, C.W.; Maddux, L.D.; Gordon, W.B. Changes in soil microbial and chemical properties under long-term crop rotation and fertilization. *Soil Sci. Soc. Am. J.* **1997**, *61*, 1672–1678. [CrossRef]

37. Sainju, U.M.; Stevens, W.B.; Caesar-Tonthat, T. Soil carbon and crop yields affected by irrigation, tillage, crop rotation, and nitrogen fertilization. *Soil Sci. Soc. Am. J.* **2014**, *78*, 936–948. [CrossRef]

38. Varvel, G.E.; Wilhelm, W.W. Soil carbon levels in irrigated western Corn Belt rotations. *Agron. J.* **2008**, *100*, 1180–1184. [CrossRef]

39. Bennett, H.H.; Chapline, W.R. Soil Erosion: A National Menace. In *U.S. Department of Agriculture Circular No. 33*; U.S. Government Printing Office: Washington, DC, USA, 1928; pp. 20–23.

40. USDA-NRCS. *75 Years Helping People Help the Land: A Brief History of NRCS*; U.S. Department of Agriculture Natural Resources Conservation Service: Washington, DC, USA, 2014. Available online: http://www.nrcs. usda.gov/wps/portal/nrcs/detail/national/about/history/?cid=nrcs143_021392 (accessed on 21 July 2014).

41. USDA-NRCS. *Soil Erosion on Cropland 2007*; U.S. Department of Agriculture Natural Resources Conservation Service: Washington, DC, USA, 2007. Available online: http://www.nrcs.usda.gov/wps/portal/nrcs/ detail/national/technical/nra/nri/?cid=stelprdb1041887 (accessed on 17 July 2014).

42. Brady, N.C.; Weil, R.R. *The Nature and Properties of Soils*, 14th ed.; Pearson Education, Inc.: Upper Saddle River, NJ, USA, 2008.

43. Liebig, M.A.; Varvel, G.E.; Doran, J.W.; Wienhold, B.J. Crop sequence and nitrogen fertilization effects on soil properties in western Corn Belt. *Soil Sci. Soc. Am. J.* **2002**, *66*, 596–601. [CrossRef]

44. Tarkalson, D.D.; Hergert, G.W.; Cassman, K.G. Long-term effects of tillage on soil chemical properties and grain yields of a dryland winter wheat-sorghum/corn-fallow rotation in the Great Plains. *Agron. J.* **2006**, *98*, 26–33. [CrossRef]

45. Sainju, U.M. Tillage, cropping sequence, and nitrogen fertilization influence dryland soil nitrogen. *Agron. J.* **2013**, *105*, 1253–1263. [CrossRef]

46. Wilhelm, W.W.; Johnson, J.M.F.; Hatfield, J.L.; Voorhees, W.B.; Linden, D.R. Crop and soil productivity response to corn residue removal: A literature review. *Agron. J.* **2004**, *96*, 1–17. [CrossRef]

47. Mahler, R.L.; Harder, R.W. The influence of tillage methods, cropping sequence, and N rates on the acidification of a northern Idaho soil. *Soil Sci.* **1984**, *137*, 52–60. [CrossRef]

48. Chen, S.H.; Collamer, D.J.; Gearhart, M.M. The effect of different ammonical nitrogen sources on soil acidification. *Soil Sci.* **2008**, *173*, 544–551. [CrossRef]

49. Herrero, M.; Thorton, P.K.; Notenbaert, A.M.; Wood, S.; Msangi, S.; Freeman, H.A.; Bossio, D.; Dixon, J.; Peters, M.; van de Steeg, J.; *et al.* Smart investments in sustainable food productions: Revisiting mixed crop-livestock systems. *Science* **2010**, *327*, 822–825. [CrossRef] [PubMed]

50. Lilienfein, J.; Wilcke, W.; Vilela, L.; Lima, S.D.; Thomas, R.; Zech, W. Effect of no-till and conventional tillage systems on the chemical composition of soils solid phase and soil solution of Brazilian Savanna soils. *J. Plant Nutr. Soil Sci.* **2000**, *163*, 411–419. [CrossRef]

51. Food and Agriculture Organization (FAO). *Salt-Affected Soils*; United Nations: Rome, Italy, 2014. Available online: http://www.fao.org/soils-portal/soil-management/management-of-some-problem-soils/salt-affected-soils/more-information-on-salt-affected-soils/en (accessed on 20 July 2014).

52. Van der Pluym, H.S.A. Extent, causes, and control of dryland saline seepage in the northern Great Plains regions of North America. Dryland Saline Seep Control. In Proceedings of the 11th International Soil Science Congress, Sub-Commission on salt affected soils, Edmonton, AL, Canada, 19–27 June 1978; pp. 48–58.

53. Black, A.L.; Brown, P.L.; Halvorson, A.D.; Siddoway, F.H. Dryland cropping strategies for efficient water use to control saline seeps in the northern Great Plains, USA. *Agric. Water Manag.* **1981**, *4*, 295–311. [CrossRef]

54. MAFRI. Soil Management Guide. Available online: http://www.gov.mb.ca/agriculture/environment/soil-management/soil-management-guide/index.html (accessed on 20 July 2014).

55. Schaible, G.D.; Aillery, M.P. *Water Conservation in Irrigated Agriculture: Trends and Challenges in the Face of Emerging Demands. EIB-99*; U.S. Department of Agriculture Economic Research Service: Washington, DC, USA, 2012.

56. Postel, S. *Pillar of Sand: Can the Irrigation Miracle Last?* W.W. Norton & Company: New York, NY, USA, 1999; p. 313.

57. Owens, H. *Tillage: From Plow to Chisel and No-Tillage, 1930–1999*; Iowa State University MidWest Plan Service: Ames, IA, USA, 2001; p. 35.

58. McCarthy, J.R.; Pfost, D.L.; Currence, H.D. Conservation tillage and residue management to reduce soil erosion. Available online: http://extension.missouri.edu/p/G1650 (accessed on 22 July 2014).

59. Stewart, B.A.; Baumhardt, R.L.; Evett, S.R. Major Advances of Soil and Water Conservation in the U.S. Southern Great Plains. In *Soil and Water Conservation Advances in the United States*; Special Publication 60; Zobeck, T.M., William, F.S., Eds.; Soil Science Society of America: Madison, WI, USA, 2010; pp. 103–130.

60. Caesar-TonThat, T.; Stevens, W.B.; Sainju, U.M.; Caesar, A.J.; Gaskin, J.F.; West, M.S. Soil-aggregating bacterial community as affected by irrigation, tillage, and cropping system in the northern Great Plains. *Soil Sci.* **2014**, *179*, 11–20. [CrossRef]

61. Baumhardt, R.L.; Blanco-Canqui, H. Soil Conservation practices. In *Encyclopedia of Agriculture and Food Systems*; Van Alfen, N., Ed.; Elsevier: San Diego, CA, USA, 2014; Volume 5, pp. 153–165.

62. CTIC. *National Crop Management Survey*; Conservation Technology Information Center: West Lafayette, IN, USA, 2012. Available online: http://www.ctic.org/CRM/ (accessed on 18 July 2014).

63. USDA-NRCS. *Soil Erosion. Conservation Resource Brief Number 0602*; U.S. Department of Agriculture Natural Resources Conservation Service: Washington, DC, USA, 2006.

64. U.S. Department of Agriculture Natural Resources Conservation Service (USDA-NRCS). *Summary Report: 2010 National Resources Inventory*; USDA-NRCS: Washington, DC, USA, 2010. Available online: http://www.nrcs.usda.gov/Internet/FSE_DOCUMENTS/stelprdb1167354.pdf (accessed on 17 July 2014).

65. Van Pelt, R.S.; Baddock, M.C.; Zobeck, T.M.; Schlegel, A.J.; Vigil, M.F.; Acosta-Martinez, V. Field wind tunnel testing of two silt loam soils on the North American Central High Plains. *J. Aeolian Res.* **2013**, *10*, 53–59. [CrossRef]

66. Blanco-Canqui, H.; Stone, L.R.; Schlegel, A.J.; Lyon, D.J.; Vigil, M.F.; Mikha, M.M.; Stahlman, P.W.; Rice, C.W. No-till induced increase in organic carbon reduces maximum bulk density of soils. *Soil Sci. Soc. Am. J.* **2009**, *73*, 1871–1879. [CrossRef]

67. Merrill, S.D.; Black, A.L.; Fryrear, D.W.; Saleh, A.; Zobeck, T.M.; Halvorson, A.D.; Tanaka, D.L. Soil wind erosion hazard of spring wheat-fallow as affected by long-term climate and tillage. *Soil Sci. Soc. Am. J.* **1999**, *63*, 1768–1777. [CrossRef]

68. Sainju, U.M. Cropping sequence and nitrogen fertilization impact on surface residue, soil carbon sequestration, and crop yields. *Agron. J.* **2014**, *106*, 1231–1242. [CrossRef]

69. Zibilske, L.M.; Bradford, J.M.; Smart, J.R. Conservation tillage-induced changes in organic carbon, total nitrogen, and available phosphorus in a semi-arid alkaline subtropical soil. *Soil Tillage Res.* **2002**, *66*, 153–163. [CrossRef]

70. Statistics Agriculture. *Chapter 5 No-till Practices Increased*; Statistics Agriculture: Ottawa, ON, Canada, 2012. Available online: http://www.statcan.gc.ca/pub/95-640-x/2012002/05-eng.htm (accessed on 18 July 2014).

71. McConkey, B.G.; Lobb, D.A.; Li, S.; Black, J.M.W.; Krug, P.M. Soil erosion on cropland: Introduction and trends for Canada. *Canadian Biodiversity: Ecosystem Status and Trends 2010*; Technical Thematic Report No. 16. Canadian Councils of Resource Ministers: Ottawa, ON, Canada, 2012. Available online: http://www.speciesatrisk.ca/resource/DOCUMENT/6178No.16_Soil%20Erosion_Jun2012_E.pdf (accessed on 17 July 2014).

72. Reeves, D.W. Cover crops and rotations. In *Crops Residue Management, Advances in Soil Science*; Hatfield, J.L., Stewart, B.A., Eds.; Lewis Publishers: Boca Raton, FL, USA, 1994; pp. 125–172.

73. Delgado, J.A.; Reeves, W.; Follett, R. Winter cover crops. In *Encyclopedia of Soil Science*; Lal, R., Ed.; Markel and Decker: New York, NY, USA, 2006; pp. 1915–1917.

74. Balkcom, K.S.; Schomberg, H.H.; Reeves, D.W.; Clark, A.; Baumhardt, R.L.; Collins, H.P.; Delgado, J.A.; Kaspar, T.C.; Mitchell, J.; Duiker, S. Managing cover crops in conservation tillage systems. In *Managing Cover Crops Profitably*, 3rd ed.; Clark, A., Ed.; Handbook Series Book 9; Sustainable Agriculture Network: Beltsville, MD, USA, 2007; pp. 44–61.

75. Keeling, W.; Segarra, E.; Abernathy, J.R. Evaluation of conservation tillage cropping systems for cotton on the Texas southern High Plains. *J. Prod. Agric.* **1989**, *2*, 269–273. [CrossRef]

76. Baumhardt, R.L.; Lascano, R.J. Water budget and yield of dryland cotton intercropped with teminated winter wheat. *Agron. J.* **1999**, *91*, 922–927. [CrossRef]

77. Unger, P.W.; Vigil, M.F. Cover crop effects on soil water relationships. *J. Soil Water Conserv.* **1998**, *53*, 200–207.

78. Nielsen, D.C.; Vigil, M.F. Legume green fallow effect on soil water content at wheat planting and wheat yield. *Agron. J.* **2005**, *97*, 684–689. [CrossRef]

79. McVay, K.A.; Radcliffe, D.E.; Hargrove, W.L. Winter legume effects on soil properties and nitrogen fertilizer requirements. *Soil Sci. Soc. Am. J.* **1989**, *53*, 1856–1862. [CrossRef]

80. Langdale, G.W.; Blevins, R.L.; Karlen, D.L.; McCool, D.K.; Nearing, M.A.; Skidmore, E.L.; Thomas, A.D.; Tyler, D.D.; Williams, J.R. Cover crop effects on soil erosion by wind and water. In *Cover Crops for Clean Water*; Hargrove, W.L., Ed.; Soil and Water Conservation Society: Ankeny, IA, USA, 1991; pp. 15–22.

81. Meisinger, J.J.; Hargrove, W.L.; Mikkelsen, R.L.; Williams, J.R.; Benson, V.E. Effect of cover crops on groundwater quality. In *Cover Crops for Clean Water*; Hargrove, W.L., Ed.; Soil and Water Conservation Society: Ankeny, IA, USA, 1991; pp. 57–68.

82. Strock, J.S.; Porter, P.M.; Russelle, M.P. Cover cropping to reduce nitrate loss through subsurface drainage in the northern U.S. Corn Belt. *J. Environ. Qual.* **2004**, *33*, 1010–1016. [CrossRef] [PubMed]

83. Malone, R.W.; Jaynes, D.B.; Kaspar, T.C.; Thorp, K.R.; Kladivko, E.; Ma, L.; James, D.E.; Singer, J.; Morin, X.K.; Searchinger, T. Cover crops in the upper midwestern United States: Simulated effect on nitrate leaching with artificial drainage. *J. Soil Water Conserv.* **2014**, *69*, 192–305. [CrossRef]

84. Lal, R.; Kimble, J.M.; Follett, R.F.; Cole, C.V. *The Potential of U.S. Cropland to Sequester Carbon and Mitigate the Greenhouse Effect*; Sleeping Bear Press, Inc.: Chelsea, MI, USA, 1998.

85. Baumhardt, R.L.; Anderson, R.L. Crop choices and rotation principles. In *Dryland Agriculture*, 2nd ed.; Agronomy Monograph No. 23; Peterson, G.A., Unger, P.W., Payne, W.A., Eds.; ASA, CSSA, and SSSA: Madison, WI, USA, 2006; pp. 113–139.

86. Lemke, R.L.; VandenBygaart, A.J.; Campbell, C.A.; Lafond, G.P.; McConkey, B.G.; Grant, B. Long-term effects of crop rotations and fertilization on soil C and N in a thin Black Chernozem in southeastern Saskatchewan. *Can. J. Soil Sci.* **2012**, *92*, 449–461. [CrossRef]

87. Hansen, N.C.; Allen, B.L.; Baumhardt, R.L.; Lyon, D.J. Research achievements and adoption of no-till, dryland cropping in the semi-arid U.S. Great Plains. *Field Crops Res.* **2012**, *132*, 196–203. [CrossRef]

88. Baumhardt, R.L.; Johnson, G.L.; Schwartz, R.C. Residue and long-term tillage and crop rotation effects on rain infiltration and sediment transport. *Soil Sci. Soc. Am. J.* **2012**, *76*, 1370–1378. [CrossRef]

89. Blanco-Canqui, H.; Holman, J.D.; Schlegel, A.J.; Tatarko, J.; Shaver, T.M. Replacing fallow with cover crops in a semiarid soil: Effects on Soil Properties. *Soil Sci. Soc. Am. J.* **2013**, *77*, 1026–1034. [CrossRef]

90. Franzluebbers, A.J. Integrated crop-livestock systems in the southeastern USA. *Agron. J.* **2007**, *99*, 361–372. [CrossRef]

91. Sainju, U.M.; Lenssen, A.W.; Goosey, H.; Snyder, E.; Hatfield, P. Dryland soil carbon and nitrogen influenced by sheep grazing in the wheat-fallow system. *Agron. J.* **2010**, *102*, 1553–1561. [CrossRef]

92. Franzluebbers, A.J.; Stuedemann, J.A. Early response of soil organic carbon fractions to tillage and integrated crop-livestock production. *Soil Sci. Soc. Am. J.* **2008**, *72*, 613–625. [CrossRef]

93. Baumhardt, R.L.; Schwartz, R.C.; Greene, L.W.; MacDonald, J. Cattle grazing effects on yield of dryland wheat and sorghum grown in rotation. *Agron. J.* **2009**, *101*, 150–158. [CrossRef]

94. Lowery, B.; Schuler, R.T. Temporal effect of subsoil compaction on soil strength and plant growth. *Soil Sci. Soc. Am. J.* **1991**, *55*, 216–223. [CrossRef]

95. Soil Science Society of America (SSSA). *Glossary of Soil Science Terms*; SSSA: Madison, WI, USA, 1997.

96. Lal, R. Axle load and tillage effects on crop yields in a Mollic Ochraquarf in northwest Ohio. *Soil Tillage Res.* **1996**, *37*, 143–160. [CrossRef]

97. U.S. Census Bureau. *International Data Base World Population: 1950–2050*; U.S. Census Bureau: Washington, DC, USA, 2013. Available online: http://www.census.gov/population/international/data/idb/worldpopgraph.php (accessed on 30 August 2014).

98. Postel, S.L. Water for Food Production: Will There Be Enough in 2025? *BioScience* **1998**, *48*, 629–637. [CrossRef]

99. Ausubel, J.H.; Wernick, I.K.; Waggoner, P.E. Peak Farmland and the Prospect for Land Sparing. *Popul. Dev. Rev.* **2013**, *38*, 221–242. [CrossRef]

100. Lal, R.; Stewart, B.A. Need for Action: Research and development priorities. In *Soil Degradation*; Advances in Soil Science 11; Lal, R., Stewart, B.A., Eds.; Springer: New York, NY, USA, 1990; pp. 331–336.

sustainability

Review

Soil Quality Impacts of Current South American Agricultural Practices

Ana B. Wingeyer [1,*], Telmo J. C. Amado [2], Mario Pérez-Bidegain [3], Guillermo A. Studdert [4], Carlos H. Perdomo Varela [3], Fernando O. Garcia [5] and Douglas L. Karlen [6]

[1] Instituto Nacional de Tecnología Agropecuaria (INTA), Estación Experimental Agropecuaria Paraná, Ruta 11, km 12,5. Oro Verde, Entre Ríos 3101, Argentina
[2] Universidade Federal de Santa Maria, Centro de Ciências Rurais, Av. Roraima 1000, Santa Maria, RS 97105-900, Brazil; telmo.amado@pq.cnpq.br
[3] Universidad de la República, Facultad de Agronomía, Garzón 780, Montevideo 12900, Uruguay; mperezb@fagro.edu.uy (M.P.-B.); chperdom@fagro.edu.uy (C.H.P.V.)
[4] Universidad Nacional de Mar del Plata, Facultad de Ciencias Agrarias, Unidad Integrada Balcarce. Ruta Nac. 226 Km 73,5. Balcarce, Buenos Aires 7620, Argentina; studdert.guillermo@inta.gob.ar
[5] IPNI Latinoamérica-Cono Sur. Av. Santa Fe 910, Acassuso, Buenos Aires B1641ABO, Argentina; FGarcia@ipni.net
[6] USDA-Agricultural Research Service (ARS), National Laboratory for Agriculture and the Environment (NLAE), 2110 University Boulevard, Ames, IA 50011-3120, USA; Doug.Karlen@ars.usda.gov
* Correspondence: wingeyer.ana@inta.gob.ar; Tel./Fax: +54-343-497-5200

Academic Editor: Marc A. Rosen

Received: 19 November 2014; Accepted: 10 February 2015; Published: 17 February 2015

Abstract: Increasing global demand for oil seeds and cereals during the past 50 years has caused an expansion in the cultivated areas and resulted in major soil management and crop production changes throughout Bolivia, Paraguay, Uruguay, Argentina and southern Brazil. Unprecedented adoption of no-tillage as well as improved soil fertility and plant genetics have increased yields, but the use of purchased inputs, monocropping *i.e.*, continuous soybean (*Glycine max* (L.) Merr.), and marginal land cultivation have also increased. These changes have significantly altered the global food and feed supply role of these countries, but they have also resulted in various levels of soil degradation through wind and water erosion, soil compaction, soil organic matter (SOM) depletion, and nutrient losses. Sustainability is dependent upon local interactions between soil, climate, landscape characteristics, and production systems. This review examines the region's current soil and crop conditions and summarizes several research studies designed to reduce or prevent soil degradation. Although the region has both environmental and soil resources that can sustain current agricultural production levels, increasing population, greater urbanization, and more available income will continue to increase the pressure on South American croplands. A better understanding of regional soil differences and quantifying potential consequences of current production practices on various soil resources is needed to ensure that scientific, educational, and regulatory programs result in land management recommendations that support intensification of agriculture without additional soil degradation or other unintended environmental consequences.

Keywords: soil degradation; erosion; soil organic matter; no-till; agricultural intensification

1. Introduction

Our global population is anticipated to be 8.1 billion in 2025 and 9.6 billion by 2050, with most of the growth occurring in developing countries [1] and urban settings [2]. In addition to population growth, the global Gross Domestic Product (GDP) is expected to grow at a rate of 2.1% year^{-1} from 2005/2007–2050 [3]. Collectively, population growth, increased per capita income, and the resultant anticipated dietary changes (*i.e.*, more meat and dairy consumption) are expected to increase global crop demand by 100%–110% by 2050 [4].

Meeting this global crop demand may be challenging because agricultural production, which grew 2.2% year^{-1} between 1987 and 2007, is projected to increase at only 1.3% yr^{-1} between 2005/2007 and 2050 [3]. This estimate was supported by Ray *et al.* [5] who studied long-term yield trends for maize (*Zea mays* L.), rice (*Oryza sativa* L.), wheat (*Triticum aestivum* L.), and soybean (*Glycine max* (L.) Merr.). Those four crops represent two thirds of the total agricultural calorie demand. Their study [5] analyzed crop yield data from 1989–2008 and projected yields to 2050 using 2008 as the baseline year. The projections indicated global average increases of 1.6%, 1.0%, 0.9% and 1.3% year^{-1} for maize, rice, wheat, and soybean, respectively. These values are well below the 2.4% year^{-1} crop yield increase needed to double current production by 2050.

Fortunately, there are several ways to increase agricultural production. Crop yields can be increased through genetic improvement and perhaps by increasing the amount and types of chemical input. Total production can be increased through more intense land use by reducing the amount of fallow, increasing the number of crops grown per year, and by cultivating new agricultural land. These options were supported by an FAO study [3] that predicted 80% of the future crop production increases will come from developing countries, with 71% of the increase coming from yield increases, 8% through higher cropping intensity and 20% by the addition of arable land.

A portion of the increasing global demand for oil seeds and cereals has been met by increasing the cropping area and converting from conventional tillage (CT) to no-till (NT) throughout Argentina, Bolivia, southern Brazil, Paraguay and Uruguay during the past 50 years. Increases in cropping area have been accompanied by changes in land tenancy, tillage practices, and greater overall productivity, but it has also contributed to a loss of crop diversity (*i.e.*, conversion of long-term perennial pastures to grain crop rotations or even monocultures). The transition from CT to NT throughout the region was initially characterized as a tremendous soil management success, but more recently it has been challenged because of the unprecedented expansion of cropland, devoted solely to soybean production, into marginal areas. Our hypothesis is that current agricultural practices are having unanticipated negative effects on soil quality in addition to the impacts associated with the loss of crop diversity.

2. Regional Soil Resource Characteristics

This review focuses on an area of approximately 195 million hectares in South America. The area includes the Pampas and Gran Chaco regions of Argentina and Uruguay, the southern highlands of eastern Paraguay, the eastern lowlands of Bolivia, as well as the Rio Grande do Sul, Santa Catarina and Parana states of Brazil. Several soil associations are found in this region. Mollisols are dominant throughout the Pampas-Chaco plains and Uruguay [6] and are among the best suited for agriculture because of their high natural fertility. Alfisols are also widespread in the Pampas-Chaco region. Alfisols are generally fertile, with high concentration of nutrient cations. Ultisols and Oxisols are the main soils in southern Brazil and eastern Paraguay; these soils have good physical qualities, but require high lime and phosphorus inputs. Vertisols are located in Entre Rios province at Argentina and Uruguay, with good fertility levels but soil physical properties that demand a careful soil management. Alluvial soils dominate the eastern lowlands of Bolivia, and also have good natural soil fertility.

The development of current agricultural practices throughout the region required the transformation of natural ecosystems into agroecosystems with reduced structural and functional complexity. The agroecosystems are also continuously evolving as the result of natural factors and human actions [7–10]. At the global scale, agroecosystems have been evolving towards oversimplification characterized by low efficiency of inputs, loss of resilience, intensification of outputs (grains plus stocks, nutrient removal and leaching and soil erosion), increased carbon consumption by the dependence on fossil energy (fertilizers, pesticides and fuel) and loss of soil quality and ecosystem services [11–13].

The capacity of agroecosystems to provide ecosystem services mainly depends on the adequate functioning of the soil [9,14]. Soil provides the media for root anchoring and healthy plant growth. In addition, soil health affects availability and transport of water, air and nutrients, the resistance to degradation and erosion, soil temperature, and water and air pollution [14,15]. Soil organic matter (SOM) constitutes one of the most affected soil components by agricultural management practices. SOM depletion is associated with alteration of important soil properties (*i.e.*, fertility, porosity, aeration-water dynamics, resistance to erosion and compaction, among others [16], and to the reduction of the capacity of soil to reorganize and restore its functionality after use or a stressful event [17]. Therefore, the capacity of soil to provide other expected environmental services (*i.e.*, supports biodiversity, regulate water partition and purification, and sequester carbon), is affected [18,19]. Other soil properties important for the adequate functioning of soil as aggregate stability and penetration resistance are impacted. Aggregate stability indicates how aggregates will react to and how porosity will be impacted by environmental events (*i.e.*, precipitation, wetting and drying cycles), while soil penetration resistance indicates the degree of difficulty for roots to grow into the soil, which is related to soil compaction and salinization processes among others that affect nutrient and water use efficiency.

3. Cropping System Changes

The primary cropping systems changed throughout the study area during the past 50 years in response to market globalization and the need to develop an internationally competitive agriculture that was beneficial for the countries, farmers, and society in general [20].

Soybean, maize and wheat have become the primary grain field crops in Argentina, Bolivia, Paraguay, Uruguay and southern Brazil (Rio Grande do Sul, Santa Catarina, and Parana states). These crops represented 63%, 19%, and 12%, respectively, of the total cropped area in 2012 according to FAOSTAT, IGBE and CONAB data [21–23]. In the last 54 years, total grain production from those three crops has grown from 14.5 million Mg (16 million tons) in 1961 to over 142 million Mg (156 million tons) in 2013 or roughly a 10-fold increase (Figure 1). Average grain yields for soybean, maize and wheat increased from 1493–2309, from 1489–5485, and from 1114–2386 kg·ha^{-1}, respectively, during this period. According to Manuel-Navarrete *et al.* [20], agricultural expansion in Argentina has been supported by the adoption of new technologies (*i.e.*, genetic resources, chemical inputs, and agricultural machinery), an increase in grain prices, a relative decrease in input costs, regional agricultural research, active farmer participation, and relatively good and stable climatic conditions [24,25].

The rates of change in cropped area, grain yield, and total production have been different for the various crops and countries within the region (Table 1). In general, soybean has expanded at the expense of other crops and through land use change [26], increasing production at an annual rate of 5.9% from 1961–2013. Most of the increase can be attributed to an increase in production area (5%), with an additional 0.8% attributed to increased grain yield. The largest recent increases in crop area and total soybean production within the study area have occurred in Argentina, Bolivia, Paraguay and Uruguay. Increases in southern Brazil have been more moderate, primarily because a significant area in that country was already devoted to soybean in 1961 (Table 1). Wheat production also increased in all the countries except Argentina, mainly due to increased grain yield in southern Brazil and Uruguay, and increased area within Paraguay and Bolivia. For maize, there was a large increase in regional production, primarily due to increased area in southern Brazil and Paraguay and higher grain yield in Uruguay and Argentina. During the 1990s and early 2000s, NT stimulated an increase in soybean/maize rotations which increased diversity, reduced insect pressure, restored SOM, and increased crop residue input and nutrient cycling.

The most significant cropping system changes throughout the region since the mid-1990s were the release and rapid adoption of glyphosate-tolerant (GT) soybean varieties and the unprecedented expansion of NT. Since 1994, soybean production has increased at an annual rate of 6.3%, primarily due to area expansion (+5.5% year^{-1}) at the expense of grasslands, maize, sunflower (*Hellianthus annuus* L.), and sorghum (*Sorghum bicolor* (L.) Moench). By 2012, soybean, maize, and wheat accounted for

67%, 20%, and 13% of the region's total cropland. With regard to yield, maize showed the highest rate of increase throughout the entire region. Wheat yield increased most rapidly in southern Brazil and Paraguay, while soybean yield increased most rapidly in Paraguay and Uruguay, but overall, yield for both crops increased at a slower pace than maize throughout the region. More recently and despite increases in potential maize yield, the soybean/maize rotation was replaced by a soybean-dominated cropping system. The driver for this change was neither agronomic nor technically based, but simply an economic one. Maize is generally a more expensive crop to grow than soybean. It is also more vulnerable to short drought during critical phenological growth stages and within the region, the commodity price was lower than for soybean.

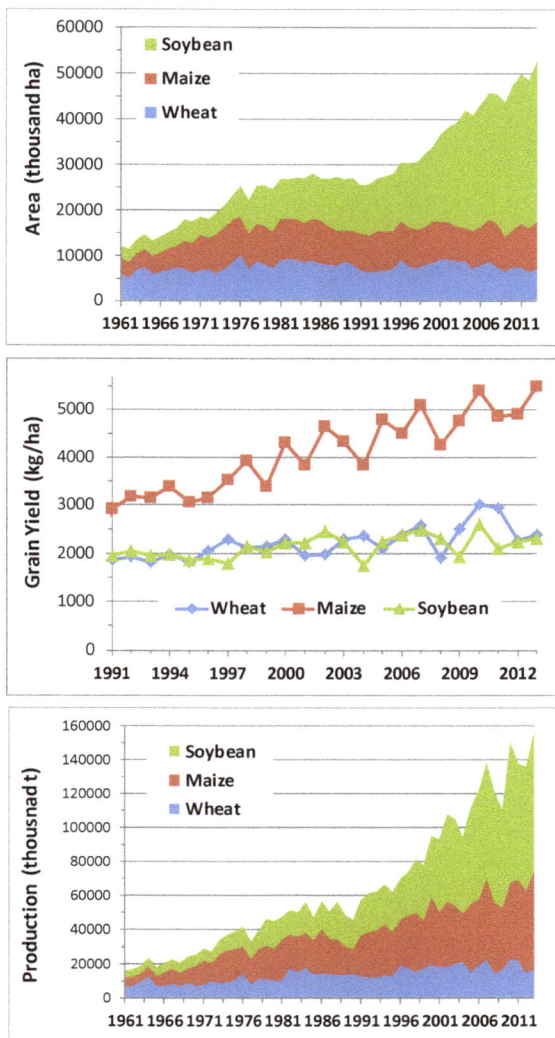

Figure 1. Fifty-year changes in planted area, grain yield, and total production of wheat, maize, and soybean in Argentina, Bolivia, southern Brazil, Paraguay, and Uruguay. Elaborated with data from FAOSTAT, IBGE and CONAB databases [21–23].

Table 1. Rates of change in area, yield, and production for wheat, maize, and soybean in Argentina, Bolivia, southern Brazil, Paraguay, and Uruguay.

Period	Area			Grain Yield			Total Production		
	Wheat	Maize	Soybean	Wheat	Maize	Soybean	Wheat	Maize	Soybean
Argentina									
1961–2013	−0.6%	1.1%	20.5%	1.3%	2.5%	1.8%	0.6%	3.6%	22.7%
1994–2013	−2.5%	3.5%	6.3%	0.8%	2.2%	1.1%	−1.7%	5.8%	7.4%
Bolivia									
1961–2013	1.7%	1.4%	16.9%	1.3%	1.3%	1.3%	3.1%	2.7%	18.4%
1994–2013	3.1%	2.3%	7.1%	2.0%	1.5%	−0.8%	5.1%	3.9%	6.2%
Southern Brazil (PR-SC-RS)*									
1961–2013	1.8%	5.0%	2.6%	3.2%	3.3%	1.2%	5.0%	8.4%	3.8%
1994–2013	5.0%	−1.7%	3.4%	3.3%	3.1%	1.0%	8.5%	1.4%	4.5%
Paraguay									
1961–2013	8.3%	4.7%	15.8%	2.0%	2.3%	1.1%	10.5%	7.1%	17.1%
1994–2013	6.0%	8.1%	7.7%	3.9%	3.2%	0.7%	10.1%	11.6%	8.4%
Uruguay									
1961–2013	0.6%	−1.6%	14.3%	2.1%	3.8%	1.9%	2.7%	2.2%	16.4%
1994–2013	5.9%	4.1%	26.4%	0.0%	6.4%	2.9%	5.9%	10.8%	30.1%
Total									
1961–2013	0.3%	2.0%	5.0%	1.4%	2.5%	0.8%	1.8%	4.6%	5.9%
1994–2013	0.3%	1.0%	5.5%	0.9%	2.4%	0.8%	1.2%	3.4%	6.3%

Elaborated with data from FAOSTAT, IGBE and CONAB databases [21–23]. * PR = Parana; SC = Santa Catarina; RS = Rio Grande do Sul States, Brazil.

Adoption of conservation tillage, including NT, has been a leading practice in South America (Table 2, Figure 2). The initial North American experience with NT (*i.e.*, Shirley Phillips and the University of Kentucky) became an important reference for the first South American experiences with NT during the early 1970s. Pioneer farmers in Brazil and Argentina recognized the importance of maintaining crop residues on the surface to protect against water erosion and to compensate for rapid residue decomposition under high temperature and moisture regimes prevalent throughout the summer cropping season. Although the adoption of NT was relatively slow from the 1970s to 1990s, it increased exponentially following the release of GT soybean varieties. The success of NT in southern Brazil and Argentina became an important reference for its widespread adoption throughout South America. Currently, NT is being used on 70%–90% of the grain crop area in Paraguay, Brazil, Argentina, Bolivia, and Uruguay. Government and farmer associations throughout the region promote adoption of NT for several reasons including its economic benefits, higher or more stable yields through improved water use efficiency, erosion control, saving on fuel and labor/time, and improved soil quality attributes (AUSID, MAGyP) [27,28].

The expansion of GT soybean and NT practices are highly correlated (e.g., 0.90 and 0.73 for Argentina and southern Brazil, respectively) and have been supported by higher soybean prices when compared to other grains within the region (Figure 3). For example, soybean grain prices increased by 196% between 1991 and 2012, compared to increases of 54%, 77%, 59%, 77%, and 96% for barley (*Hordeum vulgare* L.), maize, sorghum, sunflower, and wheat, respectively. Therefore, the main driver for increased soybean production has been the higher price when compared to other grain crops.

It is important to stress that the economic impact of soybean has not only been very positive for farmers but also for the economies of the countries in the region and society as a whole. However, to fully examine the sustainability of this cropping system change, it is important to also examine how it has affected soil resources and overall soil quality.

Table 2. Area under no-tillage in selected South American countries.

Country	Area under No-Tillage (ha) 2008/2009	Percentage of Total Cropped Area
Brazil	25,502,000	58
Argentina	25,553,000	70
Paraguay	2,400,000	90
Bolivia	706,000	72
Uruguay	655,100	82

Data source is Friedrich *et al.* [29].

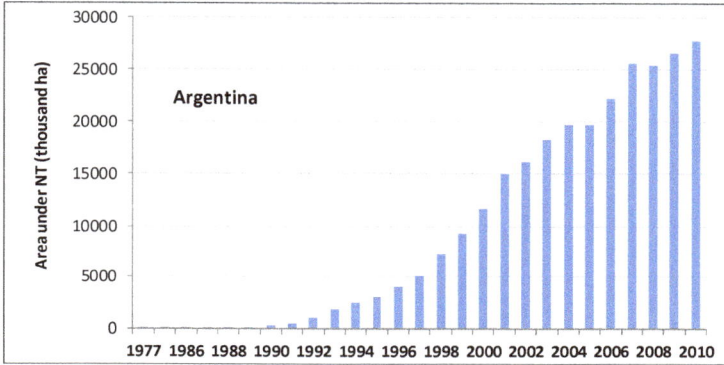

(**a**)

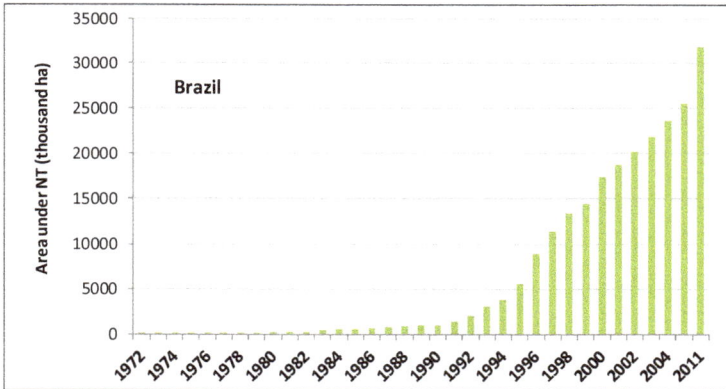

(**b**)

Figure 2. *Cont.*

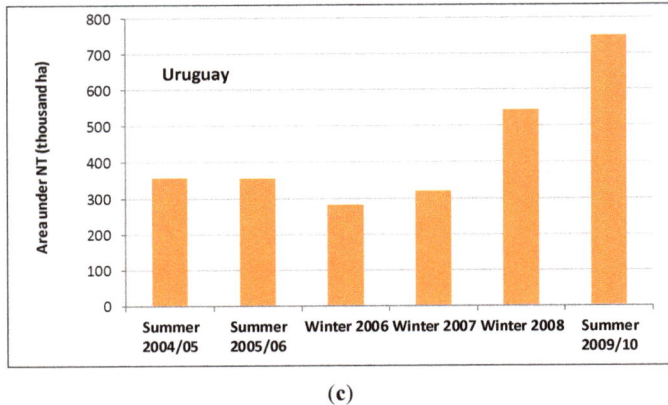

(c)

Figure 2. No-tillage adoption rates in Argentina (**a**); Brazil (**b**); and Uruguay (**c**). Elaborated with data until 2011 for Argentina [30], 2012 for Brazil [23,31], and 2010 for Uruguay [27].

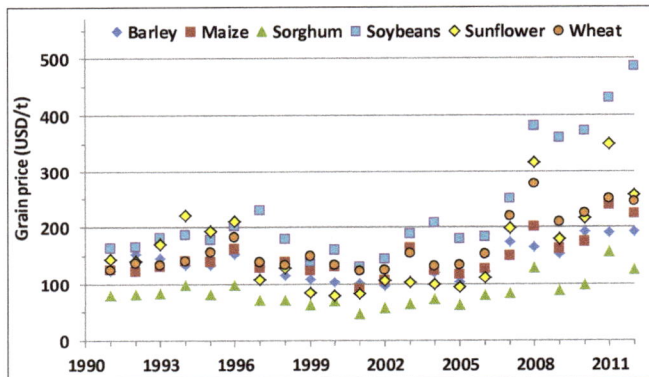

Figure 3. Twenty-year grain price averages for the primary field crops in Argentina, Bolivia, Brazil, Paraguay and Uruguay. Elaborated with data from FAOSTAT [21].

4. Soil Quality Impact of Agricultural Expansion

Assessment of soil quality indicators, including SOM content, N supply, P availability, aggregate stability, bulk density, pH, and others, has received increased attention during the last decade throughout the Pampas and extra-Pampas regions of Argentina as well as in Brazil, Uruguay, Bolivia, and Paraguay. These assessments have indicated that the cropping system changes, which have generally included removal of pasture from crop rotations, decreased crop diversity with the increased frequency of soybean, and the conversion of marginal land into cropland, are imposing a threat to soil quality [20].

In Argentina, comparisons of soil quality indicators for pristine and agricultural soils show a general reduction in SOM content and aggregate stability and an increase in bulk density with agricultural use (Table 3). Soils with less than 10 years of continuous agriculture had 83%, 62%, and 106% of the pristine SOM content, aggregate stability and bulk density, respectively, while soils 10–20 years of continuous agriculture had 64%, 48% and 116% of the pristine values, respectively (Table 3). These measurements indicate that SOM decreased by approximately 18% per decade of agricultural use.

Table 3. Aggregate stability (AS), soil organic matter (SOM) content, and bulk density (BD) of agricultural and pristine soils in the Pampas region after various years of Agriculture (YOA) and for different soil depth increments.

Zone	Soil	YOA	Depth (m)	AS	SOM	BD	Soil pH	Soil pH	Reference
				\% Relative to Pristine Conditions			Pristine	Cultivated	
Extra- pampas	O	>20	0–0.1	-	59	143	-	-	[32]
Extra-pampas	O	>20	0.1–0.2	-	83	126	-	-	[32]
Extra-pampas	O	>20	0.2–0.3	-	74	128	-	-	[32]
Pampas center	M, Argiudolls	12	>0.08	37	57	121	-	-	[33]
Pampas center	M, Argiudolls	12	0–0.08	44	56	111	-	-	[33]
Pampas center	M, Argiudolls	>20	>0.08	64	83	103	-	-	[33]
Pampas center	M, Argiudolls	>20	0–0.08	29	81	105	-	-	[33]
Pampas center	M, Argiudolls	>20	0–0.2	37	73	-	6.3	5.9	[34]
Pampas center	M, Argiudolls	>30	0–0.2	23	77	-	6.3	6.1	[34]
Pampas center	M, Argiudolls, Natralbolls	<10	>0.08	53	80	106	-	-	[33]
Pampas center	M, Argiudolls, Natralbolls	<10	0–0.08	43	72	122	-	-	[33]
Pampas center	M, Argiudolls	24	0–0.20	26	65	-	-	-	[35]
Pampas center	M, Haplustolls, Hapludolls	<10	>0.08	42	73	102	-	-	[33]
Pampas center	M, Haplustolls, Hapludolls	<10	0–0.08	39	68	110	-	-	[33]
Pampas center	M, Haplustolls, Hapludolls	>40	>0.08	37	75	112	-	-	[33]
Pampas center	M, Haplustolls, Hapludolls	>40	0–0.08	39	71	118	-	-	[33]
Pampas center, W	M, Hapludolls	13	0–0.20	40	61	-	-	-	[35]
Pampas N	I, Haplustept	4–23	0–0.025	115	75	-	6.4	7.3	[36]
Pampas NE	M, Argiudolls	>10	0–0.12	23	62	116	6.5	6.4	[37]
Pampas NE	M, Argiudolls	>20	0–0.05	57	45	109	-	-	[38,39]
Pampas NE	M, Argiudolls	>20	0.05–0.15	60	84	110	-	-	[38,39]
Pampas NE	M, Argiudolls	>20	0.15–0.3	-	86	-	-	-	[38,39]
Pampas NE	V, Hapluderts	>20	0–0.05	57	45	109	-	-	[38,39]
Pampas NE	V, Hapluderts	>20	0.05–0.15	60	84	110	-	-	[38,39]
Pampas NE	V, Hapluderts	>20	0.15–0.3	-	86	-	-	-	[38,39]
Pampas NW	M, Haplustolls	20	0–0.1	39	71	113	7.1	6.7	[40]
Pampas NW	M, Haplustolls	1–4	0–0.20	53	73	100	7.5	7.1	[41]
Pampas NW	M, Haplustolls	1–4	0.20–0.50	71	72	102	6.9	7.4	[41]
Pampas NW	M, Haplustolls	2–7	0–0.20	95	85	86	6.9	7.0	[41]
Pampas NW	M, Haplustolls	2–7	0.20–0.50	88	86	100	7.7	7.5	[41]
Pampas NW	M, Haplustolls	4–9	0–0.20	91	89	125	6.8	6.8	[41]
Pampas NW	M, Haplustolls	4–9	0.20–0.50	66	122	108	7.5	7.2	[41]
Pampas NW	M, Haplustolls	3	0–0.1	63	89	105	5.7	6.5	[42]
Pampas NW	M, Haplustolls	>10	0–0.1	37	63	120	5.7	6.6	[42]
Pampas SE	M, Argiudolls	10	0–0.20	43	83	-	-	-	[35]

Soil: O: Oxisols, M: Mollisols, V: Vertisols, I: Inceptisols.

In Uruguay, the average SOM decrease in Alfisols after 10+ years of cultivation was 15% [43]. In southern Brazil, Campos *et al.* [44] reported 23% of decline of SOM in an Oxisol when measured 30 years after the conversion of pasture to grain production with CT practices. Ferreira [45] investigated long-term NT (>20 years) in southern Brazil and reported that compared to native fields, the average SOM declined 12% and 23% for the 0–0.3 and 0–1.0 m soil depths, respectively. In addition to near-surface ($\leq$0.2 m) SOM decline, another soil quality concern is SOM decline in deeper soil layers. This can occur with the current regional cropping systems because of the difficulty in restoring SOM at those depths. The main drivers for SOM decline within the region are: elimination of perennial pasture from long-term rotations, increased soil disturbance associated with tillage, limited rooting depth due to machinery traffic and soil compaction, soil erosion, decreased crop residue input, and less crop diversity. Short-term land tenure, increases in cropland area managed by an individual farmer, and increased inputs of low C/N crop residues also contribute to land use decisions that often reduce SOM levels. Therefore, to prevent current cropping systems from further degrading soil resources, several agronomic, economic and social factors affecting soil quality need to be addressed throughout the region to better understand and improve soil and crop management.

5. Soil Quality Evaluation

The evaluation of soil quality indicators is important to avoid widespread soil degradation due to inappropriate crop management. Previous evaluations at the landscape scale have shown important changes in SOM, pH and P availability (Figure 4) [46,47], and in the soil's capacity to supply N (Figure 5) [48]. Soil quality assessment within the 0–0.2 m depth of agricultural fields throughout the Pampas and surrounding regions (n > 34,000 samples) indicated that current SOM values ranged from 1 to 83 g·C·kg^{-1}. The highest values were found in the southeast Pampas region and with gradual decline toward the north and west (Figure 4a) [46]. In the same study, low soil pH (<6.0) due to acidity was a concern only in the north Pampas region, while for the majority of the region, soil pH was within normal range for crop production (pH 6.0–7.5) [46]. Assessments of P availability (Bray P) showed low to very low ($\leq$5 mg·kg^{-1}) values throughout the Pampas region (15.2 Mha) and medium to high ($\geq$15 mg·kg^{-1}) values north of that area (12.7 Mha) [47].

A similar survey in the 1980s [49] showed that after 25–30 years of cropping, areas that had medium to high levels of P (west and north of the Pampas region) showed a major decrease in *p* availability due to unbalanced fertilization. On the other hand, areas such as the south of Pampas region which had low to very low levels of P (>10 ppm P-Bray I) had increased P availability due to long term balanced fertilization (Figure 4c) [47]. Nitrogen mineralization potential in Buenos Aires province of Argentina ranged from 12–260 mg·kg^{-1} with the majority of the fields below 65 mg·kg^{-1} [48]. Potential N supply showed a high relationship with SOC content and represented a reduction in potential N supply of approximately 50% compared to pristine soils [48]. The availability of these reports constitutes an important step for monitoring and characterizing soil quality indicators and how soil management affects it.

In Uruguay, NT was adopted to mitigate soil loss due to intensive water erosion [50]. This would be expected in improve soil quality and even though other soil and crop management changes are hypothesized to affect the quality of natural resources including soils, quantifying long-term effects has been considered very difficult because of the lack of baseline soil quality data [51]. Therefore, several studies have measured the impact of agricultural practices on soil quality indicators as compared to undisturbed conditions. For example, a 2009–2010 survey of commercial operations in Uruguay showed an average reduction in SOM of 20% [52]. The loss of potentially mineralizable N (PMN) was reported to be as high as 41.5%. These results are consistent with previous findings [53] and are presumed to be real because most PMN is contained in the biologically active SOM fraction which is easily degraded by agricultural operations. The variability of both parameters was very high, with 20% of locations reporting SOM increases between 1% and 20%, and 30% of locations reporting losses between 30% and 60%. This indicates there are important interactions between soil management, soil type, climate, and cropping system. Regarding PMN, 12% of the locations had increases between 1% and 14% while 30% of locations reported losses between 35% and 80% [52]. A different study [51] compared soils under dairy pasture with those from crop production fields and reported that both systems had 20% SOM losses. However, PMN losses from cropland soils were much higher (42%) than in pasture soils (26%). This difference is attributed to higher soil N input by legumes in dairy pastures compared to N removal by grain production.

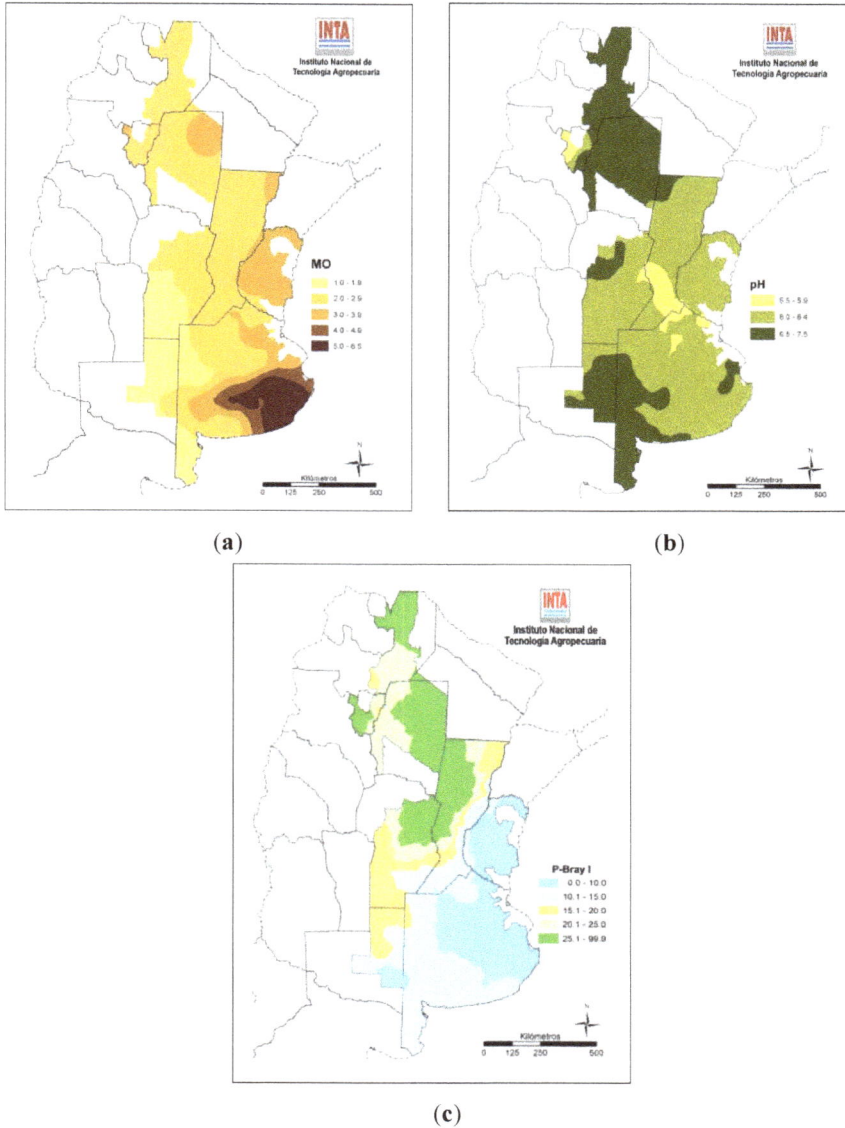

(a)

(b)

(c)

Figure 4. Median landscape values for (**a**) soil organic matter (SOM); (**b**) soil pH; and (**c**) available soil phosphorous (Bray P1) in the Pampas region of Argentina. Source: Sainz-Rozas *et al.*, 2011 and 2012 [46,47].

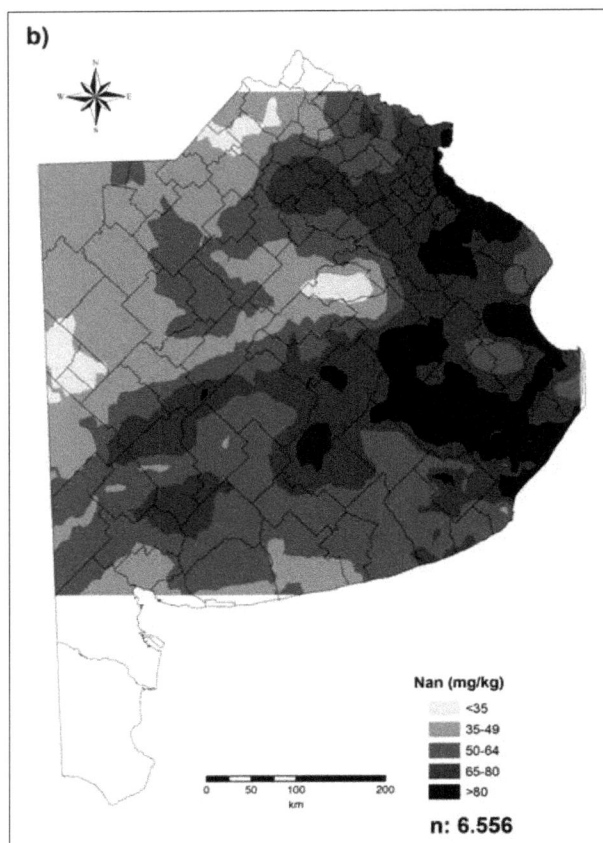

Figure 5. Surface horizon (0–20 cm) potential nitrogen supply (Nan) in cropland soils of Buenos Aires province, Argentina. *N* = number of observations. Source: Reussi-Calvo *et al.* 2014 [48].

In 2009, Mori [43] quantified several other soil quality indicators [SOM, PMN, total N (TN), exchangeable bases (EB) and water pH] in agricultural Mollisols within Uruguay (Table 4). The study showed SOM and NPM losses of 15% and 17%, respectively, for agricultural and native systems. Other negatively affected indicators were TN (−23.7%), EB (−10.5%) and pH (−7.3%). The study also demonstrated high correlation among some soil quality indicators (e.g., SOM, TN, EB and pH) (Figure 6) and therefore, soil management strategies that result in the simultaneous losses of C, N and EB are likely to cause a sharp decline in soil quality.

The use of CT in early 1970s in southern Brazil and in the 1980s in Paraguay resulted in severe (>50 Mg·ha^{-1}·year^{-1}) soil erosion due to intensive rainfall during the maize and soybean establishment period of September to December. It was estimated that soil loss for each Mg of harvested grain was 10 Mg·ha^{-1} throughout the 1970s [54,55]. The use of CT also promoted wind erosion in sandy soils and flat croplands and resulted in significant runoff from bare soil on the undulating topography of southern Brazil and Paraguay croplands [54].

Table 4. The range and average change (agricultural *versus* undisturbed conditions) in selected soil quality indicators of 15 Argiudolls from Uruguay.

Average	SOM	TN	PMN	EB	pH
Young	−16.4%	−25.0%	−21.6%	−6.0%	−5.7%
San Manuel	−14.0%	−22.7%	−12.8%	−14.4%	−8.6%
All	−15.1%	−23.7%	−16.9%	−10.5%	−7.3%
			Range		
Min	−37.0%	−41.9%	−54.3%	−38.2%	−17.1%
Max	1.8%	−1.4%	51.0%	15.8%	5.9%

Elaborated with data from Mori [43]. SOM: Soil organic matter; TN: total nitrogen; PMN: potentially mineralizable nitrogen; EB: exchangeable bases; pH: water-pH.

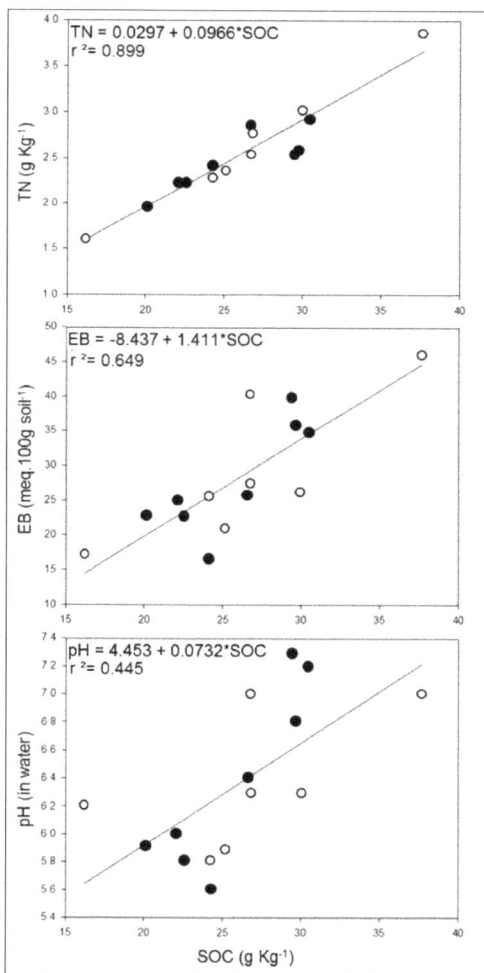

Figure 6. The statistical relationship between soil quality indicators (SOC, NT, EB and water-pH) in San Manuel (closed) and Young (open) soils from Uruguay. Regression results based on pooled data from both sites. Elaborated with data from Mori [43].

The intense soil disturbance of Oxisols by CT also resulted in decreased soil aggregate size and, as a consequence, labile fractions of SOC that were previously occluded inside aggregates were exposed to microbial processes. Soil erosion also removed the top soil carrying with it SOM and clay fractions, thus resulting in a sharp decline in soil quality. Biological oxidation of SOM was also stimulated by tillage, mixing of crop residues, and high temperature and moisture conditions during the summer. It was estimated that with CT, crop residue input as high as 16 Mg·ha^{-1}·year^{-1} would be necessary to maintain SOC content [55]. Needless to say, achieving such a level of crop residue is very difficult with grain crops. For southern Brazil and Paraguay, the first steps toward improving soil quality were associated with reducing soil disturbance by conversion from CT to NT and by increasing crop residue inputs to maintain soil protection throughout the year. Long-term experiments with agricultural grain crop rotations indicated that even under NT it is necessary to design crop systems with high amounts of crop residue input in order to balance the fast SOM decay associated with high temperature and moisture conditions in subtropical and tropical environments of southern Brazil and Paraguay [56]. Furthermore, adoption of higher crop diversity and incorporation of pastures with well-developed root systems will help restore soil structure and relieve soil compaction issues associated with the large soybean expansion area. Equilibrated soil fertilization, lime and use of soil amendments as gypsum, should also be considered as tools to sustain high crop residue input, especially on the naturally acid and low fertility soils of southern Brazil [57].

6. Developing Cropping Systems for Sustained Grain Yield, Maintenance of Soil Quality, and Environmental Protection

In southern Brazil, a recent study carried out by Ferreira [45] in six different croplands showed that there was an average SOM decline of 23% in the 0–100 cm soil layer after 25 years of CT. On-farm research showed SOM values that were similar to those observed in the few long-term experiments carried out in Rio Grande do Sul [44]. On the other side, restoration of SOM with NT in southern Brazil has been shown to be a long term process (>20 years) with increases ranging from 61%–117% (Figure 7) depending upon climate, clay content and crop rotation. With continuous soybean crops in summer and black oat (*Avena strigosa* Schieb.) or wheat in the winter, there was only a slight increase in SOM under NT in relation to CT. Conversely, crop rotations that included soybean alternated with maize as summer crops and cover crops during winter showed enhanced SOM restoration in the range of 85%–116%. In the region, it is possible to grow crops all year along since there is only a very short window of frost, and rainfall is generally well distributed. Typically, successful farmers use cover crops mixtures that include black oat + vetch (*Vicia sativa* L.) + oilseed radish (*Raphanus sativus* var. *oleiferus* Metzg.) before maize, which is then followed by oilseed radish during a short 3 month window after which wheat followed by soybean are grown. This crop rotation is designed to increase crop residue input, increase crop diversity and sustain high grain yields. In the scenario of long-term soybean as the main summer crop, special attention must be given to the use of cover crops and pastures in the short windows in order to maintain soil quality.

In Argentina, studies on SOM dynamics in Mollisols within the southeast of the Pampas region showed a quick decline in SOC (Figure 8a) and its particulate fraction (POC) when soils under pasture were converted to cropland regardless if they were managed using CT (moldboard plow) or NT management [58–60]. Crops used in the rotation had a significant impact on the rate of SOC loss [46] which was primarily associated with the amount of C returned through crop residues input [61,62]. Under continuous grain cropping, adoption of NT resulted in reduced losses, or even increases in SOC and POC in shallow soil layers (0–5 cm) regardless of fertilizer management [60,62–65], or agricultural intensification levels [66]. However, none of these studies reported improvements in SOC or POC due only to the adoption of NT or to fertilization when the 0–20 cm depth soil layer was taken into account. Nonetheless, the inclusion of a 3-year pasture in the rotation after 7–8 years of grain crops restored SOC (Figure 8a) and POC contents to original contents before cropping after pasture [58,60]. These results suggest that for the naturally rich SOM Mollisols of the southeastern Pampa, restoration and maintenance of SOM content within sustainable levels would require a combination of NT and pasture in rotation with grain crops.

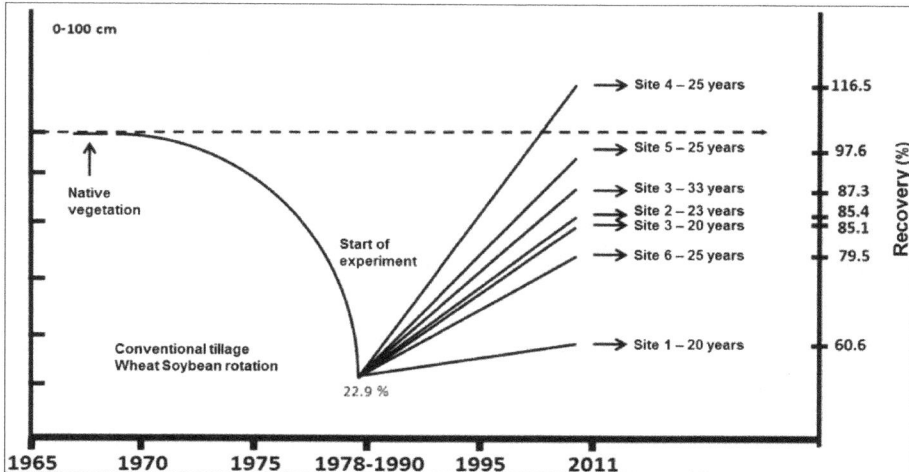

Figure 7. Soil organic matter in the 0–100 cm soil depth following the use of conventional tillage to replace native pasture with grain crops and after conversion to no-till with different soybean-based crop rotations. Source: Ferreira (2014) [45].

Direct consequences of SOC and POC loss under cropping systems include the reduction of soil N and S supplying capacity [48,67–70] and an increased dependence on N [71,72] and S [73] fertilizers to maintain crop yields. Under continuous cropping, potential N supply within 85% of farmer fields in the southeast of Pampas region was <100 mg·N·kg^{-1} while only 35% of soils under crop rotations including short term pastures were below that level [74]. Regarding aggregate stability, both physical breakdown of aggregates and loss of SOM due to soil disturbance contributed to the reduced aggregate stability of soils under agriculture [58,75,76] The use of NT delayed the reduction in aggregate stability compared to CT, but ultimately, aggregate stability reached similar values for both systems [77]. Furthermore, within the 0–20 cm soil depth, aggregate stability under continuous cropping did not improve with the conversion of CT to NT [77]. Aggregate stability was highly related to POC content [78] and particularly with POC content in macro-aggregates [79]. However, the introduction of a pasture in the rotation not only resulted in recovery of SOC and POC contents but also improved aggregate stability (Figure 8b) [58,77,78]. Based on these studies, we conclude that cropping systems need to be designed and developed regionally. For Mollisols in Argentina under temperate climate conditions, crop rotations including pastures were more critical to aggregate stability and SOC content than tillage systems or fertilization. However, for tropical and subtropical environments in southern Brazil and Paraguay, the role of minimum soil disturbance, lime and fertilization were more critical.

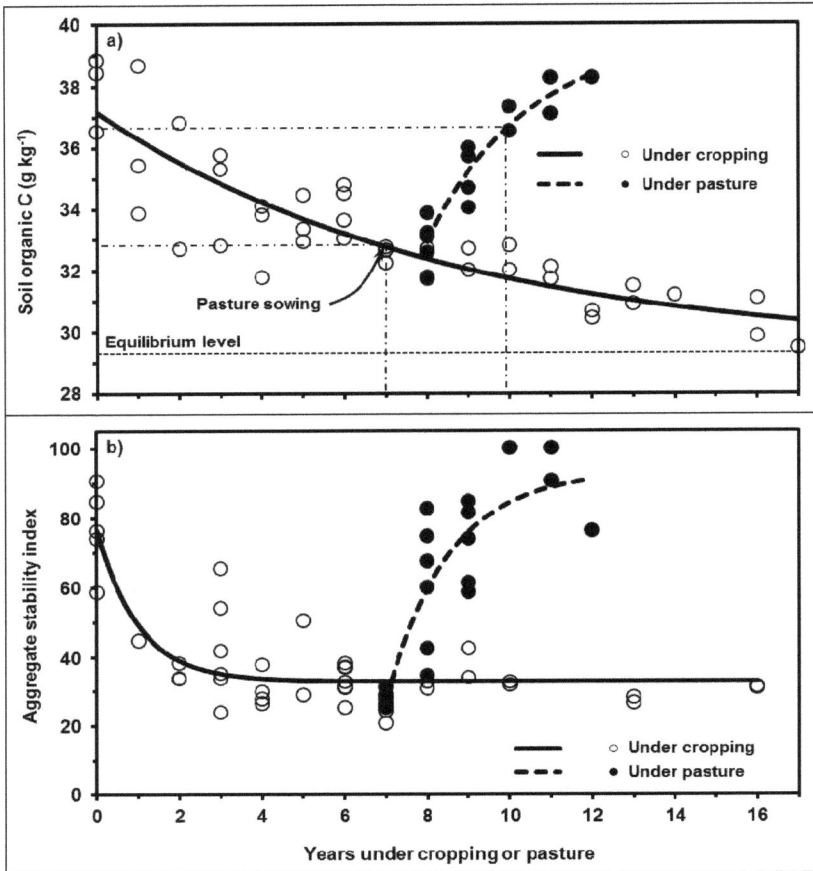

Figure 8. Changes in (**a**) soil organic C; and (**b**) aggregate stability in the arable layer of Mollisols from the southeast Pampas region of Argentina. Elaborated from Studdert *et al.* [60] and unpublished data.

In a review of long term experiments in Uruguay, Garcia Préchac *et al.* [80] reported a six-fold erosion reduction under NT compared to CT conditions. Soil losses were similar between continuous pastures and crop rotations that incorporated pasture suggesting the efficacy of pastures to reduce soil erosion was evident even in short-term rotations. Using USLE modeling, they estimated an average annual erosion rate of less than 7 Mg·ha^{-1}·year^{-1} in a Typic Argiudoll in a soybean/wheat or in a soybean/winter cover crop rotation. This is a moderate level according to Clérici *et al.* [81]. Using Century modeling [82], the same crop rotations were projected to result in long-term SOC losses.

In another study in Uruguay, Salvo *et al.* [83] quantified changes in SOC stocks in a long term study that included CT and NT and two crop rotations (continuous grain crops, *versus* a rotation with three-years of pasture and three years of grain crops). The study was carried on a Typic Argiudoll and the measurements were taken 10 years after initiation of the experiment. The results comparing NT *versus* CT, showed a SOC increase of 29% within the 0–3 cm depth under NT. Under CT, the rotation that included pastures increased SOC content by 23% at the same soil depth. The inclusion of maize under NT and pasture-crop rotations resulted in a 12% increase in SOC as compared with the treatment with only soybean and sunflower as summer crops. The evolution of SOC and other physical properties after the incorporation of winter pastures such as ryegrass (*Lolium multiflorum* L.) and oat (*Avena sativa* L.) to continuous soybeans managed under NT was investigated by Sawchik *et al.* [84]. After six years, SOC within the 0–7.5 cm layer was 17% higher on treatments with soybean and winter pasture than under continuous soybean. Infiltration rates were also higher in treatments with soybean and winter pasture compared to continuous soybean. Thus, in Uruguay, the recommendation for maintaining a stable SOC content requires inclusion of pastures and alternating C3 and C4 summer crops instead of using only C3 summer crops as soybean.

Soil management effects on soil quality in two long-term experiments carried out on Alfisols in Rio Grande do Sul were evaluated by Amado *et al.* [85]. They found that CO_2 respiration, aggregate stability and infiltration rates were the most effective soil quality indicators to discriminate between cropping and tillage systems. In general, the adoption of NT and use of cover crops increased CO_2 respiration suggesting higher biological activity (Table 5). Adoption of NT also resulted in enhanced aggregate stability, increased infiltration rates, and reduced soil erosion. In this study, the decline of SOC was associated with decreased CO_2 respiration, aggregate stability and infiltration suggesting a loss of soil quality with CT and in association with lack of crop rotation. They also reported that the soil quality kit from USDA/ARS was an efficient tool to evaluate soil quality under contrasting soil management scenarios in the region. From the two long-term experiments, it was determined that improved cropping systems should have legume cover crops or pasture in the crop rotation in order to increase both crop N and C inputs to soil. Therefore, the treatments under NT that have hairy vetch in consortium with black oats, maize in consortium with cowpea (*Vigna unguiculata* (L.) Walp.), or tropical legume cover crops such as pigeon pea (*Cajanus cajan* (L.) Millsp.), lab-lab (*Lablab purpureus* (L.) Sweet) or mucuna (*Mucuna* sp.) in consortium with maize had the best soil quality. Furthermore, in this case, mineral N fertilization applied to maize did not replace the role of a legume cover crop for improving soil quality.

Campos *et al.* [44] in a long-term experiment carried out on an Oxisol reported a linear relationship between crop residue input and SOC stock in the 0–30 cm soil depth (Figure 9). This study supports previous research suggesting an annual dry matter input of 8–10 $Mg \cdot ha^{-1} \cdot year^{-1}$ is needed to maintain or increase SOC stock under NT [55]. In addition, this experiment showed an increase of 16% in crop residue production under NT compared to CT. The increase in crop residue input under NT may be associated with the improvement in soil fertility and soil quality compared to CT [44].

Table 5. Soil respiration (CO2), aggregate stability (AS), pH, nitrate and nitrite content (N-NO3⁻ + N-NO2⁻), bulk density and infiltration rate for different management systems (MGMENT) and two soils using the Soil Quality Kit (SQK) and the reference method (REF).

MGMENT	CO2		AS		pH		N-NO3⁻ + N-NO2⁻		Bulk Density		Infiltration Rate	
	SQK	REF	SQK	REF	SQK	REF	SQK	REF	SQK	REF	SQK	REF
	kg·ha⁻¹·day⁻¹		%				kg·ha⁻¹		Mg·m⁻³		mm·h⁻¹	
Rhodic Paleudalf												
Bare fallow	6.7 b¹	15.9 b	33.4	23.9 b	5.6 a	5.5 b	1.2 b	1.2	1.72 a	1.66 a	<1 c	1
NT Fw/M	23.8 b	36 b	54.9	78.9 a	5.7 a	5.6 ab	1.9 a	3.0	1.45 ab	1.34 b	75 b	50
NT P/M	21.4 b	35.3 b	61.4	76.8 a	5.6 a	5.6 b	2.0 a	2.5	1.45 ab	1.41 b	202 a	86
NT M/Lcc	54.0 a	80.9 a	64.1	77.7 a	5.8 a	5.8 a	2.0 a	2.0	1.33 b	1.35 b	190 a	195
NV	43.5 a	97.4 a	71	96.8 a	5.2 b	5.3 c	0.9 b	1.2	1.35 b	1.37 b	35 bc	23
CV (%)	12.9	16.9	23.1	7.3	1.5	0.7	5.4	23.3	4.6	2.8	15.5	99.6
p-value	0.003	0.005	0.209	0.002	0.01	0.002	0.002	0.056	0.022	0.008	0.008	0.216
R	0.95 ***		0.86 **		0.98 ***		0.69 *		0.85 **		0.60 ns	
Typic Paleudult												
CT O/M 0N	16.2 b	26.7 e	51.5 c	52.2 b	5.4 a	5.5 ab	1.7 b	2.6 c	1.39	1.37	601 ab	392
CT O/M	17.4 b	36.6 de	55.6 bc	62.7 b	5.1 ab	5.3 ab	2.2 b	3.3 bc	1.37	1.42	570 ab	308
RT O/M	16.0 b	36.3 de	54.7 bc	63.8 b	5.4 ab	5.5 ab	5.0 b	5.1 bc	1.49	1.46	319 abc	570
NT O/M	20.5 b	51.6 cd	64.4 ab	86.9 a	5.7 a	5.7 a	5.2 b	5.9 abc	1.49	1.42	188 bc	142
NT O+V/M+CB	24.5 b	58.8 bc	64.7 ab	88.1 a	5.1 ab	5.4 ab	9.5 a	8.4 ab	1.34	1.43	461 abc	310
NT M+PB	39.9 a	73.5 ab	71.0 a	97.2 a	4.5 b	4.9 b	12.1 a	11 a	1.32	1.30	690 a	356
NV	40.4 a	87.6 a	70.0 a	93.1 a	5.1 ab	5.3 ab	1.5 b	2.1 c	1.43	1.52	40 c	6
CV (%)	18	12.5	6.5	7	5.8	4.7	27.1	32.9	6.7	6.1	42.1	108.7
p-value	0.0001	0.00002	0.0004	0.00002	0.01762	0.02846	0.00004	0.0007	0.2223	0.1278	0.0056	0.4593
R	0.85 ***		0.91		0.98 ***		0.91 ***		0.50 *		0.42 ns	

Source: Amado *et al.* [85]. [1] Means in a column followed by same letters are similar at the 0.05 probability level using Tukey. (2) ***, **, *: significant at 0.001, 0.01, and 0.05. ns: non-significant. CV (%) = variation coefficient. NV = native vegetation; CT = conventional tillage; RT = reduced tillage; NT = no tillage; P = pasture; Fw = fallow; Lcc = legume cover crop; O = oat; M = maize; V = vetch; CB = cowpea; PB = pigeon pea; 0N = no N fertilization; p-value: probability value; R: correlation coefficient.

In a study including native vegetation, CT and NT treatments from Argentina, United States and Brazil, the role of biological activity in soil C protection was investigated [86]. The soil types were Mollisol (United States), Vertisol (Argentina) and Oxisol (Brazil) with long-term tillage system adoption. Microbial biomass, evaluated by total phospholipid fatty acids (PLFA), was higher in NT than in CT for the Mollisol and Oxisol probably due to maintained permanent soil protection and high C input by crop residues (Figure 10). Biological activity was also related to the amount of macroaggregates in topsoil (Figure 11). The relationship was stronger in the Mollisol than Oxisol, presumably reflecting the presence of iron and aluminium oxides and higher tillage intensity (eight crops in three years). The presence of macroaggregates has been suggested as an important mechanism for C protection under NT systems. With the exception of the 0–5 cm depth increment in the Oxisol, CT and NT had decreased amounts of macroaggregates and increased amounts of microaggregates when compared to native vegetation (Figure 12). These results agree with aggregate stability findings for long-term tillage experiments on Mollisols from the southeastern Pampas region and Oxisols of Brazil mentioned previously. Long-term NT adoption contributed to the improvement of some soil quality indicators in the region and was crucial for reducing soil erosion. However, quality of agricultural soils in the region has been affected despite the unprecedented adoption of this soil conservation practice. Fortunately, research findings point out that soil quality recovery throughout the region is possible with an integrated soil management approach that includes NT, crop rotations that include short term pasture, and balanced fertilization.

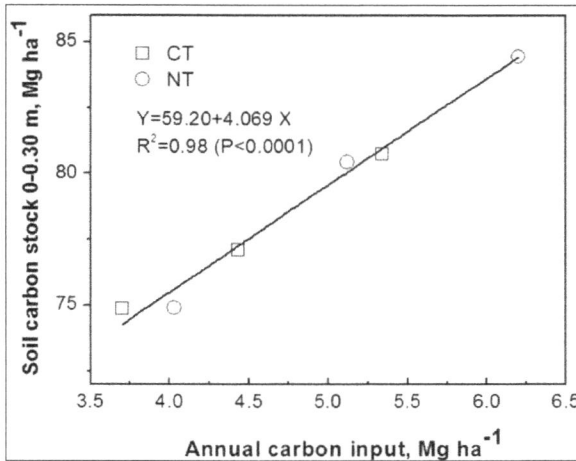

Figure 9. The relationship between annual C input and soil organic C stocks (0–0.30 m) within a subtropical Oxisol under conventional (CT) and no-till (NT) systems. Source: Campos *et al.* [44].

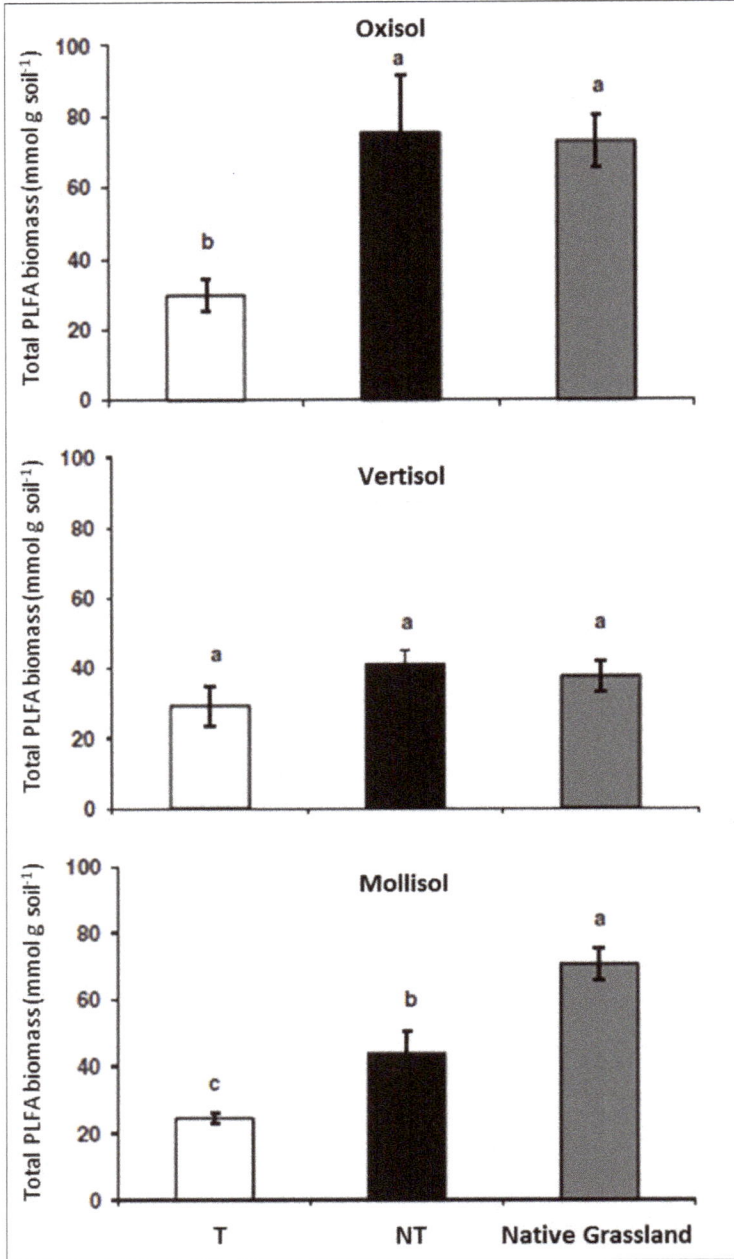

Figure 10. Mean and standard error (SE) values for microbial biomass estimated through PLFA using tilled (T), no-till (NT) and native grassland samples from the 0–0.05 m increments of an Oxisol (Brazil), Mollisol (USA) and Vertisol (Argentina). Source: Fabrizzi *et al.* [86]. Reproduced with permission from journal of Biogeochemistry.

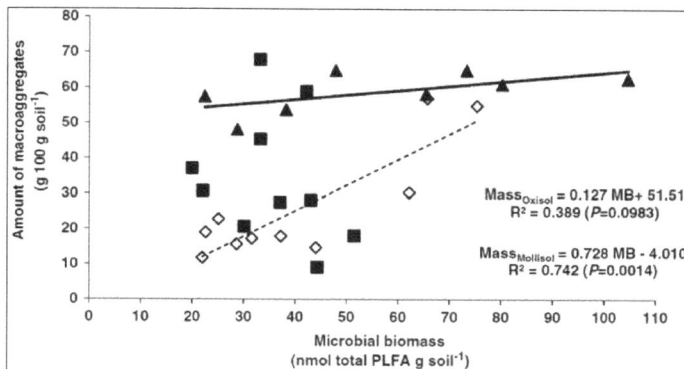

Figure 11. The relationship between microbial biomass, estimated through the PLFA technique, and the amount of macroaggregates (>250 μm) in an Oxisol (▲), Vertisol (■), and Mollisol (◊). Source: Fabrizzi *et al.* [73]. Reproduced with permission from journal of Biogeochemistry.

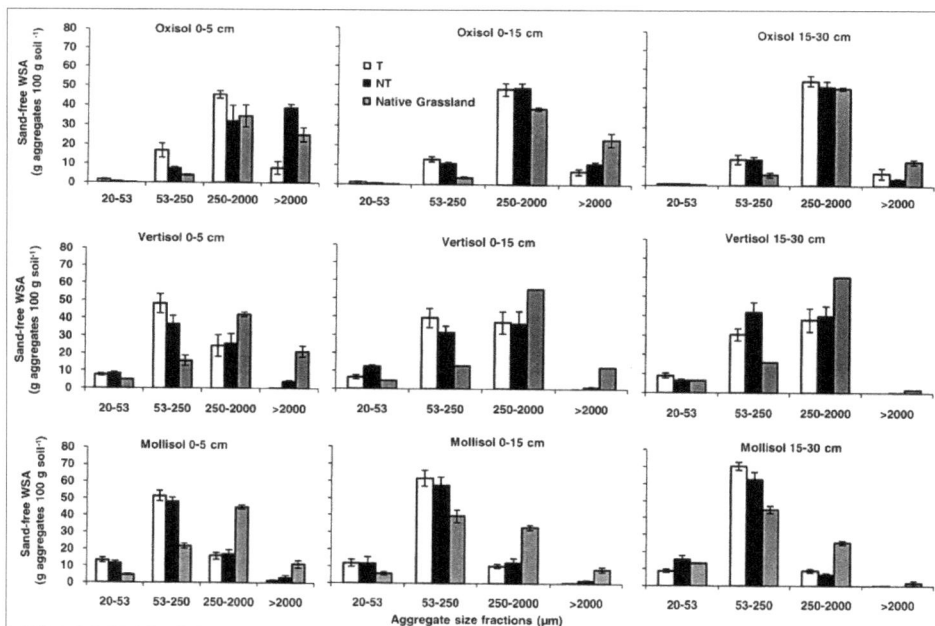

Figure 12. Distribution of sand-free water stable aggregates (WSA) under tilled (T), no-till (NT) and native grassland within the 0–5, 0–15, and 15–30 cm depth increments for the Oxisol, Vertisol, and Mollisol sites. Error bars represent the standard errors of the means. Source: Fabrizzi *et al.* [73]. Reproduced with permission from journal of Biogeochemistry.

7. Strategies for Protecting and Restoring Regional Soil Quality

Recognizing that soil is a nonrenewable resource, several legislative efforts have been initiated throughout the region to recognize, conserve, and protect both the ecosystem services and capacity for food production provided by agricultural soils. Bolivia and Uruguay have national laws of soil use and conservation under agricultural management, while in Argentina and Brazil several provinces/states have specific legislation for soil conservation. In Uruguay, farmers are currently required to develop a Land Use Management Plan with the advice of a certified agronomist for each specific field under agricultural management [87]. Each management plan is then used to determine potential soil erosion associated with the proposed management by using a software package (Erosion 6.0) based on the Universal Soil Loss Equation (USLE). For a plan to be approved, the estimated potential erosion needs to be less than the maximum allowable soil loss threshold for that soil. The Environment Protection law in Bolivia states that agricultural activities should maintain soil productive capacity and avoid soil loss and degradation [88]. To obtain and maintain a soil use license, farmers need to comply with the solicitation requirements by presenting a Management Plan and a Study of Environmental Impacts, to follow the conditions stated in the granting documentation (the Environmental Impact Declaration) of the regulatory authority, and to report the plan progress. The license is valid for 10 years if all the requirements are met. In San Luis province (Argentina), there is an enforced legislation with emphasis on peanuts (*Arachis hypogeae* L.) due to the crop's high risk for soil erosion. Similar to Uruguay, farmers that want to grow peanuts in San Luis province are required to present a five year management plan with the advice of a certified agronomist for each field under production [89]. The plans are evaluated for the potential soil loss based on soil and landscape characteristics, proposed rotations and technology. In Entre Ríos province, Argentina [90], and Parana State of southern Brazil there are specific legislative actions promoting soil conservation practices (e.g., construction of terraces). The legislation is voluntary with modest government support in Argentina, but mandatory in Brazil. The Argentinean legislation in Entre Ríos province also establishes mandatory and voluntary areas for implementation of conservation practices in the province, where farmers in mandatory conservation areas should abide to the conservation guidelines. While not widely spread in the region, these soil conservation efforts based on research findings constitute a starting point for directing the future agricultural expansion towards socially acceptable, economically viable, and environmentally sustainable systems.

8. Summary and Conclusions

This review provides an overview of research that supports our hypothesis that South America's current agricultural practices are detrimental to long-term soil quality even though NT has become a cornerstone for those practices. The analysis highlights the impact of monocultures and a general lack of biodiversity on soil degradation through wind and water erosion, SOM depletion and nutrient loss.

We found that regional economic and land tenure conditions are at odds with practices aimed at long term soil quality conservation and improvement. However, there is strong evidence that farmers embrace new practices if they understand the challenges and benefits. In the region, the adoption of NT was a voluntary reaction by farmers, agronomists and researchers in response to the unsustainable soil erosion observed under CT. The region's adoption of NT was a huge success reaching more than 54 million ha (approximately 45% of global NT area) in less than three decades without direct government financial support for the conversion from CT to NT. In the soybean cropland of the main agroecozones, NT reached adoption rates as high as 90%.

Codifying this trend, Uruguay and Bolivia, and provinces/states in Argentina and Brazil established soil conservation regulations. A couple examples include requiring soil use management plans with a crop rotation program (Uruguay and the San Luis province, Argentina) and promoting the construction of terraces (Entre Ríos province, Argentina and Parana state, Brazil). These government programs differ greatly from those in USA, Canada, and Australia, because they do not stimulate farmers economically for adopting conservation management practices and providing environmental services for the entire society.

Regional farmer adoption of NT without government support needs to be used as an example when looking for solutions to resolve soil degradation problems and the need for increased biodiversity and crop diversification. Developing and achieving adoption of alternative crop rotation systems with equivalent economic return has been a challenge for regional farmers, despite the positive and well-documented impacts of incorporating cover crops, pasture and even maize in the rotation found in agronomic research. There remains a need for multiple types of educational materials and programs that emphasize the importance of conserving the environment and enhancing soil quality. Ultimately, however, farmer compensation (direct and/or market based) for environmental services (*i.e.*, carbon sequestration, reducing nutrient runoff, and increasing biodiversity) may be required to overcome the increasing use of crop monocultures. In developed countries, consumers and retailers are requesting sustainable agricultural commodities. These demands will affect the global supply chain and producers will need to adapt to meet those requirements that often include soil quality performance indicators.

Argentina, Bolivia, southern Brazil, Paraguay and Uruguay have tremendous agricultural potential but without aggressive action and changing current crop production trends, the region will suffer soil degradation as the growing global population increases demand for food. New paradigms of agricultural production are needed to improve current soil conditions. Increased cooperation between the government, scientific community and farmers is crucial for developing effective long term solutions. Everyone's aim should be to develop more sustainable agricultural systems that balance soil quality, environmental sustainability and agricultural production while maintaining economic and social benefits for all.

Acknowledgments: We like to thank the contribution of Emilio Oyarzabal with the conception, discussion, data gathering, and writing of the drafts and revisions of this manuscript.

Author Contributions: Ana Wingeyer, Telmo Amado, Mario Pérez-Bidegain, Carlos Perdomo Varela, Guillermo Studdert and Fernando García were involved in the conception, discussion, and data gathering for this review. All authors contributed to the writing of the manuscript and gave thought to the conclusions. Doug Karlen edited the manuscript and provided critical final review. All authors read and approved the final manuscript.

Conflicts of Interest: The authors declare no conflict of interest.

References

1. United Nations. *United Nations World Population Prospects: The 2012 Revision*; United Nations: New York, NY, USA, 2013.

2. United Nations. *United Nations World Urbanization Prospects: The 2014 Revision Highlights*; United Nations: New York, NY, USA, 2014.

3. Alexandratos, N.; Bruinsma, J. *World Agriculture towards 2030/2050: The 2012 Revision*; Food and Agriculture Organization of the United Nations: Rome, Italy, 2012.

4. Tilman, D.; Balzer, C.; Hill, J.; Befort, B.L. Global food demand and the sustainable intensification of agriculture. *Proc. Natl. Acad. Sci. USA* **2011**, *108*, 20260–20264. [CrossRef] [PubMed]

5. Ray, D.K.; Mueller, N.D.; West, P.C.; Foley, J.A. Yield trends are insufficient to double global crop production by 2050. *PLoS One* **2013**, *8*. Article 6. [PubMed]

6. Duran, A.; Morrás, H.; Studdert, G.; Liu, X. Distribution, properties, land use and management of Mollisols in South America. *Chin. Geogr. Sci.* **2011**, *21*, 511–530. [CrossRef]

7. Altieri, M.A. *Agroecology: The Scientific Basis of Alternative Agriculture*; Institute of Technology Publications: London, UK, 1987.

8. Tivy, J. *Agricultural Ecology*; Longman Scientific & Technical: Harlow, UK, 1990.

9. Lal, R. Tillage effects on soil degradation, soil resilience, soil quality, and sustainability. *Soil Tillage Res.* **1993**, *27*, 1–8. [CrossRef]

10. Elliott, E.T.; Cole, C.V. A perspective on Agroecosystem Science. *Ecology* **1989**, *70*, 1597–1602. [CrossRef]

11. Addiscott, T.M. Entropy and sustainability. *Eur. J. Soil Sci.* **1995**, *46*, 161–168. [CrossRef]

12. Meadows, D.H.; Meadows, D.L.; Randers, J. *Beyond the Limits*; Chelsea Green Publishing Company: Post Mills, VT, USA, 1992; p. 205.

13. Viglizzo, E.F. La interacción sistema-ambiente en condiciones extensivas de producción. *Rev. Argent. Prod. Anim.* **1989**, *9*, 279–294.

14. Hillel, D. An overview of soil and water management: The challenge of enhancing productivity and sustainability. In *Soil Management: Building A Stable Base for Agriculture*; Hatfield, J.L., Sauer, T.J., Eds.; Am. Soc. Agron.; Soil Sci. Soc. Am.: Madison, WI, USA, 2011; pp. 3–11.

15. Reicosky, D.C.; Sauer, T.J.; Hatfield, J.L. Challenging balance between productivity and environmental quality: Tillage impacts. In *Soil Management: Building A Stable Base for Agriculture*; Hatfield, J.L., Sauer, T.J., Eds.; Am. Soc. Agron.; Soil Sci. Soc. Am.: Madison, WI, USA, 2011; pp. 13–37.
16. Weil, R.R.; Magdoff, F. Significance of soil organic matter to soil quality and health. In *Soil Organic Matter in Sustainable Agriculture*; Magdoff, F., Weil, R.R., Eds.; CRC Press: Boca Raton, FL, USA, 2004; pp. 1–43.
17. Kanal, A.; Kolli, R. Influence of cropping on the content, composition and dynamics of organic residue in the soil of the plough layer. *Biol. Fertil. Soils* **1996**, *23*, 153–160. [CrossRef]
18. Lal, R. Enhancing eco-efficiency in agro-ecosystems through soil C sequestration. *Crop Sci.* **2010**, *50*, S120–S131. [CrossRef]
19. Powlson, D.S.; Gregory, P.J.; Whalley, W.R.; Quinton, J.N.; Hopkins, D.W.; Whitmore, A.P.; Hirsch, P.R.; Goulding, K.W.T. Soil management in relation to sustainable agriculture and ecosystem services. *Food Policy* **2011**, *36*, S72–S87. [CrossRef]
20. Manuel-Navarrete, D.; Gallopin, G.C.; Blanco, M.; Diaz-Zorita, M.; Ferraro, D.O.; Herzer, H.; Laterra, P.; Murmis, M.R.; Podesta, G.P.; Rabinovich, J.; *et al.* Multi-causal and integrated assessment of sustainability: The case of agriculturization in the Argentine pampas. *Environ. Dev. Sustain.* **2009**, *11*, 621–638.
21. Food and Agriculture Organization of the United Nations Statistics Division. *Economic and Social Development Department*; FAO: Rome, Italy, 2014.
22. Instituto Brasileiro de Geografia e Estatistica (IBGE). Instituto Brasileiro de Geografia e Estatistica. Available online: http://www.ibge.gov.br/ (accessed on 15 November 2014).
23. Companhia Nacional de Abastecimiento (CONAB). Companhia Nacional de Abastecimiento. Brazil. Available online: http://www.conab.gov.br (accessed on 15 November 2014).
24. Sierra, E.M.; Hurtado, R.H.; Spescha, L. Corrimiento de las isoyetas anuales medias decenales en la región pampeana 1941–1990. *Rev. Fac. Agron. UBA* **1994**, *14*, 139–144.
25. De la Casa, A.C.; Ovando, G.G. Climate change and its impact on agricultural potential in the central region of Argentina between 1941 and 2010. *Agric. For. Meteorol.* **2014**, *195–196*, 1–11.
26. Viglizzo, E.F.; Frank, F.C.; Carreno, L.V.; Jobbagy, E.G.; Pereyra, H.; Clatt, J.; Pincen, D.; Ricard, F.M. Ecological and environmental footprint of 50 years of agricultural expansion in Argentina. *Glob. Chang. Biol.* **2011**, *17*, 959–973. [CrossRef]
27. Asociación Uruguaya pro Siembra Direta. Guía de Siembra Directa. Available online: http://www.ausid.com.uy (accessed on 15 November 2014).
28. Pognante, J.; Bragachini, M.; Casini, C. *Siembra Directa*; Actualización Técnica 58; PRECOP-INTA, MAGyP: Buenos Aires, Argentina, 2011; p. 28.
29. Friedrich, T.; Derpsch, R.; Kassam, A. Overview of the global spread of conservation agriculture. *Field Actions Sci. Rep.* **2012**, *6*, 1–7.
30. Asociación Argentina de Productores en Siembra Directa. Relevamiento de Superficie Agrícola Bajo Siembra Directa 2010. Available online: http://www.aapresid.org.ar/superficie/ (accessed on 10 October 2014).
31. Federação Brasileira de Plantio Direto na Palha. Área do Sistema Plantio Directo. Available online: http://febrapdp.org.br/area-de-pd (accessed on 10 October 2014).
32. Toledo, D.M.; Galantinni, J.A.; Ferreccio, E.; Arzuaga, S.; Gimenez, L.; Vazquez, S. Indicadores e índices de calidad en suelos rojos bajo sistemas naturales y cultivados. *Cienc. Suelo* **2013**, *31*, 201–212.
33. Ferreras, L.; Magra, G.; Besson, P; Kovalevski, E.; García, F. Indicadores de calidad física en suelos de la Región Pampeana Norte de Argentina bajo siembra directa. *Cienc. Suelo* **2007**, *25*, 159–172.
34. Urricariet, S.; Lavado, R.S. Indicadores de deterioro en suelos de la pampa ondulada. *Cienc. Suelo* **1999**, *17*, 37–44.
35. Vázquez, M.E.; Berazategui, L.A.; Chamorro, E.R.; Taquini, L.A.; Barberis, L.A. Evolución de la estabilidad estructural y diferentes propiedades químicas según el uso de los suelos en tres áreas de la Pradera Pampeana. *Cienc. Suelo* **1990**, *8*, 203–210.
36. Rojas, J.M.; Buschiazzo, D.E.; Arce, O.E.A. Parámetros edáficos relacionados con la erosión eólica en inceptisoles del Chaco. *Cienc. Suelo* **2013**, *31*, 133–142.
37. Wilson, M.; Paz-Ferreiro, J. Effects of Soil-Use Intensity on Selected Properties of Mollisols in Entre Ríos, Argentina. *Commun. Soil Sci. Plant Anal.* **2012**, *43*, 71–80. [CrossRef]
38. Novelli, L.E.; Caviglia, O.P.; Melchiori, R.J.M. Impact of soybean cropping frequency on soil carbon storage in Mollisols and Vertisols. *Geoderma* **2011**, *167–168*, 254–260.

39. Novelli, L.E.; Caviglia, O.P.; Wilson, M.G.; Sasal, M.C. Land use intensity and cropping sequence effects on aggregate stability and C storage in a Vertisol and a Mollisol. *Geoderma* **2013**, *195–196*, 260–267.
40. Sanzano, G.A.; Corbella, R.D.; García, J.R.; Fadda, G.S. Degradación física y química de un Haplustol típico bajo distintos sistemas de manejo de suelo. *Cienc. Suelo* **2005**, *23*, 93–100.
41. Barbero, M.F. Evolución del Carbono en Suelos Provenientes de Monte Bajo Siembra Directa del Área Subhúmeda Templada y Subtropical de Argentina. Doctoral Dissertation, Universidad Católica de Córdoba, Cordoba, Argentina, 2010.
42. Campitelli, P.; Aoki, A.; Gudelf, O.; Rubenacker, A.; Sereno, R. Selección de indicadores de calidad de suelo para determinar los efectos del uso y prácticas agrícolas en un área piloto de la región central de Córdoba. *Cienc. Suelo* **2010**, *28*, 233–231.
43. Mori, C. Cambios en la Abundancia Natural de 15N Debidos a la Perturbación Agrícola. Tesis de Maestría, Universidad de la Republica, Montevideo, Uruguay, 2009.
44. Campos, B.C.; Amado, T.J.C.; Bayer, C.; Nicoloso, R.S.; Fiorin, J.E. Carbon stock and its compartments in a subtropical oxisol under long-term tillage and crop rotation systems. *Rev. Bras. Cienc. Solo* **2011**, *35*, 805–817. [CrossRef]
45. Ferreira, A.O. Estoque de Carbono em Areas Pioneiras de Plantio Direto no Rio Grande do Sul. Doctoral Dissertation, Universidade Federal de Santa Maria, Santa Maria, RS, Brasil, 2014.
46. Sainz Rozas, H.R.; Echeverria, H.E.; Angelini, H.P. Organic carbon and pH levels in agricultural soils of the pampa and extra-pampean regions of Argentina. *Cienc. Suelo* **2011**, *29*, 29–37.
47. Sainz Rozas, H.; Echeverria, H.; Angelini, H. Available phosphorus in agricultural soils of the pampa and extra-pampas regions of Argentina. *RIA* **2012**, *38*, 33–39.
48. Reussi Calvo, N.; Calandroni, M.; Studdert, G.; Cabria, F.; Diovisalvi, N.; Berardo, A. Nitrógeno incubado en anaerobiosis y carbono orgánico en suelos agrícolas de Buenos Aires. *Cienc. Suelo* **2014**, in press.
49. Darwich, N.A. Niveles de fósforo asimilable en los suelos pampeanos. *IDIA* **1983**, *409–412*, 1–5.
50. Pérez Bidegain, M.; García Préchac, F.; Hill, M.; Clérici, C. La erosión de suelos en sistemas agrícolas. In *Intensificación Agrícola: Oportunidades y Amenazas Para un País Productivo y Natura*; Universidad de la Republica: Montevideo, Uruguay, 2010; pp. 67–88.
51. Morón, A.; Molfino, J.; Ibañez, W.; Sawchik, J.; Califra, A.; Lazbal, E.; La Manna, A.; Malcuori, E. La calidad de los suelos bajo producción lechera en los principales departamentos de la cuenca: Carbono y nitrógeno. In *Seminario: Sustentabilidad Ambiental de los Sistemas Lecheros en un Contexto Económico de Cambios*; INIA: Montevideo, Uruguay, 2011; pp. 41–46.
52. Morón, A.; Quincke, A.; Molfino, J.; Ibáñez, W.; García, A. Soil quality assessment of Uruguayan agricultural soils. *Agrocienc. Urug.* **2012**, *16*, 135–143.
53. Moron, A.; Sawchik, J. Soil quality indicators in a long term crop-pasture rotation experiment in Uruguay. In Proceedings of the 17th World Congress of Soil Science, Bangkok, Thailand, 14–20 August 2002.
54. Bollinger, A.; Magid, J.; Amado, T.J.C.; Neto, F.S.; Ribeiro, M.; Calegari, A.; Ralisch, R.; Neergaard, A. Taking stock of the brazilian "zero-till revolution": A review of landmark research and farmer practice. *Adv. Agron.* **2006**, *91*, 47–110.
55. Mielniczuk, J.; Bayer, C.; Vezzani, F.M.; Lovato, T.; Fernandes, F.F.; Debarba, L. Manejo de solo e culturas e sua relação com os estoques de carbono e nitrogênio do solo. *Top. Cienc. Solo* **2003**, *3*, 209–248.
56. Amado, T.J.C.; Bayer, C.; Conceicao, P.C.; Spagnollo, E.; Costa de Campos, B.H.; da Veiga, M. Potential of carbon accumulation in no-till soils with intensive use and cover crops in southern Brazil. *J. Environ. Qual.* **2006**, *35*, 1599–1607. [CrossRef] [PubMed]
57. Dalla Nora, D.; Amado, T.J.C. Improvement in chemical attributes of oxisol subsoil and crop yields under no-till. *Agron. J.* **2013**, *105*, 1393–1403. [CrossRef]
58. Studdert, G.A.; Echeverría, H.E.; Casanovas, E.M. Crop-pasture rotation for sustaining the quality and productivity of a typic argiudoll. *Soil Sci. Soc. Am. J.* **1997**, *61*, 1466–1472. [CrossRef]
59. Eiza, M.J.; Fioriti, N.; Studdert, G.A.; Echeverria, H.E. Organic carbon fractions in the arable layer: Cropping systems and nitrogen fertilization effects. *Cienc. Suelo* **2005**, *23*, 59–67.
60. Studdert, G.A.; Domínguez, G.F.; Agostini, M.A.; Monterubbianesi, M.G. Cropping systems to manage southeastern pampas' Mollisol health. Organic C and mineralizable N. In *International Symposium on Soil Quality and Management of World Mollisols*; Liu, X., Song, C., Cruse, R.M., Huffman, T., Eds.; Northeast Forestry University Press: Harbin, China, 2010; pp. 199–200.

61. Studdert, G.A.; Echeverria, H.E. Crop rotations and nitrogen fertilization to manage soil organic carbon dynamics. *Soil Sci. Soc. Am. J.* **2000**, *64*, 1496–1503. [CrossRef]

62. Dominguez, G.F.; Diovisalvi, N.V.; Studdert, G.A.; Monterubbianesi, G.M. Soil organic C and N fractions under continuous cropping with contrasting tillage systems on Mollisols of the southeastern Pampas. *Soil Tillage Res.* **2009**, *102*, 93–100. [CrossRef]

63. Diovisalvi, N.V.; Studdert, G.A.; Dominguez, G.F.; Eiza, M.J. Effect of two tillage systems under continuous cropping on organic carbon and nitrogen fractions and on the anaerobic nitrogen indicator. *Cienc. Suelo* **2008**, *26*, 1–11.

64. Divito, G.A.; Sainz Rozas, H.R.; Echeverria, H.E.; Studdert, G.A.; Wyngaard, N. Long term nitrogen fertilization: Soil property changes in an Argentinean pampas soil under no tillage. *Soil Tillage Res.* **2011**, *114*, 117–126. [CrossRef]

65. Wyngaard, N.; Echeverria, H.E.; Sainz Rozas, H.R.; Divito, G.A. Fertilization and tillage effects on soil properties and maize yield in a southern Pampas argiudoll. *Soil Tillage Res.* **2012**, *119*, 22–30. [CrossRef]

66. Martínez, J.P.; Barbieri, P.A.; Cordone, G.; Sainz Rozas, H.R.; Echeverría, H.E.; Studdert, G.A. Secuencias con predominio de soja en ambientes de la región pampeana Argentina y su efecto sobre el carbono orgánico. In Proceedings of the XXIV Congreso Argentino de la Ciencia del Suelo y II Reunión Nacional Materia Orgánica y Sustancias Húmicas, Bahía Blanca, Buenos Aires, Argentina, 5–9 May 2014.

67. Echeverria, H.; Bergonzi, R.; Ferrari, J. A model for the estimation of nitrogen mineralization in soils of southeast Buenos Aires, Argentina. *Cienc. Suelo* **1994**, *12*, 56–62.

68. Cozzoli, M.V.; Fioriti, N.; Studdert, G.A.; Domínguez, G.F.; Eiza, M.J. Nitrógeno incubado anaeróbico y fracciones de carbono en macro y microagregados bajo distintos sistemas de cultivo. *Cienc. Suelo* **2010**, *28*, 155–167.

69. Urquieta, J.N. Nitrógeno Potencialmente Mineralizable Anaeróbico en Suelos del Sudeste Bonaerense y su Relación con la Respuesta a Nitrógeno en Trigo. Tesis de Ingeniero Agrónomo, Universidad Nacional de Mar del Plata, Balcarce, Argentina, 2008.

70. Genovese, M.F.; Echeverria, H.E.; Studdert, G.A.; Sainz Rozas, H. Nitrógeno de amino-azucares en suelos: Calibración y relación con el nitrógeno incubado anaeróbico. *Cienc. Suelo* **2009**, *27*, 225–236.

71. Domínguez, G.F.; Studdert, G.A.; Echeverría, H.E.; Andrade, F.H. Sistemas de cultivo y nutrición nitrogenada en maíz. *Cienc. Suelo* **2001**, *19*, 47–56.

72. Studdert, G.A.; Echeverría, H.E. Relación entre el cultivo antecesor y la disponibilidad de nitrógeno para el trigo en la rotación. *Cienc. Suelo* **2006**, *24*, 89–96.

73. Reussi Calvo, N.I.; Echeverria, H.E.; Sainz Rozas, H. Comparison between two plant nitrogen and sulphur determination methods: Impact on wheat sulphur diagnostics. *Cienc. Suelo* **2008**, *26*, 161–167.

74. Diovisalvi, N.A.; Berardo, A.; Reussi Calvo, N. Nitrógeno anaeróbico potencialmente mineralizable: Una nueva herramienta para mejorar el manejo de la fertilización nitrogenada. In *Simposio de Fertilidad*; Fertilizar: Rosario, Santa Fe, Argentina, 2009; p. 270.

75. Studdert, G.A.; Echeverría, H.E. Soja, girasol y maíz en los sistemas de cultivo del sudeste Bonaerense. In *Bases Para el Manejo del Maíz, el Girasol y la Soja*; Andrade, F.H., Sadras, V., Eds.; INTA—Facultad de Ciencias Agrarias UNMP: Balcarce, Buenos Aires, Argentina, 2002; pp. 413–443.

76. Roldán, M.F.; Studdert, G.A.; Videla, C.C.; San Martino, S.; Picone, L.I. Distribución de tamaño y estabilidad de agregados en molisoles bajo labranzas contrastantes. *Cienc. Suelo* **2014**, in press.

77. Domínguez, G.F.; Andersen, A.; Studdert, G.A. Cambios en la estabilidad de agregados en distintos sistemas de cultivo bajo siembra directa y labranza convencional. In XXI Congreso Argentino de la Ciencia del Suelo, Potrero de los Funes, San Luis, Argentina, 13–16 May 2008.

78. Agostini, M.A.; Studdert, G.A.; Domínguez, G.F. Relación entre el cambio en el diámetro medio de agregados y el carbono orgánico y sus fracciones. In XIX Congreso Latinoamericano y XXIII Congreso Argentino de la Ciencia del Suelo, Mar del Plata, Buenos Aires, Argentina, 16–20 April 2012.

79. Studdert, G.A. Materia orgánica y sus fracciones como indicadores de uso sustentable de suelos del Sudeste Bonaerense. In XXIV Congreso Argentino de la Ciencia del Suelo y II Reunión Nacional Materia Orgánica y Sustancias Húmicas, Bahía Blanca, Buenos Aires, Argentina, 5–9 May 2014.

80. Garcia-Prechac, F.; Ernst, O.; Siri-Prieto, G.; Terra, J.A. Integrating no-till into crop-pasture rotations in Uruguay. *Soil Tillage Res.* **2004**, *77*, 1–13. [CrossRef]

81. Clérici, C.; Baethgen, W.; García Préchac, F.; Hill, M. Estimación del impacto de la soja sobre erosion y C orgánico en suelos agrícolas de Uruguay. In XIX Congreso Argentino de la Ciencia del Suelo, Paraná, Entre Ríos, Argentina, 24–29 April 2004.

82. Parton, W.J.; Stewart, J.W.B.; Cole, C.V. Dynamics of C, N, P and S in grassland soils: A model. *Biogeochemistry* **1988**, *5*, 109–131. [CrossRef]

83. Salvo, L.; Hernández, J.; Ernst, O. Distribution of soil organic carbon in different size fractions, under pasture and crop rotations with conventional tillage and no-till systems. *Soil Tillage Res.* **2010**, *109*, 116–122. [CrossRef]

84. Sawchik, J.; Pérez-Bidegain, M.; García, C. Impact of winter cover crops on soil properties under soybean cropping systems. *Agrocienc. Urug.* **2012**, *16*, 288–293.

85. Amado, T.J.C.; Conceicao, P.C.; Bayer, C.; Eltz, F.L.F. Soil quality evaluated by "soil quality kit" in two long-term soil management experiments in Rio Grande Do Sul state, Brazil. *Rev. Bras. Cienc. Solo* **2007**, *31*, 109–121. [CrossRef]

86. Fabrizzi, K.P.; Rice, C.W.; Amado, T.J.C.; Fiorin, J.E.; Barbagelata, P.; Melchiori, R. Protection of soil organic C and N in temperate and tropical soils: Effect of native and agroecosystems. *Biogeochemistry* **2009**, *92*, 129–143. [CrossRef]

87. MAGP. *Marco Jurídico de los Planes de Uso y Manejo de Suelos*; Dirección General de Recursos Naturales Renovables: Ministerio de Agricultura, Ganadería y Pesca, de la República Oriental del Uruguay. Available online: http://www.renare.gub.uy/ (accessed on 15 November 2014).

88. GSL. *Ley de Protección y Conservación de Suelos N° IX-0315–2004*; Ministerio del Campo del Gobierno: de la Provincia de San Luis, Argentina. San Luis, Argentina. Available online: http://www.campo.sanluis.gov.ar/campoWeb/Contenido/Pagina11/File/LeyProteccinyConservacindeSuelo.pdf (accessed on 15 November 2014).

89. GER. *Ley Provincial de Conservación y Manejo de Suelos N° 8318*; Secretaria de la Producción del Gobierno: de Entre Ríos, Argentina. Paraná, Argentina. Available online: http://www.entrerios.gov.ar/minpro/userfiles/files/RECNATURALES/RECURSOSNATURALES/RECURSOSNATURALES/LEGISLACIONDESUELOS/legislacion/ley_n8318.pdf (accessed on 15 November 2014).

90. EPB. *Marco Normativo Ley N° 1333 de Protección del Medio Ambiente*; Autoridad de Fiscalización y Control Social de Agua Potable y Saneamiento Básico: Estado Plurinacional de Bolivia. La Paz, Bolivia. Available online: http://www.aaps.gob.bo/?p=334 (accessed on 15 November 2014).

sustainability

MDPI

Review

The Soil Degradation Paradox: Compromising Our Resources When We Need Them the Most

Catherine DeLong [1,*], Richard Cruse [1,†] and John Wiener [2,†]

1   Department of Agronomy, Iowa State University, Ames, IA 50011, USA; rmc@iastate.edu
2   Institute of Behavioral Science, University of Colorado, 468 UCB, Boulder, CO 80309-0468, USA;
     john.wiener@colorado.edu
*   Correspondence: crdelong@iastate.edu; Tel.: +1-515-294-7850; Fax: +1-515-294-3163
†   These authors contributed equally to this work.

Academic Editor: Douglas L. Karlen
Received: 19 November 2014; Accepted: 7 January 2015; Published: 13 January 2015

Abstract: Soil degradation can take many forms, from erosion to salinization to the overall depletion of organic matter. The expression of soil degradation is broad, and so too are the causes. As the world population nears eight billion, and the environmental uncertainty of climate change becomes more manifest, the importance of our soil resources will only increase. The goal of this paper is to synthesize the catalysts of soil degradation and to highlight the interconnected nature of the social and economic causes of soil degradation. An expected three billion people will enter the middle class in the next 20 years; this will lead to an increased demand for meat, dairy products, and consequently grain. As populations rise so do the economic incentives to convert farmland to other purposes. With the intensity and frequency of droughts and flooding increasing, consumer confidence and the ability of crops to reach yield goals are also threatened. In a time of uncertainty, conservation measures are often the first to be sacrificed. In short, we are compromising our soil resources when we need them the most.

Keywords: soil erosion; salinization; land degradation; soil degradation; climate change

1. Introduction

The year 2015 has been declared the International Year of Soils by the United Nations (UN). The goal is to raise global awareness of the importance of soil for food security, climate adaptability and ecosystem functioning. Inspired by the UN's declaration, this paper serves to acknowledge the vital role that soil plays in our ecosystem, with particular emphasis on the increasingly significant role that degraded soils will play as the global population rises, and resources are stressed by climate instability [1]. The first section addresses the importance of preserving our soil resources as agricultural demand is amplified by shifting dietary expectations and an overall increase in earth's population. As agricultural demand increases, more output will be required of our soil resources, which in turn may increase the rate of soil degradation. Considering that 25% of agricultural land is already highly degraded, research addressing the ability of our land resources to meet agricultural demand on increasingly degraded soils is an area of study that demands attention [2]. The second section places this increased agricultural demand in the context of climate instability and the resultant, and already occurring, strain on our natural resources. Often, when uncertainty is looming, such as a fluctuating climate and an unclear ability to meet demand, soil conservation measures are the first to be sacrificed in order to reach yield goals. However, this near-sighted approach compromises our soil resources when we need them the most. The soil plays a critical role in buffering against climate extremes, and yet the role of degraded soils in climate models remains poorly studied. The goal of this paper is to bring into focus the increasingly important role that our soil resources, and particularly degraded soils, will play in the future. As more output is demanded of our soil, and climate volatility compromises the ability to meet this demand, maintaining healthy soils will only become more difficult, but more necessary.

2. Soil Degradation: A Global Pandemic

The expression of soil degradation is varied, but as the other papers in the journal demonstrate, it is extensive. Soil degradation is not isolated to one region, or even one continent; it is a world problem. Eleven percent of the earth's land surface is occupied by agriculture and 25% is already highly degraded according to the UN's Food and Agriculture Organization [2]. Although the expressions of soil degradation range from salinization to the overall depletion of organic matter and nutrients, perhaps the two most extensive forms are salinization and erosion.

Salinization, or the buildup of salts in soil, decreases the osmotic potential of soils so that plants are unable to take up sufficient water to meet physiological needs. Additionally, reclaiming saline soils usually requires large amounts of irrigation water which, as will be discussed later, is a limiting resource in many regions. Salinization is an ancient problem; the birthplaces of agriculture, Mesopotamia and other parts of the Fertile Crescent, were degraded by salinization to the point of abandonment. Today, 34 million ha of land are affected by salinization and some of the major hotspots are in the United States, Pakistan, Iraq and China [2]. Wood [3] estimates that, globally and at varying degrees, salinity will affect an additional 1.5 million ha of arable land each year.

Erosion is the dominant form of soil degradation [4,5]. Erosion removes the most nutrient rich and organic matter dense layer of a soil profile. In turn, this can compromise soil fertility, structure and available water holding capacity. The reach of soil erosion is global and the rate at which it is occurring is often unsustainable. In Europe, Verheijen [6] found that on tilled, arable land, soil is eroding, on average, at 3–40 times the upper tolerable rate of 1.4 t/ha annually. In sub-Saharan Africa, Vlek [7] found that 70% of farmland is degraded due to erosion. In the United States, a meta-analysis by Pimental [8] reports that soil is eroding ten times faster than regeneration rates, while in China and India the rates are 30–40 times faster than regeneration rates. Globally, Montgomery [9] estimates that conventional agriculture practices result in erosion rates that are one to two orders of magnitude greater than both the erosion rate under natural vegetation and soil regeneration rates. In layman terms, we are losing soil much faster than we can replace it.

Additionally, the majority of water-induced soil erosion estimates do not include soil lost from ephemeral gullies, or the cuts in the land that form seasonally [10]. This means that while our estimated and reported soil erosion rates have, in many cases, already reached unsustainable levels, the reported values are potentially much lower than what is actually happening on the ground. Estimated soil erosion rates normally include only sheet and rill erosion components, and are typically reported as averages over relatively large geographical areas, such as reported periodically by the United Stated Department of Agriculture in the National Resources Inventory (NRI) [10]. The NRI uses a stratified statistical sampling methodology allowing definable confidence levels to be identified regarding soil erosion estimates for each state in the United States. The NRI is a critically important and reliable resource, however, it does not illustrate soil erosion rates occurring at spatial scales that account for topographical features, management decisions, and variable rainfall.

To illustrate, as shown in Figure 1, in 2011 the Environmental Working Group utilized the Iowa Daily Erosion Project [11] to identify estimated soil erosion rates for each township in Iowa for 2007 and contrasted results to those of the NRI statewide average for that same year [12]. For the state, NRI erosion estimates were 11.6 Mg/ha for 2007, while the Iowa Daily Erosion Project estimated that at the township level more than 2.4 million ha, or close to 17% of the state, was eroding at rates *greater* than 22 Mg/ha. The maximum estimated township erosion rates were over 130 Mg/ha in 2007. Further, estimates at the township level, approximately 100 km^2, are still too coarse to adequately express the intense spatial variability of soil erosion at the field scale. Current erosion estimates, averages at large spatial scales, therefore inadvertently conceal the damage to critical crop production areas and give a false sense of security on the impacts to crop production.

As the Food and Agriculture Organization reported, soil degradation is not a theoretical problem; it is actively diminishing production capacity and compromising livelihoods at this very moment [2]. Giller [13] found that on experiments in Zimbabwe, degraded soils were less likely to respond to fertilizers because of deficiencies of Ca, Zn, N and P. In Ghana, Diao [14] asserts that land degradation associated with soil erosion will reduce agricultural income from 2006 to 2015 by approximately $4.2 billion or 5% of agriculture's gross domestic product. Due to erosion, yields have been compromised by 20% in India, China, Iran, Israel, Jordan, Lebanon and Pakistan [15]. Globally, salinity has the potential to decrease production at a cost of $11 billion per year [3]. And lastly, Pimentel [8] found that soil erosion costs the US $37.6 billion each year in productivity losses, while worldwide the estimate is close to $400 billion annually.

Soil degradation, however, is not an inevitable result of agriculture; while 25% of cultivated land is highly degraded, 10% is improving [2]. The need to maintain and improve our soil resources will only become more essential as demand for agricultural products increase, and land and water resources diminish. Ironically, rising agricultural demand and resource stress are increasing reliance on our soil resources while also driving soil degradation. The next sections of this paper will explore some of the socio-economic and environmental factors that drive soil degradation.

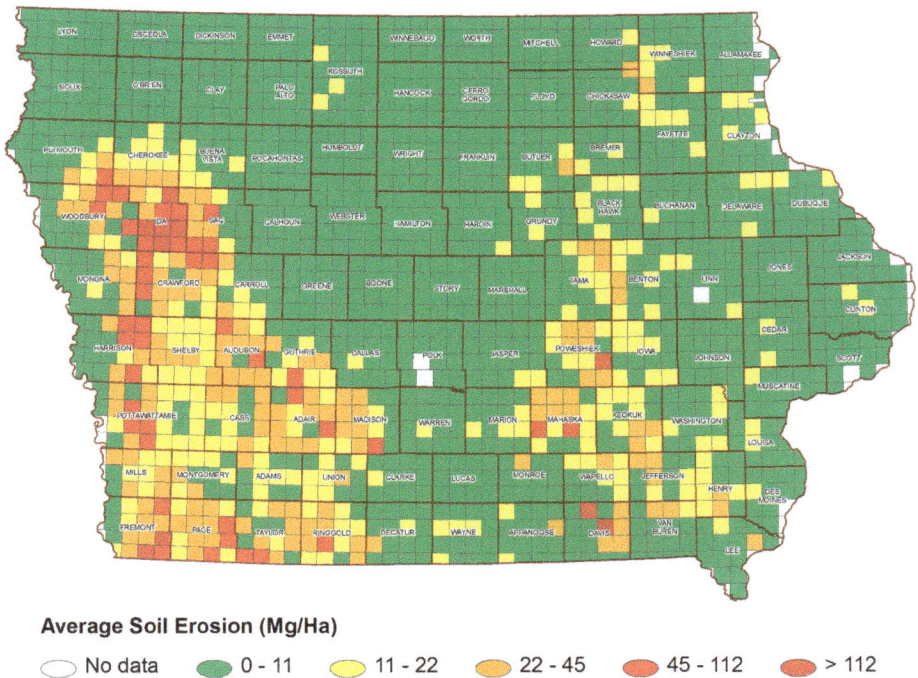

Average Soil Erosion (Mg/Ha)

◯ No data　● 0 - 11　◯ 11 - 22　● 22 - 45　● 45 - 112　● > 112

Figure 1. Average estimated sheet and rill soil erosion rates for each township in Iowa for 2007 [12].

3. Increased Demand for Agricultural Products: Rising Populations and Shifting Class Lines

In the next 40 years, it is predicted that the world population of 7.1 billion will swell by 35% [16]. This growing population will result in an increased demand for agricultural products, and intensifying demands on cropland. One-third of the food produced for human consumption is currently wasted every year, and unless there are drastic improvements in the supply chain and individuals' lifestyles, this trend will remain a constant, and any increase in population will require an increase in calories produced [17]. Not only is the population as a whole rising, but socio-economic shifts within the population are leading to an increased demand for meat and dairy products, and further demand on grain production.

Three billion people are expected to enter the middle class in the next 20 years [17], and as pointed out by Conway [18], growth in domestic product increases in unison with meat demand. What does this increased meat consumption mean for land and water resources? We will assume that the average "new" middle class will consume ~0.19 kg/capita/day of meat, or 60% of the average daily meat consumption of a US citizen [19]. Meat consumption in developing countries currently averages 0.09 kg/capita/day [20]. If three billion people are added to the middle class and meat consumption increases from 0.09 kg/capita/day to 0.19 kg/capita/day, this means an additional 300 million kg of meat must be produced daily. Assuming a 25% average protein content for meat, 75 million kg of animal protein must be produced daily to meet this need [21].

In turn, animals must consume 100–2200 kg of dry matter in order to produce one kg of protein. The conversion efficiency depends on the species of animal, environment, and quality of dry matter being fed [21]. If we assume the most efficient conversion (100 kg dry matter intake per 1 kg protein produced), our land and water resources must produce an additional 7.5 billion kg of dry matter daily. Herrero [21] indicates that the daily global dry matter consumption by livestock in 2000 was approximately 12.9 billion kg. The implication for greater feed production per unit of land area and/or expansion of land area for animal feed production is nontrivial. Currently, livestock production accounts for 23% of all agricultural water use; growth in animal production to meet this rising demand will significantly amplify water demand beyond the strain we are currently experiencing [20].

Further heightening the pressure on land resources is increased biofuel production. In the United States, ethanol, a corn-based biofuel, consumes 25% of the annual maize harvest [22]. From 2000 to 2013, this was an increase of almost 720% [23], with the amount of land dedicated to biofuel production increasing by 10% during the same period [24]. This statistic also takes into account that one-third of the calories used in ethanol production can be recycled for animal feed [25]. In Europe, by 2020 European Union Member States have predicted an increase of 4.1–6.9 million ha of land use changes associated with biofuels [26]. Although estimates for the amount of crops and land dedicated to biofuel use are varied, most studies are in agreement that the number is increasing [27–29]. Additionally, the rising number of hectares that are used for biofuels is not at the expense of uncultivated land, but rather land that is already dedicated to food or feed crops [24]. Thus, the soil is likely to be used more intensely as food production competes with fuel production on finite, and decreasing, land resources.

In the past, a Malthusian catastrophe was avoided by improved plant varieties, and increased fertilizer and irrigated water use associated with the Green Revolution [30]. However the innovations of the Green Revolution may be reaching important plateaus [27,31]. During the Green Revolution productivity growth was at ~2% per year, while today it has declined to ~1% [32]. Yield plateaus have been observed in rice in the Republic of Korea, and wheat in northwest Europe, and India. Additionally, production plateaus have been witnessed for rice and maize in China, which is currently the largest producer of these crops [33]. Despite attempts to make "drought resistant" crops, yield gains will always be tied to the availability of water sources and a fundamental principle of plant physiology that nutrients can only be taken up by the plant in solution. Sinclair [34] clearly articulates that continued crop yield increases are not likely since yield is coupled to transpiration, and a plant cannot continually and exponentially increase its water uptake. Evidence is also mounting that past yield gains have come at the cost of nutritional content as well as the crops' ability to respond to environmental stresses such as drought [35,36]. Thus a higher quantity of crops may be needed to meet the nutritional needs of a single person while the population as a whole is rising.

Further threatening food production is the reality that agricultural land is rapidly being converted to other uses for economic gain. Lambin [37] predict that urbanization will remove 1.6–3.3 Mha of prime agricultural land from production every year. In Bangladesh, land is being converted at an annual rate of 0.56%, resulting in a loss of rice production of 0.86%–1.16% annually [38]. Land use plans in Indonesia have called for as much as 42% of their high-producing rice paddy fields to be converted to other uses [39]. Rice is a major food source for over half of the world's population, and Bangladesh and Indonesia are the third and fourth highest suppliers of rice after China and India [40]. In the United States, over 9.3 million ha of agricultural land were converted to nonagricultural uses from 1982 to 2007, or about 2.5% of farmland [41]. As agricultural land is lost to urbanization, often ignored is the additional cost to our soil resources; globally 1.0–2.9 million ha of soil are degraded annually as a result of expanding cities [37]. Agricultural land area is decreasing and likely to be used more intensely due to the successive forces of rising populations, expanding cities and soil degradation. But in order to provide for the growing population on fewer hectares, we must first preserve our current soil resources. Here is yet another example of the paired fate of increased calories and soil resources; the former cannot be accomplished without maintaining the latter.

With rising food demand and a marginal ability to meet this demand due to soil degradation and land conversion, we are increasingly susceptible to production shocks and ensuing volatile food prices [42]. These "shocks" pose a significant threat to our soil resources as price volatility leads to conservation measures being undervalued and often abandoned. Generally, gross food prices have fallen over the past century and stabilized in the past three decades, but these trends have recently been interrupted by spikes in food prices [43]. Food prices, an indicator of global food availability, have been implicated as a harbinger of political unrest. Lagi [44] point out that in 2008, food riots occurred in 30 countries in North Africa and the Middle East which were experiencing high food prices. In early 2010 and late 2011, even higher food prices corresponded with riots in Mauritania, Uganda and other countries associated with the Arab Spring [44]. The root cause of these spikes in food prices is still being debated, although most researchers agree that resource competition due to biofuel production is a factor [45,46]. One thing is certain, as the demand for agricultural products increases due to rising biofuel production, growing populations and the improved economic status of billions, more will be asked of our land resources. Healthy soils will be necessary to meet these rising demands and avoid the political unrest that often accompanies food price volatility.

4. Resource Stress: Climate Change and Our Soil Resources

In 2013 the concentration of greenhouse gases in our atmosphere outpaced predictions and reached historically high levels [47]. As the concentration of these heat-trapping gases has increased, so too have global temperatures. According to the International Panel on Climate Change (IPCC), the period from 1983 to 2012 was the warmest three decades in the past 1400 years [29]. Needless to say, the effects of climate change are not likely to decrease in the near future. And with each degree of warming, the environmental repercussions are neither incremental nor linear, but exponential [29,48,49]. Impacts include volatile and increasingly intense precipitation events, lengthier droughts, sea level rise and decreasing water resources [29]. However, climate models that forecast these events rarely include the limitation of highly degraded soils in their interaction studies. The next section will detail the resource stress that is likely with climate change. It serves to highlight the importance of maintaining our soil resources as a buffer against climate extremes as well as critical to reaching yield goals.

The majority of crop models predict that global crop yields are declining and will continue to decline as a result of climate change [29,42,50,51]. Lobell [52] found that from 1980 to 2010, the warmest three decades in the past 1400 years, global maize harvest was reduced by 3.8% while wheat was reduced by 5.5%. Although there are significant regional differences in the effect of climate change on crop yields, there is particular consensus that yield losses will occur at low latitudes and tropical areas [29]. These are often the areas, as Wheeler [53] points out, where individuals are already suffering from hunger [30]. More positively, many researchers agree that increased temperatures will have a positive effect on production potential at higher latitudes, although the accompanying soil limitations at these latitudes are rarely addressed [49,50]. In fact, the most recent IPCC Report notes that scientific publications assessing climate impacts, adaptation and vulnerability have "more than doubled" between 2005 and 2010, while studies that address the intersection of climate change and soil resources, particularly the increasing prevalence of degraded soils, are minimal [28,29].

Rising atmospheric CO_2 levels result in warmer air which is able to hold a higher water vapor content. As the global hydrologic cycle intensifies so too will precipitation extremes such as extended wet and dry periods [29]. It is predicted that dry areas, such as sub-tropical regions, will become drier, and wet equatorial regions, will become wetter [54–56]. The productivity of agricultural systems is likely to be limited by environmental extremes rather than averages. For instance, plant response to extended periods of drought will be more important than gradual and incremental changes in precipitation averages. Just as the response of degraded soils is rarely addressed in climate change studies, research addressing crop response to variance in climate change, such as extended wet or dry periods, is needed.

Higher energy storm potentials and extreme precipitation events are also being observed and predicted as a result of increased energy being stored in the atmosphere as latent heat [29,57]. According to Nearing [58], as precipitation increases, soil erosion increases by a factor of 1.7, thus heightening the vulnerability of soil under future climate scenarios. A healthy soil, with high organic matter, can help buffer against precipitation extremes through increased aggregation which allows for higher infiltration and increased water and nutrient retention [59]. Soil productivity of vast areas has been reduced, and in some areas totally lost, due to soil degradation associated with erosion; a climate with an increasing frequency of extreme rainfall events will likely accelerate this process.

In order to understand the potential impact of precipitation and temperature fluctuations, we will take a closer look at the European continent. Fischer [60], using terrain suitability and current rainfall patterns, identified a broad band across Europe as having a high, relative, suitability for rain-fed agriculture. This observation is not surprising considering that central Europe has a high concentration of organic matter rich Mollisols [61]. However, as Figure 2 shows, the IPCC has pinpointed a similar band across Europe as being likely to experience a decrease in summer rainfall of more than 20% in the coming decades [62]. While growing seasonal rainfall is predicted to decrease, Figure 3 demonstrates that temperatures over the European land surface have rapidly increased since the mid to late 1980s [63] causing hotter and drier summer conditions, and an increased likelihood for crop stress. And perhaps the greatest threat to production potential, and in particular rain-fed crop potential, is the alarming rate of soil degradation in this region identified by Verheijen [6]. These analyses for the European continent serve to highlight two facts. First, we cannot properly address the risks and impacts of climate change without including soil degradation as a limiting factor. And, resource stress, such as precipitation declines, means modelers and land managers alike need to value the role that healthy soils play in buffering against climate extremes.

Through variable and extreme temperature and precipitation events, like those seen in Europe, climate change is likely to cause resource stress and crop yield fluctuations. As the global population rises and class lines shift, there is also high certainty that demand for agriculture products will increase. The combination of these two pressures, unstable crop yields and fluctuating populations, is likely to lead land managers to rely heavily on irrigation to stabilize their crop yields. Already, modern agriculture and irrigation are tightly linked. The UN's Food and Agriculture Organization states that 20% of agricultural land is irrigated, and this land provides 40% of the world's food supply [64]. Irrigated land produces yields that are, on average, two to three times higher than rain-fed land in developing countries [65]. Agriculture relies on the weather, an inherently risky position, but irrigation reduces the risk and allows higher and more stable crop yields and a more predictable income for the farmer.

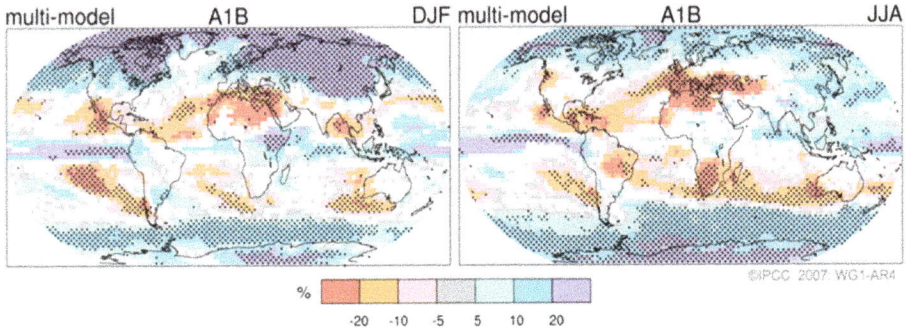

Figure 2. Projected patterns of precipitation change across the globe for the 21st century. Left side projects for the months of December, January and February. Right side projects changes for the months of June, July, and August [62].

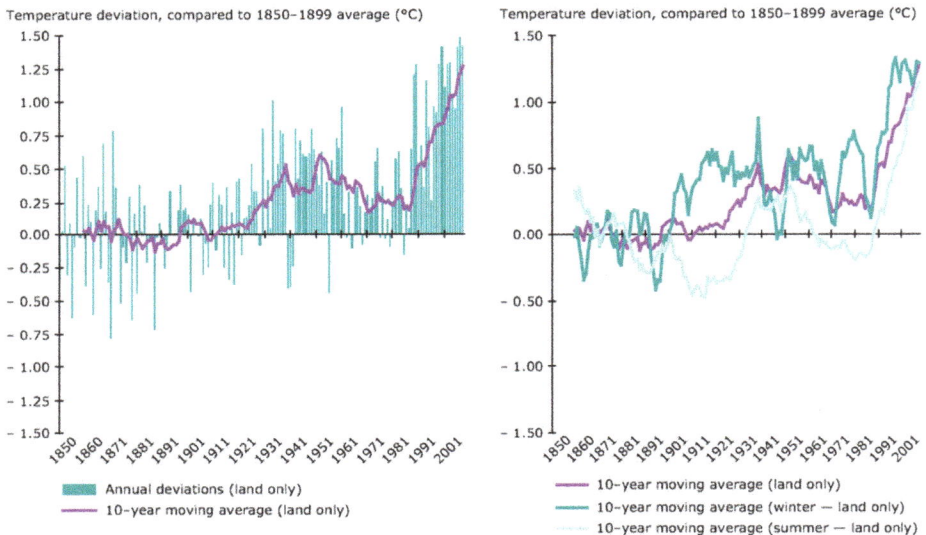

Figure 3. For the period 1850–2009, observed annual temperature (C°) deviations from the 1850–1899 average, and 10-year average for the same period over the European land surface (**left**). For the period 1850–2009, observed 10-year average temperature (C°) deviations from the 1850–1899 average for the European land surface (**right**) [63].

However, our freshwater resources are finite, and nowhere is this better illustrated than in our depleted groundwater supplies. The High Plains Aquifer in the central United States is one of the world's largest freshwater aquifers and the main source of irrigation and drinking water for the eight states that overlie it. Unsustainable pumping of the aquifer has resulted in water level declines of greater than 45 m in parts of Texas, New Mexico and Kansas [66]. In northern China, where the majority of the country's cereals, cotton, fruits and vegetables are grown, groundwater levels have declined at the rate of 0.5 to 3 m annually in the last three decades [67]. Saudi Arabia, in the past, used irrigated aquifer water for domestic wheat production, but today the vast majority of wheat is imported because of depleted groundwater supplies that were increasingly difficult to reach [68].

Globally, Konikow [69] estimates that groundwater has been depleted by approximately 4500 km³ between 1900 and 2008, with the highest rate of depletion in the last decade. Providing for a growing population with less water for crop production seems a stark reality, and one that will most likely require increased dependence on healthy soils and rain-fed agriculture.

5. Conclusions

Climate scientists are in agreement that the foreseeable future holds a higher propensity for extreme weather events, and an overall, and growing, strain on earth's resources. The main conclusion, however, is that our climate is changing and our future is increasingly uncertain. Adding to this uncertainty is the lack of inclusion given to soil, and particularly degraded soils, in climate models. Perhaps this is because soil is viewed as static, but it is more likely that studying the impact of multiple and interconnected stressors on our environment is a difficult venture. However, this is the future of our food production system, and soil degradation must be recognized as playing a dominant role or projections will not adequately represent our future.

Accelerated soil degradation is likely with increasing precipitation intensity and frequency, limiting water resources, and an increased demand for agricultural products from a growing population. Degradation is likely to accelerate as land managers respond to climate variability and increased demand by abandoning long-term soil conservation measures in order to insure yield goals for the current year. Here is the soil degradation paradox: climate variability and a growing population directly and indirectly lead to soil degradation just as healthy soils are increasingly needed to buffer against climate extremes and provide for the population.

Just as the causes of soil degradation are varied and interconnected, the solutions require cooperation, innovation and communication across many groups. The first step is for the scientific community to recognize the societal value of soil and to include it in their discussions, studies and models. In order to accomplish this, soil scientists must be included in trans-disciplinary studies, and soil scientists themselves must broaden their focus and publish their results in a language that is accessible to others. While 25% of agricultural land is highly degraded, 10% is improving [2]. Soil degradation is not an inevitability of agriculture; on the contrary, agriculture can and has improved degraded land by rehabilitating saline soils and implementing conservation measures. By being cognizant of the drivers of soil degradation and recognizing the soils critical role in providing for a growing planet and buffering against climate change, we can avoid the soil degradation paradox.

Author Contributions: Catherine DeLong was the lead author. She drafted sections and integrated sections supplied by Richard Cruse and John Wiener into the final draft. Richard Cruse was responsible for the paper topic area and focus; he also assisted in writing and editing. John Wiener supplied reference and writing support. All authors have read and approved the final manuscript.

Conflicts of Interest: The authors declare no conflict of interest.

References

1. Karlen, D.L.; Peterson, G.A.; Westfall, D.G. Soil and conservation: Our history and future challenges. *Soil Sci. Soc. Am. J.* **2014**, *78*, 1493–1499. [CrossRef]
2. Food and Agriculture Organization of the United Nations (FAO). *The State of the World's Land and Water Resources for Food and Agriculture (SOLAW)—Managing Systems at Risk*; Earthscan: New York, NY, UK, 2011.
3. Wood, S.; Sebastian, K.; Scherr, S.J. *Pilot Analysis of Global Ecosystems: Agroecosystems*; International Food Policy and Research Institute & World Resources Institute: Washington, DC, USA, 2000.
4. Troeh, F.R.; Hobbs, J.A.; Donahue, R.L. *Soil and Water Conservation*; Prentice Hall: Englewood Cliffs, NJ, USA, 1991.
5. Oldeman, L.R. Soil degradation: A threat to food security? In Proceedings of the International Conference on Time Ecology: Time for Soil Culture—Temporal Perspectives on Sustainable Use of Soil, Tutzing, Germany, 6–9 April 1997.

6. Verheijen, F.G.A.; Jones, R.J.A.; Rickson, R.J.; Smith, C.J. Tolerable *versus* actual soil erosion rates in Europe. *Earth Sci. Rev.* **2009**, *94*, 23–38. [CrossRef]

7. Vlek, P.; Le, Q.B.; Tamene, L. *Land Decline in Land-Rich AFRICA—A Creeping Disaster in the Making*; CGIAR Science Council Secretariat: Rome, Italy, 2008.

8. Pimentel, D. Soil erosion: A food and environmental threat. *Environ. Dev. Sustain.* **2006**, *8*, 119–137. [CrossRef]

9. Montgomery, D. Soil erosion and agricultural stability. *Proc. Natl. Acad. Sci. USA* **2007**, *104*, 13268–13272. [CrossRef] [PubMed]

10. USDA. Summary Report: 2010 National Resources Inventory. Available online: http://www.nrcs.usda.gov/Internet/FSE_DOCUMENTS/stelprdb1167354.pdf (accessed on 12 January 2015).

11. Cruse, R.M.; Flanagan, D.; Frankenberger, J.; Gelder, J.; Herzmann, D.; James, D.; Krajewski, W.; Kraszewski, M.; Laflen, J.; Opsomer, J.; *et al.* Daily estimates of rainfall, water runoff, and soil erosion in Iowa. *J. Soil Water Conserv.* **2006**, *61*, 191–199.

12. Cox, C.; Hug, A.; Bruzelius, N. *Losing Ground*; Environmental Working Group: Ames, IA, USA, 2011.

13. Giller, K.E.; Tittonell, P.; Rufino, M.C.; van Wijk, M.T.; Zingore, S.; Mapfumo, P.; Adjei-Nsiah, S.; Herrero, M.; Chikowo, R.; Corbeels, M.; *et al.* Communicating complexity: Integrated assessment of trade-offs concerning soil fertility management within African farming systems to support innovation and development. *Agric. Syst.* **2011**, *104*, 191–203. [CrossRef]

14. Diao, X.; Sarpong, D.B. *Cost Implications of Agricultural Land Degradation in Ghana: An Economy-Wide, Multimarket Model Assessment*; International Food Policy Research Institute: Washington, DC, USA, 2007.

15. Dregne, H.E. (Ed.) *Degradation and Restoration of Arid Lands*; Texas Technical University: Lubbock, TX, USA, 1992.

16. Department of Economic and Social Affairs, Population Division, United Nations. *World Population Prospects: The 2010 Revision, Volume I: Comprehensive Tables*; ST/ESA/SER.A/313; United Nations: New York, NY, USA, 2011.

17. United Nations Secretary-General's High-Level Panel on Global Sustainability. *Resilient People, Resilient Planet: A future Worth Choosing*; United Nations: New York, NY, USA, 2012.

18. Conway, G. One Billion Hungry: Can We Feed the World? Cornell University Press: New York, NY, USA, 2012; p. 192.

19. FAOstat. Available online: http://faostat.fao.org/site/610/DesktopDefault.aspx?PageID=610#ancor (accessed on 14 October 2014).

20. World Watch Institute. Available online: http://www.worldwatch.org/global-meat-production-and-consumption-continue-rise (accessed 14 October 2014).

21. Herrero, M.; Havlik, P.; Valin, H.; Notenbaert, A.; Rufino, M.C.; Thornton, P.K.; Blummel, M.; Weiss, F.; Grace, D.; Obersteiner, M. Biomass use, production, feed efficiencies, and greenhouse gas emissions from global livestock systems. *Proc. Natl. Acad. Sci. USA* **2013**, *110*, 20888–20893. [CrossRef] [PubMed]

22. USDA. *Table 5. Corn Supply, Disappearance, and Share of Total Corn Used for Ethanol*; US Department of Agriculture Economic Research Service, US Bioenergy Statistics: Washington, DC, USA, 2013.

23. EIA. *Monthly Energy Review, September 2013*; US Energy Information Administration: Washington, DC, USA, 2013.

24. Wallander, S.; Claassen, R.; Nickerson, C. The Ethanol Decade: An Expansion of U.S. Corn Production, 2000–09. Available online: http://www.ers.usda.gov/publications/eib-economic-information-bulletin/eib79.aspx (accessed on 8 January 2015).

25. Roberts, M.J.; Schlenker, W. Identifying supply and demand elasticities of agricultural commodities: Implications for the US ethanol mandate. *Am. Econ. Rev.* **2013**, *103*, 2265–2295. [CrossRef]

26. Bowyer, C. *Anticipated Indirect Land Use Change Associated with Expanded Use of Biofuels and Bioliquids in the EU*; Institute for European Environmental Policy: London, UK, 2010.

27. Hertel, T.W.; Ramankutty, N.; Baldos, U.L.C. Global market integration increases likelihood that a future African Green Revolution could increase crop land use and CO_2 emissions. *Proc. Natl. Acad. Sci. USA* **2014**, *111*, 13799–13804. [CrossRef] [PubMed]

28. Porter, J.R.; Xie, L.; Challinor, A.J.; Cochrane, K.; Howden, S.M.; Iqbal, M.M.; Lobell, D.B.; Travasso, M.J. Food Security and Food Production Systems. In *Climate Change 2014: Impacts, Adaptation and Vulnerability. Part A: Global and Sectoral Aspects*; Contribution of Working Group II to the Fifth Assessment Report of the Intergovernmental Panel on Climate Change; Field, C.B., Barros, V.R., Dokken, D.J., Mach, K.J., Mastandrea, M.D., Bilir, T.E., Chatterjee, M., Ebis, K.L., Estrada, Y.O., Genova, R.C., *et al.*, Eds.; Cambridge University Press: Cambridge, UK; New York, NY, USA, 2014.

29. Intergovernmental Panel on Climate Change (IPCC). *Climate Change 2014: Impacts, Adaptation, and Vulnerability. Part A: Global and Sectoral Aspects*; Contribution of Working Group II to the Fifth Assessment Report of the Intergovernmental Panel on Climate Change; Field, C.B., Barros, V.R., Dokken, D.J., Mach, K.J., Mastandrea, M.D., Bilir, T.E., Chatterjee, M., Ebis, K.L., Estrada, Y.O., Genova, R.C., *et al.*, Eds.; Cambridge University Press: Cambridge, UK; New York, NY, USA, 2014.

30. Lal, R. Climate-strategic agriculture and the water-soil-waste nexus. *J. Plant Nutr. Soil Sci.* **2013**, *176*, 479–493. [CrossRef]

31. Pingali, P.L. Green Revolution: Impacts, limits and the path ahead. *Proc. Natl. Acad. Sci. USA* **2012**, *109*, 12302–12308. [CrossRef] [PubMed]

32. Bindraban, P.S.; van der Velde, M.; Ye, L.; van den Berg, M.; Materechera, S.; Kiba, D.I.; Tamene, L.; Ragnarsdóttir, K.V.; Jongschaap, R.; Hoogmoed, M.; *et al.* Assessing the impact of soil degradation on food production. *Curr. Opin. Environ. Sustain.* **2012**, *4*, 478–488. [CrossRef]

33. Cassman, K.G.; Grassini, P.; van Wart, J. Crop yield potential, yield trends, and global food security in a changing climate. In *Handbook of Climate Change and Agroecosystems; Impacts, Adaptation, and Mitigation*; Hillel, D., Rosenzweig, C., Eds.; Imperial College Press: London, UK, 2011.

34. Sinclair, T.R. Precipitation: The thousand-pound gorilla in crop response to climate change. In *Handbook of Climate Change and Agroecosystems; Impacts, Adaptation, and Mitigation*; Hillel, D., Rosenzweig, C., Eds.; Imperial College Press: London, UK, 2011.

35. Davis, D.R. Declining fruit and vegetable nutrient composition: What is the evidence. *HortScience* **2009**, *44*, 15–19.

36. Lobell, D.B.; Roberts, M.J.; Schlenker, W.; Braun, N.; Little, B.B.; Rejesus, R.M.; Hammer, G.L. Greater sensitivity to drought accompanies maize yield increase in the U.S. Midwest. *Science* **2014**, *344*, 516–519. [CrossRef] [PubMed]

37. Lambin, E.F.; Meyfroidt, P. Global land use change, economic globalization, and the looming land scarcity. *Proc. Natl. Acad. Sci. USA* **2011**, *108*, 3465–3472. [CrossRef] [PubMed]

38. Quasem, M.A. Conversion of agricultural land to non-agricultural uses in Bangladesh: Extent and determinants. *Bangladesh Dev. Stud.* **2011**, *34*, 59–85.

39. Agus, F.; Irawan. Agricultural land conversion as a threat to food security and environmental quality. *Indones. J. Agric. Sci.* **2006**, *25*, 90–98.

40. Foreign Agriculture Service. Table 09 Rice Area, Yield, Production. Available online: http://apps.fas.usda.gov/ psdonline/psdReport.aspx?hidReportRetrievalName=Table+09+Rice+Area%2c+Yield%2c+and+Production+ &hidReportRetrievalID=893&hidReportRetrievalTemplateID=1 (accessed on 17 October 2014).

41. American Farmland Trust. National Statistics Sheet. Available online: http://www.farmlandinfo.org/ agricultural_statistics/ (accessed on 20 September 2014).

42. Nelson, G.C.; Valin, H.; Sands, R.D.; Havlik, P.; Ahammad, H.; Deryng, D.; Elliott, J.; Fujimori, S.; Hasegawa, T.; Heyhoe, E.; *et al.* Climate change effects on agriculture: Economic responses to biophysical shocks. *Proc. Natl. Acad. Sci. USA* **2014**, *9*, 3274–3279. [CrossRef]

43. Godfray, H.C.H.; Beddington, J.R.; Crute, I.R.; Haddad, L.; Lawrence, D.; Muir, J.F.; Pretty, J.; Robinson, S.; Thomas, S.M.; Camilla, T. Food security: The challenge of feeding 9 billion people. *Science* **2010**, *327*, 812–818. [CrossRef] [PubMed]

44. Lagi, M.; Bertrand, K.; Bar-Yam, Y. The Food Crises and Political Instability in North Africa and the Middle East. Available online: http://arxiv.org/abs/1108.2455 (accessed on 8 January 2015).

45. Lagi, M.; Bar-Yam, Y.; Bertrand, K.Z.; Bar-Yam, Y. The Food Crises: A Quantitative Model of Food Prices Including Speculators and Ethanol Conversion. Available online: http://arxiv.org/abs/2011arXiv1109.4859L (accessed on 8 January 2015).

46. Piesse, J.; Thirtle, C. Three bubbles and a panic: An explanatory review of recent food commodity price events. *Food Policy* **2009**, *34*, 119–129. [CrossRef]

47. World Meteorological Organization and Global Atmosphere Watch. The State of Greenhouse Gases in the Atmosphere Based on Global Observations through 2013. Available online: http://www.indiaenvironmentportal. org.in/content/399916/the-state-of-greenhouse-gases-in-the-atmosphere-based-on-global-observations-through-2013/ (accessed on 8 January 2015).

48. Schewe, J.; Heinke, J.; Gerten, D.; Haddeland, I.; Arnell, N.W.; Clark, D.B.; Dankers, R.; Eisner, S.; Fekete, B.M.; Colon-Gonzalez, F.J.; *et al.* Multimodel Assessment of water scarcity under climate change. *Proc. Natl. Acad. Sci. USA* **2014**, *111*, 3245–3250. [CrossRef] [PubMed]

49. Piontek, F.; Muller, C.; Pugh, T.A.M.; Clark, D.B.; Deryng, D.; Elliott, J.; de Jesus Colon-Gonzalez, F.; Florke, M.; Folberth, C.; Franssen, W.; *et al.* Multisectoral climate impact hotspots in a warming world. *Proc. Natl. Acad. Sci. USA* **2014**, *111*, 3233–3238. [CrossRef] [PubMed]

50. Rosenzweig, C.; Elliott, J.; Deryng, D.; Ruane, A.C.; Muller, C.; Arneth, A.; Boote, K.C.; Folberth, C.; Glotter, M.; Khabarov, N.; *et al.* Assessing agricultural risks of climate change in the 21st Century in a global gridded crop model intercomparison. *Proc. Natl. Acad. Sci. USA* **2014**, *111*, 3268–3273. [CrossRef] [PubMed]

51. Elliott, J.; Deryng, D.; Muller, C.; Freier, K.; Konzmann, M.; Gerten, D.; Glotter, M.; Florke, M.; Wada, Y.; Best, N.; *et al.* Constraints and potentials of future irrigation water availability on agricultural production under climate change. *Proc. Natl. Acad. Sci. USA* **2014**, *111*, 3239–3244. [CrossRef] [PubMed]

52. Lobell, D.B.; Schlenker, W.S.; Costa-Roberts, J. Climate trends and global crop production since 1980. *Science* **2011**, *333*, 616–620. [CrossRef] [PubMed]

53. Wheeler, T.; von Braun, J. Climate change impacts on global food security. *Science* **2013**, *341*, 508–513. [CrossRef] [PubMed]

54. Held, I.M.; Soden, B.J. Robust responses of the hydrological cycle to global warming. *J. Clim.* **2006**, *19*, 5686–5699. [CrossRef]

55. Chou, C.; Neelin, J.D.; Chen, C.-A.; Tu, J.-Y. Evaluating the Rich-Get-Richer mechanism in tropical precipitation change under global warming. *J. Clim.* **2009**, *22*, 1982–2005. [CrossRef]

56. Durack, P.J.; Wijffels, S.E.; Matear, R.J. Ocean salinities reveal strong global water cycle intensification during 1950 to 2000. *Science* **2012**, *336*, 455–458. [CrossRef] [PubMed]

57. Min, S.; Zhang, X.; Zwiers, F.; Hegerl, G. Human contribution to more-intense precipitation extremes. *Nature* **2011**, *470*, 378–381. [CrossRef]

58. Tisdall, J.M.; Oades, J.M. Organic matter and water-stable aggregates in soils. *J. Soil Sci.* **1982**, *33*, 141–163. [CrossRef]

59. Nearing, M.A.; Pruski, F.F.; O'Neal, M.R. Expected climate change impacts on soil erosion rates: A review. *Soil Water Conserv. J.* **2004**, *59*, 43–50.

60. Fischer, G.; van Velthuizen, H.; Shah, M.; Nachtergaele, F. *Global Agro-ecological Assessment for Agriculture in the 21st Century: Methodology and Results*; International Institute for Applied Systems Analysis at the FAO: Rome, Italy, 2002.

61. Liu, X.; Burras, C.L.; Kravchenko, Y.S.; Duran, A.; Huffman, T.; Morras, H.; Studdert, G.; Zhang, X.; Cruse, R.M.; Yuan, X. Overview of Mollisols in the world: Distribution, land use and management. *Can. J. Soil Sci.* **2012**, *92*, 383–402. [CrossRef]

62. IPCC. *Climate Change 2007: The Physical Science Basis*; Contribution of Working Group I to the Fourth Assessment Report of the Intergovernmental Panel on Climate Change; Solomon, S., Qin, D., Manning, M., Chen, Z., Marquis, M., Averyt, K.B., Tignor, M., Miller, H.L., Eds.; Cambridge University Press: Cambridge, UK; New York, NY, USA, 2007.

63. European Environment Agency. Mean Surface Temperature in Europe 1850–2009, Annual and by Season. Available online: http://www.eea.europa.eu/data-and-maps/figures/mean-surface-temperature-in-europe (accessed 9 October 2014).

64. Turral, H.; Burke, J.; Faures, J.-M. *Climate Change, Water and Food Security*; FAO: Rome, Italy, 2011.

65. Ruane, J.; Sonnino, A.; Steduto, P.; Deane, C. Coping with water scarcity: What role for biotechnologies? In *Climate Change, Water and Food Security*; Turral, H., Burke, J., Faures, J.-M., Eds.; FAO: Rome, Italy, 2011.

66. Konikow, L.F. Groundwater Depletion in the United States (1900–2008. Available online: http://pubs.usgs. gov/sir/2013/5079/SIR2013-5079.pdf (accessed on 20 October 2014).

67. Currell, M.J.; Han, D.; Chen, Z.; Cartwright, I. Sustainability of groundwater usage in northern China: Dependence on palaeowaters and effects on water quality, quantity and ecosystem health. *Hydrol. Process.* **2012**, *26*, 4050–4066. [CrossRef]

68. Brown, L.; Black, B.; Hussein, G.H.G. Aquifer depletion. In *Encyclopedia of Earth*; Cleveland, C.J., Ed.; Environmental Information Coalition, National Council for Science and the Environment: Washington, DC, USA, 2011.

69. Konikow, L.F. Contribution of global groundwater depletion since 1900 to sea-level rise. *Geophys. Res. Lett.* **2011**. [CrossRef]

Review

Soil Degradation and Soil Quality in Western Europe: Current Situation and Future Perspectives

Iñigo Virto [1,*], María José Imaz [1], Oihane Fernández-Ugalde [2], Nahia Gartzia-Bengoetxea [3], Alberto Enrique [1] and Paloma Bescansa [1]

[1] Escuela Técnica Superior de Ingenieros Agrónomos, Universidad Pública de Navarra, Campus Arrosadia, Pamplona 31006, Spain; mj.imaz@unavarra.es (M.J.I.); alberto.enrique@unavarra.es (A.E.); bescansa@unavarra.es (P.B.)

[2] Institute for Environment and Sustainability–Land Resources, Management Unit, Joint Research Center, Ispra 21027, Italy; oihane.fernandez-ugalde@jrc.ec.europa.eu

[3] Neiker-Tecnalia, Departamento de Conservación de Recursos Naturales, Bizkaiko Zientzia eta Teknologia Parkea 812.L, Derio 48160, Spain; ngartzia@neiker.net

* Correspondence: inigo.virto@unavarra.es; Tel.: +34-948-169166; Fax: +34-948-168930

Academic Editors: Marc A. Rosen and Douglas L. Karlen

Received: 6 November 2014; Accepted: 19 December 2014; Published: 31 December 2014

Abstract: The extent and causes of chemical, physical and biological degradation of soil, and of soil loss, vary greatly in different countries in Western Europe. The objective of this review paper is to examine these issues and also strategies for soil protection and future perspectives for soil quality evaluation, in light of present legislation aimed at soil protection. Agriculture and forestry are the main causes of many of the above problems, especially physical degradation, erosion and organic matter loss. Land take and soil sealing have increased in recent decades, further enhancing the problems. In agricultural land, conservation farming, organic farming and other soil-friendly practices have been seen to have site-specific effects, depending on the soil characteristics and the particular types of land use and land users. No single soil management strategy is suitable for all regions, soil types and soil uses. Except for soil contamination, specific legislation for soil protection is lacking in Western Europe. The Thematic Strategy for Soil Protection in the European Union has produced valuable information and has encouraged the development of networks and databases. However, soil degradation is addressed only indirectly in environmental policies and through the Common Agricultural Policy of the European Union, which promotes farming practices that support soil conservation. Despite these efforts, there remains a need for soil monitoring networks and decision-support systems aimed at optimization of soil quality in the region. The pressure on European soils will continue in the future, and a clearly defined regulatory framework is needed.

Keywords: soil quality; Western Europe; sustainable soil management

---

1. Soil Degradation in Western Europe

1.1. Geographical Diversity of Soils in Western Europe

Western Europe (WE) is a loose term for the collection of countries lying in the most westerly part of Europe. However, the definition is context-dependent as it has political and geographic connotations. From a geographical point of view, the United Nations' geoscheme divides Europe into four regions: Western, Eastern, Northern and Southern Europe [1]. In this paper, the term WE is used to refer to countries in the Western half of the continent, including the Western, Northern and Southern regions [1]. Most of the countries border with the Atlantic Ocean and/or the West Mediterranean Sea. The region includes the countries that had joined the European Union (EU) before 2000 (Table 1) and which have therefore implemented Common Agricultural Policy (CAP) regulations and other EU environmental directives affecting soils (*i.e.*, Nitrates, Water and Pesticides Framework Directives) during at least the last 15 years. Iceland, Norway and Switzerland are included because of their geographical location. Historical land government and ownership also have common traits in the region. As a result, the trends in agricultural and forest soil management, and the strategies for mitigating soil degradation are somewhat similar in these countries.

In WE, interactions between climatic, geological and topographic conditions have resulted in a large natural diversity of soils. Twenty-three of the 32 reference soil groups included in the World Reference Base [2] occur in WE. Soil groups are presented for each country in Table 1, following the Soil Atlas of Europe [3]. The most common soils across WE are Cambisols, Podzols, Leptosols, Luvisols, Fluvisols Gleysols, Regosols and Calcisols. However, distribution of the soils is uneven. Cambisols and Podzols each occupy 12%–14% of the total land in Europe. Cambisols occur in a wide variety of environments and under all types of vegetation, and they are present in almost all countries. Podzols are mainly present in the boreal and temperate zones of Northern countries (Norway, Sweden, Finland, Denmark, Scotland, N Germany and some areas of France and the Alps). Leptosols (9% of the European land) are mainly present in mountainous regions of Spain, France, Switzerland, NE Italy, and Norway. Luvisols, Fluvisols, Histosols, Gleysols, Regosols and Calcisols each occupy 5%–6% of the land. Fluvisols are common in river fans, valleys, and tidal marshes in all climate zones. Histosols, which mainly comprise organic matter, are common in boreal and sub-artic regions, Scotland and Ireland. Gleysols occur in lowland areas that have been saturated with groundwater for long periods, mainly in the United Kingdom (UK) and Ireland. Regosols are widespread in arid and semi-arid areas, as well as in mountainous regions of Portugal and Spain. Finally, Calcisols are also common in Spain. They appear in regions with calcareous parent materials and distinct dry seasons, and in dry zones where carbonate-rich groundwater appears near the surface.

Table 1. Major soil types in each country according to the Soil Atlas of Europe [3]. Soils shown in bold type are predominant in the particular country.

European Region	Country *	Main Soil Types [2]
North Western	Denmark *	Cambisol, Gleysol, Luvisol, Podzol
	Finland *	Cambisol, Gleysol, Histosol, Leptosol, Podzol
	Iceland	Andosol, Histosol, Luvisol, Podzol, Umbrisol
	Ireland *	Cambisol, Gleysol, Leptosol
	Norway	Albeluvisol, Cambisol, Leptosol, Phaeozem, Podzol
	Sweden *	Cambisol, Histosol, Leptosol, Podzol, Regosol
	United Kingdom *	Cambisol, Gleysol, Histosol, Leptosol, Luvisol, Podzol, Umbrisol
Central Western	Austria *	Cambisol, Chernozem, Fluvisol, Leptosol, Luvisol, Podzol
	Belgium *	Albeluvisol, Cambisol, Fluvisol, Luvisol, Podzol
	France *	Albeluvisol, Andosol, Calcisol, Cambisol, Leptosol, Luvisol, Podzol
	Germany *	Cambisol, Chernozem, Fluvisol, Luvisol, Podzol, Umbrisol
	Luxembourg *	Arenosol, Cambisol, Fluvisol
	Netherlands *	Fluvisol, Gleysol, Histosol, Podzol
	Switzerland	Albeluvisol, Cambisol, Leptosol, Luvisol, Podzol, Umbrisol
South Western	Greece *	Cambisol, Fluvisol, Leptosol, Luvisol, Vertisol
	Italy *	Andosol, Calcisol, Cambisol, Fluvisol, Leptosol, Luvisol, Podzol, Vertisol
	Portugal *	Cambisol, Fluvisol, Luvisol, Podzol, Regosol, Umbrisol, Vertisol
	Spain *	Calcisol, Cambisol, Fluvisol, Gypsisol, Leptosol, Luvisol, Regosol, Umbrisol, Vertisol

* Countries indicated with an asterisk are part of the European Union.

Different risks and soil degradation processes occur within the various areas of WE because of significant differences in the intrinsic properties of these types of soils and local variations in each soil group.

1.2. Historical Soil Management and Present Land-Use Patterns

Land and soil are fundamental pillars of agricultural economies and are essential for industrial and urban development. In WE, agriculture and forestry have traditionally been the most widespread types of land use and have shaped the rural landscape [4]. Their relative importance has changed throughout history. The historical relationship between these types of land use has been shaped by socioeconomic and technological changes, demographical fluctuations and environmental variability. At the beginning of the Middle Ages, more than 80% of the population in WE was working in agriculture [4]. In the 14th century, the European population was drastically reduced as a result of the Black Death, and the cultivated area thus became smaller. The agricultural area then expanded until the end of the 18th century, when a new period of contraction began due to the increased productivity per hectare. During the 19th century, the agricultural area again expanded because population growth rate exceeded the agricultural productivity per hectare. Forests expanded and contracted in the opposite fashion to agricultural land. About half of WE forests are estimated to have been cleared prior to the Middle Ages [5]. The highest rates of deforestation occurred on the land best suited for farming, especially in France, Germany and the UK. Since then, the periods of most intense deforestation have coincided with those of high economic activity. Trees were felled when grain prices rose and forest land was converted to cropland. The use of wood for construction and shipbuilding also contributed to forest degradation and eventual deforestation in France, Portugal and Spain. Wood was also needed to fuel foundries and smelters early on in the Industrial Revolution, resulting in further forest degradation and deforestation, even on land not suitable for agriculture. Old-growth, primary forests essentially disappeared from WE in these periods. In the last 150 years, forests planted to produce raw materials expanded dramatically in WE. Requirements for food and/or timber, as well as town and country planning, were therefore the drivers of land use in WE until the mid 20th century.

In the early 1960s, the European Union's CAP enforced some degree of harmonization in agriculture and influenced land use patterns in many WE countries. The CAP was designed to guarantee the supply of sufficient food for EU citizens, to support the price of agricultural products, and to provide farmers with an acceptable level of income. Implementation of the CAP together with technological progress caused a sharp increase in agricultural productivity, which led to overproduction of agricultural goods in the 1970s and 1980s. In the following decades, the CAP has been reformed several times to regulate agricultural production and stabilize agricultural markets (see Section 3.2). As a consequence, the recent history of land use patterns again shows contraction of agricultural areas and expansion of forests. This process, which is particularly notable in the extensive margins (*i.e.*, alpine regions), has somewhat counterbalanced the historical loss of forest.

Nowadays, rural European landscapes are generally still strongly linked to agriculture and forestry. In 2009, agriculture (43% of the surface) and forestry (30%) were the most common primary land use categories in the EU [6]. As shown in Table 2, the land cover pattern in WE shows a clear North-South gradient [6–9]. Semi-natural forest prevails in the Northern countries (Norway, Sweden and Finland), while agricultural land (arable land, permanent crops and grassland) dominates the rural landscape between Denmark and Southern Europe, as well as in the UK and Ireland.

The proportion of arable land that is actually cultivated varies greatly among countries. This is related to population density and traditional land use. In the tilled arable area, although the most widespread system across WE is conventional tillage, the tillage systems vary greatly between countries [10] (see below, Section 2.1). Regarding woodland, the most common function of forests in WE is for wood production, although different patterns also exist among regions and countries. The Atlantic area of WE is characterized by forest plantations, which cover more than 5 million ha in Portugal, Spain, France and the UK, producing 33 million m^3 of wood in 2012. As a region, WE is the sixth industrial roundwood producer of the world, just behind Chile and in front of New Zealand. In the WE, the proportion of industrial roundwood produced in plantations is 31% in France, 52% in Spain, 65% in UK and 99% in Portugal [5]. The most common species in planted forests in WE are *Eucalyptus* spp., *Populus* spp; *Picea sitchensis*, *Pinus radiata* and *Pinus pinaster*.

Land management in agriculture and forestry has had a strong impact on the natural environment, including soils. On the one hand, over the centuries, farming has created and maintained a variety of valuable semi-natural habitats on which a wide range of wildlife depends for survival. On the other hand, land use changes and farming practices have also had negative impacts on natural resources, such as habitat fragmentation, loss of biodiversity and soil degradation. In particular, more than half of the land in Europe has suffered some type of soil degradation in the last few decades [11].

Table 2. Land use in each country in Western Europe (as a percentage of total land area).

European Region	Country	Woodland	Cropland	Grassland	Artificial Land	Other *
North Western	Denmark	18.3	48.5	21.1	7.1	4.9
	Finland	71.8	4.9	4.4	1.6	17.4
	Iceland	0.3	0.1	2.3	0.4	96.9
	Ireland	13.2	4.7	67.1	3.9	11.2
	Norway	37.5	2.9	0.7	2.1	56.8
	Sweden	75.6	4.3	4.6	1.8	13.7
	United Kingdom	19.8	21.7	40.1	6.5	11.9
Central Western	Austria	47.5	17.7	22.9	5.8	6.1
	Belgium	24.7	27.5	32.3	13.4	2
	France	31.8	30.6	26.9	5.8	4.8
	Germany	32.9	33.1	22.5	7.7	3.8
	Luxembourg	30.5	18.3	37.1	11.9	2.2
	Netherlands	12.6	23.1	38.0	12.2	14.1
	Switzerland	31.3	11.1	24.8	7.5	25.3
South Western	Greece	37.4	23.2	11.4	3.8	24.2
	Italy	34.5	32.2	15.4	7.8	10.1
	Portugal	44.2	17.6	15.1	6.2	16.9
	Spain	36.7	28.0	13.9	3.9	17.4
Average		33.4	19.4	22.3	6.1	18.9

* Includes wetlands, shrubland, bare land, water bodies and other semi-natural areas; Sources: [6] for EU members (data for 2012); [7] for Finland (data for 2010); [8] for Norway (data for 2011); [9] for Switzerland (data for 2009).

The objective of this review paper is to examine soil degradation problems in WE, along with the different strategies and policies implemented to protect soil (with special focus on agricultural and forest soils) and future perspectives, in light of present legislation and technical support to soils. With this review, we intend to provide an up-to-date summary of soil degradation, soil management, soil quality (SQ) and SQ evaluation in the region.

1.3. Soil Degradation Issues in Western Europe

Damage to Europe's soils from modern human activities increased in the second half of the 20th century and led to irreversible losses due to a number of causes, which vary in importance and intensity across WE [12]. These include increasing demands from almost all economic sectors, mostly agriculture and forestry [13], but also households, industry, transport and tourism.

The European Environment Agency (EEA) and the Joint Research Center (JRC) of the European Commission have published numerous papers and reports describing soil degradation problems in Europe, in some cases, with special emphasis on WE (e.g., [11–13]). From a general perspective, soil degradation problems can be classified into four major groups: chemical, physical and biological degradation (including soil organic matter decline), and soil loss. Land-use changes can be considered as a cross-cutting factor that also affects soils. The following sections explain the present situation in relation to these problems in WE. It is important to note that many soil degradation problems usually occur together in many areas of WE [14].

The ENVASSO Project (Environmental Assessment of Soil for Monitoring), which involved 37 partners drawn from 25 EU Member States, represents a significant step in the identification of these problems and in the quantification of their spread and importance [15]. The aim of this section is not to repeat this information, but to provide a summary of these problems in WE, adding up-to-date information and giving significant and recent examples.

1.3.1. Chemical Degradation

The three major problems of chemical degradation in WE are soil contamination, soil salinization and acidification, and nutrient depletion [14]. The causes of these problems are varied, as are their relationships with agricultural management.

When addressing soil contamination, distinction must be made between local soil contamination (contaminated sites) and diffuse contamination over large areas [11]. For local contamination, in 2003, the European Environment Agency (EEA) reported soil contamination as a growing problem in WE, despite the existence of national and international legislation controlling sources of contamination and waste management [13]. A recent review on contaminated sites in Europe identified most of these as being located close to landfill sites, industrial and commercial installations emitting heavy metals, oil installations and military camps [16]. Mineral oil and heavy metals are the main contaminants, with the metal industries, gasoline and vehicle service stations being reported as the most frequent sources of local soil contamination [16,17]. The most recent available EEA report on the management of contaminated sites shows that the sizes of these vary widely across WE [17]. While in the UK, industrial waste treatment accounts for 31% of the contaminated sites, this is reduced to 20% in Italy and 0% in The Netherlands. On the other hand, municipal waste treatment and landfill sites account for just 1% in the Netherlands and up to 41% of contaminated sites in Switzerland.

The significance of this problem also differs between WE countries. The Netherlands, Belgium (Flanders), Denmark, France, Germany and the UK all have more than the EU average of 2.46 identified contaminated sites per 1000 inhabitants [16]. The corresponding figures are much lower in Greece, Norway, Ireland and Italy. This is clearly related to past and present industrial and commercial activities in these countries. Although reclamation and remediation of these sites has increased significantly in recent years, many potentially contaminated sites are still not clearly identified as such [11]. In addition to the references given, detailed information for most WE countries can also be found in [18].

In general, agriculture and agricultural soil management are not related to this type of contamination. Data on diffuse contamination, which is in many cases related to agriculture, are scarce and inaccurate owing to the lack of harmonized requirements for gathering this type of information in the different countries [11]. The overuse of plant protection products and fertilizers are usually highlighted as significant sources of diffuse soil contamination associated with agricultural production in WE [19]. Unlike in Eastern Europe, the use of fertilizers generally decreased in WE (in ton ha^{-1} and total ton) during the last decade (2000–2012). Much of this decrease is due to the implementation of legislation to prevent contamination of fresh water due to agricultural activities in the EU (such as the Nitrate Directive 676/1991 and the Water Framework Directive 60/2000). However, the rates of application of N and P vary widely between different regions. The highest average inputs of N and the highest share of manure to the total N fertilizer application have traditionally been observed in The Netherlands and Belgium (>300 kg/ha in 2008) [20]. On the other hand, Spain and Portugal, while using only an average of 89 and 79 kg N ha^{-1}, respectively, in 2008, used less manure, although the amounts appear to be increasing. Differences in the way the data are reported by each country make straightforward comparisons difficult. The variability within countries is also high, with irrigated agriculture accounting for much higher N and P doses than dryland areas, especially in the Mediterranean region [21].

Another source of diffuse soil contamination in agricultural soils is the use of sewage sludge. Since 1986, the use of sludge as a soil fertilizer has been regulated in the EU by a specific directive (Directive

278/1986/CEE). In relation to the risk of soil degradation, this Directive mainly focuses on heavy metal concentrations. However, since publication of the Directive, more stringent legislation has been adopted by several European countries for sludge disposal on soil, with lower limits established for heavy metals and limits also established for pathogens and organic pollutants [22]. There is currently increasing concern about the presence of emerging pollutants in sewage sludge that may contaminate soil when used in agriculture, and this should result in an up-grade of this legislation in the near future (see for instance [23]). Nonetheless, in WE, the prevailing destination for sewage sludge is recycling in agriculture [24], which accounts for 44% of the total sewage sludge production overall, although this varies in different countries. For instance, the proportion is >60% in Portugal, UK, Ireland and Spain, and <30% in Germany, Sweden, Italy, Austria and Belgium. In addition, sludge management practices vary greatly between regions within the same country [22]. These differences arise from local political, social and legal conditions, such as the adaptation of legal restrictions on toxic element concentrations. Future trends in the WE seem to be for stable or increased agricultural use as the most frequent option, although some shifts may be seen as biogas and energy production from sludge is also a current trend. In relation to the potential effect of this practice on soil contamination, several local scale studies have shown different trends depending on soil type, agricultural or forest management, *etc.* (e.g., [25]). On a regional scale, a study involving the distribution of heavy metals in European soils showed that concentrations of Cd, Cu, Hg, Pb and Zn were closely correlated with agricultural practices and some parent materials [26]. In France, diffuse contamination with heavy metals was identified close to industrial sites and also associated with sewage sludge amendments carried out in agricultural areas before the present legislation was implemented [27]. In Mediterranean Spain, Co, Cr, Fe, Mn, Ni and Zn in agricultural soils have been associated with parent rocks, while Cd, Cu and Pb have been related to human activities [28].

In WE, **soil salinization** (understood as the accumulation of soluble salts in soils as a result of human activities) is mainly of concern in the Mediterranean region, where it is most frequently caused by inadequate irrigation techniques, including the use of saline water or salinized groundwater and/or poor drainage conditions [13]. In contrast to other soil degradation problems, the thresholds and baselines of salt concentrations used to assess salinization are well defined and almost universally accepted, because of the importance of this issue for the development of irrigated agriculture (e.g., [29,30]). A map of salt-affected soils in the EU reveals that these are particularly important in Spain and Greece [31]. Coastal areas of France also have a high proportion of saline soils. However, the map does not indicate which of these areas are naturally saline or have soils with poorly soluble salts such as gypsum. There is evidence that at least some parts of the Ebro Basin in Spain and smaller regions of Italy, Greece, Portugal and France have been salinized due to improper irrigation strategies; however, data on present trends in this problem are not available on a continental scale [11].

Finally, **soil acidification** may occur as a result of atmospheric acid deposition and/or the use of acidifying amendments. No systematic national and continental-level studies on soil acidification are available for non-forested soils [11]. Acid deposition has decreased drastically in WE since 1980 ([13,32]). However, the effect of this reduction on soil acidity and acidification is not evident, because while some studies report declining levels of acid (see [11]), others indicate no or very slight reductions in acidity despite much lower rates of acid deposition.

In forest soils, **nutrient depletion** due to intensive soil management has been reported in several areas of WE [33]. The depletion depends on the level of biomass removal [34]. For instance, in weathered, acidic soils with low reserves of nutrients, stem-only harvesting in *Eucalyptus sp.* stands was found to involve the export, every 15 years, of more than 80% of the nutrients available in the soil [35]. Stem-only harvesting of *Pinus radiata* and *Pinus pinaster* in Southern Europe also involved high exports of K, Mg, P and Ca, leading to losses of 60%–100% of the soil available stores [35]. In a study in the UK, it was concluded that the removal of N, P and K in the tree biomass by whole tree harvesting was three to four times greater than by stem-only harvesting of the first rotation of Sitka spruce (*Picea sitchensis* L) [36]. It was also observed that after 23 years of growth of the next rotation of trees, the plots where whole trees were harvested had a significantly lower basal area on average [36]. Furthermore, the removal of tree stumps and coarse roots from felling sites as a source of woody biomass for bioenergy generation is being established in parts of WE such as Aquitaine (France) [37]. However, harvesting roots may be unsustainable if soil fertility is reduced, with consequences for future forest production [37,38].

1.3.2. Physical Degradation

Soil compaction has been widely studied in WE. It affects the air capacity, the permeability and the water-holding capacity of soils, as well as root development and soil biological activity, and it has therefore been observed to determine plant growth and agricultural yields [39]. The two most significant human activities responsible for soil compaction in Europe are agriculture and forestry [40]. Two major causes are identified [41]: ground pressure from machinery and/or animals, and soil management in agricultural land (including tillage systems). A complete survey of the surface affected by compaction has not yet been conducted in WE. The most important work is the elaboration of a map showing soil susceptibility to compaction [42]. This map shows that some areas of Belgium, NW France and The Netherlands are highly or very highly susceptible to compaction, although parts of England and South Scotland in the UK, and some Mediterranean areas such as the Ebro and Guadalquivir basins in Spain, and the Veneto region and some parts of Lombardy and Piedmont in Italy are also affected [13]. An earlier report declared 37% of European soils as being highly or very highly sensitive to compaction [43]; however, the map was created using pedotransfer functions, and it must therefore be interpreted with caution. In addition, many of the susceptible areas in the map correspond to peatland and other types of soil that are not cultivated or are managed with heavy machinery [44]. Nonetheless, the increasing weight of agricultural machinery, the introduction of irrigation and the use of farm equipment when soil water content is high suggest that some WE agricultural soils will be increasingly compacted to ever-greater depth [44].

Forest management can result in significant compaction problems because of the weight and size of forest machinery. In planted forests in WE, different rotation schedules involve more or less frequent use of heavy machinery. *Eucalyptus* spp., which occupy 1.5 million ha of land in the Iberian Peninsula (Spain and Portugal), are cultivated through a coppice system in short rotation forestry (10–12 year rotations), generally for three consecutive rotations. Poplar (*Populus* spp.) plantations, which are mainly found in France (230,000 ha) followed by Italy, Germany and Spain, cover between 100,000 and 125,000 ha of land [5]. This species, which has a deep root system and requires rich soils and large amounts of water [45,46], is usually also managed intensively in short rotations (12–16 years), with weed control techniques (mainly surface ploughing) used regularly during the first six years. Sitka spruce (*Picea sitchensis* L) plantations, mainly located in the UK and Ireland (1.2 million hectares) [47], are typically grown on 35–45 year rotations, but rotation lengths of 25 years have been proposed. Among pine species, the rotation lengths of *Pinus radiata* are typically between 35 and 40 years and may include both pre-commercial and commercial thinning and mechanical weed control. *Pinus radiata* plantations are mainly located in N Spain (290,000 ha) [48]. After clear-cut felling, the trunks are harvested with the help of skidders that are sometimes driven over the plantation area. The conventional method of site preparation consists of the partial removal of logging residues followed by down-slope ripping or blading, which consists of pushing the logging residues and the humus layer away from the site [49]. *Pinus pinaster*, which covers 2.6 million hectares of land in Portugal, NW Spain (Galicia) and SW France (Aquitaine), is characteristic of an Atlantic climate [50] and is well adapted to sandy soils [51]. The rotation lengths are typically between 30 years in NW Spain and 45 years in France and always include thinning and mechanical weed control. These management techniques have consequences for both soil compaction and erosion risk (see Section 1.3.4). Although the effects are assumed to be most pronounced on clayey or loamy textures [52], it has been suggested that a single pass of a harvester is enough to induce a large increase in bulk density and penetration resistance in sandy soils [53].

Finally, another important factor in relation to the soil physical status is the **loss of structural stability**, which can also favor erosion and greatly reduce soil porosity and the capacity of soil to store and conduct water. In many agricultural soils in WE, this is also related to the formation of crusts (e.g., [28]). However, this topic is seldom addressed as such on regional, national or continental scales in WE Europe. This is probably because soil structure is considered as a diagnostic soil property related not only to physical degradation, but also to chemical and/or biological problems, such as organic matter decline and salinization [54]. The risk of soil crusting has only recently been addressed in relation to potential wind erosion [55]. Sandy soils, which are common in the glacial deposits of Denmark, N Germany, the Netherlands, Scandinavia and the Baltic area, as well as in some areas of the NW of the Iberian Peninsula and SW France, are less affected by the formation of a soil surface crust.

1.3.3. Organic Matter Decay and Soil Biological Degradation

Soil organic matter (SOM), in particular organic C (SOC), has been in the spotlight of soil research for decades. At the European level, an overwhelming amount of research on SOC storage, gains and losses in soils has been conducted on different scales. However, the high variability and diversity of data make comparisons difficult [56]. A general view on the average content in SOC of European soils is that most of South Europe is covered by soils with less than 2% SOC [57]. This is related to both climate and historical land use. Many areas of France fall also below this threshold. The average SOC contents are higher in northern countries, the UK and Ireland.

The reasons for the generally observed **decline in SOC** in agricultural soils in WE Europe have been summarized [15]. These include conversion of grassland, forests and natural vegetation to arable land, deep ploughing of arable soils, intensive tillage operations, overfertilization [11], drainage, liming, fertilizer use and tillage of peat soils, crop rotations with reduced proportion of grasses, soil erosion, and wildfires. The latter two are of particular importance in Mediterranean countries [58].

At a national level, some long-term studies have reported changes in the SOC contents of agricultural soils. For instance, losses of 0.5–2 g SOC/kg soil per year were observed in England and Wales between 1973 and 2003 [59]. A large-scale inventory in Austria revealed that croplands were losing 24 g C/m² annually [60]. In S Belgium, losses of 0.12 t/ha per year were reported for croplands, but with an increase 0.44 t/ha in grasslands between 1955 and 2005 [61]. Grasslands on sandy soils in the Netherlands displayed a non-homogeneous trend, with some gains and some losses of SOC between 1984 and 2004. Continuous maize crops on the same soils systematically lost SOC in the period mentioned [62]. A slight average increment of 0.10 and 0.08 g SOC/kg soil in grasslands and arable land was reported for the same period [63]. In France, long-term observations (e.g., [64]) show decreasing stocks in many regions, because of deforestation, conversion of grassland into cropland, increasing cropping intensity or climate change. Vineyards and arable land display the lowest SOC contents overall [28]. An overall decrease in SOC was also recently observed in Bavarian cropland, although the variability was high, with some plots showing no change or a net increase between 1986 and 2007 [60]. This was also reported in France, where some intensely cultivated areas showed stable or slightly higher SOC stocks over time [28].

Despite these regional-scale studies, consistent figures for SOC stocks and how they change at European level are still scarce [65]. The interaction between SOC and climate change is an important issue that complicates predictions about SOC changes in relation to future land-use changes in WE in [66]. Recent simulations predicted an overall increase in this pool in agricultural soils in Europe, with a non homogeneous distribution [66], including C losses in the South, which could be compensated by a gain in Central and Northern regions. This model also showed pastures in the UK, Ireland, the Netherlands and France as the dominant SOC reservoirs, while permanent crops (olives, vineyards and orchards) accounted for only 3% of the total SOC stock, despite being widespread in Southern Europe. Arable land was predicted as containing 43% of the total stock of C, while it represents 53% of the total agricultural surface. In forest soils, harvesting activities and site preparation may lead to the removal of the humus layer from more than 80% of the surface [67].

Change in **soil biodiversity**, understood as the variety of all living organisms found within the soil system, is directly related to soil degradation in WE [68]. Some authors have suggested that it is essential to establish the present extent and distribution of soil biodiversity and to identify current threats [69]. The strong correlation between these threats and problems related to soil degradation problems described in the study becomes evident as the major challenges to soil biodiversity in Europe are land-use changes, intensive human exploitation of soils, soil compaction, soil erosion, soil organic matter decline and soil pollution [69]. Other issues of importance are invasive species and the use of genetically-modified crops (GMCs), climate change, salinization, desertification and wildfires [69–71]. The intensity of land exploitation has been identified, both in terms of agricultural use intensity and land-use dynamics, as the main factor affecting soil biodiversity in the EU [71]. The extent and intensity of these factors enable identification of the areas of WE most at risk. The areas at high, very high, and extremely high risk are concentrated in the UK, the Netherlands and Belgium, where almost 100% of the territory appears within these categories in different maps [69,71]. Most of Central and Northern France, Denmark and Germany also fall within these categories. This is attributed to the combined effect of intense agriculture, with a relatively large number of invasive species and an increased risk for the soils to lose organic carbon. In general, Mediterranean countries and areas of Southern Europe dedicated to intensive agriculture are at a lower risk. However, some areas of Italy and Spain under intensive land-use have been identified as being at high risk of suffering a decline in soil biodiversity. A recent modeling study of the susceptibility of European soils to antibiotic contamination from cattle also shows an uneven distribution in WE [72]. The Netherlands, Ireland and Belgium displayed by far the highest risk, while the risk was much lower in Mediterranean countries.

These characteristics refer to the relative pressure exerted to soil biodiversity, but not to the actual state of soil biota. At the national level, some WE countries have carried out a systematic evaluation of soil biodiversity in national soil monitoring networks [70]. These countries include France, Germany, the Netherlands, Switzerland, Ireland and the UK [11,70,73,74]. Diverse types of soil organisms are targeted in these studies, and data on the changes in soil biodiversity are scarce. Earthworms and soil fungi are some of the most commonly studied organisms [11]. The abundance and diversity of earthworms and other soil organisms have been found to be related to land use (dairy farms display the highest numbers and arable land the lowest among non-natural sites) and soil type [75]. In France, most of the soil biological groups exhibited lower values of abundance and community richness in cropland than in meadows [76]. Within agricultural land, the intensity of the management system also affected most biological soil properties; however, the type of tillage, fertilization and pesticide use were only related to the total microbial biomass and earthworm diversity, which were lower in sites in which fertilizer use is restricted, ploughed soils and sites with high inputs of pesticide. The use of fungicides and herbicides generally increased between 1992 and 2003 [77], but their use has decreased in most WE countries [78] following the adoption of strategies encouraging low-input or pesticide-free cultivation to reduce the risks and impacts of pesticides on the environment (EU Directive 128/2009). National action plans are being developed in most WE countries following this Directive [11]. It appears from these data that the relationship between land-use and soil biodiversity is a much-needed but still pending topic in WE Europe [74].

1.3.4. Soil Loss

The two major processes that cause soil loss are soil erosion and landslides. Although both have been clearly identified as soil degradation problems in WE [11], the nature, potential remediation and severity of both are rather different, and their extent and degree of significance in Europe also differ.

Soil erosion is a key environmental problem that has long been recognized in WE. The first extensive essay on soil conservation known to the Western world was published in Germany in 1815 [79]. The paper focused on depicting the most appropriate methods of preventing soil erosion in mountainous arable land. Since then, an enormous body of work has been developed in WE, as elsewhere, with the aim of understanding, defining, modeling, measuring and preventing soil erosion (see e.g., a comprehensive review of keynote references in N Europe in [80]). Nonetheless, soil erosion by water remains one of the most widespread forms of soil degradation in WE, especially throughout Mediterranean Europe [11,13]. Soil erosion has also traditionally been strongly associated with desertification. Desertification does not only affect soil and is now perceived as a wide-ranging issue associated with the loss of SOM, salinization and other soil degradation problems in WE [15], where it mainly affects the Mediterranean countries (Spain, Portugal, S France, Greece and S Italy).

Estimated mean annual erosion rates are higher than 6 t/ha in Italy and Portugal, between 4 and 6 t/ha in Austria and the UK, and between 2 and 4 ton/ha in Spain, France, Luxembourg, Belgium and Germany [81]. However, these data are average values of very different patterns within each country. In Portugal and Italy, for instance, some regions display average rates of >20 ton/ha and others of <2 ton/ha. This leaves few regions in WE with average erosion rates below the threshold of 1 t/ha/year, which has been repeatedly cited as the safe or tolerable value for soil erosion [11,41,82,83]. If non-arable land is excluded from the analysis, the highest rates correspond to Central-East and S Italy, S and SW Spain and N Portugal, SW France and Brittany. Of these, only 2% of the areas suffering from moderate to severe erosion are permanent grassland and pasture in the EU. The trend in the period 2000–2006 was towards no change in potential erosion rates in WE, except for some areas of Italy, Portugal, Austria and the UK (the countries with the highest average erosion rates) in which increments have been detected [81]. These data may be underestimated because of the short time period considered. It is also important to note that all these data are estimates derived from erosion models (in some cases RUSLE (e.g., [84–87]) and also PESERA (Pan–European Soil Erosion Risk Assessment [88]), and they should therefore be considered as such and not as actual measured data.

A review of the relationship between land use and actual erosion rates using measured erosion in plots at the country level confirmed the dominant influence of land use and cover on soil erosion rates in many countries in WE [89]. This study concluded that high erosion rates occurred in hilly loess-rich areas of West and Central Europe (2–10 ton ha^{-1} per year) and in agricultural areas in the piedmonts of the major European mountain ranges. Within agricultural land, cropping systems in which soil lay bare for long periods displayed the highest erosion rates. The period of year when soil is covered by crops is therefore an indicator of erosion problems [90]. The magnitude of erosion varies greatly across WE Europe, mainly due to climatic differences that induce different agronomic practices. For example, the percentage of soil covered by green crops on arable land varies from 25% to 50% in Spain, Italy and Southern France, and to more than 50% above a latitude of 46°N. Conversely, bare soil, which accounts for 20%–30% of arable land in France, Spain and Italy, represents less than 10% in the UK [90]. In areas of Southern Europe dominated by vineyards, in which the soil is often bare, erosion rates are high (12.2 ton ha^{-1} per year but with a great variability (standard deviation = 27.8 ton ha^{-1}year^{-1}). Measured erosion rates on arable and bare land were also related to slope steepness and length and to soil texture, while this was not the case for plots with permanent land cover [90].

From the above-mentioned measured data, estimated average rates were calculated per country and land use [89]. The rates were lower than those calculated using model simulations (Denmark and Italy were the only two countries in WE with rates >2 ton ha^{-1}year^{-1}). In particular, erosion rates in the Mediterranean were much lower than predicted, mainly due to the stoniness of many Mediterranean soils [84]. Areas of intense erosion were, however, found in Spain (in the Guadalquivir and Ebro river basins) and Italy (Apennines and Sicily).

In the Mediterranean region, wildfires are of particular concern in relation to soil erosion [13,83]. The characteristics of post-wildfire erosion in Mediterranean countries have been summarized, showing a strong influence of the topography, slope orientation and erratic rain distribution in the observed erosion rates at different observation points in Mediterranean WE [58]. Peak and average erosion rates were reported to be similar to or lower than those reported for fire-affected soils elsewhere or even those reported for common land uses in this region. Human-induced soil degradation since prehistoric times was suggested as the most likely factor explaining low erosion rates. Human occupation, abandonment, forestry use, *etc.*, have resulted in shallow, stony and weak soils.

In general, the loss of soil in forestry operations has also been related to a significant decrease in the water holding capacity [49]. Thus, although forest roads are essential structures that provide access to forests for wood extraction, their construction is the most destructive operation in the forest environment, causing soil compaction, increased surface run-off and soil erosion. Studies in which the degree of soil disturbance has been directly evaluated have shown that mechanized labor typically produces alteration of more than 80% of the soil surface, basically through the removal of the topsoil, compaction and soil displacement in *Pinus radiata* plantations in N Spain [67]. In the same region, the use of heavy machinery increased the erosion rate from 15 to 1600 kg ha^{-1} year^{-1}. Other types of soil erosion that can affect or have affected soils in WE are wind erosion and tillage erosion (the displacement of soil masses through intense tillage). Wind erosion has been less well studied in WE [11,81]. However, soils are highly susceptible to wind erosion in N Europe (Belgium, the Netherlands, N Germany and Denmark) and some areas of SW France and Spain [55]. This has been related to a lower tendency of soils in these regions to form crusts that would prevent wind erosion (see above). Tillage erosion has recently been studied, and although available data are scarce, studies in Europe highlight the importance of the magnitude of tillage erosion relative to water erosion, with mean annual rates in the order of 3 ton/ha for Belgium, northern France, and eastern England [82]. Tillage erosion rates were measured in an experimental field in Belgium and were found to be a more important soil redistribution factor than water erosion at present [91].

In summary, soil erosion remains a matter of concern in WE, where it is related to natural characteristics, land use and soil management. However, despite significant efforts form European institutions (e.g., [43]), harmonized data on actual erosion rates for the European continent are not yet available [11,58]. Identification of vulnerable areas that are affected by erosion is required in WE [82,87]. Current efforts on standardizing national erosion measures and estimates show discrepancies between countries, models and actual plot data [86] that must be overcome for better assessment of soil erosion and soil erosion risk in WE.

Landslides represent a serious natural hazard in many areas of WE and particularly affect the mountainous regions (Alps, Pyrenees, Apennines, and other). Although the total affected area in Europe is not known, the Italian national database states that more than 400,000 ha of land in Italy is affected [11]. From the point of view of soil degradation, landslides represent dramatic but highly localized problems of soil loss.

1.3.5. Land-Use Changes

In addition to changes within agricultural and forest systems, land take and agricultural land abandonment can also have important effects on soils.

Land take can be defined as the loss of agricultural land to non-agricultural sectors. It represents an increase in settlement areas (or artificial surfaces) over time, usually at the expanse of rural areas [92]. Land take has increased significantly in WE in the last few decades [93]. In the EU, about 1000 km^2 of land is taken for housing, industry, roads or recreational purposes every year [92]. In WE, the countries in which the largest areas of agricultural land were converted to artificial surfaces between 2000 and 2006 were Spain (104,706 ha) and France (76,000 ha). In the same period, the highest proportion of the total area of agricultural land converted (1.42%), mostly for construction work, occurred in The Netherlands [94]. A high percentage of land was converted to alternative uses in 2012 in Belgium and the Netherlands (13.4% and 12.2%, respectively), which is directly related to the high population densities in these countries [95]. Many aspects of the consequences of such change are serious. Thus, although soil sealing is the most evident consequence, land take also affects soil erosion, soil productivity, water storage and biodiversity in neighboring areas, as it interrupts the exchange between soils and other ecological compartments [92]. Soil sealing has been described as the destruction or covering of soils by buildings, constructions and layers of completely or partly impermeable artificial material. It has also been defined as a process that changes the nature of soil so that it becomes impermeable. Soil sealing is the most intense form of land take in WE and is essentially an irreversible process [92]. The effects of soil sealing are much more complex than those of other soil degradation issues because the process affects other ecological compartments (biosphere, atmosphere and hydrosphere) apart from soils [92]. In addition, the drivers and potential control of soil sealing are beyond the reach of soil conservation policies and must be addressed in an integrated approach implemented at all policy levels [11,13]. Understanding the potential services that urban and other sealed soils can provide is undoubtedly a first step [96].

Agricultural land abandonment has occurred as the result of the decline in the viability of extensive systems in some areas. This process has mainly affected farmland in marginal land, such as cold, wet and/or mountainous areas of WE [97]. In France and the UK, it is a marginal process, except in French vineyards (11% of the surface lost between 1990 and 2000). The rates of agricultural land abandonment are higher in Italy and Germany (2% of the total utilizable agricultural area, and 250,000 ha, respectively). It has recently been estimated that around 8% less land will be farmed under the new trade policies in the EU, if agricultural subsidies are further decoupled from production, which will particularly affect livestock grazing farms [98]. Regions where agriculture is limited by natural climate and/or soil limitations, such as some areas of Finland and Sweden, the Pyrenees, NW Spain and Portugal, the Massif Central and Brittany in France, the Apennines (Italy), the Alps, and other upland areas of Germany and arid zones of SE Spain, would be the most strongly affected. It has also been argued that there is reasonable evidence from trends in the drivers of abandonment that significant amounts of farmland will be abandoned in Europe over the next few decades, particularly in extensively grazed areas [99]. Eurostat identifies a higher risk of land abandonment in Southern states (Portugal, Spain, Italy and Greece) and also in some areas of N Finland and Sweden and NW Ireland [100].

The consequences of land abandonment on soil resources in WE are not yet clear. On the one hand, land abandonment has been associated with soil degradation issues such as wildfire and erosion in S Europe and biodiversity loss in N Europe [98]. On the other hand, it can provide an opportunity for restoration of natural habitats. The final balance seems to be both site and time-dependent. For instance, it has been noted that erosion problems tend to decrease and disappear as vegetation colonizes cropland following abandonment of sloping land [97]. The same study also established that the degree of intensification of agriculture before abandonment is also important: abandonment of intensively managed farmland has more positive effects as it reduces soil erosion and use of pesticides and fertilizers to a greater extent than abandonment of extensively managed farmland. In semi-arid mountainous areas of Mediterranean Europe, land abandonment has been shown to reduce soil loss and sediment delivery in surface water flows, despite slower plant colonization. When the abandoned farmland comprises terraced areas, this trend can be reversed, as small-scale landslides and erosion may increase as terraces collapse [101,102]. Slow recovery of soil properties has been noted in abandoned vineyard soils in semi-arid Spain and attributed to intense soil degradation under the vines [103]. In the Alps, small differences have been observed in the soil biological properties of abandoned land and organically fertilized meadows and grasslands [104].

Finally, a particular case of land-use change can be observed within agricultural land upon **irrigation adoption** (e.g., [105–107]). Overall, the socio-economic relevance of irrigation is considerable in WE [108]. In France, Greece and Spain together, the total area of irrigated land was 7.4 million ha in 2000, which represents an increase of 28.8% between 1990 and 2000 [109]. Across the entire European continent, irrigated land increased from 14.5 to 25.2 million ha in 2003 [110]. In 2007, more than 30% of the total area occupied by agriculture in Italy and Greece was irrigated. The proportion in other Mediterranean countries such as Spain and Portugal was close to 15%, and Northern countries such as the Netherlands and Denmark it was also >15% [111]. In 2011, irrigated land already represented 13% of the agricultural land in Central and Western Europe [112]. In Spain, 20.2% of the agricultural land was irrigated, which in absolute terms (in total, an irrigated area of 3,780,000 ha) represents the largest area of irrigated land in WE [113].

According to available data in Eurostat, the trend is for irrigation land to be stabilized in WE. In the Mediterranean region, present irrigation projects also include improvements in terms of water-use efficiency and soil conservation [108]. However, the environmental impacts of irrigation are variable and still poorly documented in many WE countries [108]. In relation to soil degradation, problems involving erosion and salinization are most often associated with irrigation in Southern Europe [108,113–115], although some cases of organic matter loss have also been reported [105]. Recent studies have shown that deficit irrigation strategies must be implemented with caution because they may lead to soil salinization and sodification problems, especially when moderately saline waters are used [116,117].

2. Mitigation of Soil Degradation in Agricultural and Forest Soils in Western Europe

Different strategies promoting management systems that attempt to reduce or mitigate soil degradation in agricultural and forest areas have been adopted in WE. The degree of implementation and the effectiveness of these strategies vary greatly between countries. In the EU, the European Parliament launched a pan-European project in 2007 to evaluate the status of soil degradation and appropriateness of relevant policy measures and the so-called "soil friendly farming practices". This project, which remains one of the most comprehensive studies of sustainable agricultural management at the EU level [118], summarized the actual and potential effect of some farming practices on soil degradation processes. The study found that actual and potential effects of farming practices on soil degradation depend on the type of degradation considered and that the implementation of farming practices is a complex and site-specific process that requires the cooperation of farmers and depends on local conditions (soil and climatic). These factors determine the potential success of changes in farming practices in the prevention and mitigation of soil degradation. A pan-European meta-analysis has recently revised the effects of some sustainable farming practices on soil chemical, physical and biological properties, also concluding that the success and adoption of such strategies is highly site-dependent [119].

2.1. Conservation Farming

Conservation farming (CF) comprises a series of techniques granting minimum mechanical soil disturbance, permanent soil cover with crop residues and diversification of crop species grown in the same field [120]. These techniques have been reported to reduce soil degradation in different agroecological situations (and in some cases are designed for this purpose). As excessive tillage and/or crop residue removal and uncontrolled grazing have been the cause of many soil degradation-associated problems, CF seems an appropriate strategy for solving such problems. Statistics on the share of conventional, conservation and zero tillage on the total arable area and the tilled arable area show that CF techniques are less frequently adopted in Europe than in other areas (especially the US and Latin America), and reduced tillage is more common than no-tillage (NT) and the use of cover crops [121]. The uptake of NT varies widely in WE; for example, in 2009 it varied from 4.5% to 10% of total arable land in Finland and Greece and from 2.5% to 4.5% in Spain and the UK [118]. Reduced tillage was practiced on 40%–55% of the arable land in Finland and the UK, and on 20%–25% of arable land in France, Germany and Portugal. Although the major driver for CF adoption in WE in recent decades has generally been the need to reduce costs in extensive agriculture [119,122], soil management and soil degradation issues have been given as reasons for both adopting and abandoning CF in different areas. In Northern Europe (including France), soil erosion, soil crusting in loamy soils and the need to increase SOM and soil trafficability are widely cited as reasons for adopting CF [123]. In the Mediterranean countries, soil water storage and water-use efficiency can be added to this list of reasons [124]. Different studies show that the effectiveness of CF in solving these problems is highly site-dependent because the contrasting soil and climate types existing in WE exert a strong influence on the success of these techniques [119,125]. Numerous studies have examined the effect of CF on soil properties across WE. Some general trends can be drawn from reviews and pan-European projects conducted on the topic in recent years.

There is some agreement about the positive effect of CF in reducing erosion in WE [119,121]. The most widely reported benefits of CF in relation to erosion are the increased soil infiltrability and/or the protective effect of crop residues on the soil surface. This is of particular importance in Southwestern Europe [125]. The capacity of reduced tillage and NT to increase SOC stocks is currently under discussion [126]. SOC increments have been reported under NT in many WE countries (see [125] for cases in UK, Spain, Portugal, France and Germany). In one study, NT increased SOC concentrations and SOC stocks by on average 3% and 7%, respectively, although differences were observed depending on the soil type and climate area [119]. Other studies reported no significant differences in the long term (e.g., in Scotland and Switzerland [125], in NE Spain after 20 years [127], and in N France in a 41-year experimental trial [128]) or inconsistent results (e.g., in Denmark [129]). This has been attributed to CF favoring stratification of SOC over accumulation in depth [130], or to NT being able to stock more SOC only when it induces higher yields and inputs of crop residues to the soil [131]. Climate and soil constraints may therefore modulate the effect of CF on crop yields and SOC accumulation in WE [125]. Other cultivation practices associated with low intensity soil management have been observed to have a greater potential for SOC accumulation than NT in WE. For instance, ley cropping systems and cover crops have been shown to perform better than straw incorporation and reduced tillage [66].

The effects of CF on physical properties have also shown variable results across WE. The soil water-retention capacity has been observed to be greater in semi-arid land under NT in Spain [132,133] and other Mediterranean countries in WE [124], but not in Germany [134]. In general, increased infiltration rates under NT have been reported [125], although this seems to be related to the type of soil and to the presence and activity of earthworms. It has been reported that NT induced soil physical degradation in some cases (higher bulk density, lower aggregate stability and permeability [119]). Soil compaction, usually understood as increased bulk density relative to that of non-CF soils, has generally been reported during the first years of NT and reduced tillage adoption, especially in Germany and Scandinavia [135,136], although it sometimes seems to decrease over time [125]. This problem is directly related to waterlogging and soil ponding in many areas in Northern and WE, for instance in Denmark and Germany [123] and Finland [137]. The positive effect of CF on the soil biological activity, especially earthworms, is generally observed all across WE [119,121,135,136,138].

The above-mentioned studies generally suggest that the adoption of NT and other CF techniques have had different consequences on WE depending on the type of soil, climate, the CF activities and the crops involved. When negative effects have been observed, these can be included in the reasons for fluctuations in the interest in CF shown by WE farmers [121]. The greatest soil limitations for CF adoption in N Europe are soil texture (sand and silt soils and heavy clays are difficult to cultivate without ploughing), soil drainage and the soil water regime (cold and wet conditions hinder organic matter incorporation into the soil and favor waterlogging) [121,122,125]. Also directly related to soils, reduced yields are also frequently cited as a reason for abandoning CF techniques, especially in Northern Europe [121]. A meta-analysis revealed that the effect of CF on crop yields in Europe is highly site-dependent, with NT tending to decrease yields to a greater extent than reduced tillage, and that the effect varies depending on climate, soil characteristics, type of crops and management under conventional farming [139]. Adoption of CF in the UK is mainly under systems of non-inversion tillage because the generally reduced yields under NT do not favor direct drilling [140]. In Mediterranean land, higher yields are commonly obtained under NT than under other tillage systems only in dry years [119,132,141].

Increases in weeds, pests or diseases, although not directly related to soil characteristics, are agronomic limitations also often cited as reasons for rejecting CF in WE. The combined use of CF techniques, especially NT, and intensive use of herbicides and other phytosanitary products (increasingly discouraged by European environmental and agricultural policies, see Section 3.2), together with the reduced acceptance of GMCs by European consumers are undoubtedly also related to the lower uptake of these techniques in WE than in other regions. Some authors acknowledge that future restrictions on the use of herbicides may deter NT adoption in WE [125]. However, others report cases of lower mobility and persistence of herbicides in soils under conservation tillage in Mediterranean Europe [124]. Finally, other circumstances not related to soil or environmental issues, such as the existence of subsidies, technical problems with weeds and soil management, scarce technical expertise, unfavorable market conditions and other socio-economic aspects are very often cited as explaining the lower use of NT and adoption of CF in WE [119,121,122,142]. For example, although CF adoption (especially the use of permanent soil cover [143]) in olive groves in Mediterranean land has proven to be efficient in reducing erosion and increasing soil fertility (e.g., [144]), it has been shown that the factors that determine the adoption of such practices include the socio-demographic characteristics of olive growers and the role of social capital as well as the characteristics of the olive groves [145].

Considering that CF is not equally suitable for all WE agrosystems [121], there is a need to define which regions are the most suitable for its implementation, which entails the need for soil databases and decision support systems [119,122]. Because of the complexity and site-specificity of CF implementation, training of farmers and the adaptation of CF techniques to local circumstances are still required in WE [118,119].

2.2. Organic Farming, Agroecology and Agroforestry

Organic farming (OF) in WE is mainly regulated by national and regional rules issued from a common legislative framework within the EU (Regulations 834/2007 and 889/2008). OF is defined therein as a system of farm management […] that combines best environmental practices, a high level of biodiversity, the preservation of natural resources […] and production […] using natural substances and processes. These regulations clearly state that OF must adopt those farming practices that aid the conservation and improvement of soils, with special emphasis on organic matter management. Since the beginning of the 1980s, agroecology has emerged as a distinct conceptual framework with holistic methods for the study of agroecosystems [146]. Agroecology, which includes organic management of soils and other approaches, has been defined as a means of protecting natural resources with guidelines for designing and managing sustainable agroecosystems. It is difficult to obtain accurate data from farmers following these principles as a whole in WE, and therefore only OF data 15% can be used to analyze its potential impact in the region.

Organic farming has developed rapidly during recent years in the EU. According to Eursotat [147], in 2012 the EU-27 had a total area of 5.8 million ha cultivated as fully-converted organic land plus 4.2 million ha under conversion, representing an increase from 2.6 million in 2003 (2.8 including fully converted and in conversion). However, in 2012, the whole OF area represented only 5.7% of the total agricultural area in the EU. In absolute terms, the countries with the most extensive fully converted areas in 2012 were Spain (1.4 million ha), Italy (0.9 million ha) and Germany (0.9 million ha). In relative terms, in 2012, the importance of the OF sector was highest in Austria (18.6%), followed by Sweden (15.8%), Italy (15%) and much further behind by Spain (7.5%), Denmark (7.3%) and Germany (5.8%). The number of studies and reports on the effect of OF on soils in WE is much fewer than for CF. Effects on the soil organic carbon and soil biodiversity have been reported, although these were not consistent in all parts of WE [118]. A meta-analysis revealed that, on average, SOC was significantly higher in organic plots than in conventional plots, although some paired studies reported no differences [148]. Another study also reported that organic farms in Europe tend to have higher SOC contents and lower nutrient losses per unit of field area [149]. However, both studies report that the differences may be lower when expressed per product unit. Gains in organic matter under OF have been reported in many Mediterranean agrosystems (e.g., in olive orchards [150] and in dryland crop rotations [151]) and attributed to the use of organic amendments (e.g., [152]).

The effects of larger organic C stocks on other soil properties reportedly vary in different sites and for different crops. Organic fertilization (relative to mineral N) yielded significant increases in

SOC, N availability, earthworm density and activity, microbial biomass and diversity in European soils [119]. Microbial biomass was also found to be significantly higher under organic than conventional management [118]. Other authors observed better structure and porosity in soils under OF than in conventionally cultivated soils in the UK, but this effect was found to be scale and time-dependent [153]. In a study evaluating the long-term effects of organic viticulture in France, the authors found that OF led to an increase in SOC, potassium content, soil microbial biomass, and nematode densities [154]. However, OF also increased soil compaction, decreased endogeic earthworm density and did not modify the soil micro-food web evaluated by nematofauna analysis. Similar values of organic matter were associated with OF and conventional management in agroforestry (*dehesa*) soils in Spain, where differences in the soil physical condition were more dependent on the soil type than on management [155].

In relation to OF and other low-intensity farming systems, changes in land allocation as a result of social and land-use policy changes has resulted in extensification of agriculture in some areas of WE, especially those where revenues from agricultural production are low and/or land costs are high and decrease because of lower demand [156]. Some of these low-intensity areas support agroforestry systems. In a modeling study of the environmental benefits in the Mediterranean and Atlantic regions of Europe, soil erosion and nitrogen leaching were found to be lower than in conventional cropping land, and carbon sequestration was enhanced [157]. In a review of the benefits of alley cropping systems combining agriculture and short rotation coppices by growing trees in agricultural sites in temperate Europe, these systems proved efficient for soil carbon sequestration, improving fertility, controlling erosion, storing water and regulating its quality, and increasing the overall productivity compared to conventional farming [158]. In a review of alley cropping agroforestry systems in Europe, other authors also reported overall increases in the soil organic C stocks [159].

The *dehesas* and *montados*, which are characteristic of Western Spain and Portugal, represent a particular case in agricultural land management. These systems, which are the most extensive agroforestry systems in Europe, integrate forestry (usually with evergreen holm oak (*Quercus ilex* L.) and cork oak (*Quercus suber* L.) with agricultural and livestock management practices [160]. They cover more than 3 million ha [161]. They are usually developed on acid, sandy, poor soils in semi-arid land, and yield significantly higher primary productivity than forestry or crops alone. Some authors found that soil fertility increased near the trees, with a significant increase in organic C, total N, available P, CEC and exchangeable calcium and potassium in *dehesa* systems in Central-Western Spain [161]. Similar systems in Sicily (Italy) have also been described as effective for maintaining relatively good soil condition [162]. They generally yield smaller erosion risks than croplands in the area, although they may be prone to severe erosion if stocking rates and/or tillage frequency become excessive [163].

2.3. Other Sustainable Crop Management Strategies

A significant number of agricultural practices that may have positive impacts on soils have been developed in the last few decades. Some of these practices have been implemented to a variable extent and with variable success in WE. Among those not specifically included within CF or OF, ridge tillage, contour farming and buffer strips, bench terraces and subsoiling have been identified [118]. Cover and catch crops, and the efficiency of crop rotations have been studied together with NT and organic fertilization [119]. These practices may be beneficial for addressing different soil degradation problems, and they have been identified as affecting the soil physical, chemical and biological parameters at different levels in WE [118,119]. As a result, they can help to address some soil degradation and related environmental issues. Their adoption in WE is variable.

Ridge tillage may favor retention of water in the soil. It is frequently used for crops such as potatoes and beet that are not suitable for CF. Reduced forms of ridge tillage have been shown to improve the soil physical quality and favor N mineralization in Belgium [164,165]. **Contour farming** involves developing cropping practices along the field contours, especially in sloping areas. It has been shown to be an effective measure for controlling erosion in England [166], especially when combined with minimal tillage [167]. When including hedgerows, as in the Armorican Massif in W France, gains in SOC were also significant both under the hedges and upslope [168].

Bench terraces, which are mainly built to prevent erosion and to enable cultivation in steep slopes, greatly modify the soil profile by leveling the slope to the contouring strips. They are very common in many mountainous areas in WE, especially in Alpine and Mediterranean regions, where they constitute a historical and traditional landscape feature [169]. Despite differences due to the heterogeneity of soil-forming factors across the area, terraced soils have some common elements. They display improved water availability, together with better nutrient conservation, which are known to increase crop yield in arid or semi-arid environments. Improved soil structure, physical quality, fertility, organic matter and porosity have also been reported for terrace soils, and they are often of better agricultural quality than the surrounding undisturbed soils [169]. However, significant nutrient losses have been reported to occur from terrace taluses when these are not properly built or not protected by vegetation [170]. Bench terrace systems require intensive maintenance to retain their sustainability [101,118] because if abandoned they may be subjected to gradual decay due to erosion processes and slope failure resulting in the loss of organic C and soil fertility [171].

As an intermediate system between contour farming and bench terraces, detention ponds can be constructed along the borders of the fields and perpendicular to the main slope. These ponds can store run-off and sediments. This system is common in many smallholdings in areas of Central and WE, and has been observed to be effective in reducing soil and nutrient loss through erosion [172].

Subsoiling is a cultivation practice generally used to loosen compacted or hardpan subsoil horizons with the aim of improving soil infiltration and/or root penetration [118]. It is widely used in agricultural soils that are prone to compaction in Central and N Europe [173] and also before planting vineyards in many wine-producing areas in WE (e.g., [174]). Its final effect depends on multiple factors, especially soil texture, the type of tools used in the operation and the crops planted.

The design and implementation of adequate **crop rotations**, and where possible, the introduction of **cover crops** have generally been observed to have positive effects on soils across Europe, especially in relation to SOM and biological soil properties [119]. However, trade-offs in terms of decreased crop yields and/or increased greenhouse gas (GHG) emissions must be considered for a complete assessment of the consequences of their implementation.

2.4. Sustainable Forest Management and Afforestation

As outlined above, there has been a significant increase in forested areas in WE in recent years. In the period 1990–2005, the gain in forest surfaces in Europe was around 10,939,000 ha [97]. The greatest increases in this period were observed in Ireland (+52%, with 10% of the total country surface under forest), Spain (+33%), Portugal (+22%), Italy (+19%), Greece (+14%), Denmark (+12%), and the UK (+9%). Afforestation has been strongly encouraged by EU funds and regulations, such as the Council Regulation 2080/92, and afforestation of agricultural land and the development of forestry activities on farms have been promoted [175].

In planted forests, the Ministerial Conference on the Protection of Forests in Europe (MCPFE) is a pan-European policy process for the sustainable management of the continent's forests. It develops common strategies for its 46 member countries, which include all of the EU Member States, other European countries and Russia. Cooperation between countries, which began in 1990, has produced guidelines, indicators and criteria for sustainable forest management (MCPFE, 2007). In the resolution of the Helsinki Conference in 1993, the signatories explicitly stated that "Human actions must be avoided which lead, directly or indirectly, to irreversible degradation of forest soils and sites, the flora and fauna they support and the services they provide" (MCPFE, 1993). In the subsequent Ministerial Conference held in Lisbon in 1998, the participants adopted six criteria for sustainable forest management from the Pan-European Criteria and Indicators for Sustainable Forest Management, and endorsed the associated indicators as a basis for international reporting and for development of national indicators (MCPFE, 1998). Among these criteria, soil conservation is mentioned several times.

In particular, SOC stocks are included as indicators for the maintenance of appropriate enhancement of forest resources and their contribution to global C cycles, soil condition is used an indicator for the maintenance of forest ecosystems health and vitality, and soil functions are used as indicators for the maintenance and appropriate enhancement of the protective functions in forest management.

Some countries in WE have designed good forest practice guidance for soil protection. In the UK, the government's approach to sustainable forestry involves specific good forestry practice requirements and guidelines for soils. In Northern Spain, specific regulations are implemented to avoid damage to soil in forestry operations [176].

In addition to national and regional guidelines, many planted forests in WE are established under different certification frameworks that in most cases include requirements in relation to soil management. Forest certification is a voluntary process conducted by an independent third party who issues a written statement or certificate guaranteeing that forest management is carried out according to standards considering ecological, economic and social aspects [177]. The two principal objectives of certification are to improve forest management and to ensure market access for products from certified forests, allowing both consumers and companies who sell forest products to play an important role in forest conservation [178]. The proportion of certified forest area decreases significantly from North to South in WE: in Ireland and the UK more than 50% of the forest surface is certified, in Portugal and Spain it is only around 6% and in France around 32% of the forest surface is certified [178].

3. Soil Quality and Future Trends in Western Europe

Unlike for air and water, environmental issues associated with soil degradation have been given marginal consideration in environmental regulations in WE. Soil protection has been addressed indirectly through measures aimed at the protection of air and water or developed within sectoral policies [13]. The most important initiative that partly redresses the lack of explicit soil protection is, undoubtedly, the proposal for the development of a Thematic Strategy for Soil Protection. Officially launched in 2002 (COM (2002) 179), this has led to a significant research effort and yielded an impressive amount of information on soil degradation in the EU. The final aim of these efforts was the implementation of a EU Directive for soil protection within the EU (*i.e.*, in most WE countries). Unfortunately, after several years of discussion between different European institutions, the proposal for a Soil Framework Directive similar to those existing for Air and Water was finally withdrawn from the European Commission agenda in May 2014 [179].

Although great efforts are being undertaken to recover this EU initiative, at present soil protection in WE mainly relies on national-level policies and indirect policies such as the Nitrates Directive and the Water Framework Directive [118], and the agri-environmental measures included in the CAP regulations, as explained below. As a result, many soil degradation issues are not completely covered by legislation at present. The only field in which soil protection is directly addressed in national laws is soil contamination and the management of contaminated sites. Between 1980 and 2006, most WE countries developed specific laws to address issues related to soil contamination and contaminated sites (see [180]). For instance, in Germany, the Soil Protection Act acknowledges the ecosystem functions of soils and states that they should be preserved over time. However, the Act focuses on and limits the threats to these functions derived from chemical contamination [181,182]. In the Netherlands, the Soil Protection Act states the importance of prevention, reduction and reversal of changes in the soil quality that imply a reduction or threat to the functional properties of soil has for humans, plants and animals [181]. The German Act mainly focuses on degradation risks associated with contamination of soil by toxic compounds.

3.1. Soil Quality and Ecosystem Services of Soils in Western Europe

The formal concept of soil quality (SQ) was developed in the second half of the last century [14], in response to the need to assess soil degradation problems from a holistic perspective [31,183,184]. Assessment of SQ is complicated by the fact that soil is a heterogeneous resource for which it is difficult to establish quality standards. Thus, SQ has not been defined by established universal criteria, but as the capacity of a given soil to function [185]. Proper soil functioning is understood as the capacity of a soil to accomplish its natural (ecosystemic), social and economic functions in a sustained way over time [186]. Defining soil functions was one of the goals of the European Thematic Strategy for Soil Protection. Five critical soil functions have been identified: production of food and other biomass; storage, filtration and transformation of minerals, water and other elements including C; supply of habitat and gene pool for a variety of organisms; acting as the physical and cultural environment for mankind (present and past); and as a source of raw materials. As the Commission's Communication (COM (2002) 179) states, most of these functions are inter-dependent and the development of some of them (raw materials, physical environment for mankind) may imply a reduction in the ability of soils to accomplish the others.

Since this Communication was launched, some efforts have been made to develop SQ monitoring systems, mainly within the EU. At the continental level, the basis for SQ and sustainability evaluation was established via definition of a common framework to assess soil functions, degradation threats and soil-use options [187]. This framework proposed a three-step evaluation in which the capacity of a given soil to accomplish a selected function is first evaluated. The existing threats for the considered soil and soil function are then determined, and finally, the capacity of the soil to accomplish the function is evaluated for different levels of pressure from the threats identified in the second step. This approach acknowledges that the results of the three steps, and especially the sensitivity of a soil to different threats, is soil- and site-dependent. This implies that the soil functional ability (number of functions that a soil can accomplish) and the soil responses to different levels of human-induced or natural threats (soil response capacity) must be evaluated to define SQ for a given soil [188]. The development of this framework requires detailed information on soil types, soil characteristics and threats to soil in each area studied. Its full development in detail therefore seems complicated. A first step is the identification of risk areas based on clearly described criteria (such as in [189]) for the identified threats to soil. Strategies for evaluating the risk of SOM decline, soil erosion, soil compaction, salinization and landslides in WE have been suggested [189]. For each of these, the authors provided the information needed to evaluate the risk of soil degradation based upon soil/topography/climate parameters in each site. For most sites, it was concluded that determining quantitative scores or thresholds requires more accurate information than is currently available.

The ENVASSO project is another important pan-European attempt to advance towards the identification of SQ indicators (SQI) and baseline values. The main aim of this project was the creation of a comprehensive, harmonized soil information system in Europe via the design and testing of an integrated and operational set of indicators [70,74]. Its output ([15,41]) includes selected indicators, threshold and baseline levels for the major soil threats identified in the European Thematic Strategy for Soil Protection (COM(2002) 179 final) and its subsequent evaluations (e.g., COM(2012)46). For each soil threat, three parameters were selected from an initial base of 290 indicators [15]. Some of the selected soil parameters are actually measured values, and others are estimated through modeling. The indicators were selected by experts, following these criteria: relevance for assessing each soil threat, ease of application, link to policy aims and applicability in a pan-European context. Baseline and threshold values were established for some of these indicators. However, it is recognized that such values may have to be established separately for different areas in Europe because of the variety of soil types and the variability in environmental conditions and land use. Table 3 summarizes the soil threats and properties suggested as indicators by the ENVASSO Project team, and which of those were finally selected as the best indicators for each threat.

The performance of those indicators was tested in different pilot areas in Europe, and the results of these tests have been reported in detail [190,191]. Complete descriptions of the protocols that should be used in each case have been published [84]. The purpose of drawing up this list of indicators was to establish a monitoring network in which changes in soil characteristics can be periodically controlled [192].

Sustainability **2015**, *7*, 313–365

Table 3. Proposed and selected soil properties for monitoring soil quality in the Environmental Assessment of Soil for Monitoring (ENVASSO) Project.

Soil Threat	Soil Indicator *	Source $	Baseline	Threshold
Soil contamination Diffuse and local	**Heavy metal content**	M	National background levels	National legislation
	Nutrient balance	M	Average national balance	Defined at a regional level
	Organic pollutant concentration	M	National background levels	National legislation
	Topsoil pH	M	Not defined	Not defined
Salinization	**Salt profile**	M	EC saturation extract < 2 dS/m	0.10% salt content or EC_e < 4 dS/m
	Exchangeable sodium percentage (ESP)	M	<5%	>15%
Soil compaction	**Density (bulk, packing and total density)**	M/E	Measured in non-compacted soils	Packing density: 1.75 g/cm^3
	Air-filled pore volume at a specified suction	M	Measured in non-compacted soils	Air-filled pore vol. at 5 KPa >10%
	Permeability	M	Not defined	Not defined
	Mechanical resistance	M	Dependent on soil structure status	Penetration resistance < 2–5 MPa
	Structure status	E	Not defined	Not defined
	Vulnerability to compaction	E	Not defined	Persistent (not recoverable)
	Drainage	E/M	Not defined	Not defined
	Precompression strength	E	Measured in non-compacted soils	<90–120 KPa
Decline in SOM	**Topsoil organic carbon content §**	M	Not defined	Not defined
	Soil organic carbon stock	M	Not defined	Not defined
	Peat stocks	E	Not defined	Not defined
	Topsoil C:N ratio	M	Not defined	Not defined
Soil biodiversity	Microbial and fungal diversity	M	Not defined	Not defined
	Earthworm diversity and fresh biomass	M	Not defined	Not defined
	Macrofauna diversity	M	Not defined	Not defined
	Collembola/Enchytraeid diversity	M	Not defined	Not defined
	Acari diversity	M	Not defined	Not defined
	Nematode diversity	M	Not defined	Not defined
	Microbial respiration	M	Not defined	Not defined
	Microbial activity (enzymes)	M	Not defined	Not defined
Erosion Water, wind and tillage erosion	**Estimated soil loss**	E	Water and tillage erosion: N Europe 0–3 ton/ha*year S Europe 0–5 ton/ha*year Wind: N&S Europe: 0–2 ton/ha*year Water: 0.5 ton/ha*year	Water and tillage erosion: N Europe 1–2 ton/ha*year S Europe 1–2 ton/ha*year Wind: N&S Europe: 2 ton/ha*year Water: 1–2 ton/ha*year
	Measured or observed soil loss	M		
Soil sealing	No soil properties as indicators			
Landslides	No soil properties as indicators			

* Indicators shown in bold type are within the three selected for each threat; $ M: measured; E: estimated or calculated; § For desertification: SOM in desertified land, salt content in desertified land and soil biodiversity in desertified land; Sources: [15,41,192].

Development of the European Soil Data Centre provides additional mechanisms for reporting information on soil and SQ data and adequate definitions of SQ, SQI and monitoring networks [85]. The spatial density of soil monitoring networks is very non-homogeneous in WE, with no or very few systematic sampling sites available for many of the indicators shown in Table 3 [192]. In fact, some of those SQI (e.g., those related to soil erosion or soil organic C) have been monitored with much higher intensity and frequency than others such as soil biodiversity [70]. The LUCAS (Land Use/Land Cover Area Frame Survey) represents the first effort to build a consistent spatial database of the topsoil (0–30 cm) cover across Europe, based on standard sampling and analytical procedures [193]. The aim of LUCAS is to gather harmonized information on land use/land cover and several soil properties, such as soil texture, organic carbon, nitrogen content, pH and cation exchange capacity. The survey also provides territorial information for the analysis of the interactions between agriculture, environment and countryside, such as irrigation and land management. LUCAS field surveys have been carried out every three years since 2006. The next LUCAS field survey is planned for 2015.

Finally, within agricultural soils, the above-mentioned Communication of the European Commission on the development of agri-environmental indicators for monitoring the integration of environmental concerns into the CAP (COM(2006) 508 final) also established SQ as a state/impact indicator set [194] (see Section 3.2). This indicator set, which has not yet been completely assessed, aims to describe the following: (i) the soil capacity for biomass production; (ii) the input required for optimal productivity; (iii) the soil response to climatic variability; and (iv) carbon storage, filtering and buffering capacity. These four issues are to be integrated in a SQ index aimed at quantifying the ability of soils to provide agri-environmental services by performing their functions and responding to external influences. The index is determined following a similar previously described approach [187]. The SQ index is calculated from four sub-indicators of similar weight, which are relevant either to the agricultural and/or to environmental performance of soil: (i) soil productivity index; (ii) soil fertilizer response rate; (iii) production stability; and (iv) soil environmental services. So far, only indicators (i) and (ii) have been calculated for most regions and countries within the EU. The productivity index (i) takes into account both the inherent soil properties, and the climate and topography of each territory. When climate is more limiting for rainfed agricultural production, soil properties supporting productivity gain more weight in the index than in areas with fewer or no climatic limitations. The index therefore does not only represent soil characteristics and should be considered as a land quality index rather than a SQ indicator. In WE, the most productive croplands are found in NW France, Belgium and the Netherlands, together with W England and Scotland. The most widespread areas of land of low productivity are in Spain and SE Italy.

The response to fertilization (ii) was calculated by assigning a fertilizer response score for each soil unit based on soil properties. The areas with a high response value matched those with high productivities in (i). Conversely, the areas with soils displaying a low response to fertilization in rainfed croplands were found in Spain, especially in the Ebro and Guadalquivir river basins. The stability of crop production will be estimated from soil characteristics that explain higher variability under limiting water and climate conditions. Finally, four soil functions will be considered for evaluating the environmental services of agricultural soils: organic C storage, the filtering capacity, the transforming capacity of the soils, and their biodiversity and biological activity. Further development of this index will be of use in land-use planning and environmental protection.

At a national level, some attempts have been made to establish standard systems for the periodic control of SQ and SQI, such as the National Soils Indicator Consortium in the UK (UKSIC) and the Soil Quality Monitoring Network (Réseau de Mesures de la Qualité des Sols, RMQS) in France. In the UK, the UKSIC has established a minimum set of SQI for broad-scale soil monitoring. This includes soil organic carbon, soil pH, heavy metals (Cu, Zn, Ni), Olsen P, potentially mineralizable N and bulk density [195]. These indicators were selected by experts and constitute a basis for periodical comparisons in a network of sampling sites across England and Wales. Scotland is at present developing its own soil monitoring system within the Scottish Soil Framework (SSF) [196].

In France, the development of the RMQS was designed as a periodic (every 10 years) collection of soil samples in a regular template (16x16 km). As in the UKSIC, soil properties (texture, organic C, nutrient contents and some other physical and chemical properties) are measured at each sampling point where soil and soil use are characterized [29,197].

Similar soil monitoring networks exist in other WE countries such as Germany and Austria. Many of these networks aim to monitor soil degradation at a national level by making regular and comprehensive comparisons of the selected indicators over time. They are not intended for field-scale application to detect main soil constraints and thus to derive soil management and conservation recommendations for particular sites [198].

One problem associated with studying SQ in such a way is the difficulty in evaluating the absolute values and observed changes in SQI. This problem arises from two sources. First, the heterogeneity of soils and soil uses makes it difficult to establish baselines and thresholds (see gaps in Table 3, and the case for SOM in [62]). Second, the same factor may have a different score depending on the soil function or ecosystem service being evaluated.

In this sense, to our knowledge, no systematic and normalized strategy for SQ evaluation at the national level equivalent to the Soil Management Assessment Framework (SMAF) in the USA [183,199,200] exists in WE. This type of evaluation focuses on the selection of a minimum data set of SQI that must include soil physical, chemical and biological attributes of soil. Different scores can be assigned to these indicators on the basis of their average values and the relationship between these values and the performance of each soil function. The scoring therefore depends on the soil type and the soil function(s) considered [183]. In contrast to SQ evaluations focused only on monitoring soil properties in time, this type of system defines different scores and quality attributes depending on the soil function and/or soil ecosystem service considered. Although these studies are complicated to carry out because they are site specific, they are very valuable for assessing SQ response to different types of agricultural management (e.g., [200]). Some initiatives are currently being developed for considering the multi-functionality of soils for land management decision (see Section 3.3 for those by Schulte *et al.* and by Volchko *et al.* [201,202]).

At regional and local levels, many studies have addressed SQ monitoring systems with a holistic approach in WE. Most of these focus on the evaluation of SQI for SQ monitoring under particular soil uses and/or under particular conditions, such as soils under OF ([203,204], forest soils under different types of management [49,205], extensive rainfed cereal crops in semi-arid land [206], Mediterranean mountain agrosystems and vineyards [207,208], and many others. Since biological SQI are generally not considered in large soil inventories or monitoring networks [209], many of these studies identify suitable biological SQI such as microbial parameters, soil fauna, earthworms and other macro invertebrates, *etc.* (e.g., [76,208,210]). The results of studies using holistic SQ evaluation systems in agricultural soils are diverse, as are the agrosystems studied. In general, different management systems are compared. To cite two examples, enhanced soil quality was observed under NT in extensive rainfed cereal systems in semi-arid Spain, but the impact of organic farming in vineyards in S France was not detectable in the overall SQ [154,206].

In summary, the development of SQ monitoring programs and SQ evaluation systems that enable accurate assessment of the soil ability to accomplish functions or ecosystem services is an ongoing and promising strategy for soil protection in WE. Two considerations are important in this framework: inclusion of the farmers' perspective and evaluation of economic trade-offs in the different evaluations [211].

3.2. Soil Protection and the EU's New Common Agricultural Policy (CAP)

Although there is a shift towards including the multi-functionality of soils into the legislation in many WE countries (e.g., The Netherlands [212], Belgium [213] and France [214]), a specific legislative framework for unpolluted agricultural soils is so far lacking. However, as most countries in WE belong to the EU, they are affected by the CAP. The CAP is based on two groups of measures or pillars. Pillar one corresponds to the legislative framework in relation to agricultural production subsidies. Pillar two includes the support policies for rural development in the EU.

Since 1999, in the so-called Cardiff process, environmental protection measures have been integrated into the CAP. This implies that the successive reforms of the CAP established a list of statutory management requirements and a reference level of good agricultural practices that should be respected by European farmers being supported by the CAP. Different requirements and reference levels have been established for different local conditions by member states or competent regional or local authorities. The cross-compliance character of these measures implies that they are mandatory for farmers receiving CAP subsidies. From the perspective of soil conservation, cross-compliance links direct payments with compliance by farmers with the obligation of keeping land in *good agricultural and environmental condition*, including standards related to soil protection (namely protecting soil from erosion and the maintenance of soil organic matter and soil structure) (EU Council Regulation 73/2009). Table 4 shows the different measures adopted in this framework for different WE countries in 2006, as compiled by GEIE Alliance Environment [215].

The CAP has also encouraged sustainable soil management by funding the provision of environmental public goods and services beyond mandatory requirements to those farmers adopting the so-called *agri-environmental measures* (AEMs). In many cases, this implies adopting agricultural activities or levels of production intensity that deliver positive environmental outcomes, while not necessarily being the first choice from the point of view of profitability. Some of these measures are related to management systems that can promote SQ. As a result, throughout its successive reforms, soil protection measures have been reinforced in the CAP and expanded to encourage organic and integrated farming, extensification, maintenance of terraces, safer pesticide use, use of certified composts, and afforestation, among others [13]. The flexibility of AEMs allowed WE countries in the EU to develop different measures or schemes to reflect different bio-physical, climatic, environmental and agronomic conditions and therefore to tailor management options to suit the characteristics of their agricultural sector. As described in a case study in Brandenburg (Germany) [182], AEMs and cross-compliance measures associated with the CAP are often the only significant official policies addressing soil conservation in agricultural land in WE [53].

Table 4. Soil-related measures adopted for maintaining arable land in good agricultural and environmental condition within the Common Agricultural Policy (CAP) framework in Western Europe (WE) countries within the EU.

Measure	Country													
	DE	AT	BE*	DK	ES	GR	FR	IE	IT	LU	NL	PT	UK	SE
Soil erosion control														
Minimum soil cover	x	x	x	x	x	x	x	x		x	x	x	x	x
Minimum land management	x		x		x	x	x		x	x	x	x	x	
Terrace conservation	x	x		x	x	x			x	x				
Other measures for erosion					x	x					x		x	
Organic matter (SOM) management														
Crop rotation	x					x	x			x			x	
Management of crop residues	x	x	x		x	x	x		x			x	x	
Other measures for SOM			x				x			x	x			x
Soil structure protection														
Use of adequate farm machinery		x			x	x			x				x	
Other measures for structure			x				x			x	x			x
Other measures														
Livestock density control	x		x	x	x	x	x	x	x	x	x	x	x	x
Grassland protection	x		x	x	x	x	x	x	x	x	x		x	x
Slope assessment	x	x			x	x		x	x				x	x
Wild vegetation control		x	x	x	x	x	x	x		x		x	x	x
Olive-grove preservation					x				x					
Other			x	x	x	x		x					x	x

* Data for Belgium include Flanders and Walonia; Source: Adapted from [215] (Data for 2006).

Within this framework, the development of agri-environmental indicators for monitoring the integration of environmental issues in the CAP was introduced in 2006 (COM 2006-508 final). As explained above, some of these indicators involve soil protection. These have been selected for monitoring farm management practices, agricultural production systems, pressures and risks to the environment and the state of natural resources. Their level of development differs greatly: while some are already operational, others are only defined and lack data. Table 5 shows these indicators and their development to date.

The changes in CAP towards more environmentally-oriented policies had different results in relation to SQ [175]. In most cases, measures included in AEMs, such as contour and reduced tillage, led to reduced erosion rates, higher biodiversity and generally improved SQ in arable land and grasslands across WE. However, promotion of set-aside, for example, may have the opposite effect in arid and semi-arid land. The difficulty in fulfilling the requirements of cross-compliance also stimulated land abandonment in some areas. The CAP has also encouraged the use of soil cover systems and crop rotations, and has contributed to the dissemination of CF [121].

The latest reform of the CAP (for the period 2014–2020) includes significant changes in relation to environmental protection: a new policy instrument of the first pillar (*greening*) is directed to the provision of environmental public goods [217]. This instrument has been designed to reward farmers for respecting three obligatory agricultural practices: (i) maintenance of permanent grassland; (ii) maintenance of ecological focus areas (land left fallow, terraces, landscape features, buffer strips and afforested areas); and (iii) crop diversification (which includes having at least three crops on the same agricultural exploitation, or including agronomic practices with minimum soil disturbance and green coverage of the soil surface in permanent crops). Implementation of these measures across the EU is expected to increase soil protection, as many of the measure directly involve soil. For instance, in Spain, ecological focus areas include set-asides, N-fixing crops, afforested surfaces and land devoted to agroforestry. The aim of this reform is also to extend and reinforce the environmental component to Pillar 2, by including agri–environmental-climate measures, OF, forestry measures and investments that are beneficial for the environment or climate (amongst others) in rural development policies.

Nevertheless, the final net effect of the new CAP on SQ in WE will depend on multiple factors, both at local and national level, and it is possible that trade-offs between conflicting agricultural sector policies will appear. For example, a policy aimed at mitigating soil erosion (achievable through CF) may conflict with another policy discouraging the use of herbicides (often critical to the initial success of CF practices [121]). Similarly, CAP measures designed to promote increased agricultural production may diverge from those developing environmental policy objectives [201].

In addition to environmental issues affecting terrestrial and aquatic systems, CAP reforms included since 2010 support climate action. The reduction of GHG emissions from farmland, when including soil management strategies and the stabilization of organic C in soils, may affect SQ in WE. The efficiency of these strategies may differ both in terms of the abatement of GHG emissions and economic costs, as shown for ten possible measures in French farms [218]. Among these measures, those directly affecting soil and soil management had positive (cover crops, hedges), very little (legume crops, agroforestry, reduced tillage) and negative (organic fertilizer application) effects in terms of net CO_2 abatement. Conversely, some agricultural practices that improve SQ have been observed to increase GHG emissions [119].

The new CAP structure offers the possibility of including climate action instruments in both Pillar 1 and Pillar 2; however, in some cases the impact of such measures is still uncertain. Nevertheless, according to [219], the new CAP will probably be one of the most important opportunities for the EU-28 to tackle the climate change issue. Implementation of CF and use of cover crops are included among the proposed measures to be adopted at farm level.

Table 5. Agri-environmental indicators for CAP monitoring in relation to soil protection in Western Europe. Indicators shown in bold type directly address soil conservation issues.

Domain	Indicator	Field of control	Related soil protection issue
Response Public policies	Agri-environmental commitments (AEC) Agricultural areas under Natura 2000	Surface under AEC (To be developed)	Soil degradation due to agricultural management (depending on each AEC in each WE country and region)
Technology and skills	Farmers' training level and environmental advicing	Farmers in training programs	—
Market signals and attitudes	Area under organic farming	(To be developed)	
Driving force Input use	Mineral fertilizer consumption Consumption of pesticides Irrigation Energy use	Mineral and organic N and P Plant-protection products Irrigable and irrigated surface Irrigation methods Source of irrigation water Use of energy by fuel type	Diffuse contamination Diffuse contamination, biodiversity decay Erosion, salinization—
Land use	**Land use change** Cropping patterns Livestock patterns	Agricultural land take Share of arable land, permanent grassland and permanent crops Share and density of major livestock types	Soil sealing Erosion, organic matter decay, compaction Compaction
Farm management Trends	**Soil cover** **Tillage practices** Manure storage Intensification/extensification Specialization Risk of land abandonment	Time when soil is under crops Share of arable land in CT, CoT and ZT * (To be developed) Inputs per factor production Number of products in farm Cessation of agricultural activities	Erosion, organic matter decay, compaction Erosion, organic matter decay, compaction, biodiversity — — Erosion, organic matter, biodiversity
Pressures and risks Pollution	**Gross nitrogen balance** **Risk of pollution by P** **Pesticide risk** Ammonia emissions Greenhouse gas emissions	Surplus N risk Surplus of P risk Pesticide consumption Ammonia emitted by agriculture GHG emitted by agriculture	Diffuse contamination Diffuse contamination Contamination, biodiversity — Soil organic matter decay
Resource depletion	Water abstraction **Soil erosion** Genetic diversity	For agricultural use Estimated water erosion (To be developed)	— Soil erosion

Table 5. *Cont.*

Domain	Indicator	Field of control	Related soil protection issue
Benefit	High Nature Value farmland	(To be developed)	—
	Renewable energy production	From agriculture and forestry	
Impact/state			
Biodiversity/habitats	Population trends in farmland birds	Monitoring 37 birds species	
Natural resources	**Soil quality**	Productivity index, fertilizer response rate, production stability and soil environmental services.	Organic matter, biodiversity, physical degradation, chemical fertility, biodiversity
	Water quality—Nitrate pollution	Nitrates in rivers and groundwater	
	Water quality—Pesticide pollution	(To be developed)	
Landscape	Landscape—state and diversity	Dominance and diversity	As in land use and soil cover.

* CT: Conventional Tillage; CoT: Conservation Tillage; ZT: zero tillage; Source: [216].

3.3. Promising Strategies for Soil Quality in Western Europe

From the above it can be concluded that SQ monitoring, assessment and protection are currently at different levels of development in WE. Some promising strategies for increasing the awareness of SQ and improving its consideration in future policies in WE include (i) the development of soil status and SQ monitoring networks at a continental scale; (ii) the inclusion of SQ and soil functionality issues in environmental and agricultural legislation; (iii) the research for accurate and, if possible, simple SQ monitoring tools; and (iv) the implementation of multi-actor and multi-target strategies for promoting and increasing SQ awareness and the effective implementation of SQ-improving management practices.

Monitoring SQ (i) is essential to measure soil degradation and to develop appropriate strategies for soil protection. This includes the creation of international networks to address critical and crosscutting soil issues. In addition to national initiatives, Europe has several projects that address these issues. For example, the above-mentioned LUCAS survey and the recently launched (2013) European Soil Partnership (ESP) are added to previous initiatives under the support of the Joint Research Center of the EU, such as the European Soil Bureau and the European Soil Data Center, from which information on soils in the EU can be retrieved at the European Soil Portal (http://eusoils.jrc.ec.europa.eu/). All these initiatives are supported by the JRC of the EU. The ESP is one of the regional partnerships of the Global Soil Partnership. The objective of this regional network is to bring together the various scattered networks and soil-related activities within a common framework, open to all institutions and stakeholders willing to actively contribute to sustainable soil management in Europe. The ESP has five main pillars of action, which include promoting sustainable management for soil protection, encouraging investment, technical cooperation, policy, educational awareness and extension in soil, promoting soil research related to productive, environmental and social development actions, enhancing the quantity and quality of soil data and information, and harmonizing methods, measurements and indicators for the sustainable management and protection of soil resources. These initiatives must account for the fact that sampling schemes suitable for inventory are not necessarily also suitable for monitoring [220]. Monitoring information on farm management practices, on how these practices affect the environment, and whether they correspond to recommended (or legislated) practices and standards may also contribute to early detection and assessment of SQ issues [221].

In relation to the **inclusion of SQ issues in legislative and assessment tools** (ii), the functional land management strategy recently proposed by Schulte and coworkers is a complete and promising model for developing policies that enable achievement of goal targets for productivity by considering and enhancing the capacity of soils to provide ecosystem services [201]. This strategy is based on optimizing five basic soil functions (biomass production, water purification, C sequestration, habitat for biodiversity and recycling of nutrients) by studying the potential of soils to supply these, as well the present and future demands by taking into account growth goals and environmental restrictions. Although the study was proposed for Irish agricultural soils, it could be expanded to other European regions (including the EU).

Another example of a decision tool that considers soil functions and that can be used to evaluate remediation alternatives for contaminated soils has been described by Volchko and coworkers [202]. This is based on the inclusion of selected ecological soil functions (basis for primary production, cycling of carbon, water, nitrogen and phosphorus) in a multi-criteria decision analysis. The degree to which these functions are fulfilled in remediated sites is determined using a minimum data set of SQI (soil texture, coarse material, organic matter, available water, pH, potentially mineralizable nitrogen, and available phosphorus), which are scored and integrated in a SQ Index, in an approach very similar to that described in the SMAF [199].

A good example of the incorporation of soil functionality criteria in legislative frameworks is the ongoing process in the above-mentioned SSF in Scotland. This framework aspires to develop the EU Soil Thematic Strategy for Scottish soils, providing a legislative framework for soil protection that accounts for the inherent soil quality and the multi-functional roles. The declared aim of this SSF is to promote the sustainable management and protection of soils consistent with the economic, social and environmental needs of Scotland. The framework identifies more than 35 actions in different fields (research, soil conservation, land management, *etc.*) linked to expected soil outcomes. Each action has a delivery date and designates the persons or bodies responsible for its accomplishment. These actions include the development of a Scottish soil-monitoring network and review of the land capacity for agriculture. The monitoring network focuses on the functions of soils related to ecosystem services. This strategy will be included in the more general strategy for land use [222], which includes for instance, the rationale for woodland expansion. In this rationale, a soil-based evaluation of land is made in order to protect sensitive soils (such as peatland soils) or high quality agricultural land, which provide essential services such as C sequestration and food production, from being converted into forest plantations. Similar systems in which the soil types (and therefore their inherent quality) are considered for management decisions have been developed in England for forestry management. These guides (e.g., Whole-tree harvesting guide [223]) determine the type of practice to be implemented as a function of soil characteristics in each forest plantation.

In this sense, the positive and negative market services provided by forests (including those related to soil protection) play a significant part in decisions about forests, but they are notoriously difficult to quantify, and people seldom agree on their value. The EU needs a policy framework that coordinates and ensures coherence of forest-related policies and allows cooperation with other sectors that influence sustainable forest management. For example, the efficient use of wood biomass as a renewable energy resource and increased utilization of domestic, renewable resources of biomass has been identified as an opportunity for many European countries to increase their energy security. Significant impacts on roundwood and wood residue markets are expected as the energy sector becomes a major consumer of wood biomass. There has already been a rapid increase in the production of energy from harvested forest residues (small diameter tree stems, branchwood and foliage) in some Nordic countries [224,225].

Another challenging aspect in many areas in the future is the **development of adequate SQI** **(iii)** [16] and in particular establishment of the relationship between their levels and soil functions for different areas and land uses across WE, as indicated for levels of soil organic matter [62]. In this sense, although much work has been done in some aspects (for instance in relation to climate change mitigation and adaptation strategies), research for developing SQ assessment tools and SQI in other aspects is still pending in WE. For example, new methodological approaches for soil biodiversity measurement are being developed [73], as well as new tools to assess land susceptibility to wind erosion [55]. New techniques such as near-infrared reflectance spectroscopy are being considered for the evaluation of SQ and soil properties [226,227].

Finally, attempts to **broaden the participating agents (multi-actor) and the objectives (multi-target)** **of new soil management strategies** that enable improved SQ **(iv)** also exist in WE. For example, the LIFE project series devoted to soil degradation problems and soil protection aims to translate science and policy into practice [228]. These measures enable the involvement of stakeholders in the launching and demonstration of new techniques and systems for sustainable soil use in the EU. There is a special need for developing cost/benefit analysis, such as that recently developed for GHG abatement in French agriculture [218].

At a different level, certification of agricultural and forest goods has been a successful strategy in some cases. In addition to public policies or official certifications, some initiatives such as integrated farming and GlobalGAP are provided to producers who apply some soil-friendly management practices, with certified labels awarded for their products. This type of labeling increases the awareness of consumers and can indirectly promote the adoption of production strategies that preserve or improve SQ.

4. Concluding Remarks and Future Threats to Soils in W Europe

The objective of this review paper was to consider soil degradation problems, the policies and strategies for soil protection and the future perspectives for SQ assessment in WE, with the aim of providing a summary of SQ problems and evaluation in the region.

The review of soil degradation showed that different population trends, economic activities, local legislative conditions and historical land use have created different types of SQ-related problems. Problems related to soil chemical, physical and biological degradation have been identified in different areas of WE. Soil losses through erosion are also significant in many regions. Many of these problems are related to agricultural and forest management, which are the predominated activities on non-urban land.

The strategies implemented in WE to overcome these problems in cultivated and forest land include conservation agriculture, organic farming and other sustainable agriculture and forest management systems. These have had results of varying success, mainly because of the site-specificity of their effectiveness. This highlights the fact that universal solutions are difficult to design and achieve for a complex problem such as soil degradation.

Moving towards strategies that consider soil functions within the framework of SQ assessment and regulation in WE would help to optimize the assessment of soil degradation problems and the search for effective strategies for sustainable soil management adapted to the characteristics of European regions. However, although promising examples exist at a local level, and a significant amount of work has been done in pan-European projects, more information than is currently available seems necessary for a national or continental-scale SQ assessment. Some platforms such as the European Soil Database, the LUCAS framework and development of the European Soil Portal will undoubtedly contribute to the harmonization of soil data and soil monitoring across Europe, for instance in the task of identifying adequate SQI that should be calibrated across the continent. Developing cost/benefit analysis and multi-actor approaches to address SQ and soil protection strategies in WE seems a promising approach.

However, so far only some problems related to soil degradation have been considered in the legislation (*i.e.*, contamination), and specific soil legislation is lacking in WE. As a result, most of the attempts involving soil protection and SQ enhancement are found in normative frameworks that affect soils only indirectly, mostly in the form of environmental-oriented restrictions to the CAP subsidies both to agricultural production and in rural development plans. Thus, the sustainable use and conservation of WE soils are not yet fully guaranteed. This is especially true for forest soils, which are not directly affected by most CAP regulations and mainly depend on certification strategies as indirect mechanisms of ensuring sustainable management.

The need for specific legislation and holistic SQ assessment strategies becomes clear in light of evidence of future threats to soils and SQ in the region. In 2012, the European Commission identified increased future soil degradation problems for European soils if several aspects are not properly addressed [229]. The greatest challenges cited were land use issues (including the increasing demand for productive soils and land take for urbanization and infrastructure areas), the preservation of SOM (especially in peatland, pastures and forest soils) and more effective use of fertilizers and organic waste that may help to optimize soil fertility without leading to SQ degradation. These three aspects are closely connected to most of the soil physical, chemical and biological degradation processes and soil loss problems described in the Soil Thematic Strategy. A review of the topics addressed suggests that, as recently noted [201], greater scientific knowledge and management of soils will be critical in meeting the challenges of food security and environmental sustainability in the forthcoming years in WE and worldwide.

Acknowledgments: The FORRISK project (Interreg SUDOE IVB, project SOE3/P2/F523) is acknowledged for partially funding this work. The authors acknowledge funding from their respective institutions for accessing publication databases. The work of three anonymous reviewers is also acknowledged.

Author Contributions: Iñigo Virto is the corresponding author of this work. He coordinated the other authors and was responsible for the final writing and edition, and for Sections 1.3, 2.3 and 3.1–3.3. María José Imaz was responsible for Sections 2.2 and all aspects related to organic farming and agroecology in the other sections. Oihane Fernández-Ugalde was responsible for Sections 1.1 and Section 1.2 and collaborated in the other sections. Nahia Gartzia-Bengoetxea was responsible for Section 2.4 and all aspects related to forest management and forestry. Alberto Enrique collaborated in Sections 2.1 and 3.1–3.3. Paloma Bescansa was responsible for Section 2.1 and assisted in the final editing process. All the authors contributed to the conclusions (Section 4) and enriched the discussion in Section 3.3.

Conflicts of Interest: The authors declare no conflict of interest.

References

1. United Nations (Statistics Division). Composition of macro geographical (continental) regions, geographical sub-regions, and selected economic and other groupings. Available online: http://millenniumindicators.un.org/unsd/methods/m49/m49regin.htm (accessed on 23 September 2014).

2. Food and Agriculture Organization of the United Nations (FAO). World reference base for soil resources 2006. In *World Soil Resources Reports 103*; FAO: Rome, Italy, 2006; p. 145.

3. European Soil Bureau Network-European Commission. *Soil Atlas of Europe*; Office for Official Publications of the European Communities: Luxembourg, 2005; p. 128.

4. Bregt, A.; de Zeeuw, K. Agriculture, Forestry and Nature, Trends and Development across Europe. In *Land Use Simulation for Europe*; Stillwell, J., Scholten, H., Eds.; Kluwer Academic Publishers: Dordrecht, The Netherlands, 2002; Volume 1, pp. 37–44.

5. Food and Agriculture Organization of the United Nations (FAO). *State of the World's Forests*; FAO: Rome, Italy, 2012; p. 46.

6. Eurostat Statistics. Land cover, land use and landscape. Available online: http://epp.eurostat.ec.europa.eu/statistics_explained/index.php/Land_cover,_land_use_and_landscape (accessed on 10 September 2014).

7. European Environment Agency (EEA). The European Environment—State and Outlook 2010 (SOER 2010)–Land Use. Available online: http://www.eea.europa.eu/soer/countries (accessed on 6 December 2014).

8. Borgen, S.K.; Hylen, G. *Emissions and Methodologies for Cropland and Grassland Used in the Norwegian National Greenhouse Gas Inventory*; Norwegian Forest and Landscape Institute: Ås, Norway, 2013; p. 45.

9. Statistique Suisse. Utilisation et couverture du sol–Données. Available online: http://www.bfs.admin.ch/bfs/portal/fr/index/themen/02/03/blank/data/01.html (accessed on 6 December 2014). (In France)

10. Eurostat, Statistics Explained. Agri-environmental indicator—Tillage practices. Available online: http://epp.eurostat.ec.europa.eu/statistics_explained/index.php/Agri-environmental_indicator_-_tillage_practices (accessed on 4 August 2014).

11. Jones, A.; Panagos, P.; Barcelo, S.; Bouraoui, F.; Bosco, C.; Dewitte, O.; Gardi, C.; Erhard, M.; Hervas de Diego, F.; Hiederer, R.; Jeffery, S.; *et al. The State of Soil in Europe—A Contribution of the JRC to the European Environment Agency's Environment State and Outlook Report—SOER 2010*; Office for Official Publications of the European Communities: Luxembourg, 2012; p. 76.

12. European Environment Agency (EEA). *Down to Earth, Soil Degradation and Sustainable Development in Europe. Environmental Issues Series No 16*; Office for Official Publications of the European Communities: Luxembourg, 2000; p. 32.

13. European Environment Agency (EEA). *Europe's Environment, the Third Assessment. Environmental Assessment Report No. 10*; Office for Official Publications of the European Communities: Luxembourg, 2003; p. 341.

14. Tóth, G.; Montanarella, L.; Rusco, E. (Eds.) *Threats to Soil Quality in Europe*; Office for Official Publications of the European Communities: Luxembourg, 2008; p. 151.

15. Kibblewhite, M.G.; Jones, R.J.A.; Montanarella, L.; Baritz, R.; Huber, S.; Arrouays, D.; Micheli, E.; Stephens, M. (Eds.) *Environmental Assessment of Soil for Monitoring Volume VI, Soil Monitoring System for Europe*; Office for Official Publications of the European Communities: Luxembourg, 2008; p. 72.

16. Panagos, P.; van Liedekerke, M.; Yigini, Y.; Montanarella, L. Contaminated sites in Europe, Review of the current situation. *J. Environ. Public Health* **2013**. [CrossRef]

17. European Environment Agency (EEA). Progress in Management of Contaminated Sites (LSI 003)-Assessment Published May 2014. Available online: http://www.eea.europa.eu/data-and-maps/indicators/progress-in-management-of-contaminated-sites-3/assessment (accessed on 5 August 2014).

18. Ferguson, C.C. Assessing risks from contaminated sites, policy and practice in 16 European Countries. *Land Contam. Reclam.* **1999**, *7*, 87–108.

19. European Environment Agency (EEA). *The European Environment—State and Outlook 2010 (SOER 2010),2012 Up-Date, Consumption and the Environment*; Office for Official Publications of the European Communities: Luxembourg, 2012; p. 67.

20. Eurostat, Statistics Explained. Agri-environmental indicator—Mineral fertilizer consumption. Available online: http://epp.eurostat.ec.europa.eu/statistics_explained/index.php/Agri-environmental_indicator_-_mineral_fertiliser_consumption (accessed on 20 September 2014).

21. Bouraoui, F.; Grizzetti, B.; Aloe, A. *Nutrient Discharge from Rivers to Seas*; Office for Official Publications of the European Communities: Luxembourg, 2008; p. 72.

22. Kelessidis, A.; Stasinakis, A. Comparative study of the methods used for treatment and final disposal of sewage sludge in European countries. *Waste Manag.* **2012**, *32*, 1186–1195. [CrossRef] [PubMed]

23. Environmental, economic and social impacts of the use of sewage sludge on land. Available online: http://ec.europa.eu/environment/waste/sludge/pdf/part_i_report.pdf (accessed on 24 November 2014).

24. Fytili, D.; Zabaniotou, A. Utilization of sewage sludge in EU application of old and new methods—A review. *Renew. Sustain. Energy Rev.* **2008**, *12*, 116–140. [CrossRef]

25. Ferreiro-Domínguez, N.; Rigueiro-Rodríguez, A.; Mosquera-Losada, R.M. Sewage sludge fertilizer use: Implications for soil and plant copper evolution in forest and agronomic soils. *Sci. Total Environ.* **2012**, *424*, 39–47. [CrossRef] [PubMed]

26. Rodríguez, L.; Reuter, H.I.; Hengl, T. A framework to estimate the distribution of heavy metals in European soils. In *Threats to Soil Quality in Europe*; Tóth, G., Montanarella, L., Rusco, E., Eds.; Office for Official Publications of the European Communities: Luxembourg, 2008; pp. 79–86.

27. Groupement d'intérêt scientifique sur les sols (GIS Sol). *L'état Des Sols de France*; Ministère de l'Écologie, du Développement Durable, des Transports et du Logement: Paris, France, 2011; p. 188. (In French)

28. Micó, C.; Recatalá, L.; Peris, M.; Sánchez, J. Assessing heavy metal sources in agricultural soils of an European Mediterranean area by multivariate analysis. *Chemosphere* **2006**, *65*, 863–872. [CrossRef] [PubMed]

29. Richards, L.A. (Ed.) *Diagnosis and Improvements of Saline and Alkali Soils*; United States Department of Agriculture (USDA): Washington, DC, USA, 1954; p. 160.

30. Abrol, I.P.; Yadav, J.S.P.; Massoud, F.I. *Salt-Affected Soils and Their Management*; FAO Soils Bulletin 39; Food and Agriculture Organization of the United Nations: Rome, Italy, 1988; p. 230.

31. Tóth, G.; Adhikari, K.; Várallyay, G.; Tóth, T.; Bódis, K.; Stolbovoy, V. Updated Map of Salt Affected Soils in the European Union. In *Threats to Soil Quality in Europe EUR 23438 EN*; Tóth, G., Montanarella, L., Rusco, E., Eds.; Office for Official Publications of the European Communities: Luxembourg, 2008; pp. 65–77.

32. European Environment Agency (EEA). Exposure of ecosystems to acidification, eutrophication and ozone (CSI 005). Available online: http://www.eea.europa.eu/data-and-maps/indicators/exposure-of-ecosystems-to-acidification-2/exposure-of-ecosystems-to-acidification (accessed on 5 August 2014).

33. Kimmins, J.P. Evaluation of the consequences for future tree productivity of the loss of nutrients in whole tree harvesting. *For. Ecol. Manag.* **1977**, *1*, 169–183. [CrossRef]

34. Madeira, M. Changes in soil properties under Eucalyptus plantations in Portugal. In *Biomass Production by Fast Growing Trees*; Pereira, J.S., Landsberg, J.J., Eds.; Kluwer Academic Publisher: Dordrecht, The Netherlands, 1989; pp. 81–89.

35. Merino, A.; Balboa, M.A.; Soalleiro, R.R.; Conzalez, J.G.A. Nutrient exports under different harvesting regimes in fast-growing forest plantations in southern Europe. *For. Ecol. Manag.* **2005**, *207*, 325–339. [CrossRef]

36. Walmsley, J.D.; Jones, D.L.; Reynolds, B.; Price, M.H.; Healey, J.R. Whole tree harvesting can reduce second rotation forest productivity. *For. Ecol. Manag.* **2009**, *257*, 1104–1111. [CrossRef]

37. Augusto, L.; Achat, D.L.; Bakker, M.R.; Bernier, F.; Bert, D.; Danjon, F.; Khlifa, R.; Meredieus, C.; Trichet, P. Biomass and nutrients in tree root systems—Sustainable harvesting of an intensively managed *Pinus pinaster* (Ait.) planted forest. *Global Chang. Biol. Bioenergy* **2014**, in press.

38. Verkerk, P.J.; Anttila, P.; Eggers, J.; Lindner, M.; Asikainen, A. The realisable potential supply of woody biomass from forests in the European Union. *For. Ecol. Manag.* **2011**, *261*, 2007–2015. [CrossRef]

39. Van den Akker, J.J.H.; Arvidsson, J.; Horn, R. Introduction to the special issue on experiences with the impact and prevention of subsoil compaction in the European Union. *Soil Tillage Res.* **2003**, *73*, 1–8. [CrossRef]

40. Le Bas, C.; Houšková, B.; Bialousz, S.; Bielek, P. Identifying risk areas for soil degradation in Europe by compaction. In *Common Criteria for Risk Area Identification according to Soil Threats; European Soil Bureau Research Report No. 20*; Eckelmann, W., Baritz, R., Bialousz, S., Bielek, P., Carre, F., Houšková, B., Jones, R.J.A., Kibblewhite, M.G., Kozak, J., le Bas, C., *et al.*, Eds.; Office for Official Publications of the European Communities: Luxembourg, 2006; pp. 35–42.

41. Huber, S.; Prokop, G.; Arrouays, D.; Banko, G.; Bispo, A.; Jones, R.J.A.; Kibblewhite, M.G.; Lexer, W.; Möller, A.; Rickson, R.J.; *et al. Environmental Assessment of Soil for Monitoring. Volume I, Indicators & Criteria*; Office for Official Publications of the European Communities: Luxembourg, 2008; p. 339.

42. Houšková, B.; Montanarella, L. The natural susceptibility of European soils to compaction. In *Threats to Soil Quality in Europe*; Tóth, G., Montanarella, L., Rusco, E., Eds.; Office for Official Publications of the European Communities: Luxembourg, 2008; pp. 23–35.

43. Van-Camp, L.; Bujarrabal, B.; Gentile, A.-R.; Jones, R.J.A.; Montanarella, L.; Olazabal, C.; Selvaradjou, S.-K. *Reports of the Technical Working Groups Established under the Thematic Strategy for Soil Protection. VOL II—EROSION*; Office for Official Publications of the European Communities: Luxembourg, 2004; p. 192.

44. Joint Research Center (JRC). Soil Themes—Soil Compaction. In European Soil Portal—Soil data and information systems. Available online: http://eusoils.jrc.ec.europa.eu/library/themes/compaction/ (accessed on 6 August 2014).

45. De Mier, A. Optimización de los sistemas de plantación y producción de chopo. In Proceedings of the I Simposio del Chopo, Zamora, Spain, 9–11 May 2001; pp. 97–105.

46. Fernández, A.; Hernanz, G. *El Chopo* (Populus sp.). In *Manual de Gestión Forestal Sostenible*; Junta de Castilla y León: Valladolid, Spain, 2004.

47. Moore, J. *Wood Properties and Uses of Sitka Spruce in Britain*; Forestry Commission: Edinburgh, Scotland, 2001; pp. 1–48.

48. Mead, D.J. *Sustainable Management of Pinus Radiata Plantations*; FAO: Rome, Italy, 2013.

49. Gartzia-Bengoetxea, N.; González-Arias, A.; Kandler, E.; Martínez de Arano, I. Potential indicators of soil quality in temperate forest ecosystems, a case study in the Basque Country. *Ann. For. Res.* **2009**. [CrossRef]

50. De la Mata, R.; Zas, R. Transferring Atlantic maritime pine improved material to a region with marked Mediterranean influence in inland NW Spain, a likelihood-based approach on spatially adjusted field data. *Eur. J. For. Res.* **2010**, *129*, 645–658. [CrossRef]

51. Bercetche, J.; Pâques, M. Somatic embryogenesis in maritime pine (*Pinus Pinaster*) Somatic Embryogenesis in Woody Plants. *For. Sci.* **1995**, *44–46*, 221–242.

52. Fisher, R.F.; Binkley, D. *Ecology and Management of Forest Soils*; Wiley: New York, NY, USA, 2000.

53. Ampoorter, A.; Goris, R.; Cornelis, W.M.; Verheyen, K. Impact of mechanized logging on compaction status of sandy forest soils. *For. Ecol. Manag.* **2007**, *241*, 162–174. [CrossRef]

54. Logson, S.; Berli, M.; Horn, R. Quantifying and modeling soil structure dynamics. In *Quantifying and Modeling Soil Structure Dynamics*; Logson, S., Berli, M., Horn, R., Eds.; Advances in Agricultural Systems Modeling, Soil Science Society of America: Madison, WI, USA, 2013; Volume 3, pp. 1–10.

55. Borrelli, P.; Ballabio, C.; Panagos, P.; Montanarella, L. Wind erosion susceptibility of European soils. *Geoderma* **2014**, *232–234*, 471–478.

56. Lugato, E.; Panagos, P.; Bampa, F.; Jones, A.; Montanarella, L. A new baseline of organic carbon stock in European agricultural soils using a modelling approach. *Global Chang. Biol.* **2014**, *20*, 313–326. [CrossRef]

57. Jones, R.J.A.; Hiederer, R.; Rusco, E.; Loveland, P.J.; Montanarella, L. Estimating organic carbon in the soils of Europe for policy support. *Eur. J. Soil Sci.* **2005**, *56*, 655–671. [CrossRef]

58. Shakesby, R.A. Post-wildfire soil erosion in the Mediterranean, Review and future research directions. *Earth Sci. Rev.* **2011**, *105*, 71–100. [CrossRef]

59. Bellamy, P.H.; Loveland, P.J.; Bradley, R.I.; Lark, R.M.; Kirk, G.J.D. Carbon losses from all soils across England and Wales 1978–2003. *Nature* **2005**, *437*, 245–248. [CrossRef] [PubMed]

60. Capriel, P. Trends in organic carbon and nitrogen contents in agricultural soils in Bavaria (south Germany) between 1986 and 2007. *Eur. J. Soil Sci.* **2013**, *64*, 445–454. [CrossRef]

61. Goidts, E.; van Wesemael, B. Regional assessment of soil organic carbon changes under agriculture in Southern Belgium (1955–2005). *Geoderma* **2007**, *141*, 341–354. [CrossRef]

62. Hanegraaf, M.C.; Hoffland, E.; Kuikman, P.J.; Brussaard, L. Trends in soil organic matter contents in Dutch grasslands and maize fields on sandy soils. *Eur. J. Soil Sci.* **2009**, *60*, 213–222. [CrossRef]

63. Reijneveld, A.; van Wensem, J.; Oenema, O. Soil organic carbon contents of agricultural land in the Netherlands between 1984 and 2004. *Geoderma* **2009**, *152*, 231–238. [CrossRef]

64. Saby, N.P.A.; Arrouays, D.; Antoni, V.; Lemercier, B.; Follain, S.; Walter, C.; Schvartz, C. Changes in soil organic carbon in a mountainous French region, 1990–2004. *Soil Use Manag.* **2008**, *24*, 254–262. [CrossRef]

65. Jandl, R.; Rodeghiero, M.; Martinez, C.; Cotrufo, M.F.; Bampa, F.; van Wesemael, B.; Harrison, R.B.; Guerrini, I.A.; de Richter, D., Jr.; Rustad, L.; *et al.* Current status, uncertainty and future needs in soil organic carbon monitoring. *Sci. Total Environ.* **2014**, *468–469*, 376–383.

66. Lugato, E.; Bampa, F.; Panagos, P.; Montanaralla, L.; Jones, A. Potential carbon sequestration of European arable soils estimated by modelling a comprehensive set of management practices. *Global Chang. Biol.* **2014**, in press.

67. Martínez de Arano, I.; Gartzia-Bengoetxea, N.; González-Arias, A.; Merino, A. Gestión forestal y conservación de suelo en los bosques cultivados del País Vasco. In Proceedings of the Reunión Nacional de Suelos XXVI, Durango, Spain, 25–27 June 2007. (In Spanishe)

68. Gardi, C.; Jeffery, S. *Soil Biodiversity*; Joint Research Center-Institute for Environment and Sustainability; EUR 23759 EN; Office for Official Publications of the European Communities: Luxembourg, 2009; p. 24.

69. Jeffery, S.; Gardi, C.; Jones, A.; Montanerlla, L.; Marmo, L.; Miko, L.; Ritz, K.; Peres, G.; Römbke, J.; van der Putten, W.H. (Eds.) *European Atlas of Soil Biodiversity*; Office for Official Publications of the European Communities: Luxembourg, 2010; p. 128.

70. Turbé, A.; de Toni, A.; Benito, P.; Lavelle, P.; Lavelle, P.; Ruiz, N.; van der Putten, W.H.; Labouze, E.; Mudgal, S. *Soil Biodiversity, Functions, Threats and Tools for Policy Makers*. Available online: http://ec.europa. eu/environment/archives/soil/pdf/biodiversity_report.pdf (accessed on 18 December 2014).

71. Gardi, C.; Jeffery, S.; Saltelli, A. An estimate of potential threats levels to soil biodiversity in EU. *Global Chang. Biol.* **2013**, *19*, 1538–1548. [CrossRef]

72. De La Torre, A.; Iglesias, I.; Carballo, M.; Ramírez, P.; Muñoz, M.J. An approach for mapping the vulnerability of European Union soils to antibiotic contamination. *Sci. Total Environ.* **2012**, *414*, 672–679. [CrossRef]

73. Gardi, C.; Montanarella, L.; Arrouays, D.; Bispo, A.; Lemanceau, P.; Jolivet, C.; Mulder, C.; Ranjard, L.; Römbke, J.; Rutger, M.; *et al.* Soil biodiversity monitoring in Europe, ongoing activities and challenges. *Eur. J. Soil Sci.* **2009**, *60*, 807–819.

74. Pulleman, M.; Creamer, R.; Hamer, U.; Helder, J.; Pelosi, C.; Pérès, G.; Rutgers, M. Soil biodiversity, biological indicators and soil ecosystem services—An overview of European approaches. *Curr. Opin. Environ. Sustain.* **2012**, *4*, 529–538.

75. Rutgers, M.; Schouten, A.J.; Bloem, J.; van Eekeren, N.; de Goede, R.G.M.; Jagers op Akkerhuis, G.A.J.M.; van der Wal, A.; Mulder, C.; Brussaard, L.; Breure, A.M. Biological measurements in a nationwide soil monitoring network. *Eur. J. Soil Sci.* **2009**, *60*, 820–832. [CrossRef]

76. Cluzeau, D.; Guernion, M.; Chaussod, R.; Martin-Laurent, F.; Villenave, C.; Cortet, J.; Ruiz-Camacho, N.; Pernin, C.; Mateille, T.; Philippot, G.; *et al.* Integration of biodiversity in soil quality monitoring, Baselines for microbial and soil fauna parameters for different land-use types. *Eur. J. Soil Biol.* **2012**, *49*, 63–72.

77. Eurostat. The Use of Plant Protection Products in the European Union. In *Eurostat Statistical Books*; Office for Official Publications of the European Communities: Luxembourg, 2007; p. 215.

78. European Environment Agency (EEA). *Environmental Indicator Report 2013*; *Natural Resources and Human Well-Being in a Green Economy*; Office for Official Publications of the European Communities: Luxembourg, 2013; p. 77.

79. Dotterweich, M. The history of human-induced soil erosion, Geomorphic legacies, early descriptions and research, and the development of soil conservation—A global synopsis. *Geomorphology* **2013**, *201*, 1–34. [CrossRef]

80. Fullen, M.A. Soil erosion and conservation in northern Europe. *Progress Phys. Geogr.* **2003**, *27*, 331–358. [CrossRef]

81. Eurostat, Statistics Explained. Agri-environmental indicator—Soil erosion. Available online: http:// epp.eurostat.ec.europa.eu/statistics_explained/index.php/Agri-environmental_indicator_-_soil_erosion (accessed on 4 August 2014).

82. Verheijen, F.G.A.; Jones, R.J.A.; Rickson, R.J.; Smith, C.J. Tolerable *versus* actual soil erosion rates in Europe. *Earth Sci. Rev.* **2009**, *94*, 23–38. [CrossRef]

83. Verheijen, F.G.A.; Jones, R.J.A.; Rickson, R.J.; Smith, C.J.; Bastos, A.C.; Nunes, P.; Keizer, J.J. Concise overview of European soil erosion research and evaluation. *Soil Plant Sci.* **2012**, *62*, 185–190.

84. Jones, R.J.A.; Verheijen, F.G.A.; Reuter, H.I.; Jones, A.R (Eds.) *Environmental Assessment of Soil for Monitoring Volume V, Procedures & Protocols*; EUR 23490 EN/5; Office for the Official Publications of the European Communities: Luxembourg, 2008; p. 165.

85. Panagos, P.; Meusburger, K.; Ballabio, C.; Borrelli, P.; Alewell, C. Soil erodibility in Europe, a high-resolution dataset based on LUCAS. *Sci. Total Environ* **2014**, *479–480*, 189–200.

86. Panagos, P.; Meusburger, K.; van Liedekerke, M.; Alewell, C.; Hiederer, R.; Montanarella, L. Assessing soil erosion in Europe based on data collected through a European network. *Soil Sci. Plant Nutr.* **2014**, *60*, 15–29. [CrossRef]

87. Bosco, C.; de Rigo, D.; Dewitte, O.; Poesen, J.; Panagos, P. Modelling soil erosion at European scale, towards harmonization and reproducibility. *Natl. Hazards Earth Syst. Sci. Dis.* **2014**, *2*, 2639–2680. [CrossRef]

88. Gobin, A.; Govers, G.; Kirkby, M.J.; le Bissonnais, Y.; Kosmas, C.; Puigdefabrecas, J.; van Lynden, G.; Jones, R.J.A. *PESERA Project Technical Annex. Contract No. QLKS-CT-1999-01323*; European Commission: Luxembourg, 1999.

89. Cerdan, O.; Govers, G.; le Bissonnais, Y.; van Oost, K.; Poesen, J.; Saby, N.; Gobin, A.; Vacca, A.; Quinton, J.; Auerswald, K.; *et al.* Rates and spatial variations of soil erosion in Europe, A study based on erosion plot data. *Geomorphology* **2010**, *122*, 167–177.

90. Eurostat, Statistics Explained. Agri-environmental indicator—Soil cover. Available online: http://epp. eurostat.ec.europa.eu/statistics_explained/index.php/Agri-environmental_indicator_-_soil_cover (accessed on 30 July 2014).

91. Van Oost, K.; van Muysen, W.; Govers, G.; Deckers, J.; Quine, T.A. From water to tillage erosion dominated landform evolution. *Geomorphology* **2005**, *72*, 193–203. [CrossRef]

92. Prokop, G.; Jobstmann, H.; Schönbauer, A. *Report on Best Practices for Limiting Soil Sealing and Mitigating its Effects*; Technical Report-2011-050; European Commission, Office for Official Publications of the European Communities: Luxembourg, 2011; p. 76.

93. European Environment Agency (EEA). *The European Environment—State and Outlook 2010 (SOER 2010)–Land Use*; Office for Official Publications of the European Communities: Luxembourg, 2010; p. 49.

94. Eurostat, Statistics Explained. Agri-environmental indicator-land use change. Available online: http://epp. eurostat.ec.europa.eu/statistics_explained/index.php/Agri-environmental_indicator_-_land_use_change (accessed on 20 July 2014).

95. Eurostat, Statistics Explained; Land cover statistics. Available online: http://epp.eurostat.ec.europa.eu/ statistics_explained/index.php/Land_cover_statistics (accessed on 23 July 2014).

96. Morel, J.L.; Chenu, C.; Lorenz, K. Ecosystem services provided by soils of urban, industrial, traffic, mining, and military areas (SUITMAs). *J. Soil Sediments* **2014**, in press.

97. Pointereau, P.; Coulon, F.; Girard, P.; Lambotte, M.; Stuczynski, T.; Sánchez-Ortega, V.; del Rio, A. Analysis of farmland abandonment and the extent and location of agricultural areas that are actually abandoned. In *Institute for Environment and Sustainability*; EUR 23411EN–2008; Anguiano, E., Bamps, C., Terres, J.-M., Eds.; JRC Scientific and Technical Reports; Office for Official Publications of the European Communities: Luxembourg, 2008; p. 204.

98. Renwick, A.; Jansson, T.; Verburg, P.H.; Revoredo-Giha, C.; Britz, W.; Gochte, A.; McCracken, D. Policy reform and agricultural land abandonment in the EU. *Land Use Policy* **2013**, *30*, 446–457. [CrossRef]

99. Keenleyside, C.; Tucker, G.M. *Farmland Abandonment in the EU, an Assessment of Trends and Prospects*; Institute for European Environmental Policy: London, UK, 2010.

100. Eurostat, Statistics Explained. Agri-environmental indicator—Risk of land abandonment. Available online: http://epp.eurostat.ec.europa.eu/statistics_explained/index.php/Agri-environmental_indicator_ -_risk_of_land_abandonment (accessed on 14 October 2014).

101. García-Ruiz, J.M.; Lana-Renalult, N. Hydrological and erosive consequences of farmland abandonment in Europe, with special reference to the Mediterranean region—A review. *Agric. Ecosyst. Environ.* **2011**, *140*, 317–338. [CrossRef]

102. Duarte, F.; Jones, N.; Fleskens, L. Traditional olive orchards on sloping land, Sustainability or abandonment? *J. Environ. Manag.* **2008**, *89*, 86–98. [CrossRef]

103. Acín-Carrera, M.; Marqués, M.J.; Carral, P.; Álvarez, A.M.; López, C.; Martín-López, B.; González, J.A. Impacts of land-use intensity on soil organic carbon content, soil structure and water-holding capacity. *Soil Use Manag.* **2013**, *29*, 547–556. [CrossRef]

104. Zeller, V.; Bardgett, R.; Tappeiner, U. Site and management effects on soil microbial properties of subalpine meadows, a study of land abandonment along a north-south gradient in the European Alps. *Soil Biol. Biochem.* **2001**, *33*, 639–649. [CrossRef]

105. Nunes, J.M.; López-Piñeiro, A.; Albarrán, A.; Muñoz, A.; Coelho, J. Changes in selected soil properties caused by 30 years of continuous irrigation under Mediterranean conditions. *Geoderma* **2007**, *139*, 321–328.

106. Denef, K.; Stewart, C.E.; Brenner, J.; Paustian, K. Does long-term center-pivot irrigation increase soil carbon stocks in semi-arid agro-ecosystems? *Geoderma* **2008**, *145*, 121–129. [CrossRef]

107. Holland, T.C.; Reynolds, A.G.; Bowen, P.A.; Bogdanoff, C.P.; Marciniak, M.; Brown, R.B.; Hart, M.M. The response of soil biota to water availability in vineyards. *Pedobiologia* **2013**, *56*, 9–14. [CrossRef]

108. Baldock, D.; Caraveli, H.; Dwyer, J.; Einschütz, S.; Petersen, J.E.; Sumpsi-Viñas, J.; SVarela-Ortega, C. *The Environmental Impacts of Irrigation in the European Union*; Environment Directorate of the European Commission, Institute for European Environmental Policy: Brussels, Belgium, 2000; p. 138.

109. Eurostat. Water Use Intensity. Available online: http://epp.eurostat.ec.europa.eu/portal/page/ portal/agri_environmental_indicators/indicators_overview/agricultural_production_systems (accessed on 31 July 2014).

110. Vlek, P.L.G.; Hillel, D.; Braimoh, A.K. Soil degradation under irrigation. In *Land Use and Soil Resources*; Braimoh, A.K., Vlek, P.L.G., Eds.; Springer Science: Berlin, Germany, 2008; pp. 101–120.

111. Eurostat, Statistics Explained. Agri-environmental indicator—Irrigation. Available online: http://epp. eurostat.ec.europa.eu/statistics_explained/index.php/Agri-environmental_indicator_-_irrigation (accessed on 10 August 2014).

112. United Nations Educational, Scientific and Cultural Organization (UNESCO). Irrigated land as percentage of cultivated land. Available online: http://www.unesco.org/new/fileadmin/MULTIMEDIA/HQ/SC/ temp/wwap_pdf/Irrigated_land_as_a_percentage_of_cultivated_land.pdf (accessed on 31 July 2014).

113. Zdruli, P. Land resources of the Mediterranean, status, pressures, trends and impacts on future regional development. *Land Degrad. Dev.* **2014**, *25*, 373–384. [CrossRef]

114. European Commission. Agriculture and environment—Agriculture and Water. Available online: http://ec.europa.eu/agriculture/envir/water/index_en.htm (accessed on 31 July 2014).

115. Martínez-Alvarez, V.; García-Bastida, P.A.; Martín-Gorriz, M.; Soto-García, M. Adaptive strategies of on-farm water management under watersupply constraints in south-eastern Spain. *Agric. Water Manag.* **2014**, *136*, 59–67. [CrossRef]

116. Aragüés, R.; Medina, E.T.; Martínez-Cob, A.; Faci, J. Effects of deficit irrigation strategies on soil salinization and sodification in a semiarid drip-irrigated peach orchard. *Agric. Water Manag.* **2014**, *142*, 1–9. [CrossRef]

117. Aragüés, R.; Medina, E.T.; Clavería, I.; Martínez-Cob, A.; Faci, J. Regulated deficit irrigation, soil salinization and soil sodification in atable grape vineyard drip-irrigated with moderately saline waters. *Agric. Water Manag.* **2014**, *134*, 84–93. [CrossRef]

118. Louwagie, G.; Gay, S.H.; Burrel, A. (Eds.) Addressing soil degradation in EU agriculture, relevant processes, practices and policies. In *Report on the project "Sustainable Agriculture and Soil Conservation (SoCo)"*; Office for Official Publications of the European Communities: Luxembourg, 2009; p. 208.

119. Spiegel, H.; Zavattaro, L.; Guzmán, G.; D'Hose, P.A.; Schlatter, N.; Ten, B.H.; Grignani, C. Impacts of soil management practices on crop productivity, on indicators for climate change mitigation, and on the chemical, physical and biological quality of soil. Available online: http://www.catch-c.eu/deliverables/D3.371_ Overall%20report_23July14.pdf (accessed on 24 November 2014).

120. Food and Agriculture Organization of the United Nations (FAO). What is conservation agriculture? Available online: http://www.fao.org/ag/ca/1a.html (accessed on 20 September 2014).

121. Lahmar, R. Adoption of conservation agriculture in Europe. Lessons of the KASSA project. *Land Use Policy* **2010**, *27*, 4–10. [CrossRef]

122. Lahmar, R.; Arrúe, J.L.; Denardin, J.E.; Gupta, R.K.; Ribeiro, M.F.S.; de Tourdonnet, S.; Abrol, I.P.; Barz, P.; de Benito, A.; Bianchini, A.; *et al*. *Knowledge Assessment and Sharing on Sustainable Agriculture. Synthesis Report*; Centre de coopération internationale en recherche agronomique pour le développement (CIRAD): Montpellier, France, 2007; p. 125.

123. De Turdonnet, S.; Nozières, A.; Barz, P.; Chenu, C.; Düring, R.-A.; Frielinghaus, M.; Kölli, R.; Kubat, J.; Magid, J.; Medvedev, V.; *et al. Comprehensive Inventory and Assessment of Existing Knowledge on Sustainable Agriculture in the European Platform of KASSA*; Centre de coopération internationale en recherche agronomique pour le développement (CIRAD): Montpellier, France, 2007; p. 55.

124. Arrúe, J.L.; Cantero-Martínez, C.; Cardarelli, A.; Kavvadias, V.; López, M.V.; Moreno, F.; Mrabet, R.; Murillo, J.M.; Pérez de Ciriza, J.J.; Sombrero, A.; *et al. Comprehensive Inventory and Assessment of Existing Knowledge on Sustainable Agriculture in the Mediterranean Platform of KASSA*; Centre de coopération internationale en recherche agronomique pour le développement (CIRAD): Montpellier, France, 2007; p. 24.

125. Soane, B.D.; Ball, B.C.; Arvidsson, J.; Basch, G.; Moreno, F.; Roger-Estrade, J. No-till in northern, western and south-western Europe, A review of problems and opportunities for crop production and the environment. *Soil Tillage Res.* **2012**, *118*, 66–87. [CrossRef]

126. Powlson, D.S.; Stidling, C.M.; Jat, M.L.; Gerard, B.G.; Palm, C.A.; Sanchez, P.A.; Cassman, K.G. Limited potential of no-till agriculture for climate change mitigation. *Nat. Clim. Chang.* **2014**, *4*, 678–683. [CrossRef]

127. Álvaro-Fuentes, J.; Plaza-Bonilla, D.; Arrúe, J.L.; Lampurlanés, J.; Cnatero-Martínez, C. Soil organic carbon storage in a no-tillage chronosequence under Mediterranean conditions. *Plant Soil* **2014**, *376*, 31–41. [CrossRef]

128. Dimassi, B.; Mary, B.; Wylleman, R.; Labreuche, J.; Couture, D.; Piraux, F.; Cohan, J-P. Long-term effect of contrasted tillage and crop management on soilcarbon dynamics during 41 years. *Agric. Ecosyst. Environ.* **2014**, *188*, 134–146.

129. Schjønning, P.; Heckrath, G.; Christensen, B.T. Threats to soil quality in Denmark. In *A Review of Existing Knowledge in the Context of the EU Soil Thematic Strategy*; DJF Report Plant Science No. 143; Faculty of Agricultural Sciences, Aarhus University: Aarhus, Denmark, 2009; p. 124.

130. Angers, D.A.; Eriksen-Hamel, N.S. Full-inversion tillage and organic carbon distribution in soil profiles, a meta-analysis. *Soil Sci. Soc. Am. J.* **2008**, *72*, 1370–1374. [CrossRef]

131. Virto, I.; Barré, P.; Burlot, A.; Chenu, C. Carbon input differences as the main factor explaining the variability in soil organic C storage in no-tilled compared to inversion tilled agrosystems. *Biogeochemistry* **2012**, *108*, 17–26. [CrossRef]

132. Bescansa, P.; Imaz, M.J.; Virto, I.; Enrique, A.; Hoogmoed, W.B. Soil water retention as affected by tillage and residue management in semiarid Spain. *Soil Tillage Res.* **2006**, *87*, 19–27. [CrossRef]

133. Fernández-Ugalde, O.; Virto, I.; Bescansa, P.; Imaz, M.J.; Enrique, A.; Karlen, D.L. No-tillage improvement of soil physical quality in calcareous, degradation-prone, semiarid soils. *Soil Tillage Res.* **2009**, *106*, 29–35. [CrossRef]

134. Vogeler, I.; Rogasik, J.; Funder, U.; Panten, K.; Schnug, E. Effect of tillage systems and P-fertilization on soil physical and chemical properties, crop yield and nutrient uptake. *Soil Tillage Res.* **2009**, *103*, 136–143. [CrossRef]

135. Tebrügge, F.; Düring, F.-A. Reducing tillage intensity—A review of results from a long-term study in Germany. *Soil Tillage Res.* **1999**, *53*, 15–28. [CrossRef]

136. Rasmussen, K.J. Impact of ploughless soil tillage on yield and soil quality, A Scandinavian review. *Soil Tillage Res.* **1999**, *53*, 3–14. [CrossRef]

137. Alakukku, L.; Ristolainen, A.; Sarikka, I.; Hurme, T. Surface water ponding on clayey soils managed by conventional and conservation tillage in boreal conditions. *Agric. Food Sci.* **2010**, *19*, 313–326.

138. Virto, I.; Bescansa, P.; Imaz, M.J.; Enrique, A.; Hoogmoed, W. Burning crop residues under no-till in semi arid land, Northern Spain. Effects on soil organic matter, aggregation and earthworm populations. *Aust. J. Soil Res.* **2007**, *45*, 414–421.

139. Van den Putte, A.; Govers, G.; Diels, J.; Gillijns, K.; Demuzere, M. Assessing the effect of soil tillage on crop growth, A meta-regression analysis on European crop yields under conservation agriculture. *Eur. J. Agron.* **2010**, *33*, 231–241. [CrossRef]

140. Morris, N.L.; Miller, P.C.H.; Orson, J.H.; Froud-Williams, R.J. The adoption of non-inversion tillage systems in the United Kingdom and the agronomic impact on soil, crops and the environment—A review. *Soil Tillage Res.* **2010**, *108*, 1–15. [CrossRef]

141. Ruisi, P.; Giambalvo, D.; Saia, S.; di Miceli, G.; Frenda, A.S.; Plaia, A.; Amato, G. Conservation tillage in a semiarid Mediterranean environment, results of 20 years of research. *Ital. J. Agron.* **2014**, *9*, 1–7. [CrossRef]

142. Basch, G.; Geraghty, J.; Streit, B.; Sturny, W. No Tillage in Europe—State of the Art. Available online: http://dspace.uevora.pt/rdpc/bitstream/10174/2999/1/No-Till%20Farming%20Systems.pdf (accessed on 18 December 2014).

143. Gómez, J.A.; Sobrinho, T.A.; Giráldez, J.V.; Fereres, E. Soil management effects on runoff, erosion and soil properties in an olive grove of Southern Spain. *Soil Tillage Res.* **2009**, *102*, 5–13. [CrossRef]

144. Rodríguez-Lizana, A.; Espejo-Pérez, A.J.; González-Fernández, P.; Ordóñez-Fernández, R. Pruning residues as an alternative to traditional tillage to reduce erosion and pollutant dispersion in olive groves. *Water Air Soil Pollut.* **2008**, *193*, 165–173. [CrossRef]

145. Rodríguez-Entrena, M.; Arriaza, M. Adoption of conservation agriculture in olive groves, evidences from southern Spain. *Land Use Policy* **2013**, *34*, 294–300. [CrossRef]

146. Wezel, A.; Casagrande, M.; Celette, F.; Vian, J.F.; Ferrer, A.; Peigné, J. Agroecological practices for sustainable Agriculture: A review. *Agron. Sustain. Dev.* **2014**, *34*, 1–20. [CrossRef]

147. Eurostat, Statistics Explained. Organic farming statistics. Available online: http://epp.eurostat.ec. europa.eu/statistics_explained/index.php/Organic_farming_statistics#Potential_for_growth (accessed on 25 October 2014).

148. Mondelaers, K.; Aertsens, J.; van Huylenbroeck, J. A meta-analysis of the differences in environmental impacts between organic and conventional farming. *Br. Food J.* **2009**, *111*, 1098–1119. [CrossRef]

149. Tuomisto, H.L.; Hodge, I.D.; Riordan, P.; Macdonald, D.W. Does organic farming reduce environmental impacts? A meta-analysis of European research. *J. Environ. Manag.* **2012**, *112*, 309–320.

150. Calero, J.; Cordovilla, M.P.; Aranda, V.; Borjas, R.; Aparicio, C. Effect of organic agriculture and soil forming factors on soil quality and physiology of olive trees. *Agroecol. Sustain. Food Syst.* **2013**, *37*, 193–214.

151. De Torres, J.; Garzón, E.; Ryan, J.; González-Andrés, F. Organic cereal/forage legume rotation in a Mediterranean calcareous soil, Implications for soil parameters. *Agroecol. Sustain. Food Syst.* **2013**, *37*, 215–230.

152. Melero, S.; Ruiz Porras, J.C.; Herencia, J.F.; Madejon, E. Chemical and biochemical properties in a silty loam soil under conventional and organic management. *Soil Tillage Res.* **2006**, *90*, 162–170. [CrossRef]

153. Papadopoulos, A.; Bird, N.R.A.; Whitmore, A.P.; Mooney, S.J. Does organic management lead to enhanced soil physical quality? *Geoderma* **2014**, *213*, 435–443.

154. Coll, P.; Le Cadre, E.; Blanchart, E.; Hinsinger, P.; Vilenave, C. Organic viticulture and soil quality. A long-term study in Southern France. *Appl. Soil Ecol.* **2011**, *50*, 37–44.

155. Parras-Alcántara, L.; Díaz-Jaimes, L.; Lozano-García, B.; Fernández Rebollo, P.; Moreno Elcure, F.; Carbonero Muñoz, M. Organic farming has little effect on carbon stock in a Mediterranean dehesa (southern Spain). *Catena* **2014**, *113*, 9–17. [CrossRef]

156. Van Meijl, H.; van Rheenen, T.; Tabeau, A.; Eickhout, B. The impact of different policy environments on agricultural land use in Europe. *Agric. Ecosyst. Environ.* **2006**, *114*, 21–38. [CrossRef]

157. Palma, J.H.N.; Grave, A.R.; Bunce, R.G.H.; Burgess, P.J.; de Filippi, R.; Keesman, K.J.; van Keulen, H.; Liagre, F.; Mayus, M.; Moreno, G.; *et al.* Modeling environmental benefits of silvoarable agroforestry in Europe. *Agric. Ecosyst. Environ.* **2007**, *119*, 320–334.

158. Tsonkova, P.; Böhm, C.; Quinkenstein, A.; Freese, D. Ecological benefits provided by alley cropping systems for production of woody biomass in the temperate region, a review. *Agrofor. Syst.* **2012**, *85*, 133–152. [CrossRef]

159. Quinkenstein, A.; Woöllecke, J.; Böhma, C.; Grünewald, H.; Freese, D.; Schneider, B.U.; Hüttl, R.F. Ecological benefits of the alley cropping agroforestry system in sensitive regions of Europe. *Environ. Sci. Policy* **2012**, *12*, 1112–1121. [CrossRef]

160. Lozano-García, B.; Parras-Alcántara, L. Land use and management effects on carbon and nitrogen in Mediterranean Cambisols. *Agric. Ecosyst. Environ.* **2013**, *179*, 208–214. [CrossRef]

161. Moreno, G.; Obrador, J.J.; García, A. Impact of evergreen oaks on soil fertility and crop production in intercropped dehesas. *Agric. Ecosyst. Environ.* **2007**, *119*, 270–280. [CrossRef]

162. Seddaui, G.; Porcu, G.; Ledda, L.; Roggero, P.P.; Agnelli, A.; Corti, G. Soil organic matter content and composition as influenced by soil management in a semi-arid Mediterranean agro-silvo-pastoral system. *Agric. Ecosyst. Environ.* **2013**, *167*, 1–11. [CrossRef]

163. Shakesby, R.A.; Coelho, C.O.A.; Schnabel, S.; Keizer, J.J.; Clarke, M.A.; Lavado Contador, J.F.; Walsh, R.P.D.; Ferreira, A.J.D.; Doerr, S.H. A ranking methodology for assessing relative erosion risk and its application to dehesas and montados in Spain and Portugal. *Land Degrad. Dev.* **2002**, *13*, 129–140. [CrossRef]

164. D'Haene, K.; Vermang, J.; Cornelis, W.M.; Leroy, B.L.M.; Schiettecatte, W.; de Neve, D.; Gabriels, D.; Hofman, G. Reduced tillage effects on physical properties of silt loam soils growing root crops. *Soil Tillage Res.* **2008**, *99*, 279–290. [CrossRef]

165. D'Haene, K.; Vandenbruwane, J.; de Neve, S.; Gabriels, D.; Salomez, J.; Hofman, G. The effect of reduced tillage on nitrogen dynamics in silt loam soils. *Eur. J. Agron.* **2008**, *28*, 449–460. [CrossRef]

166. Melville, N.; Morgan, R.P.C. The influence of grass density on effectiveness of contour grass strips for control of soil erosion on low angle slopes. *Soil Use Manag.* **2001**, *17*, 278–281. [CrossRef]

167. Quinton, J.N.; Catt, J.A. The effects of minimal tillage and contour cultivation on surface runoff, soil loss and crop yield in the long-term Woburn Erosion Reference Experiment on sandy soil at Woburn, England. *Soil Use Manag.* **2004**, *20*, 343–349. [CrossRef]

168. Walter, C.; Merot, P.; Layer, B.; Dutin, G. The effect of hedgerows on soil organic carbon storage in hillslopes. *Soil Use Manag.* **2003**, *19*, 201–207.

169. Stanchi, S.; Freppaz, M.; Agnelli, A.; Reinsch, T.; Zanini, E. Properties, best management practices and conservation of terraced soils in Southern Europe (from Mediterranean areas to the Alps), A review. *Quat. Int.* **2012**, *265*, 90–100. [CrossRef]

170. Durán-Zuazo, V.H.; Martínez-Raya, A.; Aguilar-Ruiz, J. Nutrient losses by runoff and sediment from the taluses of orchard terraces. *Water Air Soil Pollut.* **2004**, *153*, 355–373. [CrossRef]

171. Bellin, N.; van Wesemael, B.; Meerkek, A.; Vanaker, V.; Barbera, G.G. Abandonment of soil and water conservation structures in Mediterranean ecosystems. A case study from south east Spain. *Catena* **2009**, *76*, 114–121.

172. Fiener, P.; Auerswald, K.; Weigand, S. Managing erosion and water quality in agricultural watersheds by small detention ponds. *Agric. Ecosyst. Environ.* **2005**, *110*, 132–142. [CrossRef]

173. Olesen, J.E.; Munkholm, L.J. Subsoil loosening in a crop rotation for organic farming eliminated plough pan with mixed effects on crop yield. *Soil Tillage Res.* **2007**, *94*, 376–385. [CrossRef]

174. Coulouma, G.; Boizard, H.; Trotoux, G.; Lagacherie, P.; Richard, G. Effect of deep tillage for vineyard establishment on soil structure, A case study in Southern France. *Soil Tillage Res.* **2006**, *88*, 132–143. [CrossRef]

175. Stoate, C.; Bladi, A.; Boatman, N.D.; Herzon, I.; van Doorn, A.; Snoo, G.R.; Rakosy, L.; Ramwell, C. Ecological impacts of early 21st century agricultural change in Europe—A review. *J. Environ. Manag.* **2009**, *91*, 22–46. [CrossRef]

176. Gartzia-Bengoetxea, N. Risk of degradation of forest soils in the Basque Country. FORRISK, International workshop on soil degradation risks in planted forests. In Proceedings of the International Conference on Soil Degradation Risks in Planted Forests, Bilbao, Spain, 10 September 2014.

177. Bass, S. Certification. In *Encyclopedia of Forest Sciences*; Academic Press: Oxford, UK, 2004; pp. 1350–1357.

178. Gafo Gómez-Zamalloa, M.; Caparros, A.; San-Miguel Ayanz, A. 15 years of forest certification in the European Union. Are we doing things right? *For. Syst.* **2011**, *20*, 81–94.

179. Official Journal of the European Union (OJEU). Withdrawal of obsolete Commission proposals. OJEU 153/3. Available online: http://eur-lex.europa.eu/legal-content/EN/TXT/PDF/?uri=CELEX,52014XC0521(01) &from=EN (accessed on 2 August 2014).

180. Rodrigues, S.M.; Pereira, M.E.; Ferreira da Silva, E.; Hursthouse, A.S.; Duarte, A.C. A review of regulatory decisions for environmental protection, Part I—Challenges in the implementation of national soil policies. *Environ. Int.* **2009**, *35*, 202–213. [CrossRef] [PubMed]

181. Römbke, J.; Breure, A.M.; Mulder, C.; Rutgers, M. Legislation and ecological quality assessment of soil, implementation of ecological indication systems in Europe. *Ecotoxicol. Environ. Saf.* **2005**, *62*, 201–210. [CrossRef] [PubMed]

182. Prager, K.; Hagemann, N.; Schuler, J.; Heyn, N. Incentives and enforcement, the institutional design and policy mix for soil conservation in Brandenburg (Germany). *Land Degrad. Dev.* **2011**, *22*, 111–123. [CrossRef]

183. Andrews, S.S.; Karlen, D.L.; Cambardella, C.A. The soil management assessment framework, a quantitative soil quality evaluation method. *Soil Sci. Soc. Am. J.* **2004**, *68*, 1945–1962. [CrossRef]

184. Karlen, D.L.; Ditzler, C.A.; Andrews, S.S. Soil quality, why and how? *Geoderma* **2003**, *114*, 145–156. [CrossRef]

185. Karlen, D.L.; Mausbach, M.J.; Doran, J.W.; Cline, R.G.; Harris, R.F.; Schuman, G.E. Soil quality, A concept, definition, and framework for evaluation. *Soil Sci. Soc. Am. J.* **1997**, *61*, 4–10. [CrossRef]

186. Blum, W.E.H. Characterisation of soil erosion risk, an overview. In *Threats to Soil Quality in Europe*; Tóth, G., Montanarella, L., Rusco, E., Eds.; JRC Scientific and Technical Reports; Office for Official Publications of the European Communities: Luxembourg, 2008; pp. 5–10.

187. Tóth, G.; Stolbovoy, V.; Montanarella, L. *Soil Quality and Sustainability Evaluation—An Integrated Approach to Support Soil-Related Policies of the European Union*; EUR 22721 EN; Office for Official Publications of the European Communities: Luxembourg, 2007; p. 40.

188. Tóth, G. Soil quality in the European Union. In *Threats to Soil Quality in Europe*; Tóth, G., Montanarella, L., Rusco, E., Eds.; JRC Scientific and Technical Reports; Office for Official Publications of the European Communities: Luxembourg, 2008; pp. 11–19.

189. Eckelmann, W.; Baritz, R.; Bialousz, S.; Bielek, P.; Carre, F.; Houšková, B.; Jones, R.J.A.; Kibblewhite, M.G.; Kozak, J.; le Bas, C.; *et al. Common Criteria for Risk Area Identification According to Soil Threats. European Soil Bureau Research Report No.20*; EUR 22185 EN; Office for Official Publications of the European Communities: Luxembourg, 2006; p. 94.

190. Micheli, E.; Bialousz, S.; Bispo, A.; Boixadera, J.; Jones, A.R.; Kibblewhite, M.G.; Kolev, N.; Kosmas, C.; Lilja, H.; Malucelli, F.; Rubio, J.L.; Stephens, M. (Eds.) *Environmental Assessment of Soil for Monitoring, Volume IVa Prototype Evaluation*; EUR 23490 EN/4A; Office for the Official Publications of the European Communities: Luxembourg, 2008; p. 96.

191. Stephens, M.; Micheli, E.; Jones, A.R.; Jones, R.J.A. (Eds.) *Environmental Assessment of Soil for Monitoring Volume IVb, Prototype Evaluation–Pilot Studies*; EUR 23490 EN/4B; Office for the Official Publications of the European Communities: Luxembourg, 2008; p. 487.

192. Morvan, X.; Saby, N.P.A.; Arrouays, D.; Le Bas, C.; Jones, R.J.A.; Verheijen, F.G.A. Soil monitoring in Europe, a review of existing systems and requirements for harmonisation. *Sci. Total Environ.* **2008**, *391*, 1–12.

193. Tóth, G.; Arwyn, J.; Montanarella, L. The LUCAS topsoil database and derived information on the regional variability of cropland topsoil properties in the European Union. *Environ. Montioring Assess.* **2013**, *185*, 7409–7425. [CrossRef]

194. Eurostat, Statistics Explained. Agri-environmental indicator—Soil quality. Available online: http://epp.eurostat.ec.europa.eu/statistics_explained/index.php/Agri-environmental_indicator_-_soil_quality (accessed on 12 August 2014).

195. Robinson, D.A.; Cooper, D.; Emmett, B.A.; Evans, C.D.; Keith, A.; Lebron, I.; Lofts, S.; Norton, L.; Reynolds, B.; Tipping, E.; *et al.* Defra soil protection research in the context of the soil natural capital/ecosystem services framework. In *Project SP1607, Synthesis of Soil Protection Work 1990–2008*; Centre for Ecology and Hydrology: Bangor, UK, 2010; p. 131.

196. Scottish Government. *The Scottish Soil Framework*; Scottish Government: Edinburgh, UK, 2009; p. 66.

197. Jolivet, C.; Arrouays, D.; Boulonne, C.; Ratié, C.; Saby, N. Le Réseau de mesures de la Qualité des sols de France (RMQS). Etat d'avancement et premiers résultats. *Étude et Gestion des Sols* **2006**, *13*, 149–164.

198. Mueller, L.; Schindler, U.; Mirschel, W.; Shepherd, G.; Ball, B.C.; Helming, K.; Rogasi, J.; Eulenstein, F.; Wiggering, H. Assessing the productivity function of soils. A review. *Agron. Sustain. Dev.* **2010**, *30*, 601–614. [CrossRef]

199. Wienhold, B.J.; Karlen, D.L.; Andrews, S.S.; Stott, D.E. Protocol for indicator scoring in the soil management assessment framework (SMAF). *Renew. Agric. Food Syst.* **2009**, *24*, 260–266. [CrossRef]

200. Karlen, D.; Cambardella, C.; Kovar, J.L.; Colvin, T.S. Soil quality response to long-term tillage and crop rotation practices. *Soil Tillage Res.* **2013**, *133*, 54–64. [CrossRef]

201. Schulte, R.P.O.; Creamer, R.E.; Donnellan, T.; Farrelly, N.; Reamonn, F.; O'Donoghue, C.; O'hUallachim, D. Functional land management, a framework for managing soil-based ecosystem services for the sustainable intensification of agriculture. *Environ. Sci. Policy* **2014**, *38*, 45–58. [CrossRef]

202. Volchko, Y.; Norman, J.; Rosén, L.; Bergknut, M.; Josefsson, S.; Söderqvist, T.; Norberg, T.; Wiberg, K.; Tysklind, M. Using soil function evaluation in multi-criteria decision analysis for sustainability appraisal of remediation alternatives. *Sci. Total Environ.* **2014**, *485–486*, 785–791.

203. Marinari, S.; Mancinelli, R.; Campiglia, E.; Grego, S. Chemical and biological indicators of soil quality in organic and conventional farming systems in Central Italy. *Ecol. Indic.* **2006**, *6*, 701–711. [CrossRef]

204. Gómez, J.A.; Álvarez, S.; Soriano, M.A. Development of a soil degradation assessment tool for organic olive groves in southern Spain. *Catena* **2009**, *79*, 9–17. [CrossRef]

205. Zornoza, R.; Mataix-Solera, J.; Guerrero, C.; Arcenegui, V.; García-Oresnes, F.; Mataix-Beneyto, J.; Morugán, A. Evaluation of soil quality using multiple lineal regression based on physical, chemical and biochemical properties. *Sci. Total Environ.* **2007**, *378*, 233–237. [CrossRef] [PubMed]

206. Imaz, M.J.; Virto, I.; Bescansa, P.; Enrique, A.; Fernández-Ugalde, O.; Karlen, D.L. Tillage and residue management effects on semi-arid Mediterranean soil quality. *Soil Tillage Res.* **2010**, *107*, 17–25. [CrossRef]

207. Miralles, I.; Ortega, R.; Almendros, G.; Sánchez-Marañón, M.; Soriano, M. Soil quality and organic carbon ratios in mountain agroecosystems of South-east Spain. *Geoderma* **2009**, *150*, 120–128. [CrossRef]

208. Salome, C.; Coll, P.; Lard, E.; Vilenave, C.; Blanchart, E.; Hinsinger, P.; Marsden, C.; le Cadre, E. Relevance of use-invariant soil properties to assess soil quality ofvulnerable ecosystems, The case of Mediterranean vineyards. *Ecol. Indic.* **2014**, *43*, 83–93. [CrossRef]

209. Garbisu, C.; Alkorta, I.; Epelde, L. Assessment of soil quality using microbial properties and attributes of ecological relevance. *Appl. Soil Ecol.* **2011**, *49*, 1–4. [CrossRef]

210. Ruiz, N.; Mathieu, J.; Célini, L.; Rollard, C.; Hommay, G.; Iorio, E.; Lavelle, P. IBQS, a synthetic index of soil quality based on soil macro-invertebrate communities. *Soil Biol. Biochem.* **2011**, *43*, 2032–2045.

211. Andrews, S.S.; Flora, C.B.; Mithcell, J.P.; Karlen, D.L. Growers' perceptions and acceptance of soil quality indices. *Geoderma* **2003**, *114*, 187–213. [CrossRef]

212. Boekhold, S. Soil protection in The Netherlands, a changing perspective. In Proceedings of the International Conference on Land Protection, Bilbao, Spain, 22 October 2012.

213. Goidts, E. Soil protection strategy, from an integrated vision to a practical implementation. In Proceedings of the International Conference on Land Protection, Bilbao, Spain, 22 October 2012.

214. Bodenez, P. The current policy on contaminated land management in France and the possible evolutions to a broader soil protection strategy. In Proceedings of the International Conference on Land Protection, Bilbao, Spain, 22 October 2012.

215. GEIE Alliance Environnement. *Synthèse Des Évaluations Conduites Dans Le Contrat Cadre N° 30-CE-0067379/00-89 Sur Les Effets Sur L'environnement De Mesures De La PAC*; GEIE Alliance Environnement: Brussels, Belgium, 2010; p. 68.

216. Eurostat Statistics. Agri-environmental indicators. Available online: http://epp.eurostat.ec.europa.eu/statistics_explained/index.php/Agri-environmental_indicators (accessed on 10 September 2014).

217. European Union. DG Agriculture and Rural Development. Unit for Agricultural Policy Analysis and Perspectives,2013; Overview of CAP Reform 2014–2020; Agricultural Policy Perspectives Brief # 5. Available online: http://ec.europa.eu/agriculture/policy-perspectives/policy-briefs/05_en.pdf (accessed on 24 November 2014).

218. Pellerin, S.; Bamière, L.; Angers, D.; Béline, F.; Benoît, M.; Butault, J.P.; Chenu, C.; Colnenne-David, C.; de Cara, S.; Delame, N.; et al. *How Can French Agriculture Contribute to Reducing Greenhouse Gas Emissions? Abatement Potential and Cost of Ten Technical Measures*; Synopsis of the Study Report; Institut National de la Recherche Agronomique (INRA): Paris, France, 2013; p. 92.

219. European Union. DG Internal policies. Policy Department for Structural and Cohesion policies. Agriculture and Rural Development, 2014; Measures at Farm Level to Reduce Greenhouse Gas Emissions from EU Agriculture. NOTES. Available online: http://www.europarl.europa.eu/RegData/etudes/note/join/2014/513997/IPOL-AGRI_NT%282014%29513997_EN.pdf (accessed on 24 November 2014).

220. Lark, R.M. Estimating the regional mean status and change of soil properties, two distinct objectives for soil survey. *Eur. J. Soil Sci.* **2009**, *60*, 748–756. [CrossRef]

221. Piorr, H.-P. *Experiences with the Evaluation of Agricultural Practices for EU Agri- Environmental Indicators*; OECD Report. OECD (Organisation for Economic Co-operation and Development: Paris, France, 2013; p. 17. Available online: http://www.oecd.org/tad/sustainable-agriculture/44820415.pdf (accessed on 18 December 2014).

222. Dobbie, K.E.; Bruneau, P.M.C.; Towers, W. (Eds.) *The State of Scotland's Soil*. Available online: www.sepa.org.uk/land/land_publications.aspx (accessed on 24 November 2014).

223. Nisbet, T.; Dutch, J.; Moffat, A. *Whole-Tree Harvesting; A Guide to Good Practice*; Forestry Commission: Edinburgh, UK, 1997; p. 19.

224. Ling, E. Forest fuel extraction and wood ash recycling. In Proceedings of the RecAsh International Seminar, Prague, Czech Republic, 7–9 November 2005.

225. Hakkila, P. Factors driving the development of forest energy in Finland. Sustainable production systems for bioenergy, impacts on forest resources and utilization of wood for energy. *Biomass Bioenergy* **2006**, *30*, 281–288.

226. Cécillon, L.C.; Barthes, B.G.; Gómez, C.; Ertlen, D.; Genot, V.; Hedde, M.; Stevens, A.; Brun, J. Assessment and monitoring of soil quality using near-infrared reflectance spectroscopy (NIRS). *European J. Soil Sci.* **2009**, *60*, 770–784. [CrossRef]

227. Moreira, C.S.; Brunet, D.; Verneyre, L.; Sá, S.M.O.; Galdós, M.V.; Cerri, C.; Bernoux, M. Near infrared spectroscopy for soil bulk density assessment. *Eur. J. Soil Sci.* **2009**, *60*, 785–791. [CrossRef]

228. Camarsa, G.; Sliva, J.; Toland, J.; Hudson, T.; Nottingham, S.; Rosskopf, N.; Thévignot, C. *Life and Soil Protection*; Publications Office of the European Union: Luxembourg, 2014; p. 68.

229. European Commission. *The implementation of the Soil Thematic Strategy and ongoing activities; Report COM(2012) 46 final*. Available online: http://eusoils.jrc.ec.europa.eu/library/jrc_soil/policy/DGENV/COM%282012% 2946_EN.pdf (accessed on 18 December 2014).

sustainability

MDPI

Review

Wind Erosion Induced Soil Degradation in Northern China: Status, Measures and Perspective

Zhongling Guo [1], Ning Huang [2], Zhibao Dong [3], Robert Scott Van Pelt [4] and Ted M. Zobeck [5,*]

[1] College of Resource and Environmental Sciences/Hebei Key Laboratory of Environmental Change and Ecological Construction, Hebei Normal University, Shijiazhuang 050024, China; gzldhr@gmail.com

[2] Key Laboratory of Mechanics on Disaster and Environment in Western China, Lanzhou University, Lanzhou 730000, China; huangn@lzu.edu.cn

[3] Key Laboratory of Desert and Desertification, Chinese Academy of Sciences, Lanzhou 730000, China; zbdong@lzb.ac.cn

[4] Wind Erosion and Water Conservation Research Unit, Agricultural Research Service, United States Department of Agriculture (USDA), Big Spring, TX 79720 USA; Scott.VanPelt@ars.usda.gov

[5] Wind Erosion and Water Conservation Research Unit, Agricultural Research Service, United States Department of Agriculture (USDA), Lubbock, TX 79415, USA; Ted.Zobeck@ars.usda.gov

* Correspondence: Ted.Zobeck@ ars.usda.gov; Tel.: +1-806-723-5240; Fax: +1-806-723-5272

External Editors: Douglas L. Karlen and Marc A. Rosen

Received: 31 October 2014; in revised form: 20 November 2014; Accepted: 26 November 2014; Published: 4 December 2014

Abstract: Soil degradation is one of the most serious ecological problems in the world. In arid and semi-arid northern China, soil degradation predominantly arises from wind erosion. Trends in soil degradation caused by wind erosion in northern China frequently change with human activities and climatic change. To decrease soil loss by wind erosion and enhance local ecosystems, the Chinese government has been encouraging residents to reduce wind-induced soil degradation through a series of national policies and several ecological projects, such as the Natural Forest Protection Program, the National Action Program to Combat Desertification, the "Three Norths" Shelter Forest System, the Beijing-Tianjin Sand Source Control Engineering Project, and the Grain for Green Project. All these were implemented a number of decades ago, and have thus created many land management practices and control techniques across different landscapes. These measures include conservation tillage, windbreak networks, checkerboard barriers, the Non-Watering and Tube-Protecting Planting Technique, afforestation, grassland enclosures, *etc.* As a result, the aeolian degradation of land has been controlled in many regions of arid and semiarid northern China. However, the challenge of mitigating and further reversing soil degradation caused by wind erosion still remains.

Keywords: soil degradation; wind erosion; northern China

1. Introduction

Soil degradation is one of the most serious ecological and environmental problems facing the world [1]. In China, the total land area affected by soil degradation is approximately 861,600 km^2, accounting for 9.0% of the national territory [2]. It recognized that water erosion, wind erosion, salinization, acidification, and soil contamination are the main factors leading to soil degradation [3]. The most typical and serious form of soil degradation for China is soil erosion caused by wind or water. For arid and semi-arid northern China, the dominant soil degradation force involves aeolian processes [4]. Recently, the aeolian desertification survey of China revealed that the total area suffering from soil degradation caused by wind erosion covered 375,935.5 km^2 by 2010, about 44.1% of the total soil degraded land in China [5].

Wind-induced soil degradation may take several forms. The first and most visible form is total removal of the topsoil from bare fields, particularly on knolls within fields. This fertile topsoil may be transported to other areas of the field, deposited along the margins of the field, or totally lost into adjacent land. A more subtle form of wind-induced soil degradation is the winnowing of the finer, more chemically active particles from the soil. These fine soil particles have high surface area to volume ratios and thus carry disproportionate amounts of soil nutrients and organic carbon with them. Fine soil particles may be carried long distances from the source fields and even deposited in the oceans where they are lost from terrestrial ecosystems. They are also key to soil water holding capacity and soil aggregation and their loss often results in a looser, drier, sandier soil surface. Thus, a negative feedback loop is formed where wind-induced soil degradation may lead to a more vulnerable soil surface. Finally, wind-blown sand in the form of dunes or sand sheets may migrate over and bury fertile farm fields and grasslands resulting in lost ecosystem productivity.

Wind-induced soil degradation significantly affects local economic, society and ecosystems sustainability. The total economic loss due to wind-induced soil degradation has been estimated to be more than 54 billion Chinese Yuan (approximately 6.8 billion U.S. dollar) per year. About 170 million residents in arid and semi-arid northern China are threatened by wind-induced soil degradation [6]. Severe soil degradation generally decreases land productivity, which may result in local poverty, malnutrition, and disease. In turn, the poverty may push residents to over-exploit local natural resources, which may further degrade local sensitive ecosystems [7]. To combat wind erosion induced soil degradation and enhance local sustainability, the government invests about 0.024% of the annual Chinese gross domestic product (GDP) to launch a series of national policies and ecological projects [8]. Consequently, many land management practices and control techniques for different landscapes have subsequently been created. These national policies coupled with local residents' efforts, have made significant achievements in combating soil degradation in China. The expanding rate of soil degradation for some typical regions (such as Horqin Sands and Mu Us Sands) is controlled [9]. However, the campaign for reversing the trend of soil degradation still needs more efforts from government officials, local residents and research in the future.

The purpose of this paper is to review Chinese processes and experiences in combating wind erosion induced soil degradation for the last 60 years. In this paper, we focus on (1) historical and current trends of wind-induced soil degradation in northern China; (2) status of typical regions suffering from wind-induced soil degradation; (3) current land management practices and problems to combat wind-induced soil degradation; and (4) perspectives on reversing wind-induced soil degradation.

2. Historical and Current Trends of Wind-Induced Soil Degradation in Northern China

2.1. Wind-Induced Soil Degradation in Northern China

In northern China, low annual precipitation (generally less than 500 mm) and strong wind between March and June are the climatic driver to wind-induced soil degradation [4]. Wind erosivity can be used to describe how the climate affects wind erosion. Many equations of wind erosivity have been proposed [10]. The C-value from the Food and Agriculture Organization (FAO) is one of the most widely-used wind erosivity indexes [4,10]. The C-value is estimated from monthly climate data (average wind speed, potential evapotranspiration, and precipitation) [10]. Figure 1 illustrates the spatial pattern of C-value for northern China. The regions with high C-value are mainly distributed in Xinjiang, Inner Mongolia, Qinghai, Ningxia, and Gansu province, with the exception of a few coastal regions (Figure 1). These dry and windy regions with high C-value are generally suffering serious wind-induced soil degradation.

Sustainability **2014**, *6*, 8951–8966

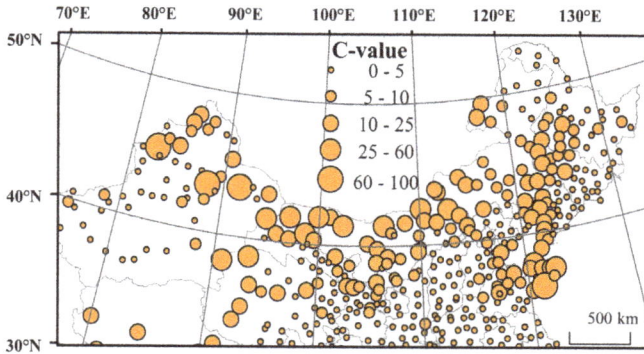

Figure 1. Spatial pattern for the wind erosivity (C-value) estimated from the Food and Agriculture Organization (FAO) method in northern China. Notes: Dr. Benli Liu (Cold and Arid Regions Environmental and Engineering Research Institute, Chinese Academy of Sciences, Lanzhou, Gansu, China) provided the figure. The authors thank him for his contribution.

In modern China, wind-induced soil degradation mitigation efforts began in the late 1950s [11]. The first national aeolian desertified land survey was conducted in 1994 with a repeating period of five years using remote sensing, the latest survey was performed during 2009. Spatially, most land with wind-induced soil degradation (aeolian desertified land) are distributed in arid and semi-arid northern China where the annual rainfall is below 500 mm [7]. As Figure 2 shows, aeolian degraded lands are mainly scattered in (1) the semi-arid agro-pastoral ecotone of northern China (about 40.5% of total aeolian degraded land) where wind erosion and sand sheet incursions are responsible for most soil degradation; (2) the semi-arid steppe in the middle of Inner Mongolia (about 36.5% of total aeolian degraded land) where reactivation of fixed dunes and shifting sand incursions create the most soil degradation; and (3) the margins of oases and lower reaches of inland rivers in the arid region (about 23% of total aeolian degraded land) where reactivated fixed dunes cause most soil degradation [7].

Generally, human activities are considered to be one of most important factor contributing to wind-induced soil degradation [7,12]. According to the statistics of wind-induced soil degradation in arid and semi-arid regions of northern China, over-cultivation, over-grazing, over-collecting fuel wood, inappropriate irrigation management, and engineering construction are responsible for 25.4%, 28.3%, 31.8%, 8%, and 1% of aeolian soil degraded land, respectively [7].

For the last 60 years, the trends of soil degradation caused by wind erosion in northern China frequently changed with human activities [7,12]. Table 1 presents the fluctuations of aeolian soil degradation during the last 60 years. In this paper, the standard titled "Classification Standard of Sandy desertification degrees" [12] was used to classify the wind-induced land degradation hazard, which is classified as slight, moderate, severe, or very severe (Table 2). The total area of aeolian degraded land changed from 296,470.4 km^2 to 375,935.5 km^2 between the 1950s and 2010. From the 1950s to 2000, aeolian degraded land rapidly expanded with an accelerating rate. The rates of increase for periods of the 1950s to 1975, 1975 to 1990 and 1990 to 2000 were 1560 km^2 per annum, 2100 km^2 per annum, 3600 km^2 per annum, respectively [13]. In contrast, from 2000 to 2010 the amount of aeolian- degraded land annually shrank. From 2000 to 2005 aeolian-degraded land decreased 1635.3 km^2 per annum and decreased 1114.4 km^2 per annum from 2005 to 2010. Analysis of wind-induced soil degradation spatial pattern for different period shows that variations of aeolian soil degraded land mainly occurred in the agro-pastoral ecotone of northern China [5]. In a word, wind-induced soil degradation has passed through two stages in arid and semiarid northern China. First, rapid wind-induced soil degradation occurred from the 1950s to the late 2000s. Second, wind-induced soil degradation has generally been prevented in many regions of arid and semiarid northern China by the

2010s, except for a few regions where arid-windy climate and more frequent human activities threaten the soil health (such as Bashang region, Minqin Oasis, Hexi Corridor, *etc.*) [7].

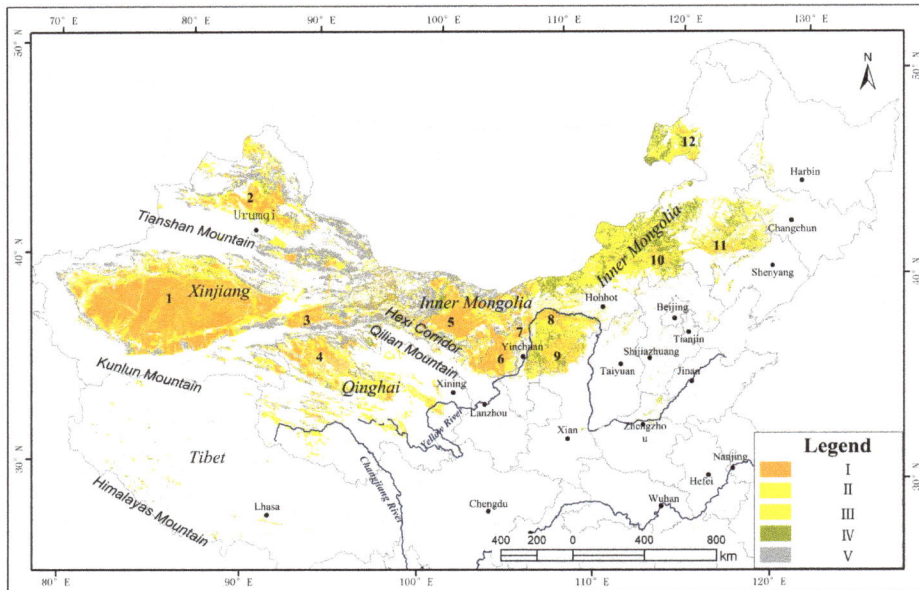

Figure 2. Spatial pattern of the aeolian soil degraded land in northern China. Notes: 1 = Taklimakan Desert, 2 = Gurbantunggut Desert, 3 = Kumtagh Desert, 4 = Qaidam Basin Desert, 5 = Badain Jaran Desert, 6 = Tengger Desert, 7 = Ulan Buh Desert, 8 = Hobq Desert, 9 = Mu Us Sandy Land, 10 = Onqin Daga Sandy Land, 11 = Horqin Sandy Land, 12 = Hulunbeir Sandy Land. I = Sandy Desert; II = aeolian degraded land with very severe and severe hazard; III = aeolian degraded land with moderate hazard; IV = aeolian degraded land with slight hazard; V = Gobi. The standard titled "Classification Standard of Sandy desertification degrees" [12] was used to classify the wind-induced land degradation hazard.

Table 1. Changes of aeolian soil degraded land from 1950s to 2010 for arid and semi-arid northern China.

Class	1950s	1975	1990	2000	2005	2010
	Area of the Class (km²)/Percentage of Total Area for the Class (%)					
Slight	–	93,886.3/29.2	109,041.6/30.7	132,795.6/34.1	129,793.4/34	127,066.2/33.8
Moderate	–	72,525.4/22.6	81,736.7/23	89,170.5/22.9	87,120.9/22.8	85,863.7/22.8
Severe	–	76,851.7/23.9	83,477.1/23.5	85,969.2/22.1	84,086.7/22	83,307.3/22.2
Very severe	–	78,204.1/24.3	81,050.7/22.8	81,785.1/21	80,543.8/21.1	79,735.9/21.2
Total	~295,000.0	321,430.4	355,268.8	389,683.7	381,507.3	375,935.5

Notes: The data were compiled from [7] and [13]. (–) = no data; (~) = approximate data; In the 1950s, the aeolian soil degraded land survey lasted several years, thus the approximate total area of soil degradation is derived from the data in 1975 [5], and the detailed soil degradation hazard data is also not available.

Table 2. Classification Standard of Sandy desertification degrees.

Class	Percentage of Blown Sand Area (%)	Percentage of Annual Increasing Area (%)	Percentage of Vegetation Cover (%)	Percentage of Annual Reduction in Biomass (%)
Slight	<5	<1	>60	<1.5
Moderate	5–15	1–2	60–30	1.5–3.5
Severe	25–50	2–5	30–10	3.5–7.5
Very severe	>50	>5	10–0	>7.5

Notes: The data were obtained from [7].

2.2. Wind-Induced Soil Degradation for Typical Regions

The rain-fed agricultural region of the agro-pastoral ecotone in northern China (APEC) is a typical region suffering from wind-induced soil degradation where changes of aeolian soil degraded land mainly occurred in the APEC for last 60 years [7]. It is generally recognized that the APEC is a transitional zone including grassland and farmland, where the soil degradation is very sensitive to climate and human activities. The trends of soil degradation for different parts of the APEC differ according to the specific locality and stressors.

The Bashang region, located in the northeast part of the APEC, is a typical region where the wind-induced soil degradation is increasing due primarily to intensified human activities. The land use dramatically changed during the 20th century. The natural landscape of the Bashang region is steppe, and cultivation of steppe grassland for grain production, often leads to severe wind-induced soil degradation. The total area of aeolian degraded land changed from 2524.0 km^2 to 4608.6 km^2 between 1975 and 1987 [7]. The degraded land increased from 4608.6 km^2 in 1987 to 6970.4 km^2 in 2000 [7]. Improper tillage practices generally leads to significant soil degradation. The fine soil material carrying much of the organic C and N, and P nutrients is the first to be winnowed from the soil and lost from the landscape as fugitive dust. After eight years of cultivation, more than 50% of the soil organic C, total N, and total P had been lost from the topsoil (0–20 cm plough layer), and the nutrients decreased 60%–79% in the topsoil after 50 years of cultivation near the town of Datan in the Bashang region [14]. Soil degradation was so severe that some of the cultivated lands were abandoned. With increasing abandonment time, the soil nutrients tended to increase (Figure 3) [14]. This indicates that the soil health can be significantly improved when intensified human disturbance such as tillage is discontinued.

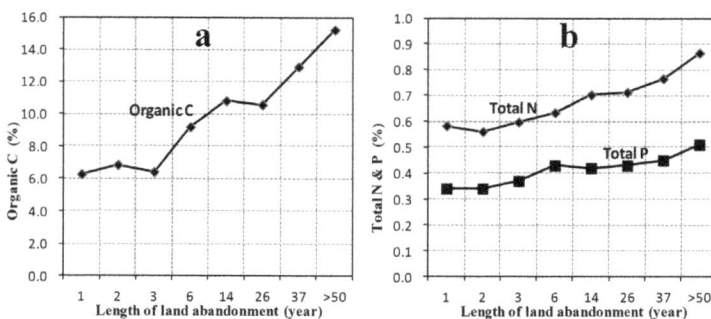

Note: the data were obtained from [14].

Figure 3. Soil properties affected by year of land abandonment for Datan Town in Bashang region. Note: the data were obtained from [14].

The trends of wind-induced soil degradation in the Horqin region is different from the Bashang region. Figure 4 presents the changes of aeolian degraded land in the Horqin region from 1959 to 2000.

The wind-induced soil degradation accelerated in the 1970s, and decelerated in the early 1980s. Before 1987, the aeolian degraded land increased. From 1959 to 1975, the total area of aeolian degraded land area increased by 9084.0 km^2. The total area increased by 9624.0 km^2 between 1975 and 1987. After 1987, the trend of increasing wind-induced soil degradation was reversed. The total area of aeolian degraded land declined from 61,008.0 km^2 in 1987 to 50,198.0 km^2 in 2000. During this period, the total area of the severe and slightly aeolian degraded land area increased 393.0 km^2 and 1749.0 km^2, respectively, but very severe and moderate aeolian degraded land area decreased 488.0 km^2 and 12,463.0 km^2, respectively [13]. The atmospheric environment has also improved with the reductions of aeolian degraded land. Figure 5 presents the variations of the indexes of atmospheric environment for Nanman Banner (County) in the Horqin region. The annual days with sandstorm and annual dusty days gradually decreased while the annual average wind speeds fluctuated between 3.2 to 3.8 m s^{-1} from the 1960s to the 1990s (Figure 5). The reduction of soil degradation with time is generally attributed to effective comprehensive artificial measures, such as grassland restoration, afforestation, enclosures, *etc.* [15].

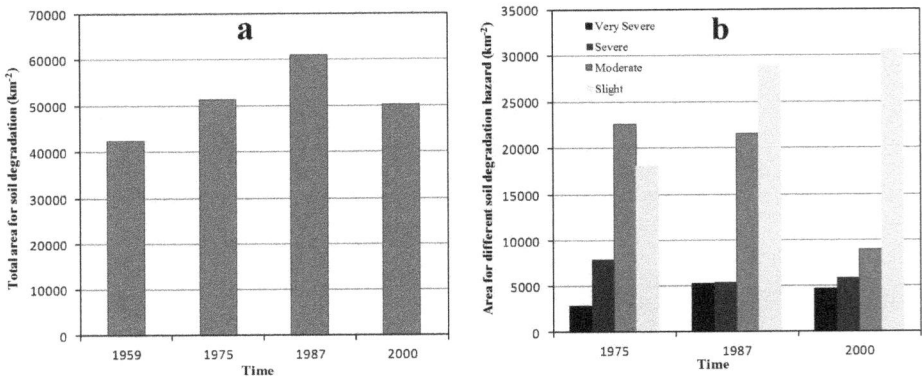

Figure 4. Changes of the (**a**) total area for aeolian degraded land and (**b**) area for different aeolian degraded hazard in Horqin region. Note: the data were obtained from [13]. The data for different soil degradation hazard were unavailable for 1959.

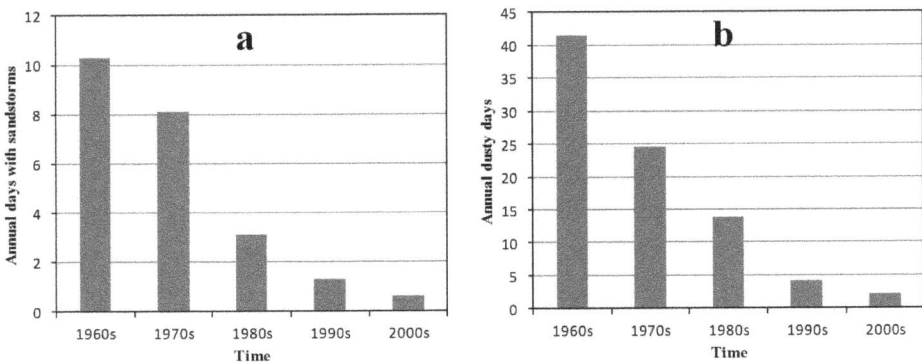

Figure 5. Variations of the (**a**) annual days with sandstorm (**b**) annual dusty day in Naiman Banner (County) in Horqin region. Note: the data were edited from [15]. The visibility is less than 1.0 km in the days with sandstorms, the visibility is between 1.0 and 10.0 km on dusty days.

3. Current Land Management Practices Status

To decrease soil loss by wind erosion and enhance local ecosystems, the Chinese government has been encouraging residents to reduce wind-induced soil degradation through policies, economic, and technical measures. On average, the government invests 0.024% of the Chinese gross domestic product (GDP) to mitigate wind-induced soil degradation. As a result, about 20% of the degraded lands have been controlled [8,16]. The measures for combating wind-induced soil degradation include national policies (projects) and land management practices at the field scale.

3.1. State Policy and Projects to Combat Wind-Induced Soil Degradation

To combat wind-induced soil degradation and further reverse the degradation trend, a series of state policies have been implemented. In 1994, the Chinese government signed the United Nation Convention to Combat Desertification (UNCCD) for promoting international cooperation [17]. To better implement the UNCCD, the Chinese Committee for Implementing the UNCCD (CCICCD) was organized [9]. The Natural Forest Protection Program (NFPP) and the National Action Program to Combat Desertification (NAPCD) were conducted during the late 1990s [18]. Based on these policies, the national strategic objectives to combat wind-induced soil degradation were divided into three stages: (1) a short-term objective (1996–2000) in which 3.2 million hectares of lands affected by wind erosion will be rehabilitated; (2) a mid-term objective (2001–2010) in which 7.5 million hectares of lands suffering from wind erosion will be rehabilitated; (3) a long-term objective (2011–2050) in which about 30.0 million hectares of wind eroded lands will be rehabilitated [19]. The three stages are closely interrelated and constitute a basic framework of wind-induced soil degradation control. With the implementations of these policies, the Chinese Government gradually recognized the importance of legislation for combating soil degradation and the Law of Combating Desertification (LCD) was enacted in 2002 [9]. In addition, a national monitoring system which consists of 43 research stations across China under the direction of the State Forestry Administration of China (SFA), the China Desert Ecosystem Research Network (CDERN), has been under development since 1978 (Figure 6) [17]. The national policies mentioned above are the guidelines to combat wind-induced soil degradation in China.

Meanwhile, a number of national ecological engineering projects have been launched. The Three-North (northwestern, northern, and northeastern parts of China) Shelterbelt Project (TNSP) (1979–2050), one of the most ambitious conservation programs in the world, was established to prevent soil degradation through extensive afforestation in arid and semiarid China (Figure 6) [20]. The project involves about 590 counties in 13 provinces, covering a total area of 4.1 million km^2, accounting for 42.4% of China's territory [19]. During the project period, 35.7 million hectares of afforestation is planned and the forest coverage will change from 4.0% to 16.0% for the project region [19]. From 1979 to 2010, about 27.9 million hectares of afforestation have already been implemented [19]. Another national afforestation project, the Grain for Green Project (GGP) (or named as Returning Farmlands to Grassland and Forest Project) (1999–2010), aims to return 147.0 million hectares of farmlands and 173.0 million hectares of grassland to forest between 1999 and 2010 [21]. The GGP began its pilot program in Sichuan, Shanxi and Gansu province in 1999 and finally extended to 1897 counties in 25 provinces of China [21]. By 2010, the GGP achieved its hectares goals, and the Chinese government restarted the GGP on 10 October 2014 [22]. In addition, the Beijing and Tianjin Sandstorm Source Control Project (BTSC) (2001-present) has been conducted for reducing the wind-induced soil loss and related sandstorms in the Beijing-Tianjin megacity belt [23]. The BTSC involves 75 counties in Beijing, Tianjin, Hebei, Inner Mongolia and Shanxi covering an area of 458,000 km^2 around the Beijing-Tianjin megacity belt (Figure 6). From 2001 to 2010, the first stage of BTSC has been implemented with an investment of 55.8 billion Chinese Yuan (approximately 9.1 billion U.S. dollar), with the result that 18.0 million hectares of land have been acquired and 2.6 million hectares of farmland have been afforested [24]. As to the benefit of the BTSC, the spatially average wind-induced soil loss decreased from 26.3 in 2001 to 18.7 t hm^{-2} a^{-1} in 2010 with the total vegetation coverage of the BTSC increased

319

from 40.9% to 49.1% [23]. However, the large-scale afforestation projects did not effectively solve the local wind-induced soil degradation [25]. The overall survival rate of planting trees during afforestation projects is only 15% in the arid and semi-arid northern China [26]. This suggests that afforestation alone could not effectively reverse the trend of soil degradation. However, planting grasses or bushes may be an effective measure to curb soil degradation for some regions [25].

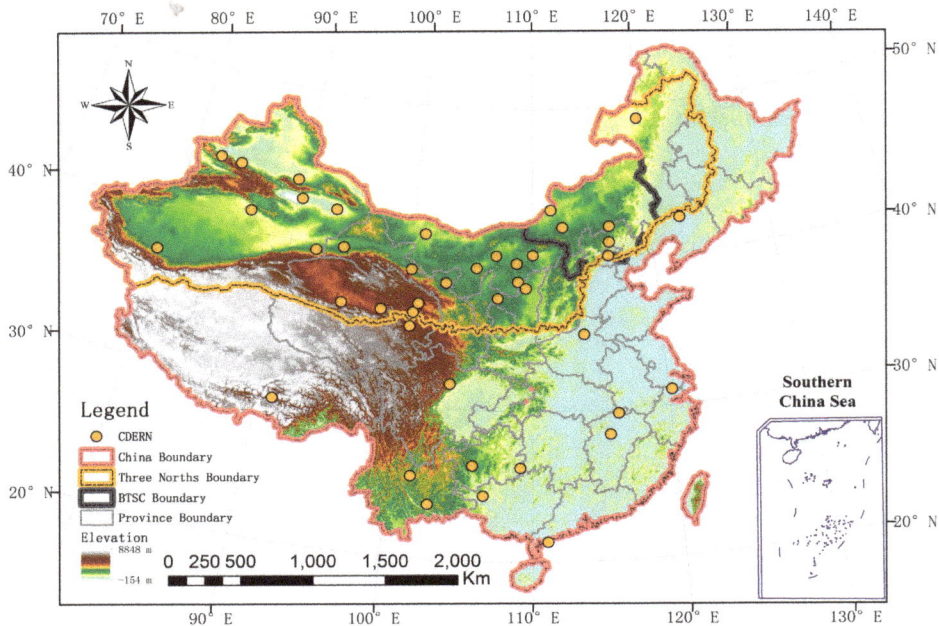

Figure 6. State projects to combat wind erosion induced soil degradation. Notes: CDERN = research stations of the China Desert Ecosystem Research Network; Three Norths = northwestern, northern, and northeastern parts of China; BTSC = Beijing and Tianjin Sandstorm Source Control Project.

Additionally, many local policies, regulations, and projects have also gradually been proposed to complement or augment the state plans. These policies and projects, at different scales, offer an ongoing strategy-system to combat wind erosion induced soil degradation in China.

3.2. Current Land Management Practices Status for Different Landscape

At the field scale, the policies and projects mentioned above have yielded many typical and classical land management practices and control techniques for different landscapes. The main landscapes suffering wind-induced soil degradation are sandy land, farmland and grassland in arid and semiarid northern China.

Sandy land is most susceptible to aeolian degradation. A typical engineering measure for sandy land erosion control is the "Straw Checkerboards Barrier" (Figure 7). This technique effectively reduce wind velocity, thus lower field sand transport rate [27]. Research has revealed that the wind velocity can be reduced by 20%–40% at a height of 0.5 m and the soil surface aerodynamic roughness could increase by 400–600 times when the height of the checkerboard barriers is 0.15–0.20 m [27]. The economic and reasonable height of "Straw Checkerboards Barrier" is 0.1–0.2 m [27]. The building materials of checkerboards barrier are flexible and include straw, shrub branches, stones, clay, and artificial plastic products (Figure 7). This classical "Checkerboards Barrier" technique is still widely used in China. More recently, an ecological technique, the "None-Watering and Tube-Protecting

Planting Technique for *Haloxylon ammodendron*", was invented and used for sandy land ecological recovery and restoration [28]. The planting technique uses plastic or sand-made tubes to nurse the seedlings of *Haloxylon ammodendron* based on the theory that the high temperature of sandy land surface layer (0–2 cm) (>50 °C) is one of the important ecological limiting factors [28]. After several years of testing, it was found that the technique can efficiently increase the percentage of average seedling survival by greater than 70% and annual growth rate by greater than 20% for *Haloxylon ammodendron*. In addition, the technique is also suitable for planting other sandy land plants [28].

The farmland scattered in arid and semi-arid northern China, especially farmland with bare surfaces, is another landscape undergoing soil degradation. Conservation tillage is generally considered to be an economical, practical and feasible wind erosion control method [29]. The efficacy of conservation tillage and its application for dry lands of northern China have been addressed since the 1970s [30]. Many reports have shown that conservation tillage could efficiently mitigate wind-induced soil loss at field scales even at regional scales in China [9,31]. The Chinese government has been encouraging residents to adopt applicable conversation practices since 2002 [30]. However, at a national level, the traditional cultivation practices, such as intensive tillage, residue removal or burning, are still common [30]. It may be a long time before local farmers accept and embrace conservation tillage. In contrast, windbreak networks for farmland have been gradually and steadily increasing due to the strong support of the Chinese government. As a part of the TNSP and BTSC, windbreak networks for farmland projects in arid and semiarid northern China have obtained continuous national investment. By 2008, the total area of farmland with windbreak networks was 533,300 hectares in the Three-North (northwestern, northern, and northeastern parts of China) region [19]. The effects of windbreak networks for farmland on controlling soil degradation is closely related to its porosity (density), orientation, height, width, distance between barrier rows, and length [32]. At a local scale, windbreak networks could lead to reductions of wind speed and turbulence intensity within a certain distance in the leeward and improve micro-agro-climate. At a regional scale, windbreak networks can increase terrain roughness, so a dense network has been suggested as the cause of a reduction in the average surface wind speed for the region [32,33]. Therefore, windbreak networks with optimal design is a feasible measure to combat wind erosion induced soil degradation for the farming regions in arid and semi-arid northern China where the water resource is sufficient to build and sustain windbreaks.

Figure 7. Various materials used for Checkerboards Barriers. (**a**) Straw; (**b**) Shrub branches; (**c**) Stones and Shrubs; (**d**) Plastic.

Grasslands are also very sensitive to wind-induced soil degradation in arid and semi-arid northern China. Enclosing degrading grassland to keep out grazing animals is considered to be a simple, economic and effective measure to maximize pastoral productivity and curb soil degradation, thus it is widely used in the rain-fed regions of northern China [34]. For example, an enclosed grassland significantly increased soil seed density and facilitate vegetation restoration in the Horqin region [35]. This study showed that seed density in the enclosed grassland increased by 15.7%, 482.5% and 728.1% for sites enclosed for two-year, six-year, and 12-year periods, respectively, and the vegetation coverage of the six-year and 12-year sites increased by 261.6% and 271.6%, respectively [35]. Nevertheless, some research also questioned the total regional benefit for Chinese households for grassland enclosures. Actually, enclosures generally do not decrease the number of livestock for a region. This measure could force more grazing animals from enclosed grassland to non-enclosed grassland, which may increase stocking rate of non-enclosed grassland. Investigations from Inner Mongolia revealed that grassland enclosures conducted at a village level actually increased soil degradation processes across vast territories while only protecting small isolated fields [36]. Planting grass on the degraded grassland (planted grassland) is an important measure to recover and further reconstruct the grassland ecosystem [37]. Planted grassland could also enhance pastoral productivity in a relative short time. Experiments conducted in Qaidam Basin have showed that the forage of the planted grassland could increase by 380% compared with the degraded rangeland [38]. The planted grassland generally requires tillage, fertilization, and irrigation. It is an expensive and water-consuming measure, which limits its wide use in arid and semi-arid northern China.

These measures and techniques for different landscapes are typical practical-experiences in preventing wind-induced soil degradation. Actually, the design, construction, and implementation of measures and techniques for preventing wind-induced soil degradation generally depend on local geographical features, regional soil degradation control experiences, laboratory experiments, and field observations [9]. It is still a challenge to develop and determine economical and efficient local measures or techniques for mitigating soil degradation.

4. Perspectives on Reversing Wind-Induced Soil Degradation

National policies and projects together with local land management practices and control techniques outline a blueprint to combat wind erosion induced soil degradation in arid and semi-arid northern China. Although significant progress and abundant achievements for preventing wind-induced soil degradation have been made to date, the campaign for reversing the trend of soil degradation still needs more efforts from government officials, local participants, and researchers in the future.

In China, the campaign for preventing wind-induced soil degradation involves more than 10 state ministries and administrations [14]. More time and effort is needed to coordinate these government branches to more effectively combat soil degradation. Accordingly, the government's decision-making may lag behind the development of soil degradation. Thus a powerful steering committee with a more effective institutional framework may be necessary to curb and further reverse soil degradation. The Chinese government could improve current policy in many ways. An efficient financial system with more investments has been proposed to fund the control projects and related research, and to further improve local residents' enthusiasm for combating soil degradation [39]. It is also a feasible way to explore international cooperation and funding.

The campaign for preventing wind-induced soil degradation also needs continuous research involvement and affordable control techniques. In the degradation-prone regions, an increasing local population with a resulting expanding economy makes the soil degradation and degradation reversal processes more complex. Meantime, research on soil degradation is discontinuous in northern China, which in turn decreases the locally effective tools to combat soil degradation [39]. Although it is generally recognized that vulnerable eco-environments and irrational human activities result in soil degradation, there is ongoing debate on the soil degradation process for different temporal or spatial scales [12,40]. Therefore, a series of spatially diverse long-term research projects on soil degradation are needed to support executive decision-making.

Wind-induced soil degradation is a physical, economical and social-related process [41]. Basically, executive policy and research involvement finally need to inspire local residents' initiatives. To effectively encourage local residents to combat wind-induced soil degradation, affordable technical supports to combat soil degradation and adequate economic rewards from preventing soil degradation are necessary.

5. Conclusions

Soil degradation due to wind erosion is a significant concern in arid and semi-arid northern China. The total area of lands suffering from wind-induced soil degradation was 375,935.5 km^2 in 2010. Most wind-induced degraded lands (aeolian desertified lands) are scattered in arid and semi-arid northern China where the annual rainfall is below 500 mm. For the last 60 years, the trends of soil degradation caused by wind erosion frequently changed with human activities. Changes of aeolian soil degraded land mainly occurred in the agro-pastoral ecotone of northern China. The wind-induced degraded lands increased from 1950s to 2000 but decreased between 2000 and 2010. The reductions of soil degradation are attributed to a series of state policies and projects. These policies include CCICCD (1994–present), NFPP (1998–present), NAPCD (2000-present). In light of these policies, a national strategic objectives were described by a short-term objective (1996–2000), a mid-term objective (2001–2010) and a long-term objective (2011–2050), respectively. Enaction of the LCD (2002–present) and construction of the CDERN (1978–present) were also launched. The national projects consisted of the TNSP (1979–2050), the GGP (1999–2010 and 2014–present) and the BTSC (2001–present). These national policies and projects yielded many land management practices and control techniques for different landscapes. These include conservation tillage and windbreak networks for farmland, checkerboards barrier, and the None-Watering and Tube-Protecting Planting Technique for sandy land, and planted grassland and grassland enclosures for grassland. Although progress and achievements for preventing wind-induced soil degradation have been made to date, more government officials, local residents and research efforts are still needed to reverse the trend of wind-induced soil degradation in the future.

Acknowledgments: This research was funded by the Natural Science Foundation of China (Grant No. 41101251, 41301291 and 41330746), Natural Science Foundation of Hebei Province, China (Grant No. D2013302034 and D2014205063). The authors would like to thank the anonymous reviewers for their careful comments.

Author Contributions: Zobeck T.M. initiated the review paper for responding the invitation from guest editor of the special issue. Guo Z., Huang N., Dong Z., Van Pelt R.S. and Zobeck T.M. drafted the manuscript, and approved the final version.

Conflicts of Interest: The authors declare no conflict of interest.

Disclaimer: Mention of trade names or commercial products in this publication is solely for the purpose of providing specific information and does not imply recommendation or endorsement by the U.S. Department of Agriculture. USDA is an equal opportunity provider and employer.

References

1. Karlen, D.L.; Andrews, S.S.; Weinhold, B.J.; Doran, J.W. Soil quality: Humankind's foundation for survival. *J. Soil Water Conserv.* **2003**, *58*, 171–179.
2. Wang, T. Progress in sandy desertification research of China. *J. Geogr. Sci.* **2004**, *14*, 387–400.
3. Lal, R. Soil degradation by erosion. *Land Degrad. Dev.* **2001**, *12*, 519–539. [CrossRef]
4. Dong, Z.; Wang, X.; Liu, L. Wind erosion in arid and semiarid China: An overview. *J. Soil Water Conserv.* **2000**, *55*, 439–444.
5. Wang, T.; Song, X.; Yan, C.; Li, S.; Xie, J. Remote sensing analysis on aeolian desertification trends in northern China during 1975–2010. *J. Desert Res.* **2011**, *31*, 1351–1356.
6. Wang, T.; Xue, X.; Luo, Y.; Zhou, X.; Yang, B.; Ta, W.; Wu, W.; Zhou, L.; Sun, Q.; Wang, X.; *et al.* Human causes of aeolian desertification in Northern China. *Sci. Cold Arid Regions* **2008**, *1*, 1–13.

7. Wang, T.; Chen, G.; Zhao, H.; Xiao, H. *Deserts and Aeolian Desertification in China*; Science Press: Beijing, China, 2011.

8. Wang, G.; Wang, X.; Wu, B.; Lu, Q. Desertification and its mitigation strategy in China. *J. Resour. Ecol.* **2012**, *3*, 97–104. [CrossRef]

9. Ci, L.; Yang, X. *Desertification and Its Control in China*; Higher Education Press: Beijing, China; Springer-Verlag: Berlin-Heidelberg, Germany, 2010.

10. Liu, B.; Qu, J.; Wagner, L.E. Building Chinese wind data for Wind Erosion Prediction System using surrogate US data. *J. Soil Water Conserv.* **2013**, *68*, 104A–107A. [CrossRef]

11. Dong, Z.; Wang, T.; Qu, J. The history of desert science over the last 100 years. *J. Desert Res.* **2003**, *23*, 1–5.

12. Wang, X.; Chen, F.; Hasi, E.; Li, J. Desertification in China: An assessment. *Earth-Sci. Rev.* **2008**, *88*, 188–206. [CrossRef]

13. Wang, T.; Wu, W.; Xue, X.; Sun, Q.; Zhang, W.; Han, Z. Spatial-temporal changes of sandy desertified land during last 5 decades in northern China. *Acta Geograohica Sinica* **2004**, *59*, 203–212.

14. Zhao, W.; Xiao, H.; Liu, Z.; Li, J. Soil degradation and restoration as affected by land use change in the semiarid Bashang area, northern China. *Catena* **2005**, *59*, 173–186. [CrossRef]

15. Zhang, G.; Wang, C.; Bian, R.; Qu, X.; Yang, X. The climate change analysis of most typical region in Horqin sandy area for nearly 50 years. *Chin. Agric. Sci. Bull.* **2012**, *23*, 287–290.

16. Fan, S.; Zhou, L. Desertification control in China: Possible solutions. *AMBIO A J. Hum. Environ.* **2001**, *30*, 384–385.

17. Wang, F.; Pan, X.; Wang, D.; Shen, C.; Lu, Q. Combating desertification in China: Past, present and future. *Land Use Policy* **2013**, *31*, 311–313. [CrossRef]

18. Zhao, J.; Wu, G.; Zhao, Y.; Shao, G.; Kong, H.; Lu, Q. Strategies to combat desertification for the twenty first century in China. *Int. J. Sustain. Dev. World Ecol.* **2002**, *9*, 292–297. [CrossRef]

19. State Forestry Administration (SFA). *National Report on the Three-North Shelterbelt Projects for Last 30 Years (1978–2008)*; China Forestry Press: Beijing, China, 2008.

20. Yang, X.; Zhang, K.; Jia, B.; Ci, L. Desertification assessment in China: An overview. *J. Arid Environ.* **2005**, *63*, 517–531. [CrossRef]

21. State Forestry Administration (SFA). *National Report on Ecological Benefits of the Grain for Green Projects (2013)*; China Forestry Press: Beijing, China, 2013.

22. Chinese Government Website. Available online: http://www.gov.cn/xinwen/201410/10/content_2 761817.htm (accessed on 17 October 2014).

23. Gao, S.; Zhang, C.; Zou, X.; Wu, Y.; Wei, X.; Huang, Y.; Shi, S.; Li, H. *Assessment on the Beijing and Tianjin Sandstorm Source Control Project*, 2nd ed.; Science Press: Beijing, China, 2012.

24. Zeng, X.; Zhang, W.; Cao, J.; Liu, X.; Shen, H.; Zhao, X. Changes in soil organic carbon, nitrogen, phosphorus, and bulk density after afforestation of the "Beijing–Tianjin Sandstorm Source Control" program in China. *Catena* **2014**, *118*, 186–194. [CrossRef]

25. Cao, S. Why large-scale afforestation efforts in China have failed to solve the desertification problem. *Environ. Sci. Tech.* **2008**, *42*, 1826–1831. [CrossRef]

26. Tong, C.; Wu, J.; Yong, S.; Yang, J.; Yong, W. A landscape-scale assessment of steppe degradation in the Xilin River Basin, Inner Mongolia, China. *J. Arid Environ.* **2004**, *59*, 133–149. [CrossRef]

27. Qiu, G.; Lee, I.; Shimizu, H.; Gao, Y.; Ding, G. Principles of sand dune fixation with straw checkerboard technology and its effects on the environment. *J. Arid Environ.* **2004**, *56*, 449–464. [CrossRef]

28. Ma, H.; Zhang, H.; Ma, L.; Ren, C.; Wang, Z.; Gao, X.; Fang, H.; He, X. None-watering and tube-protecting planting technique for *Haloxylon ammodendron* under desert and its extension. *Sci. Sin. Vit.* **2014**, *44*, 248–256.

29. Zobeck, T.M.; Halvorson, A.D.; Wienhold, B.J.; Acosta-Martinez, V.; Karlen, D.L. Comparison of two soil quality indexes to evaluate cropping systems in northern Colorado. *J. Soil Water Conserv.* **2008**, *63*, 329–338.

30. Wang, X.B.; Cai, D.X.; Hoogmoed, W.B.; Oenemad, O.; Perdok, U.D. Developments in conservation tillage in rainfed regions of North China. *Soil Tillage Res.* **2007**, *93*, 239–250. [CrossRef]

31. Guo, Z.; Zobeck, T.M.; Zhang, K.; Li, F. Estimating potential wind erosion of agricultural lands in northern China using the Revised Wind Erosion Equation and geographic information systems. *J. Soil Water Conserv.* **2013**, *68*, 13–21. [CrossRef]

32. Vigiak, O.; Sterk, G.; Warren, A.; Hagen, L.J. Spatial modeling of wind speed around windbreaks. *Catena* **2003**, *52*, 273–288. [CrossRef]

33. Bao, Y.; He, X.; Yang, J.; Sun, C.; Li, H.; Zong, P. Study on preventing effect of three farmland shelter-forest network on soil wind erosion. *Chin. J. Soil Water Conserv.* **2007**, *21*, 5–8.

34. Deng, L.; Sweeney, S.; Shangguan, Z. Long-term effects of natural enclosure: Carbon stocks, sequestration rates and potential for grassland ecosystems in the Loess Plateau. *CLEAN-Soil Air Water* **2014**, *42*, 617–625. [CrossRef]

35. Jiang, D.; Miao, R.; Oshida, T.; Zhou, Q. Effects of fence enclosure on vegetation restoration and soil properties in Horqin sandy land. *Ecol. Environ. Sci.* **2013**, *22*, 40–46.

36. William, D.M. Grassland enclosures: Catalyst of land degradation in Inner Mongolia. *Hum. Org.* **1996**, *55*, 307–313.

37. Hu, Z. The importance of artificial grassland in the development of prataculture and the control of environment in China of 21 century. *Grassl. Turf.* **2000**, *88*, 12–15.

38. Wang, Q.; Niu, D.; Jiang, W.; Ma, Y.; Li, Q. Effect and benef it of forage cultivation and livestock raising to the protection of ecosystem in the farming-pastoral region of Qiadam Basin. *Acta Agrestia Sin.* **2005**, *13*, 226–232.

39. Wang, S.; Zheng, R.; Yang, Y. Combating desertification: The Asian experience. *Internet For. Rev.* **2000**, *2*, 112–117.

40. Chen, Y.; Tang, H. Desertification in north China: Background, anthropogenic impacts and failures in combating it. *Land Degrad. Dev.* **2005**, *16*, 367–376. [CrossRef]

41. Luca, S.; Marco, Z.; Margherita, C. Territorial systems, regional disparities and sustainability: Economic structure and soil degradation in Italy. *Sustainability* **2014**, *6*, 3086–3104. [CrossRef]